ESO ASTROPHYSICS SYMPOSIA
European Southern Observatory

Series Editor: Philippe Crane

Springer-Verlag Berlin Heidelberg GmbH

Georges Meylan (Ed.)

QSO Absorption Lines

Proceedings of the ESO Workshop
Held at Garching, Germany,
21–24 November 1994

 Springer

Volume Editor

Georges Meylan
European Southern Observatory
Karl-Schwarzschild-Strasse 2
D-85748 Garching, Germany

Series Editor

Philippe Crane
European Southern Observatory
Karl-Schwarzschild-Strasse 2
D-85748 Garching, Germany

Library of Congress Cataloging-in-Publication Data

QSO absorption lines : proceedings of the ESO workshop held at
   Garching, Germany, 21-24 November 1994 / G. Meylan, ed.
      p.   cm. -- (ESO astrophysics symposia)
   Includes bibliographical references and index.
   ISBN 978-3-662-22373-4   ISBN 978-3-540-49458-4 (eBook)
   DOI 10.1007/978-3-540-49458-4

   1. Quasars--Spectra--Congresses. 2. Absorption spectra-
-Congresses. 3. Red shift--Congresses. 4. Galaxies--Congresses.
I. Meylan, G. (Georges) II. European Southern Observatory.
III. Series.
QB860.Q764 1995
523.1'15--dc20   95-31847
                                                             ·CIP

ISBN 978-3-662-22373-4

© Springer-Verlag Berlin Heidelberg 1995
Originally published by Springer-Verlag Berlin Heidelberg New York in 1995
Softcover reprint of the hardcover 1st edition 1995

Typesetting: Camera ready by author/editor
SPIN: 10481224   42/3142-543210 - Printed on acid-free paper

Preface

The controversial question of whether the majority of the narrow absorption lines observed in QSO spectra represent cosmological intervening systems or ejecta from the QSO themselves is settled. QSO absorption line spectroscopy, initially a mere technique, has matured into an essential extragalactic research tool for understanding the content of the Universe at redshifts between 0 and 4, and beyond.

The only previous important meeting devoted to "QSO Absorption Lines" was held in May 1987 at the Space Telescope Science Institute in Baltimore, Maryland, U.S.A. Since that time, nearly a decade ago, research has been extremely active in this now well–established field of astrophysics. Theoretical studies and simulations have taken advantage of the constant progress in computer technology, and during these last few years, the observational results have benefited largely from the new facillities offered by the Hubble Space Telescope in the UV wavelength range and the Keck Telescope for high–resolution spectroscopy.

These proceedings contain the papers presented at the ESO workshop on "QSO Absorption Lines" held during November 21 – 24, 1994 in Garching bei München, Germany. The papers summarize the current state of our understanding of this field and review the new observational results and theoretical developments. The different sections address the following topics: damped Lyman–alpha systems, abundances, absorption lines from gas in our Galaxy, metal rich systems at low and intermediate redshifts, BAL QSOs and quasar environment, Lyman–alpha systems at low and hight redshifts, large–scale structure, the UV background, and formation of absorbing systems.

It is a great pleasure to acknowlegde the essential input of the scientific organizing committee: Jacqueline Bergeron, Patrick Petitjean, Peter Shaver, and Joseph Wampler. Simon White is warmly thanked for chairing the discussions of the poster sessions. The organisational success of this conference was mostly due to the now legendary efficiency of Christina Stoffer from ESO. And above all, I wish to express my gratitude to all the participants for contributing to an interesting and stimulating workshop: it is they who made the workshop happen, and this is also true for these proceedings!

European Southern Observatory   Georges Meylan
Garching, May 1995

Contents

PART III   ABUNDANCES

PART IV   ABSORPTION LINES FROM GAS IN OUR GALAXY

PART V   METAL–RICH SYSTEMS AT LOW AND
         INTERMEDIATE REDSHIFTS

PART VI   BAL QSOS AND QUASAR ENVIRONMENT

PART VII LYMAN–ALPHA SYSTEMS AT LOW AND HIGH REDSHIFTS

PART VIII   LARGE–SCALE STRUCTURE

PART XI WORKSHOP SUMMARY AND CONCLUSION

LIST OF PARTICIPANTS

Name	Institution
ARAGÓN-SALAMANCA, A.	Inst. of Astronomy, Cambridge, UK aas@mail.ast.cam.ac.uk
BAJTLIK, Stanislaw	Copernicus Astronomical Center, Warsaw, Poland bajtlik@camk.edu.pl
BECHTOLD, Jill	Steward Observatory, Tucson, USA jbechtold@as.arizona.edu
BERGERON, Jacqueline	ESO, Garching, Germany jbergero@eso.org
BERGVALL, Nils	Astron. Obs., Uppsala, Sweden nils.bergvall@astro.uu.se
BOISSÉ, Patrick	Ecole Normale Supérieure, Paris, France boisse@ensapa.ens.fr
BOKSENBERG, Alec	Royal Greenwich Obs., Cambridge, UK boksy@mail.ast.cam.ac.uk
BORGEEST, Ulf	Hamburg Sternwarte, Hamburg, Germany uborgeest@hs.uni-hamburg.de
BÖRNER, Gerhard	MPI für Astrophysik, Garching, Germany grb@mpa-garching.mpg.de
BOWEN, David	STScI, Baltimore, USA bowen@stsci.edu
BRINKMANN, Wolfgang	MPI für extraterrestrische Physik, Garching, Germany wpb@mpe-garching.mpg.de
BUNKER, Andrew	Department of Astrophysics, Oxford, UK a.bunker1@physics.oxford.ac.uk
BURBIDGE, Margaret	University of California, San Diego, USA mburbidge@ucsd.edu
CAMPUSANO, Luis	Universidad de Chile, Santiago, Chile luis@calan.das.uchile.cl
CARBALLO, Ruth	University of Cantabria, Santander, Spain carballo@ccucvx.unican.es
CARSWELL, Robert	Inst. of Astronomy, Cambridge, UK rfc@mail.ast.cam.ac.uk
CAULET, Adeline	ST-ECF, Garching, Germany acaulet@eso.org
CHARLTON, Jane	Penn State University, A. and A. Dept., University Park, USA charlton@astro.psu.edu

CHERNOMORDIK, Victor	Scientific Council on Cybernetics, Academy of Sciences, Moscow, Russia leo@pccross.msk.su
CHIBA, Masashi	Tohoku Astron. Inst., Sendai, Japan chiba@astroa.astr.tohoku.ac.jp
CHURCHILL, Christopher	University of California, Santa Cruz, USA cwc@lick.ucsc.edu
CLEMENTS, David	ESO, Garching, Germany dclement@eso.org
COMBES, Francoise	Obs. de Paris - DEMIRM, Paris, France bottaro@obspm.fr
CRISTIANI, Stefano	Dipart. di Astronomia, Padova, Italy cristiani@astrpd.astro.it
DAVIDSEN, Arthur	Johns Hopkins Univ., Baltimore, USA afd@pha.jhu.edu
DEHARVENG, Jean-Michel	Laboratoire d'Astronomie Spatiale, Marseilles, France jmd@astrsp-mrs.fr
DE LA FUENTE, Antonio	ESA-VILSPA, Madrid, Spain af@vilspa.esa.es
DINSHAW, Nadine	Steward Observatory, Tucson, USA ndinshaw@as.arizona.edu
DI SEREGO ALIGHIERI, S.	Osservatorio Astrofisico di Arcetri, Firenze, Italy sperello@arcetri.astro.it
DISNEY, Mike	University of Wales, Cardiff, UK M.Disney@astro.cf.ac.uk
DOBRZYCKI, Adam	Center for Astrophysics, Cambridge, USA adam@kurdel.harvard.edu
D'ODORICO, Sandro	ESO, Garching, Germany sdodoric@eso.org
DOROSHKEVICH, Andrei	Keldysh Inst. of Applied Mathematics, Moscow, Russia dorr@tacsg1.tac.dk
DRINKWATER, Michael	AAO, Coonabarabran, Australia mjd@aaocbn.aao.gov.au
FALL, Mike	STScI, Baltimore, USA fall@stsci.edu
FARDAL, Mark	Univ. of Colorado, CASA, Boulder, USA fardal@shapley.colorado.edu
FERNÁNDEZ-SOTO, Alberto	University of Cantabria, Santander, Spain fsoto@ccucvx.unican.es
FRANCIS, Paul	Univ. of Melbourne, Parkville, Australia pjf@tauon.ph.unimelb.edu.au

FRAYER, David　　　　　　　　　NRAO, Charlottesville, USA
　　　　　　　　　　　　　　　　dfrayer@nrao.edu
FRIAÇA, Amancio　　　　　　　Royal Greenwich Obs., Cambridge, UK
　　　　　　　　　　　　　　　　amancio@mail.ast.cam.ac.uk
FRIED, Josef　　　　　　　　　MPI für Astronomie, Heidelberg, Germany
　　　　　　　　　　　　　　　　fried@mpia-hd.mpg.de
FRITZE - V. ALVENSLEBEN, U.　Landessternwarte, Heidelberg, Germany
　　　　　　　　　　　　　　　　ufritze@gwdg.de
GIACCONI, Riccardo　　　　　　ESO, Garching, Germany
　　　　　　　　　　　　　　　　rgiaccon@eso.org
GIALLONGO, Emanuele　　　　　Osservatorio Astronomico di Roma.
　　　　　　　　　　　　　　　　Monteporzio, Italy
　　　　　　　　　　　　　　　　giallongo@roma2.infn.it
GOLINKIN, Anatoli　　　　　　　Inst. of Radio Astron., Kharkov, Ukraine
　　　　　　　　　　　　　　　　rai@ira.kharkov.ua
GREEN, Richard　　　　　　　　NOAO, Tucson, USA
　　　　　　　　　　　　　　　　green@noao.edu
GROOTE, Detlef　　　　　　　　Hamburg Sternwarte, Hamburg, Germany
　　　　　　　　　　　　　　　　dgroote@hs.uni-hamburg.de
GUILLEMIN, Patrick　　　　　　Institut d'Astrophysique, Paris, France
　　　　　　　　　　　　　　　　guillemin@iap.fr
IMPEY, Chris　　　　　　　　　Steward Observatory, Tucson, USA
　　　　　　　　　　　　　　　　cimpey@as.arizona.edu
JAKOBSEN, Peter　　　　　　　ESA/ESTEC, Noordwijk, The Netherlands
　　　　　　　　　　　　　　　　pjakobse@astro.estec.esa.nl
JENKINS, Edward　　　　　　　Princeton University Obs., Princeton, USA
　　　　　　　　　　　　　　　　ebj@astro.princeton.edu
KAPER, Lex　　　　　　　　　ESO, Garching, Germany
　　　　　　　　　　　　　　　　lkaper@eso.org
KARLSSON, K.-G.　　　　　　　Mid Sweden Univ., Härnösand, Sweden
　　　　　　　　　　　　　　　　KG.Karlsson@nth.mh.se
KHARE, Pushpa　　　　　　　　Utkal University, Bhubaneswar, India
　　　　　　　　　　　　　　　　pk@iopb.ernet.in
KHERSONSKY, Valery　　　　　Univ. of Pittsburgh, A. and A. Dept.,
　　　　　　　　　　　　　　　　Pittsburgh, USA
　　　　　　　　　　　　　　　　vkk@phyast.pitt.edu
KNEISSL, Rüdiger　　　　　　　MPI für Astrophysik, Garching, Germany
　　　　　　　　　　　　　　　　ruk@mpa-garching.mpg.de
KÖHLER, Susanne　　　　　　　Hamburg Sternwarte, Hamburg, Germany
　　　　　　　　　　　　　　　　st2b304@hs.uni-hamburg.de
LANZETTA, Kenneth　　　　　　New York State Univ., Stony Brook, USA
　　　　　　　　　　　　　　　　lanzetta@sbastc.ess.sunysb.edu
LE BRUN, Vincent　　　　　　　Institut d'Astrophysique, Paris, France
　　　　　　　　　　　　　　　　vlebrun@iap.fr

LEIBUNDGUT, Bruno — ESO, Garching, Germany
bleibund@eso.org

LEVSHAKOV, Sergei — Physico-Technical Inst.,
St. Petersburg, Russia
lev@astro.pti.spb.su

LIEBSCHER, Dierck — Astrophysik. Inst., Potsdam, Germany
deliebscher@aip.de

LINDER, Suzanne — Penn State University, A. and A. Dept.,
University Park, USA
slinder@astro.psu.edu

LINDNER, Ulrich — Univ.-Sternwarte, Göttingen, Germany
ulindner@eden.uni-sw.gwdg.de

LIPMAN, Keith — Inst. of Astronomy, Cambridge, UK
kl@cast0.ast.cam.ac.uk

LÓPEZ, Sebastian — Hamburg Sternwarte, Hamburg, Germany
st3b314@hs.uni-hamburg.de

LOWENTHAL, James — University of California, Santa Cruz, USA
james@lick.ucsc.edu

LOXEN, Johannes — Univ.-Sternwarte, Göttingen, Germany
jloxen@gwdg.de

LU, Limin — Caltech, Pasadena, USA
ll@troyte.caltech.edu

MADAU, Piero — STScI, Baltimore, USA
madau@stsci.edu

MATHUR, Smita — Center for Astrophysics, Cambridge, USA
smita@cfa.harvard.edu

MCMAHON, Richard — Inst. of Astronomy, Cambridge, UK
rgm@mail.ast.cam.ac.uk

MEIKSIN, Avery — University of Chicago, Chicago, USA
meiksin@oddjob.uchicago.edu

MEYLAN, Georges — ESO, Garching, Germany
gmeylan@eso.org

MIRALDA ESCUDÉ, Jordi — Inst. for Advanced Study, Princeton, USA
jordi@guinness.ias.edu

MO, Houjun — MPI für Astrophysik, Garching, Germany
hom@mpa-garching.mpg.de

MOLARO, Paolo — Osservatorio Astronomico, Trieste, Italy
molaro@oat.trieste.it

MØLLER, Palle — STScI, Baltimore, USA
moller@stsci.edu

MÜCKET, Jan — Astrophysik. Inst., Potsdam, Germany
jpmuecket@aip.de

MURAKAMI, Izumi — C.I.T.A., Toronto, Canada
murakami@cita.utoronto.ca

NORMAN, Colin — STScI, Baltimore, USA
norman@stsci.edu

OCH, Susanne — ESO, Garching, Germany
soch@eso.org

PERRY, Judith — Inst. of Astronomy, Cambridge, UK
jjp@mail.ast.cam.ac.uk

PETITJEAN, Patrick — Institut d'Astrophysique, Paris, France
petitjean@iap.fr

PETTINI, Max — Royal Greenwich Obs., Cambridge, UK
pettini@mail.ast.cam.ac.uk

PFENNIGER, Daniel — Observatoire de Genève, Switzerland
pfennige@scsun.unige.ch

PRIETO, Almudena — MPI für extraterrestrische Physik, Garching, Germany
alm@rosat.mpe-garching.mpg.de

RAO, Sandhya — Univ. of Pittsburgh, A. and A. Dept., Pittsburgh, USA
rao@phyast.pitt.edu

RAUCH, Michael — OCIW, Pasadena, USA
mr@ociw.edu

REES, Martin — Inst. of Astronomy, Cambridge, UK
mjr@mail.ast.cam.ac.uk

REIMERS, Dieter — Hamburg Sternwarte, Hamburg, Germany
dreimers@hs.uni-hamburg.de

RIEDIGER, Rüdiger — Technische Universität, Berlin, Germany
rschmidt@aip.de

ROBERTSON, Gordon — Sydney University, Sydney, Australia
jgr@astrop.physics.su.oz.au

SANZ, José Luis — ESA-VILSPA, Madrid, Spain
jls@vilspa.esa.es

SAUCEDO, Julio — Steward Observatory, Tucson, USA,
jsaucedo@as.arizona.edu

SAVAGE, Blair — Univ. of Wisconsin, Dept. of Astronomy, Madison, USA
savage@madraf.astro.wisc.edu

SAVAGLIO, Sandra — Universita della Calabria, Italy
savaglio@fis.unical.it

SCHNEIDER, Peter — MPI für Astrophysik, Garching, Germany
peter@mpa-garching.mpg.de

SCIAMA, Dennis — SISSA, Trieste, Italy
sciama@tsmi19.sissa.it

SEALEY, Katrina — University of New South Wales, Kensington, Australia
kms@newt.phys.unsw.edu.au

SHAVER, Peter
 ESO, Garching, Germany
 pshaver@eso.org
SHCHEKINOV, Yuri
 Rostov State Univ., Rostov-on-Don, Russia
 yuris@riphys.rnd.su
SIEBERT, Joachim
 MPI für extraterrestrische Physik,
 Garching, Germany
 jos@mpe-garching.mpg.de
SMETTE, Alain
 Kapteyn Astron. Inst., Groningen,
 The Netherlands
 asmette@astro.rug.nl
STEIDEL, Charles
 MIT, Cambridge, USA
 steidel@trystero.mit.edu
STENGLER-LARREA, Erik
 Royal Greenwich Obs., Cambridge, UK
 eas@mail.ast.cam.ac.uk
STOCKE, John
 Univ. of Colorado, CASA, Boulder, USA
 stocke@hyades.colorado.edu
STORRIE-LOMBARDI, Lisa
 Inst. of Astronomy, Cambridge, UK
 lsl@mail.ast.cam.ac.uk
SZUSZKIEWICZ, Ewa
 SISSA, Trieste, Italy
 ewa@tsmi19.sissa.it
THIMM, Guido
 ESO, Garching, Germany
 gthimm@eso.org
TRIMBLE, Virginia
 Univ. of Maryland, College Park, USA
 vtrimble@astro.umd.edu
TRIPP, Todd
 Univ. of Wisconsin, Dept. of Astronomy,
 Madison, USA
 tripp@madraf.astro.wisc.edu
TURNSHEK, David
 Univ. of Pittsburgh, A. and A. Dept.,
 Pittsburgh, USA
 turnshek@vm2.cis.pitt.edu
TYTLER, David
 University of California, San Diego, USA
 tytler@cass155.ucsd.edu
ULRICH, Marie-Helene
 ESO, Garching, Germany
 mulrich@eso.org
VILKOVISKIJ, Emmanuel
 Fessenkov Astrophysical Institute,
 Almaty, Kazakhstan,
 vilk@afi.academ.alma-ata.su
VLADILO, Giovanni
 Osservatorio Astronomico, Trieste, Italy
 vladilo@oat.ts.astro.it
VOGEL, Susanne
 Hamburg Sternwarte, Hamburg, Germany
 st2b306@hs.uni-hamburg.de
WAMPLER, Joe
 ESO, Garching, Germany
 jwampler@eso.org
WANG, Boqi
 Johns Hopkins Univ., Baltimore, USA
 boqi@pha.jhu.edu

WARREN, Steve   Imperial College, London, UK
                            s.j.warren@ic.ac.uk
WEBB, John   University of New South Wales,
                            Kensington, Australia
                            jkw@edwin.phys.unsw.edu.au
WEYMANN, Ray   OCIW, Pasadena, USA
                            rjw@ociw.edu
WHITE, Simon   MPI für Astrophysik, Garching, Germany
                            swhite@mpa-garching.mpg.de
WIEDEMANN, Günter   ESO, Garching, Germany
                            gwiedema@eso.org
WIKLIND, Tommy   Onsala Space Obs., Onsala, Sweden
                            tommy@oso.chalmers.se
WOLFE, Arthur   University of California, San Diego, USA
                            awolfe@ucsd.edu
WOMBLE, Donna   Caltech, Pasadena, USA
                            dsw@astro.caltech.edu
WOODGATE, Bruce   NASA Goddard Space Flight Center,
                            Greenbelt, USA
                            woodgate@uit.gsfc.nasa.gov

PART I

INTRODUCTION

QSO Absorption Lines: Progress, Problems and Opportunities

Ray Weymann

Carnegie Observatories, Pasadena CA 91101, USA

Abstract. Some brief impressions are given of the changes that have taken place in our view of the nature of QSO absorption lines over the last two decades. Some comments are next made on what is, in my opinion, one of the most interesting current problems, as well as the opportunities that the new generation of space and ground-based facilities, together with detailed numerical modeling, afford for its solution.

1 Introduction: Some Historical Perspectives

Since it is just a little less than 15 years ago that Carswell, Smith and I wrote a review on QSO absorption lines (Weymann, Carswell & Smith 1981) that review provides a convenient vantage point from which to gauge the progress we have made since then. In general it has been my experience that I cannot understand anything I have written after it is more than a few months old, but that review article was an exception. Indeed, I even agree with much of what was said in it, so let me start with a brief review of that review. We began by speculating that in the decade of the 1970s more large telescope time was devoted to QSO absorption lines than any other topic, part of the motivation being to "resolve the intense controversy...over the origin of QSO absorption lines." Although QSO absorption line spectroscopy is clearly a major activity at many large observatories, I sense that it is not quite the dominant observational activity that it was then, and that at the present time as much telescope time is now devoted to high redshift galaxy and "large scale" structure problems as to QSO absorption line spectroscopy. The "intense controversy" referred to was of course the question of whether the majority of the narrow highly displaced lines represented cosmologically intervening systems or ejected material and/or redshifts arising from new physics. Today, almost all astronomers familiar with this field would say that this controversy has been conclusively resolved. (At least for the heavy majority of narrow displaced lines—there are some new indications that a small number of narrow systems might be remnants of ejection episodes; Moller 1994).

With this issue largely behind us, QSO spectroscopy is now developing into the powerful tool for extragalactic research that we all tried to convince telescope allocation committees 15 years ago that it should be. We are now using this tool to probe the properties of uncondensed matter at very high redshifts and at very large impact parameters from galaxies in a manner which cannot be done with any other tool.

To continue just a bit longer with the review of this review, we found it useful to classify QSO absorption lines as follows:

A) Broad Absorption Lines.

B) Complexes of narrow lines ("associated absorption") lying within about 3000 km/sec of the emission line redshift.

C-1) Intervening systems with high enough column densities ($\sim$ 15.5 $<$ log $N_{HI} <$ 22) and/or high enough heavy element abundances that lines from heavy elements could be detected with the S/N and resolution then feasible.

C-2) Intervening systems in which no heavy element lines were detected with the current technology. (The "Lyα forest"—I believe this review may have been where that phrase was first introduced.)

Evidence is now becoming quite strong that the distinction between the "Lyα Forest" and heavy element systems is not a very meaningful distinction but rather simply reflects differences in various properties (*e.g.*, clustering) as a function of column density. Indeed, there is evidence (Tytler 1994) that some degree of metal enrichment may be an ubiquitous property of all absorbing clouds, at least out to redshifts greater than about 3 to 4.

Aside from this point, the classification described above still seems a reasonable one.

But there have been some major developments which lay almost entirely in the future at the time the review was written. Among these one might mention:

1) The properties and significance of the damped Lyα systems and their interpretation as galactic disks or proto–galactic disks. The potential of these systems for studying the detailed history of nucleosynthesis, the conversion of baryonic matter into stars and the dynamical properties of these disks has been pioneered by Wolfe and collaborators and is summarized by Wolfe (1994).

2) The use of the "proximity effect" to measure the intergalactic ionizing background radiation. Although we anticipated that this effect should exist, its development as a powerful tool to actually measure the strength of the ionizing background radiation, (*e.g.*, as described by Bechtold 1994) lay entirely in the future.

3) The successful launch of Space Telescope has of course opened up the low redshift regime for the exploration of the Lyα and C IV lines and we are now obtaining the first information on the evolution of the Lyα lines at these redshifts, as described by Boksenberg (1994) as well as the first information on the relation of these features to galaxies, as described by Lanzetta (1994), Bergeron (1994) and Stocke (1994).

The steady improvement in sensitivity of astronomical spectroscopic instrumentation is of course responsible for much of the gain in our knowledge of QSO absorption lines. Compared to the Mt. Wilson Coude Spectrograph with which I took my first spectrum (of α Ori in 1960), the gain is astonishing: Dunham (1956), describing this system, writes that "spectra of stars of the 6th...magnitude can be photographed... in about six hours under favorable seeing conditions" with a resolution of about 4 km s^{-1}. At this conference, work is reported based on spectra of QSOs at almost this resolution but over 10 magnitudes fainter!

2 Some Current Problems and Future Opportunities

Among the numerous current problems which are discussed at this workshop, I want to address here only one: Namely, what are the nature of the low redshift Lyα absorbers, their relationship to the high redshift Lyα absorbers and their relationship to galaxies.

It has often been suggested that there are "two populations" of absorbers. This suggestion has been raised in at least three different contexts: i) In distinguishing between high redshift systems which do, or do not, show accompanying heavy element absorbers ii) In distinguishing between Lyα absorbers of comparable column density at high redshift vs. those at low redshift and iii) In distinguishing between the low redshift Lyα absorbers which appear to be closely associated with galaxies, and those which do not.

As noted above, recent results (Tytler 1994) indicate that C IV lines are seen down to moderately low (log $N_{HI} \approx 14.5 \, cm^{-2}$) column densities, and it appears that the metallicity of these systems may be no lower than that at higher column densities (Cowie et al. 1995). There is thus very likely a continuity of properties from the higher column density systems to the lower column density systems, with apparently more tendency for higher amplitude velocity (and presumably spatial) clustering at the higher column densities.

In the case of context (ii), the current evidence suggests a slower rate of evolution at low redshifts than at high redshifts for the Lyα clouds and possibly a greater tendency for clustering at the lower redshifts (Boksenberg 1994). This, and the tendency for many of the low redshift Lyα absorbers to be associated with galaxies, suggests that possibly two different types of clouds might be involved, with a relatively unclustered rapidly evolving population dominating at high redshifts and a more stable population, possibly associated with the outer envelops of galaxies themselves, dominating at low redshifts.

Finally, at low redshifts, there are still somewhat conflicting results concerning the degree of association of the Lyα absorbers and galaxies, as well as the significance of this association. Lanzetta (1994) finds that luminous galaxies nearly always produce readily observable (e.g., with FOS) Lyα absorbers for impact parameters $\leq$ about 160 kpc h_{100}^{-1}. He also finds a correlation between the strength of the line and the impact parameter; see also Bergeron (1994). On the other hand, the less extensive material based upon the higher resolution GHRS data indicates that most of the Lyα lines—which are generally weaker than the lines detected by FOS, and which are about 4 times more numerous per unit redshift interval—are not nearly as strongly associated with luminous galaxies (Morris *et al.*1993). This is corroborated by Stocke (1994) who also notes that there are some absorbers which are found in voids. The likely resolution of this disagreement is that, as noted above in the context of the high redshift systems, there is a dependence of the degree of association of the absorbers upon column density. This possibility is reinforced by the new data presented by Lanzetta (1994)

In any event, it is still an interesting question to consider the origin and properties of these low redshift clouds. The fact that the covering factor for the production of moderate column density Lyα clouds for impact parameters $\leq$ 160 kpc from luminous galaxies is $\geq$ 1 is generally stated by saying that "luminous galaxies have huge halos". What exactly does this statement imply about the physical properties of the absorbing clouds? Does it imply, for example that the clouds consist of the gas bound gravitationally in very low luminosity dwarfs orbiting the galaxy, or, perhaps expelled in winds from such dwarfs, or, perhaps confined by ram pressure associated with motion through galactic halo gas, or, perhaps gas confined by the thermal pressure of very hot halo gas.

The last possibility has been considered in some detail by Mo (1994a ; see also Mo 1994b and the discussion by Murray *et al.*1993) and his analysis, summarized below, has some implications which may be explored by future HST observations, especially with the high resolution ($\sim$ 12 km s^{-1}) mode of STIS with its broad spectral coverage.

There is still only very meagre direct spectroscopic evidence for *extended* hot halo gas in galaxies (*i.e.*, to the dimensions implied by the impact parameters referred to above.) There is definitely evidence for very high ionization gas in our own galaxy, though not out to these dimensions (Savage 1994), and O VI as well as Ne VII is seen in FOS spectra of intermediate redshift systems (Bergeron 1994, Reimers 1994) though it is not clearly established yet whether these systems are collisionally ionized (indicative of a very hot gas) or ionized by hard UV radiation. There is fairly unambiguous evidence for hot X-ray emitting gas which is apparently "intra–group" gas in small groups of mostly early–type galaxies (Mulchaey *et al.* 1995a) as well as hot halo gas surrounding ellipticals and S0s (*e.g.,* Fabbiano 1995 and references therein.) Interestingly, however, the groups studied by Mulchaey *et al.* do *not* show detectable emission for groups dominated by spirals, possibly because their lower characteristic virial velocities imply temperatures which produce the bulk of the x-ray emission below the response of ROSAT (Mulchaey 1995b).

Consider uniform clouds pressure–confined by hot gas in the outer part of a roughly isothermal halo in hydrostatic equilibrium with the gravitational potential of a luminous galaxy. Various constraints can be imposed upon the locus of allowed regions in the $(n, T)_{hot}$ plane at some large distance (say $\sim$ 100 kpc) from the center of the galaxy. Assuming the ionization in the cloud is due to the background UV ionizing radiation (as opposed to local sources associated with the galaxy) then at low redshifts the characteristic size of the cloud will be

$$R_{kpc} \approx 24 \times N_{14} \times J^* \times (T_c^*)^{11/4}/(P')^2. \tag{1}$$

where N_{14} is the neutral hydrogen column density in units of 10^{14}, J^* is the value of the background radiation field at the Lyman limit in units of 5.0×10^{-23}, T_c^* is the cloud temperature in units of 3×10^4, and P' is the product of the hot halo gas density and temperature, $n_h \times T_h$.

Mo identifies several mechanisms by which clouds may be destroyed, and if such destruction occurs on a fairly short time scale ($\ll t_{Hub}$) an ample source of material and an efficient means of producing the clouds would have to be found. One such mechanism is gravitational collapse (which might lead to the production of a star cluster). A second is evaporation by thermal conduction if the clouds are too small or the gas too hot. Other possibly significant destruction mechanisms are tidal disruption, cloud—cloud collisions and the Kelvin–Helmoltz instability. Mo concludes that the typical destruction time scale is likely to be of order 10^9 years, provided the cloud radii are not much less than about 10 kpc (in which case the time scale for destruction by the KH instability becomes progressively shorter.)

In addition, there is the question of the origin of the hot halo itself and how its temperature is maintained. The halo is presumably formed by shock heating during the collapse phase of the galaxy itself: If the cooling time for the gas is not short compared to a Hubble time, the hot gas will remain as a relic of the collapse phase. If the cooling time were shorter than the Hubble time a continuous source of heating—either internal, or external (e.g., accretion) would have to be present. However, Mo also sets a comparable empirical constraint on the physical conditions of hot halo gas by the total x-ray luminosity which would be emitted by hot, dense, halo gas. (The following caveats to these arguments should be mentioned, however: If it is supposed that at some impact parameter, e.g., 100 kpc, the density and temperature of the gas are such that the cooling time is comparable to the Hubble time, then, if the density varies as roughly $n_h \sim r^{-2}$, then at e.g., 30 kpc, the cooling time will be very short compared to a Hubble time and a heating source would have to found anyway to avoid the collapse of the halo. Additionally, as mentioned above, if the temperature is too low, the x-rays may be too soft to have been detected, even though the total luminosity is quite high.)

Finally, there is the question of the origin of the clouds in the first place— if the cooling time of the halo is indeed greater or comparable to the Hubble time, then the clouds cannot have cooled via thermal condensation unless there were a mechanism which previously raised the cloud gas to higher densities. Mo suggests that gravitational confinement by minihalos provides such a mechanism, and that the gas inside the minihalos is stripped by the ram pressure of the larger parent galaxy during merger processes leaving the gas pressure–confined by the parent galaxy hot halo.

Considering all the foregoing, the most likely regime in the outer halos of galaxies for the existence of pressure–confined clouds seems to be a region running from about $\log n_h \approx -6$ for T_h slightly above 10^6 (corresponding to virial velocities ~ 200 km s^{-1}) down to about $\log n_h \approx -5$ for temperatures of about $T \approx 3 \times 10^5$ (corresponding to virial velocities of about 100 km s^{-1}.) In particular, the much hotter and comparatively dense gas giving rise to the x-ray emission in giant ellipticals and in small groups dominated by early type galaxies referenced above, would seem a rather inhospitable environment for the clouds. This is an expectation that can readily be tested when the identification of the morphological types of the galaxies in the vicinity of the Lyα clouds becomes complete.

The model also has two other attractive features: It accounts at least qualitatively for the anti–correlation between line strength and impact parameter as discussed by Lanzetta (1994) and Bergeron (1994). Quantitatively, for clouds of roughly the same mass, one expects the column densities of the clouds to decrease with impact parameter roughly as $r^{-10/3}$. Moreover, if we express the column density in terms of the total gas mass of the cloud instead of the radius, then equation (1) takes the form

$$ N_{14} \approx (M/4.5 \times 10^7 M_\odot)^{1/3} \times (P')^{5/3}/(J^* \times (T_c^*)^{29/12}). \tag{2} $$

The insensitivity of the H I column density to the cloud mass and its sensitivity to the confining pressure has been noted by Webb and Barcons (1991) in a somewhat different context. Here, it suggests why it may be that very weak lines have thus far been found to be not strongly correlated with galaxies: The pressure in the vicinity of luminous galaxies or groups of galaxies may be too high for very low column density clouds to exist. The fiducial value of $J_\nu = 5 \times 10^{-23}$ used by Mo is almost certainly too high for the value at *zero* redshift (Vogel & Weymann 1995), unless there are strong local sources of UV radiation able to escape into the halo. However, the background radiation is expected to rise rapidly with increasing redshift and equation (2) suggests, that, other things being equal, smaller column densities may be found near galaxies at moderate redshifts than at the very lowest redshifts.

If the clouds are moving in the galaxy halo with roughly the virial velocity, then some lines of sight will frequently intersect more than one cloud and complex velocity structure should be seen, but this requires resolutions higher than the FOS data to be seen. Such complexity is notable for its rarity in the relatively meagre data obtained with the GHRS G160M grating, but this data mostly involves lines of sight which do *not* involve 100–kpc scale impact parameters. It is not entirely clear what should be expected in any case—the process which strips the gas from minihalos will, as the structure is shredded by K-H instabilities, also be brought to rest with respect to the halo gas before it finally dissipates and more detailed simulations along the lines initiated by Murray *et al.* (1993) should be carried out. Observationally, more lines of sight with small impact parameters should be investigated with GHRS–G160M and STIS.

Finally, an additional intriguing possibility exists for direct detection of the hot halo gas in Lyα, provided the temperature of the halo gas doesn't much exceed 3×10^5: At an impact parameter of 100 kpc and gas density of $\sim 10^{-5}$ a broad (~ 70 km s^{-1} Doppler width) weak Lyα line should be readily detectable with high S/N GHRS or STIS spectra. Such a weak broad line should be readily distinguishable from any cooler Lyα clouds embedded in the halo. As emphasized by Bergeron, if such a halo has even very modest metal enrichment, than O VI would be readily detectable as well, though its unambiguous association with the hot halo gas may be a little more difficult.

Although the model of pressure–confined clouds in hot halos of luminous galaxies has several attractive features, it almost certainly cannot account for *all* the low redshift Lyα lines, especially those at the lowest column densities, some of which appear to be located well away from any galaxies (except possibly those of extremely low optical luminosity.) One of the lessons of the detailed N–body + hydrodynamical calculations carried out at high redshifts by Cen *et al.* (1994) and more recently by Hernquist, Katz, Weinberg and Miralda–Escude (1995), is that the dynamical processes at work which produce the high redshift Lyα absorption are varied and complex, and in retrospect the simple spherical quasi–static models invoked to explain them seem rather naive. It should be a high priority task to see that such simulations are extended to zero redshift.

In connection with the remark at the outset concerning the incredible gain over the last 4 decades in the sensitivity with which astronomical spectra can be obtained, it is also worth noting that the gain in computational speed far, far exceeds even this gain. As a graduate student I still vividly remember performing many–digit multiplications with a "high speed" Marchant calculator with a computation time of a few seconds: The current generation of parallel processors represent gains of about 9 orders of magnitude in speed! While we do not quite yet have the spatial resolution and the inclusion of all the detailed physics that one would wish for in order to definitively compare the simulations with observations, this should become available within the coming decade. Coupled with the new generation of ground–based and space telescopes there is every expectation that the use of QSO absorption lines will in fact fully realize the potential for studying the evolution of structure in the universe that we had hoped it would when our review of this subject was written in 1981.

References

Bechtold, J. 1994, paper presented at this Workshop.

Bergeron, J. 1994, paper presented at this Workshop.

Boksenberg, A. 1994, paper presented at this Workshop.

Cen, R., Miralda-Escude, J., Ostriker, J.P., & Rauch, M. 1994, ApJ (Letters) 437, 9.

Cowie, L.L., Songaila, A., Kim, T-S., & Hu, E.M. 1995, submitted to AJ.

Dunham, T. Jr. 1956, in *Vistas in Astronomy* ed. Beer, A. (London: Pergamon Press), vol 2 p. 1223.

Fabbiano, G. 1995, Harvard–Smithsonian Center for Astrophysics Preprint # 4055, and references therein.

Hernquist, L., Katz, N., Weinberg. D., & Miralda–Escude, J. 1995, in preparation.

Lanzetta, K. 1994, paper presented at this Workshop.

Mo, H.J. 1994a, MNRAS 269, L49.

Mo. H.J. 1994b, poster paper presented at this Workshop.

Moller, P. 1994, paper presented at this Workshop.

Morris, S.L., Weymann, R.J., Dressler, A., McCarthy, P.J., Smith, B. A., Terrile, R.J., Giovanelli, R. & Irwin, M. 1993, ApJ 419, 524.

Mulchaey, J.S., Davis, D.S., Mushotzky, R.F., & Burstein, D. 1995a, submitted to ApJ.

Mulchaey, J.S. 1995b, private communication.

Murray, S.D., White, S.D.M., Blondin, J.M., & Lin, D.N.C. 1993, ApJ 407, 588.

Reimers, D. 1994, paper presented at this Workshop.

Savage, B.D. 1994, paper presented at this Workshop.

Stocke, J. 1994, paper presented at this Workshop.

Tytler, D. 1994, paper presented at this Workshop.

Vogel, S., & Weymann, R.J. 1995, in preparation.

Webb, J., & Barcons, X. 1991, MNRAS 250, 270.

Weymann, R.J., Carswell, R.F., & Smith, M.G. 1981, Ann. Rev. Astron & Astroph 19, 41.

Wolfe, A. 1994, paper presented at this Workshop.

PART II

DAMPED

LYMAN-ALPHA SYSTEMS

Evolution of the Neutral Gas and Metal Content of Damped Lyman Alpha Systems

A. M. Wolfe

Department of Physics and CASS, University of California, San Diego, 0111, La Jolla CA, 92093 USA

Abstract. Surveys for damped Lyα absorption systems indicate that the neutral gas content of the universe increases with redshift at $z < 3.3$. The evolution may be due to gas consumption by star formation in protogalaxies. The redshift dependence of the neutral gas content at $z > 3.3$ places restrictions on various models for structure formation. The metal content of the damped Lyα systems is discussed. Accurate spectra obtained with the HIRES echelle spectrograph on the Keck 10 m telescope are used to obtain accurate metallicities of damped Lyα systems for the first time. The small sample reveals a wide range of metallicities. The kinematic structure of the gas is consistent with a thick, rapidly rotating disk.

1 Introduction

The damped Lyα systems comprise a population of high-redshift H I layers that dominate the neutral gas content of the universe in the redshift interval $z = [0,4]$. In this paper I discuss evidence that the damped systems are the progenitors of nearby ordinary galaxies. In § 2 I discuss new evidence that the comoving mass density of neutral gas decreases with decreasing redshift at $z < 3.3$. Section § 2 discusses evidence that the gas content evolves because of star formation. I also discuss how the gas evolution restricts the cold + hot dark models for structure formation. In § 3 I discuss recent evidence concerning the metal content of the damped Lyα systems. Evidence obtained from the Keck 10 m telescope is presented concerning the metal abundance and kinematic structure of the gas which gives rise to damped Lyα absorption. Concluding remarks are given in § 4.

2 Evolution of Neutral Gas Content

The evidence for evolution is given in Figure 1 which plots $\Omega_g(z)$, the comoving density of neutral gas, in units of current critical density, versus redshift. The data points are inferred from a statistical sample of 62 damped Lyα systems found along a redshift path $\Delta z = 324$ in which systems with $N(\text{H I}) \geq 2 \times 10^{20}$ cm^{-2} could have been detected at a confidence level exceeding 10σ. The primary data consists of (a) low-resolution surveys for candidate damped Lyα absorption features and (b) intermediate-resolution follow-up

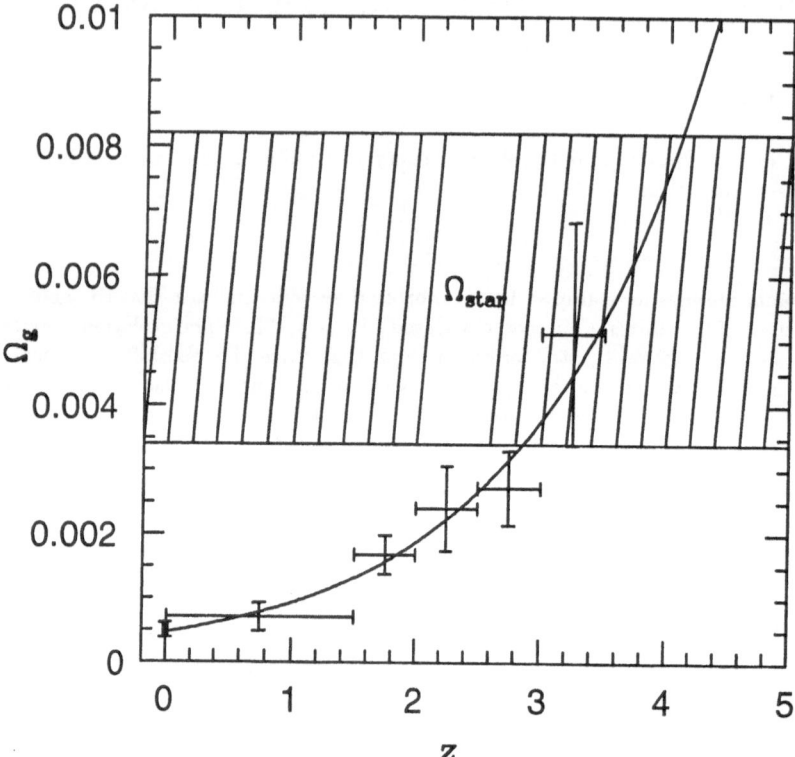

Fig. 1. Comoving mass density of neutral gas in damped Lyα systems *vs* redshift z for cosmology with $q_0 = 0.5$ and $\Lambda = 0$. Absorption redshift z_{abs} and $N(\text{H I})$ pairs from the statistical sample and the selection function $g(z)$ in Wolfe et al. 1995a were used to derive $\Omega_g(z)$. Vertical error bars are 1σ errors and horizontal bars indicate bin sizes. Data point at $z = 0$ derived from 21 cm emission by nearby galaxies (Rao & Briggs 1993). Redshift bins of $z = 0.008 - 1.5$, $1.5 - 2.0$, $2.0 - 2.5$, $2.5 - 3.0$, and $3.0 - 3.5$ are adopted. Horizontal lines are the $\pm$ 1σ contours for estimates of Ω_{star}, the cosmological density of visible matter in nearby galaxies. Solid curve is maximum likelihood fit to exponential function. The fit is extended beyond $z = 3.5$ for illustrative purposes only.

spectroscopy required to test whether or not a candidate is in fact a damped Lyα line arising from gas with $N(\text{H I}) \geq 2 \times 10^{20}$ cm^{-2}, the column-density threshold of the survey.

The data in Figure 1 show a factor of 10 decrease in $\Omega_g(z)$ between $z \approx 3.3$ and $z = 0$. The decrease is caused by the decline of the high column-density systems with $N(\text{H I}) \geq 10^{21}$ cm^{-2}. By contrast there is no evidence for evolution of systems with $2 \times 10^{20} < N(\text{H I}) < 10^{21}$ cm^{-2}. Figure 1 shows another interesting phenomenon. The decrease of $\Omega_g(z)$ measured from damped Lyα lines extends naturally to the data point at $z = 0$ which is inferred from 21 cm emission by nearby galaxies (Rao & Briggs 1993).

2.1   Gas Consumption by Star Formation

So far, there have been two attempts to interpret these results. The first interpretation is from the "..what you see is what you get.." school; namely, the decrease in $\Omega_g(z)$ is caused by gas consumption by star formation (Lanzetta, Wolfe, & Turnshek 1995; Wolfe et al. 1995a). On the other hand Mike Fall (see this volume) assumes the decline in $\Omega_g(z)$ to be an apparent effect mainly caused by obscuration of background QSOs from dust in foreground damped Lyα systems. I will only discuss the star formation model

The evidence for star formation is as follows. First, Figure 1 shows that $\Omega_g(z)$ in the highest redshift bin is in good agreement with Ω_{star}, the cosmological mass density of visible stars in current galaxies. As a result the damped Lyα systems contain sufficient baryons in the form of cold neutral gas to form all the visible stars we see today. Second, the link between our measurements of $\Omega_g(z)$ and the mass density of gas contributed by nearby spirals is evidence for star formation, because the current gas content of galaxies is the residue of star formation during the past Hubble time. It is interesting that ordinary spirals and not dwarf galaxies dominate the present cosmic density of neutral gas, because ordinary spirals also dominate Ω_{star}. Third, empirical expressions for gas consumption by star formation in nearby galaxies predict the consumption rate to increase as a non-linear function of $N(\text{H I})$ (Schmidt 1959) which explains why the decline in gas is most rapid at high column densities. A model where damped Lyα systems are assumed to be randomly oriented exponential disks in which gas consumption follows the Schmidt (1959) law is in good agreement with the observed evolution of $f(N, z)$, the frequency distribution of H I column densities (Wolfe et al. 1995a).

2.2   Evolution at $z > 3.5$

Does the increase of $\Omega_g(z)$ with z continue at redshifts higher than the highest redshift data point in Figure 1? There is tentative evidence by Storre-Lombardi et al. (this volume) that $\Omega_g(z)$ does *not* keep increasing at higher redshifts, and that it may in fact decrease with increasing z at $z > 3.5$. The latter result is tentative because only a tiny fraction of the damped Lyα candidates have been confirmed. Nevertheless, the data of Storre-Lombardi et al. is probably sufficiently accurate to rule out the type of exponential increase of $\Omega_g(z)$ with z shown as the the solid curve in Figure 1. This is a maximum-likelihood fit to the data, which has been extrapolated to higher redshifts. In any case, comparison with Ω_{bbns}, the baryon density required by big-bang nucleosynthesis to explain the light-element abundances, reveals that the baryons in damped Lyα systems account for about 10 % or more of the total baryonic content of the universe.

The $\Omega_g(z)$ relation has also been used to constrain hierarchical theories of galaxy formation. Specifically, cold dark matter models in which the power spectrum of density fluctuation amplitudes is constrained at large scales by the COBE result generate too much power on small scales (Ma & Bertschinger 1994). The introduction of hot dark matter helps, since for a fixed large-scale normalization much of the small-scale power is removed But this also delays the formation of galaxies and of damped Lyα systems, if they are assumed to be collapsed baryonic material in the potential wells of normal galaxies. The latter assumption has been adopted by various authors to constrain the mixed cold+hot dark matter models. Although there are differences in details, the consensus view is that the cold+hot model cannot account for the $\Omega_g(z)$ observed at high redshifts unless the damped systems at $z > 3$ are low mass dwarf galaxies (Klypin et al. 1995).

3   Metal Content

So far I have focused on what the Lyα absorption lines in damped Lyα systems tell us about the neutral gas content of high-redshift galaxies. I will now describe what we are learning from the metal lines.

We have started an ambitious program using the Keck 10 m telescope and its high-resolution echelle spectrograph, HIRES (Vogt 1992), to obtain accurate velocity profiles of various low-ion transitions in damped Lyα systems. The principal goals are to

(a) record the emergence of metals in galaxies,

(b) trace their abundances relative to hydrogen from $z \approx 4.5$ to the present,

(c) determine the kinematic state of the absorbing gas.

Because the damped systems are likely to be drawn from the bulk of the galaxy population at high redshifts, the evolutionary history of *these* metallicities and kinematics should be more representative of the overall galaxy population than for similar quantities measured in the solar neighborhood.

3.1   Abundance Methodology

Element abundances are determined from resonance absorption lines arising in damped Lyα systems. We concentrate on weak, unsaturated lines arising in low-ionization species such as Zn^+, Cr^+, and Ni^+. These are the dominant ionization states in gas with $N(\text{H I}) \geq 2 \times 10^{20}$ cm^{-2}. The total column density of the gas is given by $N(\text{H I})$ which is well determined from the damping profiles. As a result ionization corrections to ratios such as $N(Zn^+)/N(H^0)$ are not required to compute the Zn abundance. The gas phase abundance of Zn is especially important, since Zn is relatively undepleted by grains in the ISM (Pettini et al. 1994). Moreover Zn should trace Fe, the most widely used indicator of metallicity, because [Zn/Fe] $\approx$ 0 for stars with $-$ 3.0 $<$ [Fe/H] < 0, where the logarithmic abundance ratio of elements X and Y, [X/Y]$\equiv$ log(X/Y)$-$log(X/Y)$_\odot$. While previous determinations of [Zn/H] at intermediate resolution (Pettini et al. 1994) have limited accuracy, because Zn II 2062.66 is blended with Cr II 2062.23, these transitions are easily resolved at the 4 $-$ 8 km s^{-1} resolution of HIRES. Weak, unsaturated lines such as Zn II 2026 and Ni II 1741 are important for another reason; namely, their velocity profiles trace the kinematics of the highest column-density gas. We rely on stronger transitions such as Si II 1527 and Fe II 1608 to determine the kinematics of the lower column-density gas.

3.2 Abundances

We have obtained accurate metal abundances for 3 damped Lyα systems. Results for the $z = 2.309$ absorber toward PHL 957 have already been published (Wolfe et al. 1994), and have been submitted for publication in the case of the $z = 2.468$ damped system toward Q0201+36 (Prochaska & Wolfe 1995). Here I concentrate on the $z = 1.7764$ damped system toward Q1331+17 (Wolfe et al. 1995b). We used HIRES to obtain spectra with signal-to-noise ratios per pixel of $\approx 50 - 100$ at resolution $\Delta v \approx 6$ km s^{-1}. We resolved Zn II 2062.664 from Cr II 2062.234, features that are severely blended in all previous intermediate-resolution spectra. In Figure 2, profiles for 6 selected transitions detected in the damped system are transformed to a Doppler velocity scale in which $v = 0$ km s^{-1} corresponds to $z = 1.7764$. The profiles are arranged in order of increasing saturation. The following is a brief description of the velocity structure.

(1) The low ions Zn$^+$ and Si$^+$ give rise to the weak absorption lines Zn II 2026 and Si II 1808. In both transitions the bulk of the gas is confined to a narrow (FWHM ≈ 25 km s^{-1}) velocity structure near $v = 0$ km s^{-1}. Both transitions also show a weak absorption wing between $v = 25$ and 60 km s^{-1} (the narrow feature at $v = 50$ km s^{-1} in the Zn II profile is Mg I 2026.47). The wing contains lower column density gas not seen at negative velocities.

(2) Owing to the higher abundance of Fe, the Fe II 1608 and Fe II 2374 lines saturate near $v = 0$ km s^{-1} and are also strong in the positive velocity wing. The stronger Si II 1527 transition is heavily saturated across the same velocity interval over which the weaker lines are detected.

(3) The highly ionized gas traced by the C IV 1548 transition has a velocity structure which is completely disjoint from the low-ion gas. This feature exhibits strong absorption between $v = -$ 200 and $+$ 230 km s^{-1}. This is the broadest C IV absorption line in our sample of damped systems investigated with HIRES.

Element abundances are determined by fitting the velocity profiles with multiple Gaussian velocity components using VPFIT, a least squares technique developed by R. C. Carswell and colleagues. Each component is as-

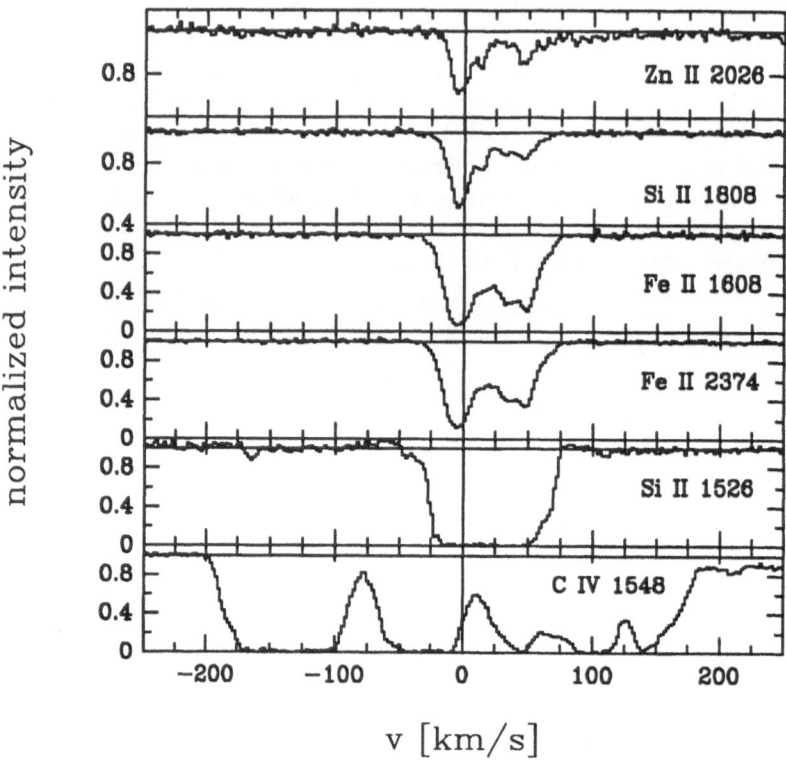

Fig. 2. Velocity profiles for metal lines in the $z = 1.7764$ damped Lyα system toward Q1331+17. The HIRES data was acquired at a resolution with FWHM $\approx$ 6 km s^{-1}. The velocity $v = 0$ corresponds to $z = 1.7764$.

signed a velocity centroid, ionic column density, and Doppler parameter. The column densities are deduced independently using the apparent depth technique in which the column density per unit velocity interval is inferred directly from the line profile (Savage & Sembach 1991). Column densities deduced from the two methods agree to within 10 %. The results of a 7 component fit are shown in Figure 3. Here I by plot the gas-phase abundance *vs* condensation temperature, T_c, a practice adopted by workers studying the ISM (Jenkins 1987). The idea is to determine whether or not the abundance pattern in damped Lyα systems shows the anti-correlation between [X/H] and T_c expected if element depletion by grain formation in stellar atmospheres occurs.

If O and Zn are relatively undepleted as in the ISM, then Figure 3 shows the $z = 1.7764$ damped system toward Q1331+17 to be metal poor. Specifical-

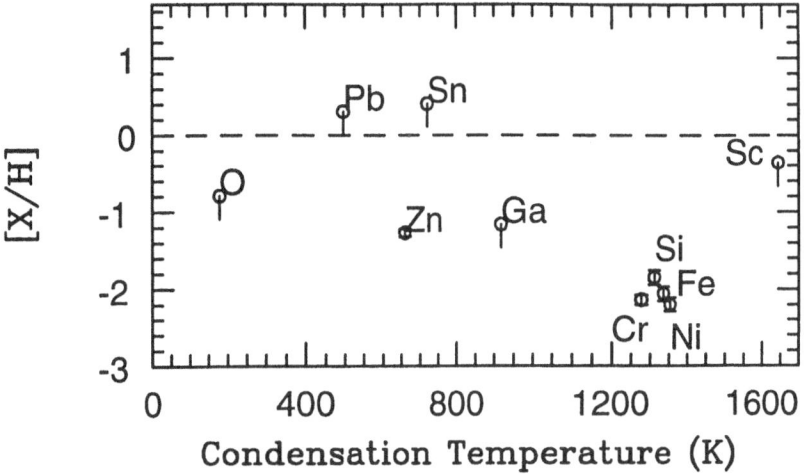

Fig. 3. Element abundance *vs* condensation temperature in the $z = 1.7764$ damped Lyα system toward Q1331+17. $[\mathrm{X/H}] = \log(\mathrm{X/H}) - \log(\mathrm{X/H})_\odot$. Thus dashed line represents solar abundance. Condensation temperature is temperature at which 50% of atomic species condenses into solid phase. Circles with error bars derived from ionic column densities determined from profile fitting and $N(\mathrm{H\ I}) = 1.5 \times 10^{21}\,\mathrm{cm}^{-2}$. Circles with negative vertical lines are 3σ upper limits on abundances.

ly, $[\mathrm{Zn/H}] = -1.27\pm0.06$ and $[\mathrm{O/H}] < -0.79$. Therefore, $[\mathrm{O/Zn}] < 0.48$ which is consistent with element enrichment by type II and type Ia supernovae, i.e., the data do not distinguish between a halo and a late disk abundance pattern. The limits on the S process elements Pb and Sn are not yet meaningful since sub-solar abundances are expected when the metallicity is low. The *overall* abundance pattern is consistent with grain depletion, since Si, Cr, Fe, and Ni have lower gas-phase abundances than Zn, as predicted.

Are the abundances patterns in the Q1331+17 damped system typical? Our sample is too small to answer that question. But it is clear that the metallicities determined with Zn exhibit a wide scatter. For example the other damped systems in our sample have $[\mathrm{Zn/H}] = -1.55$ and > -0.3. Moreover, neither system shows convincing evidence for a grain depletion pattern in the $[\mathrm{X/H}]$ *vs* T_c diagram. From their intermediate-resolution study of 17 damped systems Pettini et al. (1994) find a mean $[\mathrm{Zn/H}] = -1$ at $z \approx 2$. If confirmed by high-resolution spectroscopy, this conclusion has important consequences for the chemical evolution of the damped Lyα galaxies. The problem is illustrated in Figure 1 which shows $\Omega_g(z)$ to decrease from Ω_{star}

at $z = 3.3$ to one half this value at $z = 2$. Therefore, half the stars in galaxies descended from damped Lyα systems would have [Fe/H] = -1 in contrast to the solar neighborhood where less than 0.02 of the stars have [Fe/H] $<$ -0.6. This cosmic "G dwarf problem" suggests the damped Lyα systems do not evolve directly to chemically enriched galactic disks.

4   Kinematics

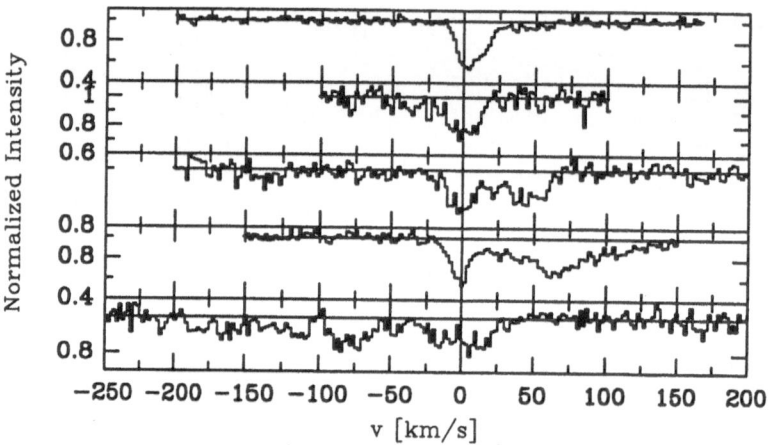

Fig. 4. Ni II 1741 velocity profiles in 5 damped Lyα systems. The profiles were acquired with HIRES with velocity resolution of 6 and 8 km s^{-1}.

I now turn to kinematics of the gas. The velocity profiles of the weak metal lines exhibit an asymmetry that may be crucial in understanding the nature of the damped systems. In Figure 4 I compare the Ni II 1741 profiles for 5 systems I observed with HIRES. In every case the highest column-density component is either at the red or blue edge of the velocity profile. This is evidence for systematic rather than random motions of the absorbing gas. The most natural explanation for the profile shape is passage of the sightline to the QSO through a rotating disk of discrete clouds in which (a) the rotation speed, Θ, is constant and (b) the number density of clouds decreases exponentially with cylindrical radius R and distance from the plane, $|z|$; i.e., $n_{cld}(R, z) = n_{cld}(0,0)\exp[-(R/R_d+|z|/h)]$.

Figure 5 shows there are two effects causing the profile asymmetry. The first is a radial effect illustrated in the left face-on panel. The figure shows how v_{los}, the line-of-sight velocity, decreases with increasing $|y|$ where y is the coordinate along the line of sight. Because the cloud density also decreas-es with increasing $|y|$, the dominant component of the resultant profile is at $v_{los} = \Theta\sin(i)$ and the weaker components are at $v_{los} < \Theta\sin(i)$; i.e., the

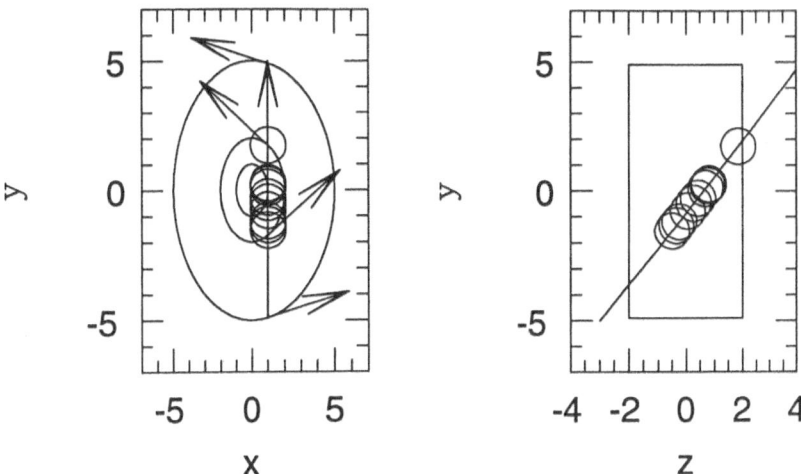

Fig. 5. Schematic diagram showing shadow of inclined sightline projected onto plane of disk, where i is angle of inclination. (a) Left panel gives face-on projection. Sightline is line $x = 1$, where all dimensionless coordinates are in units of radial scale-length R_d. Ellipses are compressed circles with radii $r = 1, 2$, and 5. Arrows represent rotation velocities tangential to the 3 circles. Small circles are cloud locations drawn with random-number generator from exponential distribution. (b) Right panel gives edge-on view showing where sightline intersects disk midplane.

strongest component is at the red boundary of the profile. The second effect is illustrated in the right edge-on panel. In this case $i = 54°$ and the sightline intersects the disk midplane at $y = -1$. Because the cloud distribution peaks at $z = 0$, clouds are also concentrated at $y = -1$. Therefore, the cloud locations along the sightline are drawn from the product of the radial and z distributions. Since $h < R_d$, the resulting distribution is peaked at $y = -1$ with an asymmetric tail leading toward $y = 0$. This is why most of the clouds are between $y \approx -1.2$ and 0.

To compute the resultant velocity profiles I transferred radiation through the cloud configuration in Figure 5. Each cloud was assigned a Gaussian velocity distribution with $b = 6$ km s^{-1} and the same ionic column density. I assumed $\Theta = 250$ km s^{-1} and $h = 0.5R_d$: a thick disk is required so that velocity spreads of ~ 100 km s^{-1} can be reproduced by typical inclination angles. The computed profile was then smoothed by the instrumental resolution of 8 km s^{-1}. For optically thin lines such as Ni II 1741 the dominant component was on the red edge of the profile. When the sightline was changed to penetrate midplane away from $y = 0$, at $y = -4$, the asymmetry was reproduced, but the dominant component was at the blue edge of the profile at $v_{los} < \Theta\sin(i)$. A series of Monte Carlo experiments showed that in about 80 % of randomly placed sightlines the resultant profile was asymmetric. Similar experiments with clouds drawn randomly from a uniform velocity distribution produced profile asymmetries in only 10 % of the cases. Therefore, the rotating disk model is consistent with our limited data set.

5 Conclusions

The evidence cited so far supports the hypothesis that the damped Lyα systems are galaxy progenitors. The main difficulty with this idea is the cosmic "G dwarf" problem (see § 3). I suggest two possible solutions. First, the determination of $\Omega_g(z = 3.3)$ is based on the detection of only 4 damped systems in a redshift path $\Delta z = 12.2$. Recent tentative results based on a redshift path twice as large suggest that $\Omega_g(z = 3.3)$ is lower (Storre-Lombardi et al. [this volume]). If this result is confirmed, the fraction of gas consumed between $z = 3.3$ and $z = 2$ would be much less than 0.5 which would lower the fraction of stars in damped Lyα galaxies that are metal poor. Second, studies of nearby galaxies suggest half the visible mass is in disks and half in bulges and spheroids (Schechter & Dressler 1987). Suppose most of the gas consumed by $z = 2$ goes into forming stars in galactic bulges rather than disks. Abundance studies of nearby spirals indicate that bulges contain a higher fraction of metal poor stars than disks; i.e., bulges do not have a G dwarf problem. Therefore, it is possible that the metallicity distribution of $z > 2$ damped Lyα systems may be similar to that of galactic bulges, an idea that we intend to test with a larger sample of HIRES spectra of damped Lyα systems.

References

Jenkins, E. B.1987, in Interstellar Processes, ed. D. J. Hollenbach & H. A. Thronson (Dordrecht:Reidel),533

Klypin, A., Borgani, S., Holtzman, J. & Primack, J. 1995, ApJ, in press.

Lanzetta, K.M., Wolfe, A.M. & Turnshek, D.A., 1995, ApJ, 440, 435

Ma, C.-P., & Bertschinger, E. 1994, ApJ, 429, 22

Pettini, M., Smith, L.J., Hunstead, R.W. & King, D.L., 1994, ApJ, 426, 79

Prochaska, J. X. & Wolfe, A. M. 1995, ApJ, submitted

Rao, S., & Briggs, F. H. 1993, ApJ, 419, 515

Savage, B. D., & Sembach, K. R. 1991, ApJ, 379, 245

Schechter, P. L., & Dressler, A. 1987, AJ, 94, 563

Schmidt, M. 1959, ApJ, 129, 243

Vogt, S. S. 1992, in ESO Conf. and Workshop Proc. 40, High Resolution Spectroscopy with the VLT, Ed. M.-H.Ulrich (Garching:ESO),223

Wolfe, A. M., Fan, X.-M, Tytler, D., Vogt, S. S., Keane, M. J., & Lanzetta, K. M. 1994, ApJ, 435, L101

Wolfe, A. M., Lanzetta, K. M., Foltz, C. B., & Chaffee F. 1995a, ApJ, submitted

Wolfe, A. M., Oren, A. L., Carswell, R. C., Songaila, A., Cowie, L. L., Vogt, S. S., & Keane M. J. 1995b, in preparation.

Consequences of Dust in Damped Lyman-Alpha Systems

S. Michael Fall[1] and Yichuan C. Pei[2]

[1] Space Telescope Science Institute, Baltimore, MD 21218, USA
[2] The Johns Hopkins University, Baltimore, MD 21218, USA

Abstract. Dust has been detected in damped Lyα systems at high redshifts by two independent methods: the reddening of background quasars and the gas-phase depletion of Cr relative to Zn. Both methods give a typical dust-to-gas ratio roughly 10% of that in the Milky Way. Here, we review some of the consequences of these observations. First, the dust in damped Lyα systems causes obscuration and hence incompleteness in optically selected samples of quasars. We estimate that up to 70% of the bright quasars at $z = 3$ are missing from such samples. Corrections for this bias help to reconcile the ionizing UV background inferred from the proximity effect with that expected from quasars alone. Second, since the existing samples of damped Lyα systems were obtained from the spectra of optically bright quasars, they too must be incomplete. This affects the distribution of HI column densities $f(N, z)$ and, especially, the comoving density $\Omega_{\mathrm{HI}}(z)$. The bias therefore has implications for the histories of gas consumption, star formation, and metal production in the damped Lyα systems. In particular, it solves a puzzle that has arisen in attempts to understand the chemical evolution in the damped Lyα systems (the "G-dwarf problem"). Third, the dust is likely to absorb most of the Lyα photons produced in the damped Lyα systems and thereby account for the null results of most searches for Lyα emission from them.

1 Introduction

The damped Lyα systems represent our best hope of tracing the evolution of normal galaxies at intermediate and high redshifts ($z \gtrsim 1$). They contain most of the cool, neutral gas in the Universe and are usually interpreted as the progenitors of present-day galactic disks. Thus, it has been a high priority of recent quasar absorption-line studies to determine as many of their properties as possible and over as wide a range in redshift as possible. The articles in these Proceedings by Wolfe and Pettini et al. describe what is known about the HI and metal content of the damped Lyα systems. Here, we concentrate on their dust content. This has consequences for a surprisingly wide range of topics, including the obscuration of quasars, cosmic chemical evolution, and searches for Lyα emission from primeval galaxies.

2   Evidence for Dust

A direct indication that the damped Lyα systems contain dust comes from
the reddening of background quasars (Fall, Pei, & McMahon 1989; Pei, Fall,
& Bechtold 1991). This is shown in Figure 1 — a comparison between the
spectral indices of quasars that do and do not have damped Lyα systems
in the foreground (the "damped" and "control" subsamples). The reddening
is small (typically, $\Delta\alpha \approx 0.4$) but significant at more than 4σ. This is a
statistical version of the familiar method for determining the reddening of
a star, with the spectra of quasars in the control subsample playing the
role of an unreddened stellar spectrum. From the typical reddening and an
assumed shape of the extinction curve, one can infer the typical dust-to-gas
ratio in the damped Lyα systems. We use the notation $k \equiv 10^{21}(\tau_B/N)\text{cm}^{-2}$,
where τ_B is the extinction optical depth in the rest-frame B band, and N
is the column density of HI. Galactic-type dust is probably ruled out by the
absence of strong extinction near 2175Å in the rest frames of the damped
Lyα systems, whereas LMC and SMC-type dust are both acceptable from
this point of view. The observed reddening then implies $k \approx 0.1$ at $\bar{z} = 2.3$.
The overall uncertainty in this estimate is about a factor of two (up or down).
For reference, the Milky Way has $k \approx 0.8$.

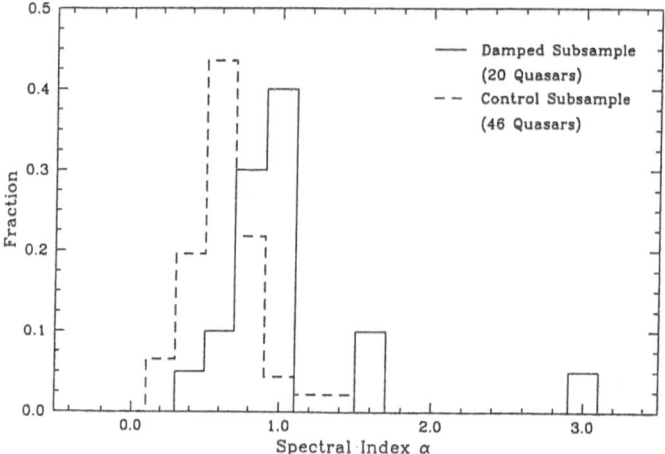

Fig. 1. Histograms of the spectral indices of quasars with and without damped Lyα
systems in the foreground (adapted from Pei et al. 1991).

Another indication that the damped Lyα systems contain dust comes from
the gas-phase abundances of Zn, Cr, and other trace elements (Meyer, Welty,
& York 1989; Meyer & Roth 1990; Pettini, Boksenberg, & Hunstead 1990;
Bergeron & Petitjean 1991; Meyer & York 1992; Pettini et al. 1994; Wolfe

et al. 1994; Steidel et al. 1995). In the local interstellar medium, Zn is only weakly depleted with respect to the Solar abundance, typically by factors of two or less, whereas Cr is depleted by factors of 50 or more. In comparison, the damped Lyα systems generally have lower abundances of Zn and higher abundances of Cr relative to Zn. This indicates that the metallicities and dust-to-gas ratios are lower than those in the Milky Way today. From a thorough analysis of the available data, Pettini et al. (1994) derive a mean metallicity of $Z \approx 0.1$ at $\bar{z} = 2.2$ and a dust-to-gas ratio entirely consistent with the reddening discussed above. The assumption here is that, apart from overall scalings in the metallicity and dust-to-gas ratio, the interstellar medium is the same in the damped Lyα systems as it is in the Milky Way (i.e., same abundance patterns and grain compositions). If this were not true, the derived values of Z and k could be misleading (Pei et al. 1991). Eventually, it should be possible to determine how the metallicity and dust-to-gas ratio vary with redshift, but the existing samples are not yet large enough to reveal such trends.

3 Obscuration of Quasars

The dust in intervening galaxies causes a bias in any optically selected sample of quasars. This is illustrated schematically in Figure 2. Here, Q1 and Q5 are not obscured at all, Q2 and Q4 are obscured lightly enough to be included in the sample, while Q3 and Q6 are obscured so heavily that they are excluded from the sample. It should be clear from Figure 2 that the number of "missing" quasars depends primarily on two factors: the fraction of the sky covered by galaxies and the typical optical depth through a galaxy, both of which may be functions of redshift. There have been several attempts to quantify this bias (Ostriker & Heisler 1984; Wright 1986; Boissé & Bergeron 1988; Heisler & Ostriker 1988; Fall & Pei 1989; Wright 1990; Fall & Pei 1993; Zuo & Phinney 1993). The results of these studies differ in large part because they are based on different assumptions about galaxies at high redshifts (e.g., that they are the same as galaxies at low redshifts). Our approach, and that of subsequent studies, has been to rely as heavily as possible on the observed properties of the damped Lyα systems. Thus, we combine the distribution of HI column densities $f(N, z)$ with the typical dust-to-gas ratio k to obtain the distribution of optical depths $\rho(\tau, z)$. From this and the observed luminosity function of quasars $\phi_o(L, z)$, we compute the true luminosity function $\phi_t(L, z)$ and hence the number of missing quasars.

The main complication in the procedure outlined above is that the existing samples of damped Lyα systems are also biased by obscuration. The reason for this is that they were obtained from the spectra of quasars that appear bright at optical wavelengths (roughly $B \lesssim 19$; see Wolfe et al. 1986; Lanzetta et al. 1991). This selection criterion eliminates some of the quasars that are obscured by dust in foreground absorbers and hence some of the absorbers themselves (Q3 and Q6 in Figure 2). Thus, we must distinguish between the true and observed distributions of HI column densities, denoted by $f_t(N, z)$ and $f_o(N, z)$. The former pertains to all lines of sight (Q1–Q6 in Figure 2), while the latter pertains to unobscured and lightly obscured lines of sight (Q1, Q2, Q4, and Q5 in Figure 2). Correspondingly, we must distinguish between the true and observed comoving densities of HI. These are given by

$$\Omega_{\mathrm{HI}}(z) = \frac{8\pi G m_H}{3 c H_0} \int_0^\infty dN N f(N, z),$$

(1)

with $\Omega_{\mathrm{HI}t}$ computed from f_t and $\Omega_{\mathrm{HI}o}$ computed from f_o. Since the comoving densities are dominated by absorbers with highest column densities, where f_t and f_o differ most [see equation (3) below], there can be large differences between $\Omega_{\mathrm{HI}t}$ and $\Omega_{\mathrm{HI}o}$.

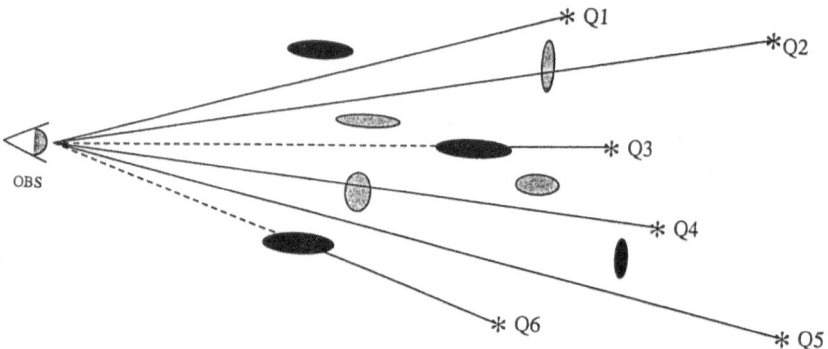

Fig. 2. Schematic illustration of the obscuration of quasars by dust in intervening galaxies.

The relation between ϕ_t and ϕ_o is inextricably coupled to the relation between f_t and f_o, presenting, in the general case, an intractable problem. This can, however, be solved in closed form under the following simplifying assumptions (Fall & Pei 1993): (a) the observed luminosity function of bright quasars has power-law form, $\phi_o(L, z) = \psi_o(z) L^{-\beta-1}$, (b) the sample of absorbers was obtained from a magnitude-limited sample of bright quasars, and (c) all absorbers at a given redshift have the same dust-to-gas ratio $k(z)$. In this case, the true luminosity function of bright quasars also has power-law form, $\phi_t(L, z) = \psi_t(z) L^{-\beta-1}$, with

$$\psi_t(z) = \psi_o(z) \exp\left\{ \int_0^z dz' \frac{(1 + z')}{(1 + 2 q_0 z')^{1/2}} \int_0^\infty dN [f_t(N, z') - f_o(N, z')] \right\},$$

(2)

$$f_t(N, z) = f_o(N, z) \exp[\beta \tau(N, z)],$$

(3)

$$\tau(N, z) = k(z)(N/10^{21} \mathrm{cm}^{-2}) \xi[\lambda_e/(1 + z)].$$

(4)

Here, $\xi(\lambda)$ is the ratio of the extinction at a wavelength λ to that at the effective wavelength λ_e of the limiting magnitude of the sample. Assumptions (a) and (b) above are valid for quasars brighter than $B \approx 19$ and for the existing samples of damped Lyα systems. Assumption (c) is harder to justify, although we have verified that equations (2)–(4) are reasonably accurate even if there is a dispersion in the dust-to-gas ratio provided $k(z)$ is then interpreted as the logarithmic mean dust-to-gas ratio (see the Appendix of Fall & Pei 1993).

We have used equations (2)–(4) to correct for the effects of obscuration on the comoving density of quasars and their contribution to the ionizing UV background. Figure 3 shows the results, which were obtained as follows. For the observed luminosity function of quasars, we have adopted the double power-law model presented by Pei (1995), and for the effective UV opacity of the Lyα forest, we have adopted the model presented by Madau (1991). The solid curve in the left-hand panel of Figure 3 shows the comoving density of observed quasars brighter than $M_B = -26$, while the solid curve in the right-hand panel shows the contribution of observed quasars to the mean intensity of Lyman-limit radiation. The data points in the right-hand panel are estimates of J_L from the proximity effect. The three dashed curves in each panel represent the corresponding true quantities, corrected for obscuration by the factor $\psi_t(z)/\psi_o(z)$ from models A, B, and C of Fall & Pei (1993). These are based on the distribution of HI column densities at $\overline{z} = 2.4$ from Lanzetta et al. (1991) and the typical dust-to gas ratio at $\overline{z} = 2.3$ from Pei et al. (1991), with interpolations at lower redshifts and extrapolations at higher redshifts. The wide range spanned by models A–C is mainly a consequence of the weak constraints on $\Omega_{\mathrm{HI}t}(z)$ (discussed above) and, to a lesser extent, the uncertainties in $k(z)$. From these calculations, we conclude that up to 70% of the bright quasars at $z = 3$ could be missing from optically selected samples. This is not enough obscuration to cause the turnover in the observed comoving density of quasars at $z \approx 3$, but it does help to account for the ionizing UV background inferred from the proximity effect.

4 Cosmic Chemical Evolution

The abundances of heavy elements in the damped Lyα systems provide some potentially valuable information about the history of star formation in these objects. To exploit this information, it is necessary to turn to models of chemical evolution. Dust plays a role in such studies because it biases estimates of $\Omega_{\mathrm{HI}}(z)$ and hence inferences about the rate of gas consumption in the damped Lyα systems. We have constructed some illustrative models of chemical evolution that include self-consistent corrections for this bias (Pei & Fall 1995). Here, we give a brief summary of the main assumptions and conclusions. Our work is motivated in part by the "G-dwarf problem" identified by Lanzetta, Wolfe, & Turnshek (1995). These authors neglected obscuration and adopted

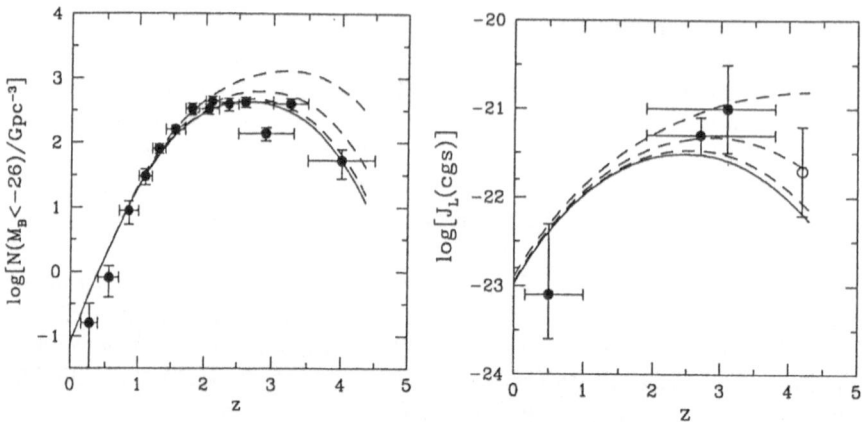

Fig. 3. Effects of obscuration on the comoving density of quasars brighter than $M_B = -26$ (left) and the contribution of quasars to the mean intensity of background radiation at the Lyman limit (right) for $q_0 = 0.5$. The data points in the left-hand panel are from Hartwick & Schade (1990, $0.1 \leq z \leq 3.3$) and Warren, Hewett, & Osmer (1994, $2.0 \leq z \leq 4.5$). The data points in the right-hand panel are from Espey (1993, $z = 2.7$), Kulkarni & Fall (1993, $z = 0.5$), Bechtold (1994, $z = 3.1$, $\Delta v = -1500$ km s^{-1}), and Williger et al. (1994, $z = 4.2$).

a "closed-box" model of chemical evolution. From the observed $\Omega_{\mathrm{HI}}(z)$, they concluded that half of the stars in the descendants of the damped Lyα systems formed before $z \approx 2.5$, with metallicities below $0.1 Z_\odot$. This is puzzling because there are very few metal-poor stars in the disk of the Milky Way. As a solution, Lanzetta et al. proposed that most of the damped Lyα systems at $z \gtrsim 2.5$ are the progenitors of present-day galactic spheroids. Our models, while idealized in several respects, demonstrate that obscuration is a critical factor in such considerations.

The natural variables with which to describe the chemical evolution in large comoving volumes of the Universe are the comoving densities of gas and stars, Ω_g and Ω_s, and the mean abundance of heavy elements in the gas Z, including dust. In the approximation of instantaneous recycling (and $Z \ll 1$), these are related by

$$\dot{\Omega}_g + \dot{\Omega}_s = \dot{\Omega}_f, \tag{5}$$

$$\Omega_g \dot{Z} - y\dot{\Omega}_s = (Z_f - Z)\dot{\Omega}_f, \tag{6}$$

where y is the IMF-averaged yield, and the dots denote differentiation with respect to proper time. When all the baryons in the Universe are considered, the "source" terms on the right-hand sides of equations (5) and (6) must, of course, vanish. Here, however, we are concerned only with the damped Lyα systems, for which the source terms represent the inflow or outflow of gas with metallicity Z_f at a rate $\dot{\Omega}_f$. To allow for a wide range of possibilities, we consider a closed-box model ($\dot{\Omega}_f = 0$), a model with inflow of metal-free gas ($\dot{\Omega}_f = +\nu\dot{\Omega}_s$, $Z_f = 0$), and a model with outflow of metal-enriched gas ($\dot{\Omega}_f = -\nu\dot{\Omega}_s$, $Z_f = Z$). Following Lanzetta et al. (1995), we ignore HII and H_2 but include He; thus, as an approximation, we set $\Omega_g = 1.3\Omega_{HIt}$.

To link the obscuration to the chemical evolution in the models, we assume that the dust-to-gas ratio is proportional to the metallicity:

$$k(z)/k(0) = Z(z)/Z(0). \tag{7}$$

This is equivalent to the assumption that the dust-to-metals ratio $k(z)/Z(z)$ is constant and hence that a fixed fraction of the heavy elements condenses into dust grains. Equation (7) is also consistent with the empirical values of k and Z in the damped Lyα systems at $\bar{z} = 2.3$. Given $f_o(N, z)$ and $k(z)$, we can compute $\Omega_{HIo}(z)$ and $\Omega_{HIt}(z)$ from equations (1), (3), and (4). However, we also require that $\Omega_g(z)$ and $Z(z)$ satisfy equations (5), (6), and (7). These coupled sets of equations — for obscuration and chemical evolution — are solved simultaneously by iteration. We choose a function $f_o(N, z)$ that reproduces the HI statistics of damped Lyα systems at $0 < z \leq 3.5$ from Lanzetta et al. (1991) and Lanzetta et al. (1995). All the other properties of the models are then determined by four input parameters: the present metallicity $Z(0)$, the present dust-to-gas ratio $k(0)$, the initial comoving density of gas $\Omega_{g\infty}$, and the relative inflow or outflow rate ν. We adopt $Z(0) = Z_\odot$ and $k(0) = 0.8$, the values in the local interstellar medium, and experiment with different combinations of $\Omega_{g\infty}$ and ν.

The results from one set of models are displayed in Figure 4; these have $\Omega_{g\infty} = 4 \times 10^{-3}h^{-1}$ and $\nu = 0.5$. Panel (a) shows the distribution of HI column densities at $\bar{z} = 2.4$; panel (b) shows the evolution in the comoving density of HI; panel (c) shows the evolution in the mean metallicity in the gas; and panel (d) shows the evolution in the comoving rate of star formation. The labels C, I, and O refer, respectively, to the closed-box, inflow, and outflow models. The solid curves represent true quantities, while the dashed curves in panels (a) and (b) represent the corresponding observed quantities; the dashed curves in panels (c) and (d) show the effects of neglecting obscuration. Evidently, all three models with obscuration (C, I, and O) match the observed mean metallicity $Z \approx 0.1Z_\odot$ at $\bar{z} = 2.2$. In contrast, the same models, but without obscuration, have metallicities $4 - 8$ times higher than the observed value. The models with obscuration are also consistent with the average properties of present-day galaxies, including the local comoving densities of gas and stars. We have checked that the success of the models does not require "fine tuning" of the input parameters $\Omega_{g\infty}$ and ν. In particular, we also find agreement between the models and the observations when $\Omega_{g\infty}$ is adjusted downward for consistency with the HI statistics of damped Lyα systems at $2.8 \leq z \leq 4.4$ from Storrie-Lombardi et al. (described in these Proceedings).

One of the main predictions to emerge from our models of chemical evolution is that the damped Lyα systems experience rapid star formation at

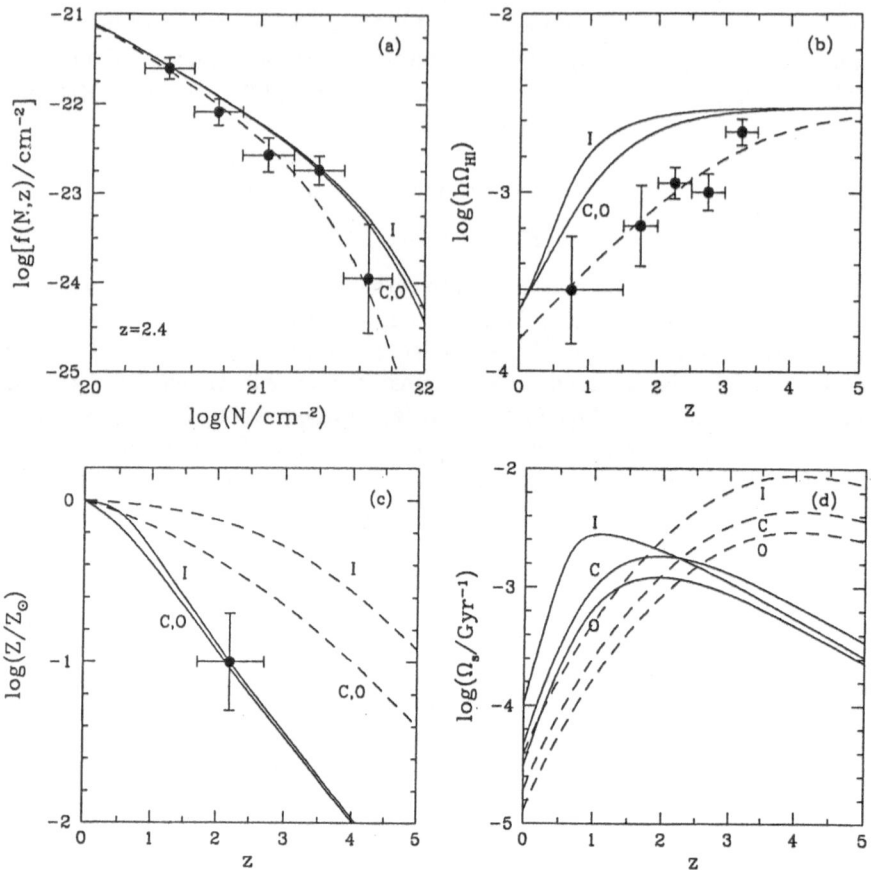

Fig. 4. Results from models of cosmic chemical evolution with $\Omega_{g\infty} = 4 \times 10^{-3} h^{-1}$, $\nu = 0.5$, and $q_0 = 0.5$. The curves are explained in the text. The data points in panels (a), (b), and (c) are from Lanzetta et al. (1991), Lanzetta et al. (1995), and Pettini et al. (1994), respectively.

relatively low redshifts. This, in fact, is the reason for the rapid rise in the mean metallicity with decreasing redshift. In comparison, models without obscuration have lower rates of star formation at $z \lesssim 2$ and correspondingly lower rates of metal production at these redshifts. This difference occurs because $\dot{\Omega}_s$ is proportional to $-\dot{\Omega}_g$ and hence either to $-\dot{\Omega}_{HIt}$ or $-\dot{\Omega}_{HIo}$, depending on whether or not obscuration is included. In the models with obscuration, $\dot{\Omega}_s \propto -\dot{\Omega}_{HIt}$ has a peak at $z \approx 1$, whereas in the models without obscuration, $\dot{\Omega}_s \propto -\dot{\Omega}_{HIo}$ has a peak at $z \approx 3 - 4$. Although the mean metallicity in the gas evolves in nearly the same way in the inflow, outflow, and closed-box models with obscuration, there are differences in the present distributions of stellar metallicities. For the parameters adopted in Figure 4, the fraction of stars with metallicities below $0.1Z_{\odot}$ is 25% in the closed-box and outflow models and only 10% in the inflow model. Decreasing $\Omega_{g\infty}$ and/or increasing ν in the inflow model would further reduce the fraction of metal-poor stars. Thus, if there is a G-dwarf problem at all, it is much less serious than claimed by Lanzetta et al. (1995).

Our models of chemical evolution entail some definite predictions about the effects of obscuration on optically selected samples of quasars and damped Lyα systems. From equation (2), we compute the ratio of the true and observed numbers of bright quasars ψ_t/ψ_o, and by integrating $f_t(N, z)$ and $f_o(N, z)$ over $N \geq 2 \times 10^{20}$ cm^{-2}, we obtain the ratio of the true and observed numbers of damped Lyα systems n_t/n_o. These are plotted against redshift in Figure 5 for the same models as in Figure 4. Evidently, ψ_t/ψ_o increases monotonically from unity at $z = 0$ to $1.4 - 1.5$ at $z = 3$, indicating a fairly modest amount of obscuration; only about 30% of the quasars are missing at $z = 3$. This is near the middle of the range of obscuration found in our previous work (i.e., similar to that in model B of Fall & Pei 1993; summarized here in Section 3). More interesting in the present context are the effects of obscuration on samples of damped Lyα systems. In our models, n_t/n_o increases from 1.8 at $z = 0$ to a maximum of $2.1 - 2.8$ at $z \approx 1$ and then decreases to 1.2 at $z = 3$. This occurs because, while the dust-to-gas ratio (and metallicity) decrease with increasing redshift, the number of absorbers per unit redshift and their rest-frame extinction have the opposite behavior. We therefore expect the existing samples of damped Lyα systems to be substantially incomplete at $z \approx 1$. As a consequence of this bias, the properties of the observed systems (luminosity, metallicity, etc.) might differ systematically from those of the population as a whole.

5 Attenuation of Lyα Emission

Another consequence of dust is that it would attenuate any Lyα emission within the damped Lyα systems. The resonant scattering of Lyα photons in HI regions causes them to execute random walks in both position and frequency. In the absence of dust, all of the Lyα photons would eventually escape in a prominent, double-peaked emission line (Harrington 1973). The Lyα luminosity would then be proportional to the ionization rate and hence to the luminosity of massive stars and/or an AGN. Such emission would be relatively easy to detect, even at $z = 3$, if the star formation rate and stellar IMF were similar to those in the Milky Way. This has been the motivation for numerous searches for Lyα emission from damped Lyα systems at known locations and from "primeval" galaxies in blank-sky fields (which might, of course, be the same things). Only a few of these searches have been successful, and even then, the Lyα emission can often be attributed to a nearby AGN (see Charlot & Fall 1993 and references therein and the article by Pettini et al. in these Proceedings). In most cases, Lyα emission has not been observed at the expected level. This means either that the star formation rates in the

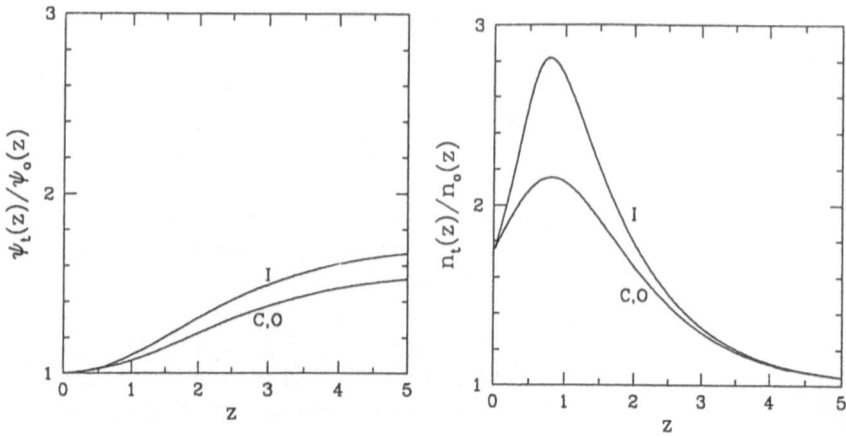

Fig. 5. Ratios of the true and observed numbers of bright quasars (left) and the true and observed numbers of damped Lyα systems (right), as functions of redshift, for the same models as in Fig. 4.

damped Lyα systems are lower than those in present-day galaxies, that the stellar IMF is different, or that the Lyα photons are absorbed by dust.

The circuitous paths taken by Lyα photons in HI regions increase the likelihood that they will be absorbed by dust grains before escaping. The condition for substantial attenuation in a homogeneous slab of HI is

$$k \gtrsim 0.01(N_\perp/10^{21} \text{ cm}^{-2})^{-4/3}(\sigma_V/10 \text{ km s}^{-1})^{2/3}, \qquad (8)$$

where $N_\perp$ is the face-on column density and σ_V is the one-dimensional velocity dispersion (Charlot & Fall 1991). Since the typical dust-to-gas ratio in the damped Lyα systems exceeds the "critical" value given above (for reasonable values of $N_\perp$ and σ_V), we expect strong attenuation of the Lyα line. This plausibly accounts for the weak Lyα emission from most damped Lyα systems and the null results of blank-sky searches for Lyα emission from primeval galaxies. However, equation (8) should be considered only a rough guide because the attenuation also depends on several other factors, including the unknown topology of the interfaces between the HI and HII regions. Fewer Lyα photons are absorbed when the circumstellar HII regions are not completely surrounded by HI (Neufeld 1991). These issues are discussed in more detail in the papers by Charlot & Fall (1993) and Chen & Neufeld (1994).

6 Conclusions

From the work reviewed here, we draw the following conclusions:

(1) The typical dust-to-gas ratio in the damped Lyα systems at $\bar{z} = 2.3$ is $k \approx 0.1$. This follows both from the reddening of background quasars and the gas-phase depletion of Cr relative to Zn.

(2) As a result of obscuration by dust in damped Lyα systems, up to 70% of the bright quasars at $z = 3$ could be missing from optically selected samples. Thus, quasars alone might account for all of the ionizing UV background inferred from the proximity effect.

(3) Models that link obscuration to the chemical evolution in the damped Lyα systems match the observed mean metallicity $Z \approx 0.1 Z_\odot$ at $\bar{z} \approx 2.2$. In this case, there is little or no G-dwarf problem.

(4) Models of chemical evolution that include obscuration have rapid star formation at $z \approx 1$. They also indicate that the existing samples of damped Lyα systems are substantially incomplete at $z \approx 1$.

(5) The enhanced absorption of Lyα photons by dust can account for the weak or absent Lyα emission from most damped Lyα systems.

References

Bechtold, J. 1994, ApJS, 91, 1

Bergeron, J., & Petitjean, P. 1991, A&A, 241, 365

Boissé, P., & Bergeron, J. 1988, A&A, 192, 1

Charlot, S., & Fall, S.M. 1991, ApJ, 378, 471

Charlot, S., & Fall, S.M. 1993, ApJ, 415, 580

Chen, W.L., & Neufeld, D.A. 1994, ApJ, 432, 567

Espey, B.R. 1993, ApJ, 411, L59

Fall, S.M., & Pei, Y.C. 1989, ApJ, 337, 7

Fall, S.M., & Pei, Y.C. 1993, ApJ, 402, 479

Fall, S.M., Pei, Y.C., & McMahon, R.G. 1989, ApJ, 341, L5

Harrington, J.P. 1973, MNRAS, 162, 43

Hartwick, F.D.A., & Schade, D. 1990, ARA&A, 28, 437

Heisler, J., & Ostriker, J.P. 1988, ApJ, 332, 543

Kulkarni, V.P., & Fall, S.M. 1993, ApJ, 413, L63

Lanzetta, K.M., Wolfe, A.M., & Turnshek, D.A. 1995, ApJ, 440, 435

Lanzetta, K.M., Wolfe, A.M., Turnshek, D.A., Lu, L.M., McMahon, R.C., & Hazard, C. 1991, ApJS, 77, 1

Madau, P. 1991, ApJ, 376, L33

Meyer, D.M., & Roth, K.C. 1990, ApJ, 363, 57

Meyer, D.M., Welty, D.E., & York, D.G. 1989, ApJ, 343, L37

Meyer, D.M., & York, D.G. 1992, ApJ, 399, L121

Neufeld, D.A. 1991, ApJ, 370, L85

Ostriker, J.P., & Heisler, J. 1984, ApJ, 278, 1

Pei, Y.C. 1995, ApJ, 438, 623

Pei, Y.C., & Fall, S.M. 1995, ApJ, in press

Pei, Y.C., Fall, S.M., & Bechtold, J. 1991, ApJ, 378, 6

Pettini, M., Boksenberg, A., & Hunstead, R.W. 1990, ApJ, 348, 48

Pettini, M., Smith, L.J., Hunstead, R.W., & King, D.L. 1994, ApJ, 426, 79

Steidel, C.C., Bowen, D.V., Blades, J.C., & Dickinson, M. 1995. ApJ, 440, L45

Warren, S.J., Hewett, P.C., & Osmer, P.S. 1994, ApJ, 421, 412

Williger, G.M., Baldwin, J.A., Carswell, R.F., Cooker, A.J., Hazard, C., Irwin, M.J., McMahon, R.G., & Storrie-Lombardi, L.J. 1994, ApJ, 428, 574

Wolfe, A.M., Fan, X.-M., Tytler, D., Vogt, S.S., Keane, M.J., & Lanzetta, K.M. 1994, ApJ, 435, L101

Wolfe, A.M., Turnshek, D.A., Smith, H.E., & Cohen, R.D. 1986, ApJS, 61, 249

Wright, E.L. 1986, ApJ, 311, 156

Wright, E.L. 1990, ApJ, 353, 411

Zuo, L., & Phinney, E.S. 1993, ApJ, 418, 28

Are Studies of Dust in Damped Lyman-α Systems Biased by Extinction ?

Patrick Boissé

Radioastronomie, Ecole Normale Supérieure, 24,rue Lhomond F-75231 Paris, France

Abstract. We discuss whether the low dust to gas ratio and the absence of molecules in damped Lyα absorbers is a true characteristic of the latter or rather the result of a selection effect induced by dust extinction.

In many respects, dust is an important component of interstellar matter. First, it plays an important role in star formation processes and second, when associated to intervening galaxies, it may strongly affect our view of the distant universe (Ostriker and Heisler, 1984). Since damped Lyα systems are thought to be due to distant intervening galaxy disks, they are prime candidates to investigate dust in high redshift galaxies.

In recent years, several studies have been devoted to this topic (see especially Fall and Pei, 1993 and references therein). A small excess of reddening has been found for quasars with damped systems (Pei et al., 1991). The 2175Å feature has been searched for but was never seen (e.g. Boissé and Bergeron, 1988) whereas it is ubiquitous in our Galaxy and also detected (although with a smaller strength or width) in the LMC, M31 (Hutchings et al., 1992) and M101 (Rosa and Benvenuti, 1994). Similarly, molecules associated with damped systems have been searched for but with very little success. Several stringent limits have been obtained (Lanzetta et al., 1989) which indicate a fraction of gas in molecular form much lower than in our Galaxy.

We now discuss whether this apparent contradiction may be understood either as evidence that the composition of interstellar matter is really different in high z galaxies or as the result of a selection effect. Indeed, intervening dust induces extinction which could bias the samples of QSOs observed spectroscopically in favor of objects having no damped absorbers in front of them or having damped absorbers but with a low dust to gas ratio (hereafter k, following Fall et al.). Fall et al. have investigated this question in much detail (see also Borgeest, these proceedings) and conclude that the bias is difficult to estimate because the dispersion in the dust to gas ratio is unknown. Such a dispersion is likely to be large for at least two reasons: i) damped Lyα absorbers may be found in various states of chemical evolution, ii) radial abundance gradients are probably present and the scatter in impact parameters implies a large spread of k values.

Let us estimate how strong the bias would be against the detection of Galactic interstellar type material. In our Galaxy, the fraction of molecular gas begins to be substantial at $N(HI) \geq 10^{21} cm^{-2}$. Such an amount of matter at $z_a \approx 2.5$ implies an extinction $A \approx 0.5$ for a background QSO in the B band (using standard ratios). In the two samples considered by Fall et al. (1993), the brightest quasars have $m_v = 16.5$ while the faintest have $m_v = 19 - 20$, value beyond which the spectra would have a too low S/N. It is then clear that in the plane (m(QSO),N(HI)), where m(QSO) is the QSO magnitude *before* attenuation by dust, the range corresponding to detectable Galactic type dust is very limited. Smaller k values will be favored because the associated extinction is weaker and the fraction of QSOs that remain observable larger. Another effect is that k is likely to correlate with N(HI); indeed, it is in the internal regions that both N(HI) and the metal abundances are large (however, in the *observed* QSOs, this effect may be masked since for large N(HI), only small k values are allowed). This acts against the detection of molecules because they lie preferentially in the inner disk. Further, their spatial distribution in the disk is very far from uniform both at small and large scales (see e.g. the CO map of M51 by Garcia-Burillo et al., 1993) which results in a very low effective cross-section.

We also note that the largest N(HI) values observed ($\approx 10^{22} cm^{-2}$) are close to those at which one expects extinction to induce a cut-off in the *observed* N(HI) distribution if k lies in the range 0.1 - 1. If absorptions with $N(HI) > 10^{22} cm^{-2}$ and low k existed, they could be recognized very easily in the Lyα forest. However, the high end of the N(HI) distribution is still poorly sampled by available data and the absence of such absorbers could also be a true property of the population: better statistics are needed.

The arguments developed above altogether indicate that we cannot conclude on the absence of Galactic type material ($k \approx 0.8$, 2175Å feature, presence of molecules) within damped Lyα absorbers. Observations allowing to probe a larger domain in the space (m(QSO), N(HI), k) are needed to reach firm conclusions. In particular, observations of fainter QSOs not selected according to their optical colors would allow to probe the area where extinguished QSOs potentially lie and may lead to the detection of gas more representative of star forming complexes which would be of great interest for comparison with interstellar matter at $z \approx 0$.

References

Boissé P., Bergeron J., 1988, A&A, 192, 1.
Fall S.M., Pei Y.C., 1993, ApJ, 402, 479.
Garcia-Burillo S., Guélin M., Cernicharo J., 1993, A&A, 274,123.
Hutchings J.B. et al., 1992, ApJ, 400, L35.
Lanzetta K.M., Wolfe A.M., Turnshek D.A., 1989, ApJ, 344, 277.
Ostriker J.P., Heisler J., 1984, ApJ, 278,1.
Pei Y.C., Fall S.M., Bechtold J., 1991, ApJ, 378,6.
Rosa M.R., Benvenuti P., 1994, A&A, 291, 1.

QSO Absorption by Nearby Galaxies

M. Trewhella, M. Disney and J. Davies

University of Wales College of Cardiff, Department of Physics and Astronomy, PO Box 913, Cardiff CF2 3YB, UK

Abstract. We investigate the obscuration of the distant universe by absorption in foreground spiral galaxies. We have used the most conservative and most radical views highlighted by the recent NATO conference on the opacity of spiral disks. We find that a significant amount of the universe will be obscured at redshifts important to QSO astronomers. This implies that any studies at high redshift should take account of this absorption.

1 Introduction

The purpose of this paper is to draw the attention of QSO astronomers to the debate now raging on the amount of smoke (dust) in nearby spirals, and the possible implications of large amounts of such smoke for the observation of higher redshift QSOs.

The long standing and comfortable idea that spirals are very largely transparent (Holmberg 1958) has been shown to be inconclusive (Disney et al. 1989), and spirals could just as well be very optically thick (Valentijn 1990, Burstein et al. 1991). A recent NATO workshop (Davies 1994) was held to try to clear up some of the disagreements of this contentious issue. It is quite likely, but by no means certain, that most spirals are opaque ($\tau_B > 1$) out to between 1 and 3 scale lengths, high enough to affect observers of distant QSOs.

2 Blocking Factor

The blocking factor $f(z)$ is the ratio of the total cross section of absorbers to the total sky area. For absorbers (galaxies) with local ($z{=}0$) number density $n_G(0)$ and absorption cross section $A_G(0)$ we have($\Lambda = 0, q_0 = 0.5$):

$$f(z) = \frac{2A_G(0)n_G(0)}{3H_0}(1+z)^{3/2} \tag{1}$$

Various authors (Beckman 1994, Wright 1990, Ostriker 1984) have used E-q.(1) together with different estimates of these parameters to argue that the blocking factor approaches 1 at redshifts between 2 and 10. Figure (1) shows (for $q_0 = 0.05$ and 0.5) the likely range in f(z) where high corresponds to $n_G(0) = 0.04 Mpc^{-3}, R_{op} = 4$ scale length, low corresponds to

$n_G = 0.01 Mpc^{-3}, R_{op} = 1$ scale length. The most probable form for f(z) is shown by the medium graph where $n_G = 0.02 Mpc^{-3}$ and $R_{op} = 2$ scale lengths. Various complications to this include extinction law, evolution, clustering and patchiness.

3 Discussion

The calculations shown above, together with the increasing amounts of smoke allowed by modern interpretations, show that smoke could be a major player in the observation of distant QSOs and primeval galaxies. We would like to emphasize the point that observations of QSOs primarily found by optical techniques can estimate only the minimum part played by smoke (if you find a QSO optically then it cannot have been heavily absorbed) Most absorbed QSOs may be too faint or too red to appear in finder surveys. The report by Webster et al. (1994) that half of the flat spectrum 1Jy (5GHz) VLA sources are red, with a $(B - K) \geq 5$, may be evidence of just that.

Fig. 1. The blocking factor

References

Holmberg E., 1958, *Mrf. Lunds. astr. Obs.*, Ser. 2, No. 136
Disney M.J., Davies J.I., Phillips S., 1989, *MNRAS*, 239 ,205
Valentijn E.A., 1990, *Nature* 346, 153
Burstein D., Haynes M.P., Faber S.M., 1991, *Nature*, 353, 515
ed. J.I. Davies, 1994 *The Opacity of Spiral Discs*, Kluwer (NATO conference)
Beckman J., 1994, in *The Opacity of Spiral Discs* see Davies, 1994
Wright E.L., 1990, *Ap.J.*, 353, 411
Ostriker J.P., 1984, *Ap.J.*, 402,479
Webster. 1994, IAU verbal report.

The Mass Spectrum of Giant Hydrogen Clouds Associated with Damped Lyα Absorbers

V.K. Khersonsky and D.A. Turnshek

Department of Physics and Astronomy, University of Pittsburgh, USA

Abstract. Comparison of the observed properties of damped Lyα absorbers (D-LAs) seen in QSO spectra with different structures in the interstellar medium of our own and nearby galaxies shows that these absorbers may be interpreted as giant hydrogen clouds (GHCs) in intervening galaxies along the lines-of-sight. We show that the observational data on the redshift and column density distribution of the DLAs may be used to study the mass spectrum of GHCs and its evolution with cosmic time. The derived mass spectrum may be represented by a power-law relation with rapid cosmic evolution in the exponent. The evolution may be interpreted in terms of global star formation processes in young galaxies.

1 Introduction

A widely accepted interpretation of high-redshift DLAs is that they are gaseous HI layers which are the progenitors of disk galaxies. To some extent this interpretation is motivated by the fact that spiral galaxies seen locally are sites which could give rise to damped Lyα absorption. However, a more general approach applicable at high redshift is to assume that the DLAs are simply GHCs which can be found in different types of young gaseous galaxies, independent of whether they are spirals or irregulars, the progenitors of spirals, irregulars or ellipticals, or possibly just dwarf galaxies which might be more numerous in the past. We propose that available observational data be interpreted in such a framework. One might then hope or speculate that future observational tests (based on images or studies of kinematics, ionization, metal abundances, dust content, molecular content, etc.) could be used to distinguish between the various possible absorbing sites.

A comparison of the observed properties of DLAs with GHCs in the MWG and nearby galaxies [see, for example, Wolfe (1993) for a review of DLA properties and Blitz (1991) and Kulkarni & Heiles (1988) for a review of GHC properties] shows (a) that the properties of DLAs are similar to those of GHCs in galaxies and (b) that the properties of GHCs are almost identical for different types of gas-containing galaxies — large spirals, irregulars, and dwarfs. Rather than being dictated by galaxy type, the observed properties of GHCs mainly appear to be related to turbulent processes in the interstellar medium which are responsible for generating the whole spectrum of GHC masses and sizes. As a consequence, the interstellar medium of a galaxy is very patchy and the neutral gas distribution consists of many clumps of different size and shape.

The mass spectrum is one of the most important properties of the GHC population. It plays a dominant role in the evolution of the cloud population itself, in terms of star formation, the generation of heavy elements and, therefore, the general evolution of galaxies.

In this contribution we show how the mass spectrum can be derived from available observational data and we investigate its main features.

2   The Mass Spectrum of GHCs

The basic goal is to derive the mass spectrum of GHCs from the observational data on the two-dimensional distribution of DLAs in column density and redshift, $d^2\mathcal{N}(z, N_{HI})/dzdN_{HI}$. If we assume that GHCs are spherical, that they have radius, R, and that they have a mean number density of neutral hydrogen, n_{HI}, then the observed two-dimensional distribution can be related to the mass spectrum of GHCs by the integral equation

$$\frac{d^2\mathcal{N}(z, N_{HI})}{dzdN_{HI}} = f(z, N_{HI})$$

$$= \frac{c\pi(1 + z)N_{HI}}{2H_0\sqrt{1 + \Omega z n_{HI}^2}} \int_{M_{min}(N_{HI})}^{\infty} n_{GHC}(z, M)dM.$$

In this equation, $n_{GHC}(z, M)$ is the mass spectrum or number density of GHCs per comoving volume, averaged over the surface of sky at redshift z, and $M_{min}(N_{HI}) = (\pi\mu N_{HI}^3/6n_{HI}^2)M_\odot$ where μ is the mean atomic weight. The left side of this equation, i.e., the function $f(z, N_{HI})$, can be derived from the observational data (Lanzetta et al. 1991; Rao & Briggs 1993, R-B; Lanzetta, Wolfe & Turnshek 1995, LWT; Rao, Turnshek & Briggs 1995, RTB). These data are shown in Fig.1 as DLA column density distributions for different redshift intervals. We approximate this observed two-dimensional function using a power-law relation of the form $\log[f(z, N_{HI})] = a(z) + b(z)\log(N_{HI}/N_{min})$ where $a(z)$ and $b(z)$ can be represented by polynomials in the redshift interval $z = 0$ - 3.5 and $N_{min} = 10^{12}$cm^{-2}. Fig.2 shows the derived relation for $f(z, N_{HI})$.

By differentiating both parts of the integral equation with respect to N_{HI}, one can obtain the mass spectrum of GHCs in the form of a power-law relation,

$$n_{GHC}(z, M) = Q(z)\left(\frac{M_\odot}{M}\right)^\delta, Q(z) = 10^{a(z)}\frac{1 - b(z)}{2\Lambda(z)}\kappa^{\delta-1}, \delta = \frac{4 - b(z)}{3}.$$

In this result κ and $\Lambda(z)$ are dimensionless parameters where $\kappa = \pi\mu N_{min}^3/6n_{HI}^2 M_\odot$ and $\Lambda(z) = \pi c(1+z)M_\odot N_{min}^3/2H_0\sqrt{1 + \Omega z n_{HI}^2}$. It should be noted that observations (e.g. in the 21 cm absorption line) generally show that kinematic (and therefore spatial) substructure is present inside each D-LA. Thus, some parts of a GHC must be denser than other parts, but we ignore this. For a mean number density of $n_{HI} \approx 1$ cm^{-3}, the derived mass spectrum of GHCs associated with DLAs is shown in Fig 3.

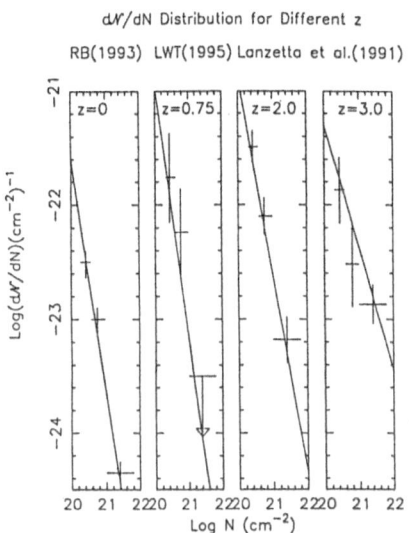

Fig. 1. Distribution of DLAs in column density at different redshifts. Crosses represent binned data. Solid lines show our approximations.

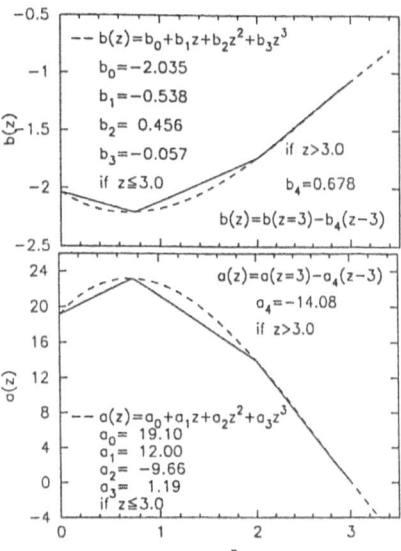

Fig. 2. The parameters of $f(z, N_{HI})$ which approximate the data. Solid lines connect the coefficients derived from Fig. 1 data. Dashed lines show the polynomial approximations.

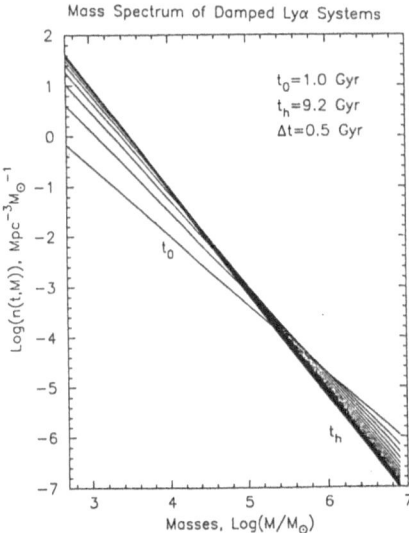

Fig. 3. Evolution of the mass spectrum of GHCs with cosmic time.

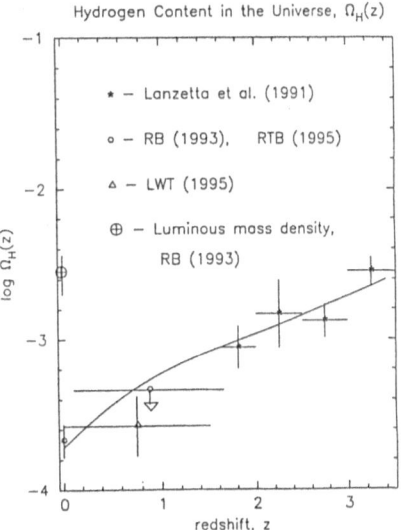

Fig. 4. Derived evolution of Ω_{HI} with redshift.

In Fig.4 we also show the evolution of the cosmological mass density parameter for neutral hydrogen, $\Omega_{HI}(z) = (1/\rho_{cr}) \int_{M_{min}}^{M_{max}} M n_{GHC}(z, M) dM$, where ρ_{cr} is critical density and M_{max} is assumed to be equal to $2 \times 10^8 M_\odot$, as calculated with the derived mass spectrum. A comparison of the derived curve with the observational data clearly shows that the derived mass spectrum matches the main integral features of the cosmological evolution of neutral hydrogen gas. This result can be used to study different aspects of the evolution of the interstellar medium in galaxies.

3   Conclusions

The derived mass spectrum of GHCs shown in Fig.3 exhibits some interesting features which we summarize below:

1. The mass spectrum of the GHCs caused by the DLAs is described by a power-law relation which covers several orders of magnitude in mass. This suggests that there is a universal mechanism for forming the hierarchy of these hydrogen clouds in galaxies. The most probable mechanism of this kind is interstellar turbulence.

2. The interstellar medium in galaxies evolves rapidly. The slope of the mass spectrum systematically increases with decreasing redshift between $z \approx 3$ and $z \approx 1$. This is a manifestation of powerful star formation processes. Star formation first destroys the most massive clouds and, as a result, forms many smaller mass clouds.

3. Therefore, there exists mass transfer between the high mass and low mass parts of mass spectrum. This mass transfer can be quantitatively related to the differential star formation rate, $\psi(t, M_\star)$, which is a key parameter in galactic evolution.

This research was supported by a grant from the Space Telescope Science Institute, which is operated by AURA, Inc., under NASA contract NAS5-26555.

References

Blitz, L. 1991, in *The physics of Star Formation and Early Stellar Evolution*, eds. Lada C.J. & Kylafis N.D. (Kluwer Academic Publishers), p.3

Kulkarni, S.R., & Heiles, C. 1988, in *Galactic and Extragalactic Radio Astronomy*, eds. Verschuur G.L. & Kellerman K.L. (Springer-Verlag, N-Y), p.95

Lanzetta, K.M., Wolfe, A.M., & Turnshek, D.A. 1995, ApJ, February 20 (LWT)

Lanzetta, K.M., Wolfe, A.M., Turnshek, D.A., Lu, L., McMahon, R.G., & Hazard, C. 1991, ApJS, 77, 1

Rao, S., & Briggs, F.H. 1993, ApJ, 419, 515 (RB)

Rao, S., Turnshek, D.A. & Briggs, F.H. 1995, ApJ, in press (RTB)

Wolfe, A.M. 1993, in *Relativistic Astrophysics and Particle Cosmology*, eds. Akerlof C.W. & Srednicki M.A., (New York: New York Acad. of Sci.), p. 281

A Damped Lyα Survey at $z < 1.65$

Sandhya Rao[1], David Turnshek[1], and Frank Briggs[1,2]

[1] University of Pittsburgh, Pittsburgh, PA 15260, USA
[2] Present address: Kapteyn Astronomical Institute, Gröningen, The Netherlands

Abstract. We have used a new approach to study the evolutionary properties of damped Lyα systems at $z < 1.65$. The approach takes advantage of the fact that all damped Lyα systems studied so far are found to have corresponding MgII absorption. From known MgII absorption-line statistics, we determine an upper limit to the number of damped Lyα systems per unit redshift, n_{DLya}, at $<z> \approx 0.8$ that is consistent with the trend at higher redshift as well as the number density at the current epoch. The data exhibit a trend established by three different observing techniques over three redshift regimes, providing important evidence that this trend actually traces the evolution of neutral gas in galaxies. Implications for the evolution of $n_{DLya}(z)$ and the cosmological neutral hydrogen gas mass density, $\Omega_{HI}(z)$, are discussed.

1 Introduction

QSO absorption lines are proving to be powerful probes for studying the evolution of galaxies (Blades et al. 1988). For over a decade numerous observations have addressed the statistical properties of various classes of absorption systems. The evolution of damped Lyα (dLyα) systems has proven to be of particular interest since at high redshift most of the neutral gas in the universe arises in such systems (Wolfe 1988).

The Lyα line is redshifted from the UV into the optical regime for redshifts $z > 1.65$, and ground-based optical surveys have produced data that constrain the evolutionary trends of the dLyα systems (Wolfe et al. 1986; Lanzetta et al. 1991). However, the redshift range $z < 1.65$ is particularly important. It corresponds to a cosmic look-back time interval spanning 62% of the age of the Universe for $q_0 = 0$ (77% for $q_0 = 0.5$). For redshifts $z < 1.3$, Bahcall et al. (1995) report that no dLyα systems have as yet been found in the HST QSO absorption-line key project survey. On the other hand, Lanzetta, Wolfe, & Turnshek (1995) used IUE data to select dLyα candidates and have provided some constraints on the evolutionary properties of the dLyα systems at low redshift. In any case, since the occurrence of a dLyα system is relatively rare and the number of QSOs studied in UV spectroscopic surveys is small, studies of the statistical properties of dLyα systems at low redshift are difficult.

Here we discuss new observational results on the incidence of dLyα absorbers at redshifts $z < 1.65$ and combine them with existing results in the redshift interval $1.65 < z < 4.0$ in order to study the evolution of the dLyα absorbers over this entire redshift range.

2   The MgII Selection Method

Since observations from the ground can detect MgII ($\lambda 2800$) systems with redshifts $0.11 < z < 2.2$, their statistical properties are well known (cf. Steidel & Sargent 1992, SS92) and observations of a sample of these systems offer the best opportunity for a follow-up dLyα study. Since all known dLyα systems exhibit low-ionization metal-line transitions (cf. Turnshek et al. 1989; Lu et al. 1993; Wolfe et al. 1993; Lu & Wolfe 1994), we can search for dLyα absorption by studying only those objects which exhibit MgII absorption in their optical spectra, thereby weeding out objects which essentially have no chance of showing dLyα in the UV. A similar point of view was adopted by Briggs & Wolfe (1983) for evaluating of the incidence of 21 cm absorption line systems selected for MgII absorption. These methods are more efficient than the normal spectroscopic survey procedures (Rao 1994). The number of dLyα systems per unit redshift, $n_{\rm DLya}(z)$, can then be constructed from known $n_{\rm MgII}(z)$ statistics by determining the fraction of dLyα absorption lines in a MgII-selected sample.

3   The Incidence of dLyα Systems with $z < 1.65$

After exclusions based on selection criteria, an unbiased sample was constructed from a list of 366 known (as of June 1994) MgII absorption systems. UV data existed for 43 MgII systems; the presence of a damped line at the expected wavelength was ruled out for 37 of these using archival I-UE and HST-FOS spectra. See Rao, Turnshek, & Briggs (1995, RTB) and Rao (1994). Four dLyα candidates were identified and two dLyα lines with $N(HI) \geq 2 \times 10^{20}$ cm^{-2} were found in recently obtained HST spectra (Steidel 1994). Therefore, we assume that at most 6 dLyα systems are present in the sample of 43 MgII systems. This fraction along with $n_{\rm MgII}(z)$ for Mg II rest equivalent width $W_r(\lambda 2796) > 0.3$Å (SS92) gives $n_{\rm DLya}(z \approx 0.8) < 0.12$, with a 1σ upper limit of 0.18. If none of the four candidates turns out to be damped, then $n_{\rm DLya}(z \approx 0.8) = 0.04^{+0.05}_{-0.01}$. Note that the upper limit on $n_{\rm DLya}(z \approx 0.8)$, shown in Figure 1a, follows a decreasing trend of $n_{\rm DLya}(z)$ with decreasing redshift, especially if the point at $z = 0$ is included. Figure 1b shows the corresponding redshift dependence of the cosmological mass density of neutral gas associated with the dLyα systems, $\Omega_{HI}(z)$.

The statistical parameters $n(z=0)$ and $\Omega_{HI}(z=0)$ were determined using well established results from direct observations of nearby galaxies (Rao & Briggs 1993, RB; Rao 1994). These measurements are much more precise and less ambiguous than those at high redshift, and hence, play an important role in constraining the evolution of galaxies. The determination of these parameters for $z < 1.65$, a time span of more than half the age of the Universe, is also crucial for interpreting the nature of the dLyα systems as well as for constraining the evolution of galaxies. The result that $n_{\rm DLya}(z)$ in this redshift interval conforms to a trend established by three different observing techniques is strong evidence that this trend actually traces the evolution of

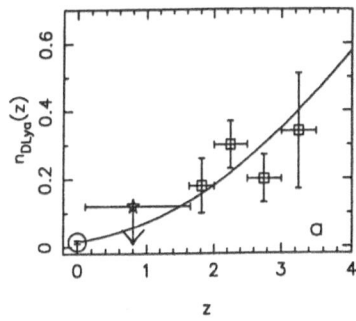 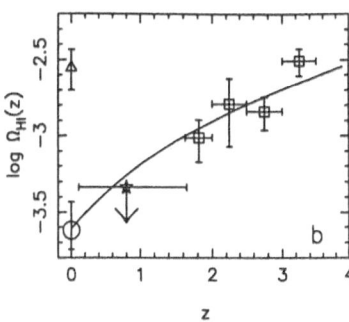

Fig. 1. (a) Squares are from Lanzetta et al. (1991), the circle is derived from s-tatistics of nearby galaxies (RB), and the star is derived from the MgII-selection method (RTB). The solid line is equation (1). (b) Symbols are as in (a); the triangle is the luminous mass at $z = 0$. The solid line is the best fit power law (Rao 1994).

neutral gas in galaxies. It lends further support to the conclusion of previous studies that dLyα systems are the progenitors of present-day galaxies. This result justifies the use of a single expression to describe the evolution of dLyα systems from redshift 4 to the present epoch. Using the data at $z > 1.65$ in combination with the constraint at $z = 0$ we derive an expression for $n(z)$ that is valid from the present epoch to a redshift of $z \approx 4$, i.e.,

$$n_{\text{DLya}}(z) = (0.015 \pm 0.004)(1 + z)^{2.27 \pm 0.25} \tag{1}$$

(RTB). This parameterization is shown in Figure 1a. The upper limit at $< z \gg 0.8$ was not used to derive this power law. For now, the result obtained in this redshift interval is simply used as strong justification for fitting a single expression over the entire redshift range $0 < z < 4$. The power-law index $\gamma = 2.27 \pm 0.25$ clearly indicates that dLyα absorbers undergo evolution, either in size or in number density, from $z \approx 4$ to 0. This result is significant at the 5σ level for $q_0 = 0$ and at the 7σ level for $q_0 = 0.5$.

Until recently, inferences about the nature of the dLyα systems have been made from observations at widely separated epochs, i.e., $z > 1.65$ and $z \approx 0$. Our results bridge this large gap in time and show that the evolution of their number density per unit redshift and neutral gas mass can be traced from redshifts of $z \approx 4$ to 0. However, given that $\Omega_{HI}(z \approx 4)$ is approximately equal to the luminous mass of all galaxies at the present epoch, it is not yet clear whether the dLyα systems trace only the spirals (Wolfe 1988) or whether they trace the evolution of neutral gas in all galaxies. There is increasing evidence that some elliptical galaxies might be merger remnants (Schweizer & Seitzer 1992; Hernquist 1993; Dressler et al. 1994). It is, therefore, quite likely that the decrease in n_{DLya} with time is due not only to the depletion of high-column-density gas due to processes such as star formation, but also to the decrease in the number density of galaxies due to mergers.

As discussed in Rao (1994), the decline in Ω_{HI} with time cannot be a consequence of star formation alone. If it were, then the star formation rate at the current epoch predicted from the observed rate of decline of Ω_{HI} would be a factor of ~ 30 lower than the observed star formation rate in nearby galaxies. The neutral gas content of galaxies is influenced not only by star formation, but also by depletion processes such as ionization of HI by hot stars as well as the metagalactic EUV radiation field and the formation of molecular hydrogen. The HI content can also be replenished by the recombination of ionized hydrogen as well as by infall of intergalactic gas. Thus, the HI content is expected to evolve over time through a complex mixture of processes.

Finally, an effect that might also be important is the obscuration of background QSOs by dust in the parent galaxies of the dLyα systems (Fall & Pei 1993). For example, it is possible that selection effects caused by dust might make the observed values of n_{DLya} and Ω_{HI} at $z\approx0.8$ lower than the "true" values. Indeed, if the observed n_{DLya} and Ω_{HI} values at low redshift were deemed too low relative to the $z = 0$ point, this might be evidence for dust.

Acknowledgements. SR is grateful to Chuck Steidel for sharing results prior to publication. This research was supported by an archival research program grant AR4921.01-92A from the Space Telescope Science Institute, which is operated by AURA, Inc., under NASA contract NAS5-26555.

References

Bahcall, J. N., et al. 1995, in press

Blades, J. C., Turnshek, D., & Norman, C. A. 1988, QSO Absorption Lines: Probing The Universe (Cambridge University Press)

Briggs, F. H., & Wolfe, A. M. 1983, ApJ, 268, 76

Dressler, A., Oemler Jr., A., Sparks, W. B., & Lucas, R. A. 1994, ApJ, 435, L23

Fall, S. M., & Pei, Y. C. 1993, ApJ, 402, 479

Hernquist, L. 1993, ApJ, 409, 548

Lanzetta et al. 1991, ApJS, 77, 1

Lanzetta, K. M., Wolfe, A. M., & Turnshek, D. A. 1995, in press

Lu, L., Wolfe, A. M., Turnshek, D. A., & Lanzetta, K. M. 1993, ApJS, 84, 1

Lu, L., & Wolfe, A. M. 1994, AJ, 108, 44

Rao, S. M. & Briggs, F. H. 1993, ApJ, 419, 515 (RB)

Rao, S. M. 1994, Ph.D. Thesis, University of Pittsburgh

Rao, S. M., Turnshek, D. A., & Briggs, F. H. 1995, in press (RTB)

Schweizer, F., & Seitzer, P. 1992, AJ, 104, 1039

Steidel, C. C., & Sargent, W. L. W. 1992, ApJS, 80, 1 (SS92)

Steidel, C. C. 1994, private communication

Turnshek et al. 1989, ApJ, 344, 567

Wolfe, A. M., Turnshek, D. A., Smith, H. E., & Cohen, R. D. 1986, ApJS, 61, 249

Wolfe, A. M. 1988, in QSO Absorption Lines: Probing the Universe, eds. Blades, Turnshek, and Norman (Cambridge University Press), p. 297

Wolfe, A. M., Turnshek, D. A., Lanzetta, K. M., & Lu, L. 1993, ApJ, 404, 480

High Redshift Lyman Limit and Damped Lyman-Alpha Absorbers

L.J. Storrie-Lombardi[1,2], R.G. McMahon[1], M.J. Irwin[3], C. Hazard[1,4]

[1] Institute of Astronomy, Madingley Road, Cambridge CB3 0HA UK
[2] current address: UCSD-CASS, Mail Code 0111, La Jolla, CA 92093 USA
[3] Royal Greenwich Observatory, Madingley Road, Cambridge CB3 0HE UK
[4] University of Pittsburgh, Pittsburgh, PA USA

Abstract. We have obtained high signal:to:noise optical spectroscopy at 5Å resolution of 27 quasars from the APM z>4 quasar survey. The spectra have been analyzed to create new samples of high redshift Lyman-limit and damped Lyman-α absorbers. These data have been combined with published data sets in a study of the redshift evolution and the column density distribution function for absorbers with $\log N(HI) \geq 17.5$, over the redshift range $0.01 < z < 5$. The main results are:

(i) Lyman limit systems: The data are well fit by a power law $N(z) = N_0(1+z)^\gamma$ for the number density per unit redshift. For the first time intrinsic evolution is detected in the product of the absorption cross-section and comoving spatial number density for an $\Omega = 1$ Universe. We find $\gamma = 1.55$ ($\gamma = 0.5$ for no evolution) and $N_0 = 0.27$ with >99.7% confidence limits for γ of 0.82 & 2.37.

(ii) Damped Lyα systems: The APM QSOs provide a substantial increase in the redshift path available for damped surveys for $z > 3$. Eleven candidate and three confirmed damped Lyα absorption systems, have been identified in the APM QSO spectra covering the redshift range $2.8 \leq z \leq 4.4$ (11 with $z > 3.5$). Combining the APM survey confirmed and candidate damped Lyα absorbers with previous surveys, we find evidence for a turnover at z~3 or a flattening at z~2 in the cosmological mass density of neutral gas, Ω_g. The Lyman limit survey results are published in Storrie-Lombardi, et al., 1994, ApJ, 427, L13. Here we describe the results for the DLA population of absorbers.

1 Introduction

How and when galaxies formed are questions at the forefront of work in observational cosmology. Absorption systems detected in quasar spectra provide the means to study these phenomena up to z~5, back to when the Universe was less than 10% of its present age. While the baryonic content of spiral galaxies that are observed in the present epoch is concentrated in stars, in the past this must have been in the form of gas. Damped Lyα absorption (D-LA) systems have neutral hydrogen column densities of $N(HI) > 2 \times 10^{20} cm^{-2}$. They dominate the baryonic mass contributed by HI. The principal gaseous component in spirals is HI which has led to surveys for absorption systems detected by the DLA they produce (Wolfe, Turnshek, Smith & Cohen 1986 [WTSC]; Lanzetta et al. 1991 [LWTLMH]; Lanzetta, Wolfe & Turnshek 1995

[LWT95]). We extend the earlier work on Lyman limit systems and DLAs to higher redshifts using observations of QSOs from the APM z>4 QSO survey (Irwin, McMahon & Hazard 1991), These data more than triple the redshift path surveyed at z>3 and allow the first systematic study up to z=4.5.

2   APM Damped Lyman-Alpha Survey at z~4

We have obtained 27 high S/N spectra at 5Å resolution at the William Herschel Telescope. The spectra were analyzed starting 3000 km s^{-1} blueward of z(emission). The analysis was stopped when the S/N ratio became too low to detect a Lyα line with W(rest)$\geq$5Å. This point was typically caused by the incidence of a Lyman limit system. Features with W(observed) $\gtrsim$ 25Å were selected with an automated procedure. Most of these are blends of the dense Lyα forest features present at high redshift. The equivalent width and FWHM were measured interactively and N(HI) was estimated for features with W or FWHM >30Å. Of the 34 measured, 15 have estimated N(HI)$\geq 2 \times 10^{20}$ cm^{-2} covering $2.8 \leq z \leq 4.4$. Only one candidate has estimated N(HI)$\geq 10^{21}$ cm^{-2}. High resolution spectroscopy of 4 candidates has confirmed 3 as damped (log N(HI)$\geq$20.3). The sensitivity of the survey with redshift was determined using the method developed by Lanzetta (see LWT95). The function $g(z)$ is calculated, giving the number of lines of sight along which a damped system at a redshift z could be detected. Figure 1a shows the sensitivity of the APM survey alone and in combination with the WTSC and LWTLMH surveys.

3   Evolution of the Number Density per Unit Redshift

The candidate and confirmed DLA systems from the APM sample and previous surveys (WTSC; LWTLMH; LWT95) have been combined to study the evolution of the number density per unit redshift for 0.01 < z < 4.7. Fit with the customary power law $N(z) - N_0(1 + z)^\gamma$, a population with no intrinsic evolution in the product of the absorption cross-section and comoving spatial number density will have $\gamma = 1/2$ ($\Omega = 1$) or $\gamma = 1$ ($\Omega = 0$). A maximum likelihood fit to the data with z>1.5 yields $N(z) = 0.03(1 + z)^{1.5\pm0.6}$, consistent with no intrinsic evolution even though the value of γ is similar to that found for the Lyman limit systems where evolution is detected at a significant level. However, there is redshift evolution evident in the higher column density systems with an apparent decline in $N(z)$ for z>3.5. These results are displayed in Figure 1b. The combined data set is plotted as dashed lines with the above fit. The results for only the absorbers with log N(HI)$\geq$ 21 are shown as solid lines. The z<1.5 bin is taken from LWT95.

4   Evolution of Ω_g – Baryons in Neutral Gas

The mean cosmological mass density contributed by damped Lyα absorbers can be estimated as

$$\langle \Omega_g \rangle = \frac{H_0 \mu m_H}{c \rho_{crit}} \int_{N_{min}}^{\infty} N f(N) dN$$

as defined in LWTLMH (equations 17-18), giving the current mass density in units of the current critical density. The errors in Ω_g are difficult to estimate because the column density distribution function, $f(N)$, is not known. LWTLMH utilised the standard error in the distribution of N(HI) which yields zero error if all the column densities in a bin are the same. We have estimated the fractional variance in Ω_g as $\sum_{i=1}^{p} N_i^2 / (\sum_{i=1}^{p} N_i)^2$ which yields $\sqrt{n}$ errors if all the column densities included in a bin are equal. This method yields larger errors. The results for Ω_g are shown in Figure 2 for $q_0=0$ and $q_0=0.5$ ($H_0=50$). The $z<1.5$ bin is taken from LWT95. The $z>1.5$ solid bins utilise the data from WTSC, LWTLMH, and the APM survey. The dotted bins exclude the APM data. The inclusion of the APM survey data for $z > 3$ lowers the value previously found for $3 < z < 3.5$ and indicates a possible turnover for $z > 3.5$. The results are also consistent with a relatively constant value of Ω_g for $z > 2$ as the error bars are still very large at high redshift. Larger samples of bright $z > 4$ quasars are needed.

5   Summary

The QSOs from the APM survey more than triple the $z > 3$ redshift path for DLA surveys. Fourteen candidate DLA systems have been identified in the APM spectra covering $2.8 \leq z \leq 4.4$ (11 with $z > 3.5$), with 3 confirmed. Combining these data with the previous surveys and fitting a single power law for $z>1.5$ gives N(z)= $.03(1+z)^{1.5\pm0.6}$, marginally consistent with no evolution models. Evolution is evident in the highest column density absorbers with the incidence of systems with log N(HI)$\geq$21 apparently decreasing for $z \gtrsim 3.5$. We find evidence for a turnover or flattening in the cosmological mass density of neutral gas, Ω_g at high redshift. The more gradual evolution of Ω_g than previously found helps alleviate the 'cosmic G-dwarf problem' (LWT95), i.e. if a large amount of star formation has taken place between $z=3.5$ and $z=2$, a much larger percentage of low metallicity stars should exist than is detected. It is also consistent with the suggestion by Pettini et al. (1994) that the wide range in DLA metallicities measured at the same epoch indicates that at $z\sim2$ they are observed prior to the bulk of star formation in the disk. Storrie-Lombardi and Wolfe have undertaken a programme using the Keck Telescope to confirm the candidate systems from the APM survey and obtain accurate column densities for determining N(z) and Ω_g.

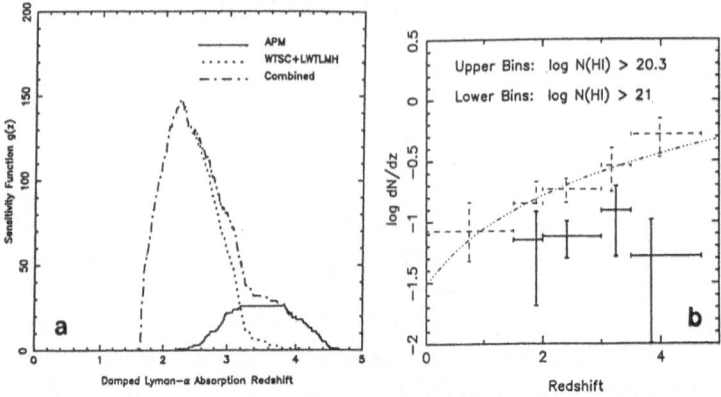

Fig. 1. (a) The sensitivity function, $g(z)$, of the DLA surveys. This gives the number of lines of sight along which a damped system at redshift z could be detected. (b) The number density of DLA per unit redshift, $N(z)$, vs. z(absorption). The dashed bins show N(z) for all the damped systems and the solid bins for systems with $N(HI) \geq 10^{21}$ cm^{-2}. A single power law fit to the sample for z>1.5 gives $N(z) = .03(1 + z)^{1.5 \pm 0.6}$.

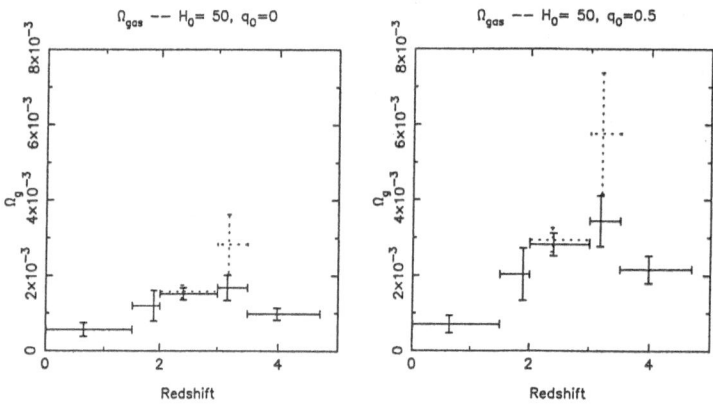

Fig. 2. The mean cosmological mass density in neutral gas, Ω_g, contributed by DLA absorbers for $0.01 \leq z \leq 4.7$ for $q_0 = 0$ and $q_0 = 0.5$ ($H_0 = 50$). The z<1.5 bin is taken from LWT95. The z>1.5 solid bins utilise the WTSC, LWTLMH, and APM survey data. The dotted bins exclude the APM data.

References

Irwin, M.J., McMahon, R.G. & Hazard, C., 1991, in *ASP Conf. Series, Vol. 21*, ed. Crampton D., (San Francisco: ASP), 117

Lanzetta, K.M., et al., 1991, ApJS, 77, 1 (LTWLMH)

Lanzetta, K.M., Wolfe, A.M. & Turnshek, D.A., 1995, ApJ, 440, 435 (LWT95)

Pettini, M., Smith, L.J., Hunstead, R.W. & King, D.L., 1994, ApJ, 426, 79

Storrie-Lombardi, L.J., McMahon, R.G., Irwin, M.J. & Hazard 1994, ApJ, 427, L13

Wolfe, A.M., Turnshek, D.A., Smith, H.E. & Cohen, 1986, ApJS, 61, 249 (WTSC)

Moderate Resolution Studies of Damped Lyα Systems at High Redshift

Julio Saucedo[1,2] and Jill Bechtold[1]

[1] Steward Observatory, University of Arizona, Tucson AZ 85721
[2] CIF, Universidad de Sonora, Hermosillo, Sonora, México

Abstract. We present a summary of the results obtained through curve of growth analysis of metal lines associated with damped Lyα systems at high redshift, z_{abs}=1.77 to 3.02. For this summary we have included our recent study of 6 QSOs with the MMT Red Spectrograph, as well as results from three other similar studies at intermediate resolution. We find that the average abundances relative to hydrogen of O I, C II, Si II, Fe II, and Al II in damped Lyα Systems resemble those of the Galactic Halo.

Damped Lyα Systems (DLAS) are known to be excellent targets for abundance studies. This is due to the fact that the column density of neutral hydrogen can be accurately determined from the damping profile. This information can be combined with column densities derived from curve of growth (COG) analysis to determine metal abundances, provided one can establish that the metal absorption lines are produced in the same region as the damped component. At intermediate resolution, however, the lines are frequently blended or saturated, which makes abundance studies questionable. In this work we use the results of COG analysis as lower limits to the abundances.

A summary of the results from COG analysis in DLAS at intermediate resolution is shown on Fig 1. There we plot logN(X i)/N(H I)), where N(X i) is the column density of one of the six low-ionization species studied (OI, CII, Si II, Fe II, Al II, Mg II), and N(H I) is the column density of H I. The data for DLAS are shown with open polygons. The error bars on the right are the average logarithmic errors for each species. This figure allows us to make a comparison with well-studied regions of the Galaxy. The filled polygons represent abundances for the Galactic Halo (in the direction of HD 36402) reported by Savage and Jeske (1981). The starred symbols are observations of the Galactic Halo toward 3C273 from Savage et al. (1993). The half filled polygons present recent results obtained for the ζ Ophiuchi clouds (denoted by A and B in in the figure). The observations for ζ Ophiuchi were carried out by Savage et al. (1992), and by Cardelli et al. (1994) using the HST. Component A (at -15 km/s) is thought to be due to the dense cloud close to ζ Ophiuchi, whereas component B (at -27 km/s) samples the local ISM. Although there is a lot of scatter in the figure, we can see that the average abundances of DLAS are similar to the values found in the Halo.

Since the Doppler velocity parameters (b-values) obtained from the COG analysis are high, the abundances are likely underestimated. In other words these results set the abundance of the Halo as a lower limit. On the other hand, the upper limits for weak lines reported by several studies (for instance Pettini et al. 1994) impose strong upper limits to the abundances of DLAS. Therefore it is reasonable to characterize the abundances of DLAS as similar to those in the Halo.

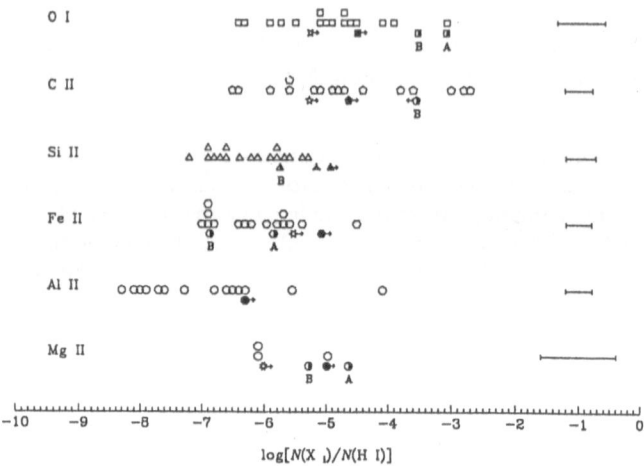

Fig. 1. Abundances (relative to hydrogen) for low-ionization species in DLAS. Unfilled polygons represent data from: Turnshek et al. (1989), Lu et al. (1993), Wolfe et al. (1995), and Saucedo and Bechtold (1995). Average logarithmic errors for each species are shown at right. Filled polygons are from the Halo toward HD36402 (Savage and Jeske 1981). Starred symbols are from the Halo toward 3C273 (Savage et al. 1993). Partially filled polygons are abundances for clouds A and B in ζ Ophiuchi from Savage et al. (1992), and from Cardelli et al. (1994) (for Si II).

Acknowledgements. This research was supported by NSF grants RII-880660, AST-9058510, and a gift from Sun Microsystems. J.S. acknowledges support from a CONACYT fellowship.

References

Cardelli, J. A., Sofia, U. J., Savage, B. D., et al., 1994, ApJ, 420, L29
Lu, L., Wolfe, A. M., Turnshek, D. A., & Lanzetta, K. M. 1993, ApJS, 84, 1
Pettini, M, Smith L. J., Hunstead, R. W., & King, D. 1994, ApJ, 426,79
Rauch, M., Carswell, R. F., Robertson, J. G., et al., 1990, MNRAS, 242, 698
Saucedo, J., & Bechtold, J. 1995, ApJS, in press
Savage, B. D., & Jeske, N. A. 1981, ApJ, 244, 768
Savage, B. D., Cardelli, J. A., Sofia, U. J. 1992, ApJ, 401, 706
Turnshek, D. A., Wolf, A. M., Lanzetta, K. M., et al., 1989, ApJ, 344, 567
Wolfe, A. M., Turnshek, D. A., Lanzetta, K. M., & Lu, L. 1993, ApJ, 404, 480

Near-Infrared Hα Searches for Companion Galaxies to Damped Ly-α Systems at $z > 2$

A.J. Bunker[1], S.J. Warren[2], P.C. Hewett[3] and D.L. Clements[4]

[1] University of Oxford, Astrophysics Dept., Keble Road, Oxford OX1 3RH, UK
[2] Imperial College, Blackett Laboratory, Prince Consort Road, London SW7 2BZ
[3] Institute of Astronomy, Madingley Road, Cambridge CB3 OHA, UK
[4] ESO, Karl-Schwarzschild-Straße 2, D-85748 Garching-bei-München, Germany

Abstract. Despite many years of work, searches for distant galaxies based on the expected Lyman-α emission have proved almost universally unsuccessful. The likely explanation for this is the presence of dust in the emitting objects which will strongly absorb the Lyman-α photons, particularly with resonant scattering from H I greatly extending the escape path length. Narrow-band observations in the near infrared, looking for the redshifted Hα line, can avoid many of these difficulties, and enable the true star formation rate (*SFR*) to be inferred. By targeting our observations on known $z > 2$ damped Lyman-α systems we hope to detect companion galaxies of the absorbers. In this poster we present encouraging preliminary results using this new strategy for finding distant galaxies.

1 Observations

The Hα line lies in the K-band at redshifts $2.08 < z < 2.66$, epochs where star formation may be at a peak according to many current theories. Our technique is to image the region around a damped Lyman-α absorption line system in both the K' continuum and a narrow-band filter at the wavelength of Hα emission at the damped system redshift. Objects that are bright in the narrow band but faint or undetected in the broad band are candidate high-redshift galaxies.

As a pilot study, observations of a field around the quasar PHL957 were conducted over 3 nights in November 1993 at the ESO 2.2-m, using the IRAC2B infrared camera (this uses a 256^2 NICMOS-3 array). Conditions were photometric with seeing about 1 arcsec. Total integration times were 3 hours in K' and nearly 7 hours in a narrow-band filter centred on 2.177 μm. This corresponds to the wavelength of Hα at the redshift $z = 2.31$ of a damped absorption system seen in the quasar's spectrum. The field of view of the camera is over 2 arcmin (0.52 arcsec per pixel), corresponding to $\sim 0.5\,h^{-1}$ Mpc at the redshift of the absorber ($h = \mathrm{H_o}/100\,\mathrm{km\,s^{-1}\,Mpc^{-1}}$). The width of the filter (5000 km s^{-1}) comfortably encompasses typical velocity dispersions of galaxy clusters.

2   Results and Conclusions

In this first study we surveyed 4.9 arcmin^2 to a 4σ limiting narrow-band flux of $f = 2.7 \times 10^{-16}$ erg cm^{-2} s^{-1} (aperture $r = 3$ pixels, 1.56 arcsec), corresponding to $m_{na} = 19.3$. Our broad-band K' frame goes 0.8 magnitudes deeper. The graph plots the the colour (difference between broad and narrow-band magnitudes) against narrow-band magnitude for all 30 objects with $S/N > 4$ found in the PHL957 field. The dot-dashed lines are lines of constant Σ, which is the number of standard deviations of the excess flux in the narrow-band relative to the broad-band. Also shown are lines of constant rest-frame equivalent width. Candidate high-redshift galaxies are objects with equivalent widths $EW_{rf} > 75$ Å, and $\Sigma > 2$. The line $\Sigma = 2$ corresponds to a SFR of $11\ h^{-2}\ M_\odot$ yr^{-1}, using the prescription of Kennicutt (1983, ApJ, 272, 54).

A previously known companion galaxy (C1), discovered first in Lyman-α (Lowenthal et al. 1991, ApJ, 377, L73), is detected at 4.5σ in the narrow-band. This object shows a clear colour excess ($\Sigma = 3.3$) indicating that we have detected the Hα emission from this object. We note that a tentative (2.5σ) spectroscopic detection of Hα has also been made (Hu et al. 1993, ApJ, 419, L13). Two other objects show smaller but significant ($> 2.5\sigma$) excesses, and are thus candidate objects at $z = 2.3$.

The first observations using our new technique have successfully detected a known galaxy at $z = 2.3$. This technique is insensitive to dust absorption in the distant galaxies, and we believe that it represents a very promising new way to find star-forming galaxies at large redshifts. We are extending these observations with the UK Infrared Telescope to the regions around other damped absorbers. Results are reported in Bunker et al. 1995 MNRAS, in press.

Table 1. Properties of galaxy C1

Offset rel. to QSO	42.6″W, 24.1″N
Projected Dist.	190 h^{-1} kpc
Redshift z_{em}	2.313
m_{narrow}	19.22 (4.5σ)
m_{broad} (K')	20.99 (1.9σ)
Col. ($m_{br} - m_{na}$)	1.77
Col. sig. Σ	3.3
$EW_{rf}^{H\alpha+[N\ II]}$ /Å	1190(2σ > 220)
$EW_{rf}^{Ly\alpha}$ /Å	140
$f_{H\alpha}$/erg cm^{-2} s^{-1}	$(2.1 \pm 0.6)10^{-16}$
$f_{Ly\alpha}$/erg cm^{-2} s^{-1}	$(5.6 \pm 0.1)10^{-16}$
V mag.	~ 23.6
SFR/h^{-2} $M_\odot$ yr^{-1}	$18 \pm 5(q_0 = 0.5)$
Lyα / Hα ratio	2.7
Reddening	$E_{B-V} = 0.16$

Fig. 1. Colour - magnitude diagram

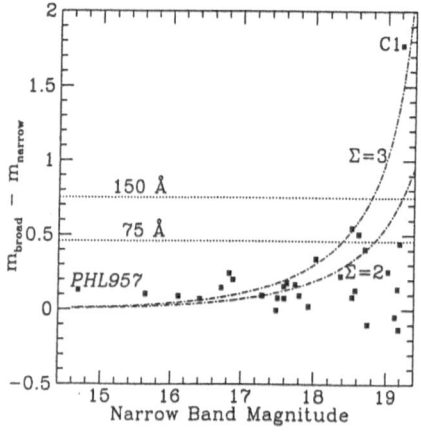

Lyman α Emission from Galaxies at High Redshifts

Max Pettini[1], Richard W. Hunstead[2], David L. King[1], and Linda J. Smith[3]

[1] Royal Greenwich Observatory, Madingley Road, Cambridge, CB3 0EZ, UK
[2] School of Physics, University of Sydney, NSW 2006, Australia
[3] Department of Physics and Astronomy, University College London, Gower Street, London WC1E 6BT, UK

Abstract. We summarise the results of a deep search for Lyman α emission from star-forming regions associated with damped Lyman α absorption systems and conclude that the Lyman α luminosity of high redshift galaxies is generally less than 10^{42} erg s^{-1}. We also present a newly discovered case, in the field of the QSO Q2059−360, where the emission is unusually strong, possibly because the damped system is close in redshift to the QSO.

1 Introduction

Damped Lyman α systems at $z \simeq 2 - 3$ arise in galaxies which are generally metal-poor and, in some cases, may have experienced only a few episodes of star formation (Pettini et al. 1995). It is therefore reasonable to expect that the fields of QSOs with damped systems may be good candidates in searches for Lyman α emission from high-redshift H II regions, particularly as the QSO light is completely extinguished by the damped Lyman α absorption line over a redshift interval which can span up to several tens of Å.

2 Lyman α Emission from Damped Absorbers

Over the last few years we have conducted such a search with high-resolution slit spectroscopy of 21 damped systems in 18 QSOs. Two noteworthy features of our approach are the use of photon-counting detectors, which maintain the photon statistics even at the very low count rates measured in the cores of the damped lines (essentially just the dark sky background) and the long integration times (between $\sim 10\,000$ and $\sim 40\,000$ s). Consequently, we are able to reach some of the lowest flux limits reported, corresponding to Lyman α luminosities $L_{\mathrm{Ly}\alpha}(3\sigma) \leq 1 \times 10^{42}$ erg s^{-1} ($H_0 = 50$ km s^{-1} Mpc^{-1}, $q_0 = 0.5$). On the other hand, a drawback of slit spectroscopy is that it samples only a limited area of sky; it is thus possible to miss the emission if the absorbing galaxy is displaced by more than ~ 1 arcsecond (the typical slit width used) from the QSO.

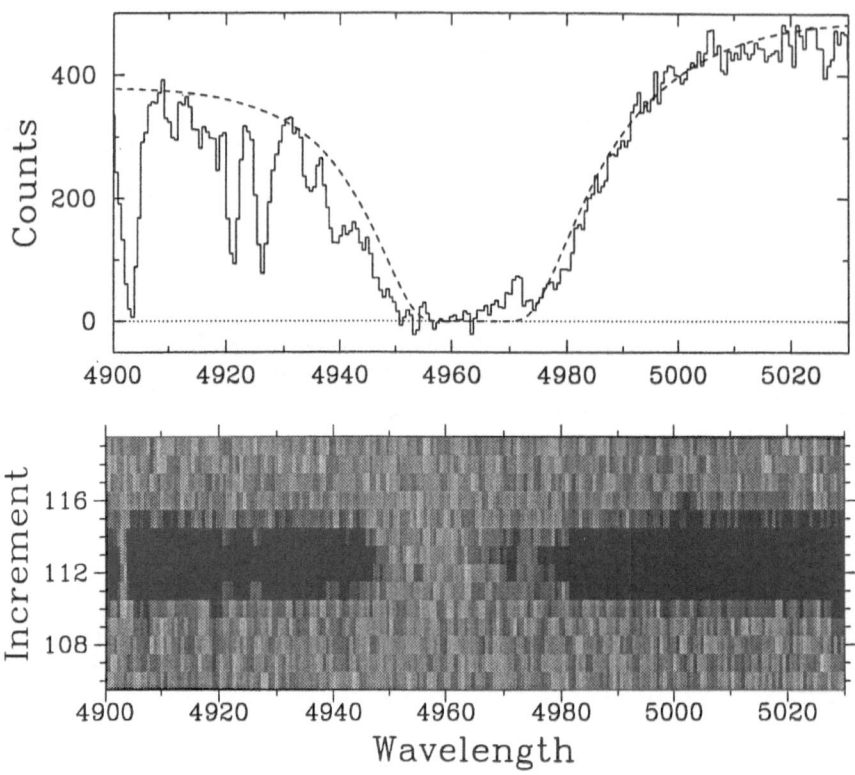

Fig. 1. *Top*: Observed profile of the $z_{abs} = 3.0831$ damped Lyman α line in the $z_{em} = 3.13$ QSO Q2059−360; the x-axis gives the wavelength in Å. The broken line shows the theoretical profile for a neutral hydrogen column density $N(H^0) = 5 \times 10^{20}$ cm^{-2}, centred at the redshift of Lyman β and metal absorption lines in the same system. *Bottom*: Grey-scale ($\pm 3\sigma$ of the sky signal) reproduction of the corresponding portion of the two-dimensional spectrum; each spatial increment is 0.75 arcsec, or ~ 5 kpc at $z = 3.08$. The spectrum was recorded with 8000 s integration using a Tektronix blue-sensitive CCD at the cassegrain focus of the Anglo-Australian telescope. The resolution is 1.4 Å *FWHM*; the data were obtained over two nights in September 1994, when the seeing was $\sim 1 - 1.5$ arcsec *FWHM*.

The principal result is that high-redshift galaxies are *not* strong Lyman α emitters; Lyman α emission is below, or close to, the detection limit in nearly all cases surveyed. We have found only two definite detections, associated with the $z_{abs} = 2.4651$ system in Q0836+113 – reported by Hunstead, Pettini and Fletcher (1990) and confirmed by new observations obtained in 1994 – and with the $z_{abs} = 3.0831$ system in Q2059−360 − shown in Fig.1. In addition, we have two or three cases of marginal detections.

From Fig.1 it is evident that there is excess signal in the base of the damped line, which we interpret as Lyman α emission redshifted by ~ 470 km s^{-1} from the absorption system. The integrated flux is $F_{Ly\alpha} = 7 \times 10^{-17}$ erg cm^{-2} s^{-1}, corresponding to $L_{Ly\alpha} = 5 \times 10^{42}$ erg s^{-1}. At the relatively coarse spatial resolution of our observations the emission appears to be approximately coincident with the QSO position. However, the velocity difference between emission and absorption suggests that the Lyman α emitting region is not associated directly with the damped system, but more likely arises in a galaxy which is a member of a cluster also including the absorber.

In this respect Q2059−360 is a similar case to the field of Q0528−250 where Moller and Warren (1993) found emission at $z_{em} = 2.811$. It may not be a coincidence that two out of the only three known detections of Lyman α emission in the core of a damped Lyman α line are absorption systems close to the QSO redshift. It remains to be seen whether these are environments favouring large bursts of star formation, or whether the Lyman continuum radiation from the QSO itself is responsible for the unusually strong emission feature.

3   Discussion

The limits reached in our survey, $L_{Ly\alpha} \le 1 \times 10^{42}$ erg s^{-1}, would correspond, *if the Lyman α line were not attenuated*, to Hα luminosities $L_{H\alpha} \le 1 \times 10^{41}$ erg s^{-1}, comparable to the luminosity of the brightest H II regions in the compilation by Kennicutt (1988) and less than the typical integrated Hα luminosity of present-day spiral galaxies (Kennicutt and Kent 1983).

The lack of significant Lyman α emission from damped Lyman α galaxies is in line with the generally null results of other Lyman α searches (e.g. Pritchet 1994; Lowenthal et al. 1995) and has motivated a considerable amount of theoretical work. The observations confirm current ideas that: *(i)* even massive bursts of star formation are probably bright in Lyman α for only a relatively brief period (Charlot and Fall 1993); *(ii)* Lyman α line radiation can be quenched very effectively by small amounts of dust, due to the large optical depths involved (Chen and Neufeld 1994); and *(iii)* Lyman α photons in any case probably escape from the parent galaxy along preferred directions determined by the inhomogeneous nature of the interstellar medium (Neufeld 1991). A local example may be provided by I Zw 18 which, despite being the most metal-poor H II galaxy known (Skillman and Kennicutt 1993) with an optical spectrum dominated by strong emission lines, has *no* detectable Lyman α emission (Kunth et al. 1994).

The continuing improvement in infrared detectors and the encouraging results of pilot observations (Hu et al. 1993; Bunker et al. 1995) suggest that, in future, searching for Balmer lines at $z > 2$ may be the most promising s-trategy for assessing the relative importance of these processes and measuring the star-formation rate of high-redshift galaxies.

References

Bunker, A.J., Warren, S.J., Hewett, P.C., Clements, D.L., 1995, MNRAS, in press

Charlot, S., Fall, S.M., 1993, ApJ, 415, 580

Chen, W.L., Neufeld, D.A., 1994, ApJ, 432, 567

Hu, E., Songaila, A., Cowie, L.L., Hodapp, K.W., 1993, ApJ, 419, L13

Hunstead, R.W., Pettini, M., Fletcher, A.B., 1990, ApJ, 356, 23

Kennicutt, R.C., 1988, ApJ, 334, 144

Kennicutt, R.C., Kent, S.M., 1983, AJ, 88, 1094

Kunth, D., Lequeux, J., Sargent, W.L.W., Viallefond, F., 1994, A&A, 282, 709

Lowenthal, J.D., Hogan, C.J., Green, R.F., Woodgate, B., Caulet, A., Brown, L., Bechtold, J., 1995, ApJ, submitted

Moller, P., Warren, S.J., 1993, A&A, 270, 43

Neufeld, D.A., 1991, ApJ, 370, L85

Pettini, M., King, D.L., Smith, L.J., Hunstead, R.W., 1995, This volume

Pritchet, C.J., 1994, PASP, 106, 1052

Skillman, E.D., Kennicutt, R.C., 1993, ApJ, 411, 655

PART III

ABUNDANCES

Abundances and Nature of the Absorbers

Patrick Petitjean

Institut d'Astrophysique de Paris - CNRS, 98bis Boulevard Arago, F-75014 Paris, France

Abstract. Observations of low redshift ($z < 0.5$) Lymanα lines by HST and detection of galaxies in the neighbourhood of the corresponding clouds have cast a new light and a shadow on the question of what is the nature of the Lymanα forest and in particular what is the connection between the absorbing gas and galaxies. Metallicity determination is a powerful tool towards clarifying the issue. The particularity of absorption line systems is that they are observed all over the accessible redshift range $0 < z < 5$. We therefore need an overall picture describing how the gas evolve. Recent N-body simulations with improved treatment of photoionization and cooling of the baryonic material show that Lymanα clouds could trace the filamentary structures of the dark matter.

1 Introduction

It has been convincingly demonstrated that low-redshift ($z \sim 0.6$) metallic absorption line systems are associated with galaxies. About sixty galaxy/MgII system pairs are known by now (Bergeron & Boissé 1991, Steidel 1995). The corresponding sample of absorption system/galaxy pairs is unique in the sense that information is known about the stellar content of the galaxy, the extent of the gaseous halo and, in favourable cases, the physical characteristics of the gas (temperature, ionization state, abundances). The best case up to now is the system at $z_{\mathrm{abs}} = 0.7913$ toward PKS 2145+06 (Bergeron et al. 1994) in which a large number of lines have been detected by HST (Bahcall et al. 1993), high resolution observation of the MgII absorption doublet is available (Petitjean & Bergeron 1990) and the associated galaxy is detected (Bergeron & Boissé 1991) at a projected distance of $55h_{50}^{-1}$ kpc. Reliable abundances have been derived and are surprisingly high ($0.5\ Z_{\odot}$) for such large distance to the centre of star formation. This demonstrates how powerful absorption line studies can be for investigating galactic structure and its evolution. Therefore it can be foreseen easily that development of the subject in the very next years, especially with the advent of 10-m class telescopes, will give unique information about when and how galaxies form.

The determination of metallicities in absorption line systems is a difficult task since one needs good determination of the column densities of HI and several ionization stages of the different elements to perform the ionization correction. It is therefore necessary to select carefully the systems in which the lines are neither strongly blended nor badly saturated. When the HI column density is very large as for damped systems, hydrogen can be considered neutral and the ionization correction is negligible. Moreover most of the elements are neutral or in their singly ionized state as OI, CII or SiII. These ions can rarely be used however since their lines are heavily saturated so the derived column densities have large uncertainties. Instead one has to search for weak lines of species such as ZnII and CrII (Meyer and York 1987a, Pettini et al. 1990, Bergeron & Petitjean 1991, Pettini et al. 1994).

In cases where the HI column density is not that large a photoionization code can be used (Bergeron & Stasińska 1986, Steidel 1990, Petitjean et al. 1992, Gruenwald & Viegas 1993). In the following I will give a few examples showing how it is possible to derive reliable information about the physical state of the gas and especially abundances. The overall aim however is to construct a consistent picture of what is the structure and nature of absorbers, and to do this the complete set of information in the field has to be used.

2   Abundances in Excess of Solar in Associated Systems

It has been suggested that systems at $z_{abs} \sim z_{em}$ could be associated with the QSO itself either because they arise in galaxies hosting the QSO or because they are the signature of ejected material. Weymann et al. (1979) surveyed a sample of 46 intermediate redshift QSOs and found an excess of CIV doublets per unit velocity within 5000 km s^{-1} from the emission redshift. Subsequently such an excess has not been found in surveys of optically selected QSOs (Young et al. 1982, Sargent et al. 1988). Foltz et al. (1986) suggested that a possible explanation for the discrepancy might be that strong $z_{abs} \sim z_{em}$ CIV systems are more often found in radio–loud QSOs. High incidence of such systems in radio loud quasars has been confirmed by Anderson et al. (1987) although no correlation between the presence of associated systems and radio properties is found to be statistically significant. Combining a recent large survey of radio–loud quasars with data from previous studies, Barthel (1988) finds an excess of strong $z_{abs} \sim z_{em}$ systems ($w_r > 3$ Å) in both radio–quiet and radio–loud QSOs.

BAL systems are undoubtedly closely related to the AGN phenomenon. Although BAL arise only in radio–quiet QSOs (Stocke et al. 1992, see also Weymann et al. 1991, Hamann et al. 1993), the observational facts may suggest that quasars with BAL or associated systems are drawn from the same population. There should be then a smooth transition from the narrow ejected associated systems to the broad absorption lines (BAL) which could correspond to recently ejected gas.

One way to address this question is to study metallicities in associated systems. In a recent work, Petitjean et al. (1994) have used high quality high resolution ($R = 15000$) data to derive abundances in nine systems observed in two QSOs, Q0424-131 and Q0450-131. Since the lines are most often saturated the uncertainties on the column densities are large except for a few

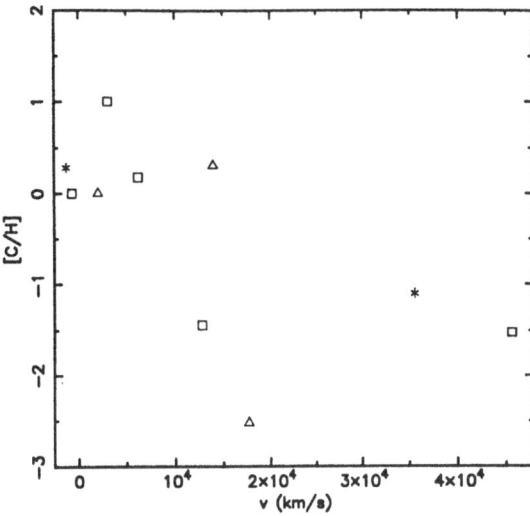

Fig. 1. Logarithm of the carbon abundance relative to solar versus ejection velocity from the QSO.

subcomponents, detected in SiII, SiIII, SiIV, CIV and NV, which are well detached from the bulk of the absorption. Moreover the HI column density is strongly constrained by the wing of the HI absorption feature. This favorable characteristics are coupled with a detailed fitting analysis of the system taking into account a consistent description of the ionization equilibrium using photoionization models. In particular the Doppler parameters are taken as $b = b_{turb} + b_{th}$ where the first term represents the turbulent motions in the gas and is the same for all elements; the second term represents the thermal broadening consistently tied up to the temperature given by photoionization models. This must be done this way since, for abundances larger than solar, the temperature is very sensitive to the abundances. Also in a few systems the zero level has been adjusted when it was clear, even from the 2D spectrum, that the lines although saturated do not go to zero.

The results for the nine systems are shown in Fig. 1 where the logarithm of the carbon abundance relative to solar is plotted versus the velocity difference of the system and the QSO. It can be seen that the systems within 10000 km s^{-1} from the QSO have abundance in excess of solar and thus are probably associated with the QSO. The other systems have typical abundances of Lyman limit intervening systems, in the range 10^{-2}-10^{-1} solar.

It has been often argued against the idea of associated systems being ejected by the QSO that the lines are too narrow for the gas to be part of

some kind of wind. However narrow ($b \sim 20$ km s^{-1}) FeII absorption lines
are detected in the strong BAL system in Q0059-2735 (Wampler et al. 1995)
showing that gas at low temperature ($T < 10^4$ K) and with small amount of
turbulent motions can be found embedded in a BAL flow. The fact that the
FeII lines do not go to the zero continuum level implies that the corresponding
clouds do not cover the source of continuum and therefore should be very
small. This is consistent with the high density ($n > 10^6$ cm^{-3}) needed to
explain collisional excitation of atomic levels situated up to 4.5 eV above the
ground state from which absorption lines are detected. Moreover structures
are seen in the CIV, AlIII, AlII, SiII, SiIII, FeII and FeIII broad absorptions.
Several distinct regions of very similar ionization state but different velocities
are present. Detailed photoionization modeling suggests that the ionization
parameter (ratio of the density of ionizing photons to the gaseous density) is
very high ($\sim 10^{-1}$); that the carbon abundance is about solar and iron may
be enhanced by possibly a factor of ten.

3   Inhomogeneities in Intervening Systems

Abundance determination within intervening systems gives unique information about history of the metal enrichment. This can be done in LLS or
damped systems. The latter are well suited to this purpose since the HI column density is well determined, the ionization correction is negligible and
weak lines from ZnII and CrII can be searched for (Meyer & York 1987a).
The results spread several orders of magnitude, with abundance relative to
solar in the range [Zn/H] = 10^{-3}-1 (Pettini et al. 1994). This may not be
surprising since these systems can certainly arise through disks, in the proximity of star forming regions, and halos of galaxies (Petitjean et al. 1992).
Abundance determination in Lyman Limit Systems requires to model the
ionization state of the gas. Since it is probable that at least at high redshift
photo-ionization is an important source of ionization, photo-ionization codes
are used (Bergeron & Stasińska 1986, Steidel 1990, Petitjean et al. 1992). To
constrain the models however a large amount of information must be gath-
ered and is sufficient up to now only in a few cases. It seems clear however
that abundances in LLS systems are of the order of 10^{-2}, 10^{-1}, 0.5 $Z_\odot$ at
redshift 3 (Steidel 1990), 2 (Petitjean et al. 1992) and 0.5 (Bergeron et al.
1994) respectively. This may strongly constrain the gas enrichment history.

In this respect an important issue is the presence of OVI absorptions
recently detected in six systems toward QSO HE1700+6416 at redshifts between 0.7 and 2.4 (Reimers et al. 1992) and in four out of five CIV systems
detected by HST (Bergeron et al. 1994). It has been convincingly detected also in a composite spectrum formed of 73 CIV systems at $z \sim 2$ (Lu & Savage
1993). It seems that the presence of an OVI phase is a common feature of CIV
systems. Although photo-ionization models can reproduce the OVI column
densities (Vogel & Reimers 1993, Bergeron et al. 1994, Riediger & Petitjean

1995), the derived dimensions of the OVI phase are uncomfortably large (up to 100 kpc for an homogeneous cloud). This is due to the low value of the density required to obtain a high enough ionization parameter. The conclusion is that either the ionizing flux is higher than what is generally believed or the OVI phase is of higher density but collisionally ionized. To decide between both hypothesis, high spectral resolution observations of OVI absorption lines are required in order to determine the width of the lines which should be of the order of $b \sim 10$ and 30 km s^{-1} in case of photo-ionization or collisional ionization respectively.

An important consideration is that the medium we observe is clumpy (Petitjean & Bergeron 1990, 1994). It is therefore probable that the line of sight passes through several phases of different densities and ionization states. The different regions produce very different absorption signatures which nonetheless could be coincident in velocity or unresolved at the resolution usually used. This is the case in particular when OVI is observed. In order to explain the column densities of all observed ions from MgII to OVI a model with two phases of different densities must be considered. Indeed the ionization parameter U (ratio of the density of ionizing photons to the total hydrogen density) in the MgII phase is found to be in the range 3×10^{-4} to 3×10^{-3} whereas values at least equal to $2\text{-}3 \times 10^{-2}$ are needed to explain the observed $N(\text{NV})/N(\text{CIV})$ and $N(\text{OVI})/N(\text{CIV})$ column density ratios. Therefore, for the same ionizing flux, the densities should differ by one order of magnitude.

This is particularly well illustrated in the LLS systems toward QSO HE1700+6416 in which CIII,IV, NIII,IV,V, OIII,IV,V,VI and most important HeI absorption lines have been observed by HST (Reimers et al. 1992). Models with constant density fail to explain the large HeI/HI ~ 0.03 ratio since ionization parameter of the order of 10^{-2} are required to reproduce the large OVI column density (Vogel & Reimers 1993). Moreover, since a large flux of photons of energy larger than 54.4 eV is needed to sufficiently ionize OV in OVI, a break in the ionizing flux spectrum at such energy (possibly produced by intervening HeII, see Madau 1992) is ruled out. However models with two phases of density differing by about one order of magnitude can reproduce the whole set of column densities within the present time uncertainties. The high density region produces the low excitation ions in particular HI and HeI. The ratio $N(\text{HeI})/N(\text{HI})$ is thus easily reproduced. Most of the absorption by twice ionized elements (CIII, NIII, OIII) originates in this phase. The low density region is highly ionized and explains the NV and OVI column densities plus part of the absorption of the three times ionized elements. A break in the ionizing spectrum can be accommodated since the ionization parameter can be made larger by decreasing the density. The model is then mostly constrained by the CIV, NIV and OIV column densities (Riediger & Petitjean 1995).

4   The Lyα Forest

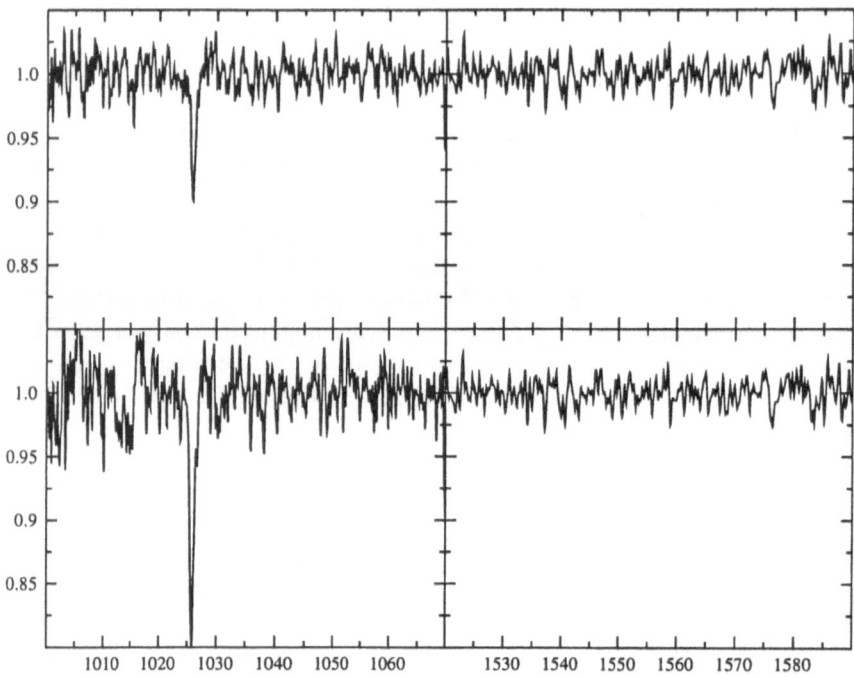

Fig. 2. OVI (left panel) and CIV (right panel) wavelength ranges in composite spectra from HST data. The upper panels correspond to Lyα systems with $w_r > 0.4$ Å (see Sect.4.1).

Recent observations by HST have shown that the observed number of Lyα lines at $z \sim 0.3$ is five to ten times larger than what a simple low-redshift extrapolation of the high-redshift z-dependence would suggest. The number density of lines per unit redshift is 100±25 for $N(HI) > 10^{13}$ cm^{-2} (Morris et al. 1991) and 15±2 for $W > 0.32$ Å (Bahcall et al. 1993) which corresponds roughly (depending also on b) to $N(HI) > 10^{14}$ cm^{-2}. This low-z excess in the observed number density of lines can be explained by a nonzero cosmological constant. Alternatively, a significant fraction of the low-redshift population could represent relics of a population of clouds which, at high redshift, were associated with galaxies. Indeed the association of galaxies with *some* of the Lyα lines has been demonstrated recently (Lanzetta et al. 1994, Le Brun et al. 1995).

4.1 Abundances

The idea that *part* of the Lyα forest could be associated somehow with galaxies is supported by the fact that, at high redshift, *part* of the Lyα forest could contain metals unseen up to now because the corresponding weak lines, especially from CIV, have been undetected due to lack of sensitivity (see Petitjean 1995). Indeed Meyer & York (1987b) found weak CIV absorption corresponding to several Lyα systems and more recently high S/N ratio observations with the Keck telescope have revealed CIV absorption associated with a few individual Lyα lines with $N(\text{HI}) \sim 10^{14.5}$ cm^{-2}. The corresponding abundances might be as high as $10^{[C/H]} = 10^{-2.5}$ $Z_{\odot}$ (Tytler 1995). On the other hand upper limits had been obtained up to now. Chaffee et al. (1986) studied two high column density ($N(\text{HI}) = 10^{16.3}$ and $10^{16.7}$ cm^{-2}) Lymanα clouds and derived [C/H] $\sim$ -3.5. Using the composite spectra technique, Norris et al. (1983) have claimed detection of OVI, but this has not been confirmed by Williger et al. (1989). Lu (1991) has reported tentative detection of CIVλ1548 implying [C/H]$\sim$-3.2. Tytler & Fan (1994) give [C/H] < -2.

We have used the same method to search for metal lines in low redshift Lyα systems using the data obtained in the course of the HST Key Program "Absorption Line Systems" (Bahcall et al. 1993). An important point is that it is possible to search for CIV *and* OVI. Indeed, depending on the shape of the ionizing flux, CIV can be highly ionized in typical photo-ionized Lyα gas. In case CV dominates the carbon ionization state, thus preventing any detection of CIV absorption, OVI could be detectable. The individual spectra have been shifted to the rest frame of the Lyα line and coadded. Great care has been taken to reject spectra where strong lines from other systems accidentally coincide with CIV or OVI. Known metal line systems are not included. We have used two samples: S1 contains all the possible systems and S2 only those with $w_r(\text{Ly}\alpha) > 0.4$ Å. For CIV and OVI, S1C, S2C, S1O and S2O contain 133, 55, 65 and 26 systems respectively. It can be seen in Fig.2 that no absorption (except Lyβ) is found down to a 3σ limit of $w_r \sim 10$ mÅ for OVI and 8 mÅ for CIV. The Lyβ equivalent width is 65 and 170 mÅ for S1O and S2O respectively. From this it is very difficult to derive abundances because of the uncertain ionization correction. However, using a standard model, we can be confident that [C/H] and [O/H] are smaller than -2.3. This is similar to what is derived in high redshift systems.

Given the small sample available, it is not possible to go further. In particular it would be of great interest to try and derive abundances in the Lyα systems that have been observed to be associated with galaxies. The question whether the whole Lyα forest or only part of it, and may be the one associated with galaxies, contains metals is still open.

4.2   A New Picture

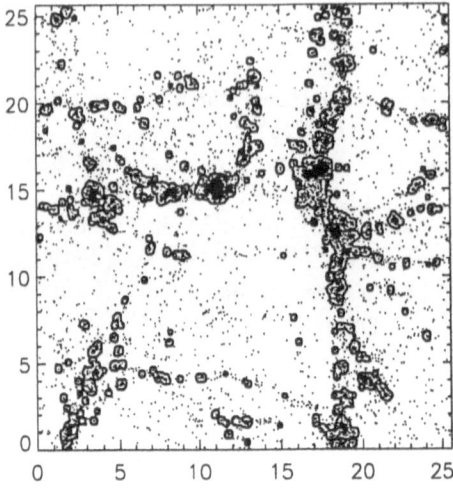

Fig. 3. Projected space distribution of cold dark matter particles in a slice of size $25 \times 25 \times 2$ Mpc^3 at redshift $z = 0$. Contours for constant neutral hydrogen column density $N(\text{HI}) = 10^{13}$ cm^{-2} are overplotted.

Since observational evidence for direct association of most Lyα absorptions with bright galaxies is not unequivocal, we have suggested (Petitjean et al. 1995) that they arise in gas associated more generally with the potential wells of rich filamentary structures that are ubiquitous in gravitational evolution dominated by dark matter (see also Cen et al. 1994). To obtain a consistent picture within the framework of models for large-scale structure, we have studied the spatial distribution of Lyα gas and its time evolution using N-body simulations. To this purpose we have improved and adapted a particle-mesh (PM) code (see see Kates et al. 1991, Klypin & Kates 1991)) to include cooling and the effects of photoionization (Mücket et al. 1995). The baryonic material follows the dark matter and is heated by an ionizing flux. We take the ionizing flux intensity proportional to the rate at which material cools below 5000 K in the simulation. This simulates an ionizing flux, produced as a consequence of gas collapsing in the dense regions, and then playing the role of a regulator preventing most of the gas in intermediate density regions from collapsing. The flux-intensity redshift dependence is thus self-consistently determined and turns out to be very similar to what is known (Miralda-Escudé & Ostriker 1990). It is found to be a power law $(1+z)^\beta$ with $\beta = -2$ and 2 for $z \geq 5$ and $z \leq 2.5$ respectively. Between

$z = 2.5$ and 5, the flux is approximately constant. The only free parameter is the normalization: We assume $F_o = 10^{-21}$ erg s^{-1} cm^{-2} sr^{-1} Hz^{-1} at $z = 5$.

The model is able to reproduce the number-density redshift-dependence which is characterized by a flat part at low redshift and a steep increase beyond $z = 2$ where the ionizing flux is at maximum (Mücket et al. 1995). The number density of lines with $N(\text{HI}) > 10^{13}$ cm^{-2} is 105 at $z = 0$ which is consistent with the observed number. The ability to reproduce this number, for an ionizing flux intensity within observational constraints, represents an important success of the scenario.

Fig. 3 shows the projected spatial distribution at $z = 0$ of the dark matter particles in a 25×25 Mpc^2 slice of depth 2 Mpc. The structures seen form the usual network of filaments connected by nodes where galaxies and groups of galaxies are to be found. The neutral hydrogen column density is computed through the slice, and contours of constant neutral hydrogen column density $N(\text{HI}) = 10^{13}$ cm^{-2} are shown. It can be seen that the Lyα gas traces the gravitational potential wells.

Testing the picture presented here – that the Lyα forest traces the filamentary network of the dark matter – will require new observations. In particular high-quality data should be obtained for the Lyα forest in QSO pairs with very different separations to probe scales from a few tens of kiloparsecs to several megaparsecs. The probability distribution of coincidences at different scales would provide strong constraints on the model. On the other hand, a detailed investigation of the metal content of the Lyα forest at high redshift using 10m-class telescopes and at low redshift using shift and stack methods are clearly needed. With the advent of 10m-class telescopes, 21 mag QSOs will be observed routinely. This will allow the mapping of the Universe in three dimensions up to redshift larger than 3. This may be the only way to study large scale structures at such high redshift and their evolution in time providing a powerful tool to test cosmological models.

Acknowledgements. This work has benefited from active collaborations with Jacqueline Bergeron, Jan Mücket, Michael Rauch and Jo Wampler. I would like to thank Bob Carswell for his constant enthusiasm and Ray Weymann for fruitful discussions.

References

Anderson S.F., Weymann R.J., Foltz C.B., Chaffee F.H.Jr. (1987): AJ 94, 278

Bahcall J.N., Bergeron J., Boksenberg A., et al. (1993): ApJS 87, 1

Bahcall, J.N., Januzzi, B.T., Schneider, D.P., et al. (1991): ApJL 377, 5

Barthel P.D., 1988. In P.A. Shaver, E.J. Wampler & A.M Wolfe (eds) ESO Mini-Workshop on Quasar Absorption Lines. ESO, Garching, p.79

Bergeron, J., Boissé, P. (1991): A&A 243, 344

Bergeron, J., Petitjean, P. (1991): A&A 241, 365

Bergeron, J., Stasińska, G. (1986): A&A 169, 1

Bergeron, J., Petitjean, P., Sargent, W.L.W., et al. (1994): ApJ 436, 33

Cen R., Miralda-Escudé J., Ostriker J.P., Rauch M., 1994, ApJL 437, L9

Chaffee, F.H., Foltz, C.B., Bechtold, J., Weymann, R.J. (1986): ApJ 301, 116

Foltz C.B., Weymann R.J., Peterson B.M., et al. (1986): ApJ 307, 504

Gruenwald R., Viegas S.M. (1993): ApJ 415, 534

Hamann F., Korista K.T., Morris S.L. (1993): ApJ 415, 541

Kates R.E., Kotok E.V., Klypin A.A. (1991): A&A 243, 295

Klypin A.A., Kates R.E. (1991): MNRAS 251, 41P

Lanzetta, K.M., Bowen, D.V., Tytler, D., Webb, J.K. (1994): preprint

Le Brun V., Bergeron J., Boissé P. (1995): this volume

Lu, L. (1991): ApJ 379, 99

Lu, L., Savage, B.D. (1993): ApJ 403, 127

Madau, P. (1992): ApJ 389, L1

Meyer, D.M., York, D. (1987a): ApJL 319, L45

Meyer, D.M., York, D. (1987b): ApJL 315, L5

Miralda-Escudé, J., Ostriker, J.P. (1990): ApJ 350, 1

Morris, S.L., Weymann, R.J., Savage, B.D., Gilliland, R.L. (1991): ApJL 377, 21

Mücket J., Petitjean P., Kates R.E. (1995): this volume

Norris, J., Hartwick, F.D.A., Peterson, B.A. (1983): ApJ 273, 450

Petitjean, P. (1995): Proceedings of the ESO workshop "Science with VLT", Garch-
    ing (Springer, Heidelberg), in press

Petitjean, P., Bergeron, J. (1990): A&A 231, 309

Petitjean, P., Bergeron, J. (1994): A&A 283, 759

Petitjean, P., Bergeron, J., Puget, J.L. (1992): A&A 265, 375

Petitjean, P., Mücket, J., Kates, R. (1995): A&AL in press

Petitjean, P., Rauch, M., Carswell, R.F. (1994): A&A 291, 29

Pettini, M., Boksenberg, A., Hunstead, R.W. (1990): ApJ 348, 48

Pettini, M., Smith L.J., Hunstead R.W., King D.L. (1994): ApJ 426, 79

Reimers D., Vogel S., Hagen H.J., et al. (1992): Nature 360, 561

Riediger, R., Petitjean, P. (1995): this volume

Sargent W.L.W., Boksenberg A., Steidel C.C. (1988): ApJS 68, 539

Steidel, C.C. (1990): ApJS 74, 37

Steidel, C.C. (1995): this volume

Stocke J., Morris S., Weymann R., Foltz C. (1992): ApJ 396, 487

Tytler, D. (1995): this volume

Tytler, D., Fan, X. (1994): ApJL 424, L87

Vogel S., Reimers D. (1993): A&AL 274, L5

Wampler, E.J., Chugai, N.N., Petitjean, P. (1995): ApJ in press

Weymann R.J., Morris S.L., Foltz C.B., Hewett P.C. (1991): ApJ 373, 23

Weymann R.J., Williams R.E., Peterson B.M., Turnshek D.A. (1979): ApJ 234, 33

Williger, G.M., Carswell, R.F., Webb, J.K., Boksenberg, A., Smith, M.G. (1989):
    MNRAS 237, 635

Young P., Sargent W.L.W., Boksenberg A. (1982): ApJS 48, 455

The Chemical Evolution of Damped Lyman α Galaxies

Max Pettini[1], David L. King[1], Linda J. Smith[2] and Richard W. Hunstead[3]

[1] Royal Greenwich Observatory, Madingley Road, Cambridge, CB3 0EZ, UK
[2] Department of Physics and Astronomy, University College London, Gower Street, London WC1E 6BT, UK
[3] School of Physics, University of Sydney, NSW 2006, Australia

Abstract. Measurements of element abundances in damped Lyman α systems are providing new means to investigate the chemical evolution of galaxies, particularly at early times. We review progress in this area, concentrating on recent efforts to extend the range of existing surveys to both higher and lower redshifts.

1 Introduction

One has only to glance at the programme of this workshop, and compare it with that of the last QSO Absorption Line meeting which took place at the Space Telescope Science Institute in April 1987, to realise the increasing amount of attention which damped Lyman α systems have attracted in the last few years. I would like to echo the words by Ray Weymann earlier today, in recognising that this is largely due to the efforts by Artie Wolfe and his collaborators who, in a series of large-scale studies, identified several lines of evidence to suggest that in this class of absorbers we may well be seeing the high-redshift progenitors of galaxies like our own. If we accept this as a working assumption, we can then bring the full power of QSO absorption line spectroscopy to bear in determining many of the physical properties of galaxies caught, as it were, in their infancy.

2 Heavy Element Abundances at Redshift $z = 2$

My collaborators and I have been involved for some years now in a programme aimed at studying the chemical evolution of high-redshift galaxies through measurements in particular of the metallicity, dust content, and star formation in the gas giving rise to damped Lyman α systems. The first stage of our survey was completed and published earlier this year (Pettini et al. 1994); I shall use those results as a starting point for my talk today, which concentrates on recent work to extend the survey over a wider redshift interval.

The filled circles in Fig.1 show the abundances of zinc measured in 17 damped Lyman α systems by Pettini et al. (1994). Pettini, Boksenberg and

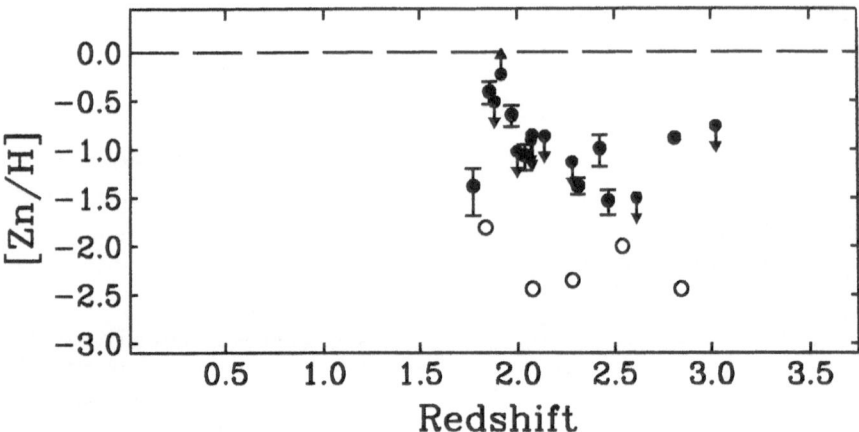

Fig. 1. Values of the abundance of Zn in 17 damped Lyman α systems reported by Pettini et al. 1994 (filled circles). [Zn/H] is plotted on a logarithmic scale relative to the solar value; thus the broken line at 0.0 corresponds to the solar abundance and [Zn/H] = −1.0 indicates an underabundance of Zn by a factor of 10. Downward pointing arrows are upper limits appropriate to cases where the Zn II lines have not been detected. Also shown for comparison are metal abundances measured from very high resolution echelle observations of damped Lyman α systems (open circles), as follows (in order of increasing redshift): [O/H] in the z_{abs} = 1.83856 system in Q1101−264 (Pettini et al. 1992); [Si/H] in the z_{abs} = 2.07623 system in Q2206−199 (Pettini and Hunstead 1990); [Fe/H] in the z_{abs} = 2.27936 system in Q2348−147 (Pettini, Lipman and Hunstead 1995); [Fe/H] in the z_{abs} = 2.53788 system in Q2344+124 (Lipman, Pettini and Hunstead 1995); and [Fe/H] in the z_{abs} = 2.8443 system in Q1946+769 (Lu et al. 1995). Two of the five open circles (the points for Q2206−199 and Q2348−147) have corresponding [Zn/H] upper limits.

Hunstead (1989) first drew attention to several advantages in using Zn as a tracer of metallicity; as these are now widely recognised, we only need to summarise them briefly here. Zn is among the few heavy elements which show little affinity for dust; this, coupled with the fact that it is mostly s-ingly ionised in H I regions, makes it likely that the 'missing' fraction of Zn – either in solid form or in unobserved ionisation stages – is small. For most known damped Lyman α systems, multiplet 1 of Zn II, $\lambda\lambda 2025, 2062$ is red-shifted into a region of the optical spectrum which can be readily observed. Furthermore, Zn is a relatively rare element ([Zn/H]$_\odot$ = 3.8 × 10^{-8}; Aller 1987); consequently the doublet lines are usually sufficiently weak that line saturation can be assessed – and the column density deduced – with an accuracy which is adequate for the present purposes. In terms of galactic chemical evolution, the metallicity measured by the Zn abundance is analogous to that traced by Fe; the nucleosynthetic origins of these two elements presumably have much in common, since [Zn/Fe] shows no departure from solar values in Galactic stars of all metallicities, down to the lowest values measured.

Returning to Fig.1, two main results are indicated. First, at redshifts between $z \simeq 2$ and 3, the galaxies giving rise to the absorption are mostly chemically young systems – most of the points in the plot lie well below the solar [Zn/H]. As pointed out by Lanzetta (1993), if the damped Lyman α lines account for most of the baryons at these redshifts, then the column density-weighted metallicity we find, $Z_{DLA} = 1/10Z_\odot$, can be interpreted as the typical 'cosmic' metallicity reached by the universe $\sim$ 13 Gyr ago ($H_0 = 50$ km s^{-1} Mpc^{-1}, $q_0 = 0.01$).

Second, there is a large dispersion in the degree of metal enrichment attained by different galaxies at essentially the same epoch. About half of the filled circles in Fig.1 are upper limits, corresponding to non-detections of the Zn II doublet; in some of these cases we *know* that the true abundances lie well below the limits of the Zn II observations. For metallicities $Z_{DLA} < 1/50Z_\odot$, the Zn II lines become vanishingly small but, on the other hand, we enter a regime where the ultraviolet resonance lines of the more abundant astrophysical elements become accessible to abundance studies through high-resolution observations with echelle spectrographs. Five such measurements are included as open circles in Fig.1 . There may be uncertainties of 0.3-0.5 dex in comparing these data with the values of [Zn/H], since some of the elements in question (Si and O in particular) may be present in non-solar proportions relative to Zn. Nevertheless, the inclusion of these points in the figure does highlight the fact that at $z \simeq 2$ damped Lyman α systems span more than two orders of magnitude in metallicity and that systems with $Z_{DLA} < 1/100Z_\odot$ are not uncommon.

Such a wide range is unlikely to be due to radial abundance gradients similar to those seen in present-day spiral galaxies. Simulations show that, within the inner regions of galaxies where the column density of neutral gas is sufficiently high to produce a damped Lyman α line ($N(H^0) \geq 2 \times 10^{20}$ cm^{-2}) randomly placed sight-lines would on average sample an abundance spread of only a factor ≈ 2 . While it is possible that the interstellar medium at these high redshifts is poorly mixed, the observations suggest that the process of chemical enrichment started at different times and probably proceeded at different rates in the galaxies picked out by the damped systems.

3 Extension to Higher Redshifts

The results in Fig.1 have been interpreted by some (e.g. Wolfe 1993) as evidence for a rapid build-up of metals in the universe between $z = 3$ and 2 . Such an effect may go hand-in-hand – at least qualitatively – with the significant consumption of gas apparently indicated by the evolution, over approximately the same redshift interval, of the H I distribution of damped Lyman α lines (Lanzetta, Wolfe and Turnshek 1995). In our view, however, published data are insufficient to measure with confidence trends in either metallicity *or* H I content with redshift. Indeed, the recent finding that a large sample of very high redshift ($z_{em} > 4$) QSOs includes significantly fewer systems with $N(H^0) > 10^{21}$ cm^{-2} than expected (Storrie-Lombardi et al. 1995) suggests that the H I redshift evolution may have been overestimated. As far as the Zn measurements in Fig.1 are concerned, while it is true that near-solar abundances are found only at $z \simeq 2$, the redshift distribution of the data in Fig.1 is too uneven to assess the significance of this result.

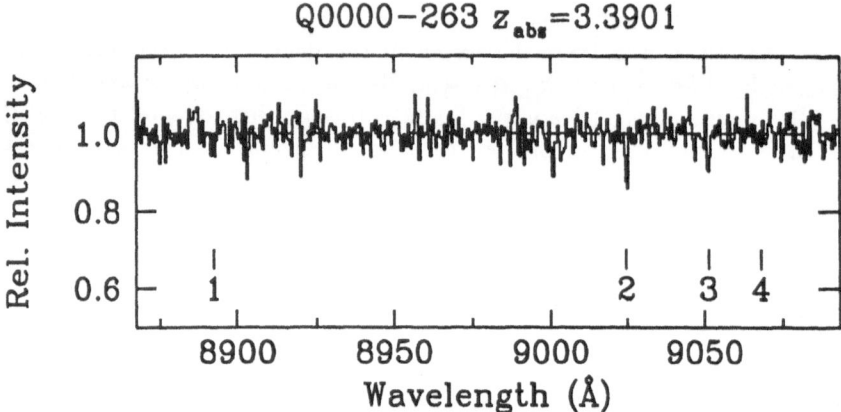

Fig. 2. Portion of the AAT spectrum of Q0000−263 encompassing the redshifted Zn II and Cr II lines in the $z_{abs} = 3.3901$ absorption system. The tick marks indicate the expected positions of the absorption features, whether they have been detected or not. Line 1: Zn II $\lambda2025.484$; line 2: Cr II $\lambda2055.596$; line 3: Cr II $\lambda2061.575+$ Zn II $\lambda2062.003$ (blended); and line 4: Cr II $\lambda2065.501$. The spectrum shown here is the sum of many exposures with a Tektronix CCD, amounting to a total integration time of 58 000 s; the resolution is 1.1 Å *FWHM* and the final S/N $\simeq 30$.

To remedy the situation, we have extended the survey in 1994 with observations of an additional 7 damped Lyman α systems, mostly at redshifts $z_{abs} \geq 2.5$. The full sample for which Zn abundance determinations (or upper limits) are now available consists of 24 damped systems in 20 QSOs, and includes more than one third of the total number of confirmed damped absorbers. The new data were obtained during a series of observing runs at the 3.9 m Anglo-Australian telescope at Siding Spring Observatory, Australia, and the 4.2 m William Herschel telescope at La Palma, Canary Islands. The instrumental set-ups were similar to those used by Pettini et al. (1994); the spectral resolution ranged from 0.7 to 1.2 Å *FWHM*.

The observations of the high-redshift $z_{em} = 4.111$ bright QSO Q0000−263, reproduced in Fig.2, are among the most sensitive in the survey. By adding together several nights' data, we achieved a moderately high signal-to-noise ratio (S/N $\simeq 30$) in the difficult region near 9000 Å, to where the Zn II and

Cr II lines associated with the $z_{abs} = 3.3901$ damped system are redshifted. No Zn II absorption is detected, implying a 3σ upper limit $[Zn/H] < -1.76$, or less than 1/60 of solar! This limit is ~ 5 times more sensitive than that placed by Savaglio, D'Odorico and Moller (1994). Interestingly, we have a clear detection of Cr II absorption; we deduce $[Cr/H] = -2.46$, nearly 300 times below the solar abundance. This is one of the few examples now known where the Cr II lines are stronger than those of Zn II, implying little – if any – depletion of Cr onto dust grains. In Pettini et al. (1994) we predicted that such cases would be discovered at the lowest metallicities, on the basis of an apparent trend in our data of increasing $[Cr/Zn]$ with decreasing Z_{DLA}.

In general, however, the observations we attempted are at the limit of what can be accomplished with 4 m-class telescopes, due to the decreasing response of CCDs and increasing sky emission at wavelengths longwards of ~ 7000 Å. Consequently, we were not always able to reach as high a sensitivity to the Zn II lines as we would have wished.

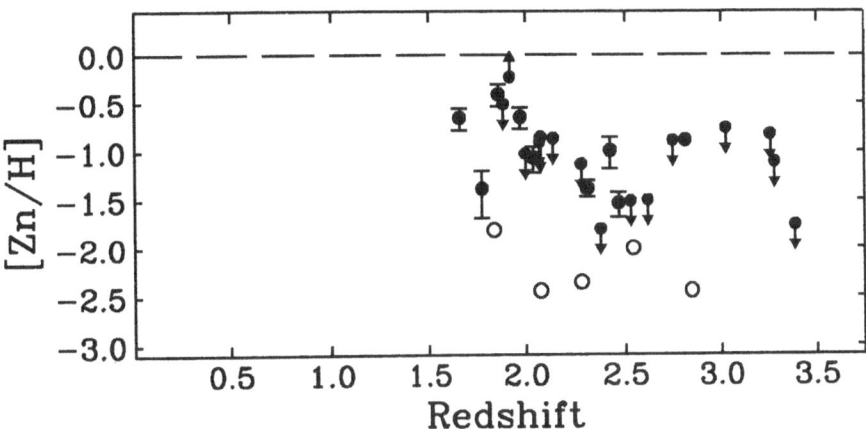

Fig. 3. Same as Fig.1, but with the addition of seven new measurements of $[Zn/H]$ obtained in 1994, mostly for systems at $z_{abs} \geq 2.5$.

As can be seen from Fig.3, we have found no new detections of Zn II at $z_{abs} \geq 2.5$. The only system showing measurable amounts of Zn at these redshifts remains the $z_{abs} = 2.8122$ absorber in Q0528−250 (Meyer, Welty and York 1989) which is at a higher redshift than the QSO itself and is probably atypical of the sample as a whole in other ways (Moller and Warren 1993). Despite their limitations, the available data therefore seem to suggest that abundances at $z \geq 2.5$ are generally lower than 1/10 of solar, the typical value at $z \simeq 2$. *All* damped Lyman α systems sampled at $z \geq 2.5$ are metal-poor, while at $z \simeq 2$ at least some galaxies have apparently attained a significant degree of metal-enrichment, comparable to that of spiral galaxies today. It may well be, then, that the interval between $z = 3$ and 2 does signal the epoch of galaxy formation, if we take the term to indicate the period when the first major episodes of star formation took place in galaxies.

In this scenario the damped system in Q0000−263 may be a good example of a primeval galaxy. The absorber has been imaged in the stellar ultraviolet continuum near the Lyman limit by Steidel and Hamilton (1992) and it appears to be a luminous galaxy ($L_B \approx 3L_*$) of dimensions $10 - 20 \, h^{-1}$ kpc. Knots which may be sites of star formation are seen in a recent *Hubble Space Telescope* image (Giavalisco 1995, private communication). And yet the interstellar medium of this galaxy – at least in the region which happens to lie in our line of sight to Q0000−263 – has apparently undergone little, if any, chemical enrichment. The metallicity deduced from Cr II and confirmed by observations of other elements by Vladilo et al. (1995), $Z_{DLA} = -2.5$, is the same as that which apparently applies to clouds in the Lyman α forest, as recently discovered by Tytler (1995). One possible interpretation is that this value may represent a 'base level' of metallicity on which the process of galactic chemical evolution subsequently builds up, although some stars in our Galaxy must have formed from more pristine gas, since stars with abundances as low as $Z \simeq -4$ are known (e.g. McWilliam et al. 1995).

Before moving on, we point out that, if the Lyman α forest does indeed trace an intergalactic medium which at $z \simeq 3$ has been already 'polluted' with heavy elements, abundance measurements of a wide complement of elements in damped systems with metallicities as low as $Z_{DLA} = -2.5$ may offer vital chemical clues to the stellar populations which produced this initial enrichment of the universe at early epochs.

4   Metal Abundances at Intermediate Redshifts

The next step in constructing a full picture of the chemical evolution of galaxies is to follow the build-up of metals to recent epochs. The difficulty here is not so much in detecting the Zn II lines (with modern blue-sensitive CCDs it is relatively straightforward to search for these down to redshifts $z_{abs} \simeq 0.6$), but rather in identifying even a modest sample of damped Lyman α systems. Cosmological effects, combined with the intrinsic evolution of the H I content, result in significantly smaller numbers of candidates than at higher redshifts (Lanzetta, Turnshek and Sandoval 1993; Bahcall et al. 1993); these difficulties are compounded by the fact that *HST* observations are then required to confirm the candidates and measure $N(\mathrm{H}^0)$.

Nevertheless there is a strong incentive to extend this work to $z < 1$, particularly given the spectacular success in identifying QSO absorbers at these redshifts by direct imaging (Steidel 1995). The morphological information provided by the images, together with the physical properties of the absorbing gas deduced from QSO absorption line spectroscopy, form a particularly powerful combination for unraveling the nature and evolutionary status of the absorbing galaxies.

To date there are only three measurements of metallicity in damped systems at intermediate redshifts, at $z_{abs} = 0.6922$ in 3C 286 (Meyer and York 1992), $z_{abs} = 0.8596$ in PKS 0454+039 (Steidel et al. 1995), and $z_{abs} = 1.3718$ in Q0935+417 (Meyer, Lanzetta and Wolfe 1995). Surprisingly, in all three cases the abundances deduced are approximately one order of magnitude below solar and *not* significantly higher than the typical metallicity at $z = 2$ ($[Zn/H] = -1.2$, -0.9 and -0.7 in 3C 286, PKS 0454+039 and Q0935+417 respectively. The last two values of $[Zn/H]$ have been increased by 0.14 over those reported by the original authors for consistency with the values of oscillator strength and solar abundance adopted in the rest of this survey).

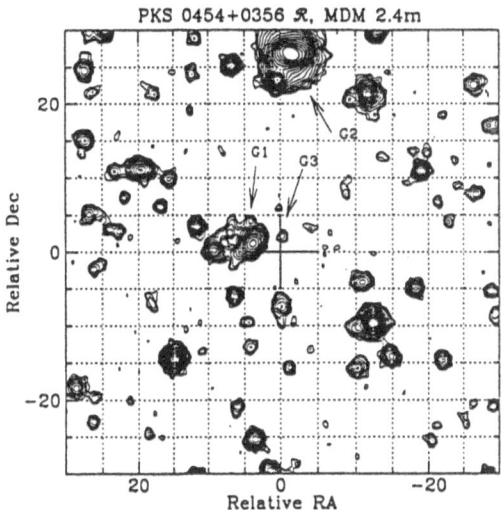

PKS 0454+0356 $\mathcal{R}$, MDM 2.4m

Fig. 4. (*Reproduced from Steidel et al. 1995*). R band image of the field of PKS 0454+039 obtained with a 7600 s-long exposure at the 2.4 m Hiltner telescope at the MDM Observatory on Kitt Peak. The *FWHM* of the seeing profile on the final co-added image is 0.89 arcsec. The image of the QSO has been subtracted by modeling the PSF of bright stars in the field. The galaxies labelled G1 and G2 are at redshifts $z = 0.072$ and 0.201 respectively. G3 is the most likely candidate for the damped absorber at $z_{abs} = 0.8596$.

The galaxies likely to be responsible for the absorption in 3C 286 and PKS 0454+039 have been identified from deep CCD images obtained in good seeing conditions and their characteristics found to be consistent with the low metal abundances measured. The absorber in front of 3C 286 is probably a luminous galaxy ($M_B = -20.8$, or $\approx 0.8 L*$), but of *low surface brightness* (Steidel et al. 1994). These authors deduced a peak surface brightness in the rest-frame B band $\mu_B(0) \simeq 23.6$ mag arcsec^{-2}, close to the median value of the sample of LSB galaxies studied by McGaugh and Bothun

(1994) and significantly lower than the Freeman value for 'normal' spirals, $\mu_B(0) \simeq 21.65 \pm 0.3$ mag arcsec^{-2}. Chemical enrichment probably proceeds at a slower pace in LSB galaxies which, even at the present epoch, have low metal abundances (McGaugh 1994); it is not surprising, then, to find [Zn/H] $= -1.2$ in one such galaxy at $z = 0.6922$.

The field of PKS 0454+039 is shown in Fig.4. The most likely candidate for the $z_{abs} = 0.8596$ absorber is the object labelled G3. If the identification is correct, the galaxy is located $\approx 9h^{-1}$ kpc ($q_0 = 0.5$) from the QSO sight-line and has an absolute magnitude $M_B = -19.0$ ($\approx 0.15L*$). Again, the fact that this is a relatively underluminous galaxy provides a plausible explanation for the low metallicity deduced by Steidel et al. (1995).

5   Bringing It All Together...

We bring the above results together in Fig.5, where we plot the full set of abundance measurements on a timescale compatible with stellar ages, and compare them with Fe abundances of F and G dwarf stars in the disk of the Milky Way from the landmark paper by Edvardsson et al. (1993). The differences are very obvious. At $z > 2$, corresponding to lookback times of more than ~ 12.5 Gyr, both the *distribution* and the *mean* of the abundances measured in damped systems resemble more closely the values found in stars in the halo, rather than the disk, of our Galaxy (Pettini et al. 1994). Presumably at this epoch most galaxies had not yet collapsed to form a thin disk. Furthermore, the only two galaxies observed at a lookback time approaching the age of the Sun (~ 5 Gyr) are evidently on very different evolutionary tracks from that of the Milky Way.

Nevertheless, these findings are not necessarily inconsistent with our working assumption that in the damped Lyman α systems we are seeing the progenitors of present-day spiral galaxies. The extensive survey of Mg II absorbers (of which the damped systems are presumably a subset) by Steidel and collaborators (Steidel, Dickinson and Persson 1994; Steidel 1995) has led to the first determination of the luminosity function of field galaxies picked out by absorption cross-section at $z \simeq 0.6$. The luminosity function is Gaussian, centred near $M_B \simeq -20.5$ (for $H_0 = 50$ km s^{-1} Mpc^{-1}). The key question, which can only be addressed with a larger sample of damped systems at intermediate redshift, is whether the distribution of metallicities can be understood in terms of the absorbers' luminosity function, appropriately modified by the scaling of absorption cross-section with luminosity (Steidel, Dickinson and Persson 1994), and the present-day dependence of metallicity on galaxy luminosity (Skillman, Kennicutt and Hodge 1989).

Another factor contributing to the differences evident in Fig.5 between the chemical evolution of the damped systems and that of the Milky Way disk may be the increasing bias introduced by dust as the metal content of the universe grows with time (Fall 1995). In these models, our view of

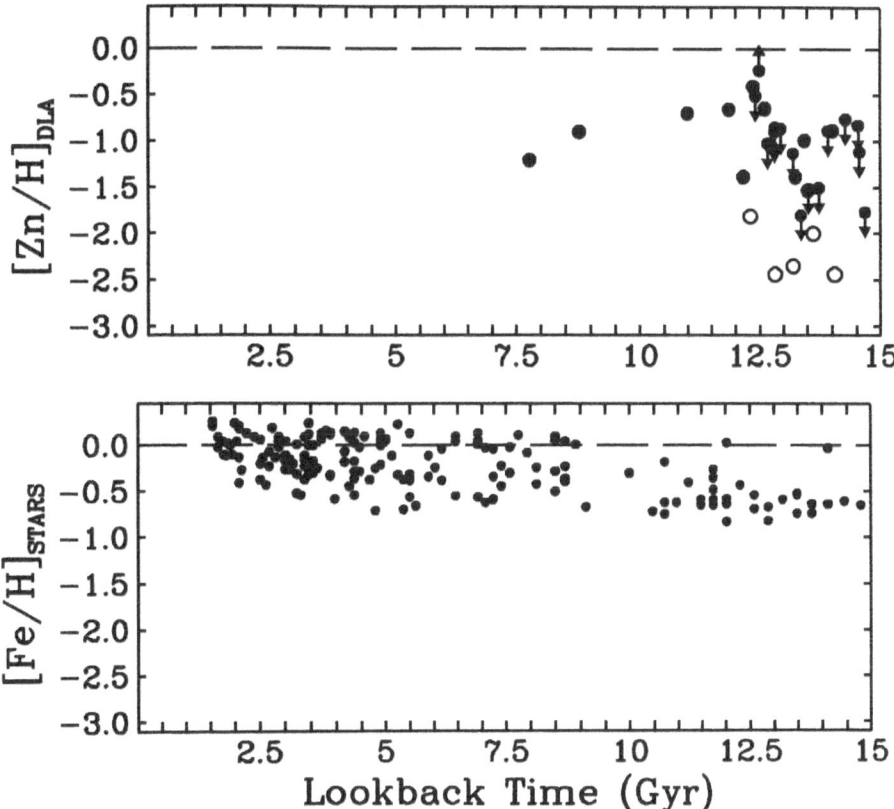

Fig. 5. *Top Panel:* Available measurements of metallicity in damped Lyman α galaxies are plotted as a function of lookback time for $H_0 = 50$ km s^{-1} Mpc and $q_0 = 0.01$, this being the set of cosmological parameters which is least discordant with stellar ages. The symbols have the same meaning as in Fig.1. *Bottom Panel:* Metallicities of 182 disk stars with measured iron abundances and ages from the large-scale study by Edvardsson et al. (1993).

distant galaxies becomes progressively more skewed in favour of metal-poor systems with decreasing redshift. QSOs which lie behind metal-rich – and presumably dusty – galaxies are reddened by amounts which, while too small to hide the QSOs completely, are nevertheless sufficient to exclude them from magnitude limited samples such as those from which the current damped Lyman α surveys are drawn.

To conclude, it is clear that the technique of QSO absorption line spectroscopy has matured significantly in recent years, to the point where it can be used effectively to probe the chemical evolution of galaxies over cosmologically long timescales. It is a powerful new tool at our disposal in an area of study which until now has been based mainly on observations of different stellar populations in the Milky Way and of H II regions in nearby galaxies.

Several aspects of the early stages of chemical enrichment in galaxies are already emerging from this work and, with the new generation of 8-10 m class telescopes, we can confidently expect significant progress towards building a full picture.

References

Aller, L.H. (1987). Spectroscopy of Astrophysical Plasmas, ed. A. Dalgarno and D. Layzer (Cambridge University Press), 89

Bahcall, J. et al. (1993). ApJS, 87, 1

Edvardsson, B., Andersen, J., Gustafsson, B., Lambert, D.L., Nissen, P.E., Tomkin, J. (1993). A&A, 275, 101

Fall, S.M. (1995). This volume

Lanzetta, K.M. (1993). The Environment and Evolution of Galaxies, ed. J.M. Shull and H.A. Thronson, Jr. (Kluwer, Dordrecht), 237

Lanzetta, K.M., Turnshek, D.A., Sandoval, J. (1993). ApJS, 84, 109

Lanzetta K.M., Wolfe, A.M., Turnshek, D.A. (1995). ApJ, in press

Lipman, K., Pettini, M., Hunstead, R.W. (1995). This volume

Lu, L., Savage, B.D., Tripp, T.M., Meyer, D.M. (1995). ApJ, in press

McGaugh, S.S. (1994). ApJ, 426, 135

McGaugh, S.S., Bothun, G.D. (1994). AJ, 107, 530

McWilliam et al. (1995). ApJ, submitted

Meyer, D.M., Lanzetta, K.M., Wolfe, A.M. (1995). In preparation

Meyer, D.M., Welty, D.E., York, D.G. (1989). ApJ, 343, L37

Meyer, D.E., York, D.G. (1992). ApJ, 399, L121

Moller, P., Warren, S.J. (1993). A&A, 270, 43

Pettini, M., Boksenberg, A., Hunstead, R.W. (1990). ApJ, 348, 48

Pettini, M., Hunstead, R.W. (1990). Australian J. Phys., 43, 227

Pettini, M., Hunstead, R.W., Smith, L.J., King, D.L. (1992). First Light in the Universe: Stars or QSOs?, ed. B. Rocca-Volmerange, M. Dennefeld, B. Guiderdoni and J. Tran Thanh Van (Editions Frontières, Gif-sur-Yvette), 97

Pettini, M., Lipman, K., Hunstead, R.W. (1995). ApJ, in press

Pettini, M., Smith, L.J., Hunstead, R.W., King, D.L. (1994). ApJ, 426, 79

Savaglio, S., D'Odorico, S., Moller, P. (1994). A&A, 281, 331

Skillman, E.D., Kennicutt, R.C., Hodge, P.W. 1989, ApJ, 347, 875

Steidel, C.C. (1995). This volume

Steidel, C.C., Bowen, D.V., Blades, J.C., Dickinson, M. (1995). ApJ, in press

Steidel, C.C., Dickinson, M., Persson, S.E. (1994). ApJ, 437, L75

Steidel, C.C., Hamilton, D. (1992). AJ, 104, 941

Steidel, C.C., Pettini, M., Dickinson, M., Persson, S.E. (1994). AJ, 108, 2046

Storrie-Lombardi, L.J., McMahon, R.G, Irwin, M.J., Hazard, C. (1995). This volume

Tytler, D. (1995). This volume

Vladilo, G., D'Odorico, S., Molaro, P., Savaglio, S. (1995). This volume

Wolfe, A.M. (1993). Proc. 16th Texas Symposium/3rd Symposium on Particles, Strings and Cosmology, ed. B. Sadoulet and J. Silk (Academic Science, New York)

Chemical Evolution of (Proto-) Galactic Disks and Metal Abundances of Damped Lyα Absorbers

Uta Fritze – v. Alvensleben

Landessternwarte, Königstuhl, D – 69117 Heidelberg, Germany

Abstract. Detailed chemical and spectrophotometric modeling of nearby galaxies allows to derive star formation (SF) histories of spiral galaxy disks as a function of Hubble type. Transforming galaxy ages to redshifts for any given cosmology (H_0, Ω_0, z_f) yields – among others – redshift evolutions of chemical abundances and element ratios. Our models explicitly account for SNI contributions to ^{56}Fe, etc. Comparison with curve of growth determined abundances of metal line systems associated with damped Lyα absorbers shows that

- the weak redshift evolution of abundances of damped Lyα systems is consistent with expectations from the slowly varying star formation rates of galactic disks
- the abundance scatter of damped Lyα systems at fixed redshift is consistent with the range of star formation histories from *Sa* through *Sd* disks.

1 Introduction

Abundances have by now been derived from curve-of-growth determined ionic column densities combined with known HI column densities for a substantial number of metal systems associated with damped Lyα absorbers over the redshift range $0.7 \lesssim z \lesssim 3.5$. From the very beginning of this kind of analyses till today, results kept puzzling many observers:

- as opposed to the case of narrow CIV or MgII absorbers, no clear redshift evolution of the damped Lyα abundances has shown up, and,
- the abundance spread among absorption systems at given redshift is large and – at least – comparable to any overall redshift evolution (see e.g. Pettini, *this volume*).

While the dynamical evolution of (proto-)galactic disks – held responsible for damped Lyα absorption – has been studied by Schiano *et al.* (1990), and their number density evolution e.g. by White *et al.* (1993) and Rao (*this volume*), no detailed modeling of their chemical evolution has been reported.

Our approach is the following: we use our evolutionary synthesis code modeling at the same time the spectrophotometric and the chemical properties of various types of galaxies from the onset of star formation (SF) until their present age. For SF histories that give detailed agreement with spectral properties of *Sa* through *Sd* galaxies, we present model predictions for a series of element abundances and compare them to damped Lyα observations.

2   Star Formation in Spiral Disks

Starting from a gas cloud of given initial mass M_{init} and abundances and using various pieces of stellar input physics (stellar yields and remnant masses for PNe, SNe I, II, stellar evolutionary tracks, a library of stellar spectra, ...), we calculate on the one hand the synthetic galaxy spectra (including gaseous emission lines and stellar metal indices) and on the other hand ISM abundances of ^{12}C, ..., ^{56}Fe, taking into account SNI contributions from carbon deflagration white dwarfs in binary systems à la Matteucci (1990). In a first approach, we restrict ourselves to a very simple 1–zone model and do not include any dynamical effects. All aspects of our model galaxies are calculated as a function of time, basic parameters are the initial mass function (IMF) and the time evolution of the star formation rate.

While it has been known since many years that SF in spiral disks proceeds on a longer timescale than in (collapsing) spheroidal systems as E galaxies or halos (see e.g. Kennicutt 1983, Sandage 1986), it is possible only today with the availability of Kennicutt's (1992) atlas of high-resolution galaxy spectra to really constrain the SF histories (i.e. time evolution of SF rates) of various kinds of spiral galaxies. Detailed comparison of our model galaxy spectra with Kennicutt's fiducial galaxies yielded characteristic timescales t_* for SF as defined by $\int_0^{t_*} \Psi(t) \cdot dt = (1 - \frac{1}{e}) \cdot M_{init}$, with $t_* = 2, 3, 10$, and 16 Gyr for spirals of type Sa, Sb, Sc, and Sd.

Spectrophotometric analyses of a large sample of, chemical abundances and element ratios of some, nearby galaxies, as well as our interpretation of CIV and MgII halo lines all led us to prefer a Scalo IMF over e.g. a Salpeter one (Fritze – v. Alvensleben et al. 1991). In the following, we assume that this same IMF is also realised in (proto-)galactic disks.

Our models allow for a detailed analysis of the ejection rate of every individual element in terms of stellar mass, nucleosynthetic site, time and galaxy type. For lack of space, these results will be presented elsewhere (Fritze – v. Alvensleben & Fricke, in preparation).

While evolutionary and cosmological corrections depending on Hubble type are necessary to understand the spectral appearance of high redshift galaxies, a direct 1–to–1 transformation between galaxy age and redshift – uniquely determined by H_0, Ω_0, Λ_0 and an assumed redshift of galaxy formation z_f – is all we need to model the redshift evolution of their chemical properties.

Comparison of our spectrophotometric models with redshift surveys (Fritze – v. Alvensleben 1989), and of the redshift evolution of CIV- and MgII- (halo-)systems (Fritze – v. Alvensleben et al. 1989) allowed to constrain the cosmological parameters to: $50 \lesssim H_0 \lesssim 70$, $\Omega_0 \sim 1$, $\Lambda_0 = 0$, and $5 \lesssim z_f \ll 10$.

According to these constraints, we specifically chose $(H_0, \Omega_0, \Lambda_0, z_f) = (50, 1.0, 0, 5)$ for our comparison with observed abundances of damped Lyα absorbers.

3   Abundances of Damped Lyα Absorbers

The most important among the many caveats connected with the derivation of abundances in damped Lyα systems is that narrow saturated lines may be hidden due to inadequate spectral resolution, so that, in this case, abundances may only be lower limits. High resolution observations revealing a complex velocity structure led to a 2 – component model for the associated metal systems: a quiescent component (Doppler parameter $b \lesssim 10 - 20$ km s^{-1}) of low ionisation with a disk-like structure that also seems to give rise to the observed HI, and a turbulent component ($b \gg 20$ km s^{-1}), predominantly of high ionisation, that may be due to HVCs in the halo. Only in cases of $b \lesssim 20$ km s^{-1} can curve-of-growth determined column densities $N(X_i)$ be safely transformed into metal abundances $[X_i/H]$.

Because abundances in the literature are given for a large variety of atomic line strengths f and for different solar abundances, we have taken published equivalent widths and redone the curve-of-growth analysis with a homogeneous and up-to-date set of atomic data and solar abundances of Anders & Grevesse 1989).

4   Preliminary Results

• Over all of the observationally accessible redshift range, $0.7 \lesssim z \lesssim 3.5$, models with SF histories as given above provide overall agreement with abundances derived for damped Lyα absorbers, not only in global metallicity Z as depicted in Fig. 1, but also for all of the individual elements available.
The same models that successfully describe high redshift (proto-)galactic disks smoothly join into ISM abundance observations on nearby galaxies. Thus, it is possible to understand the chemical evolution of spiral galaxies over virtually all of the Hubble time (at a redshift 3.5 galaxies have an age of 0.5 Gyr in our cosmology).

• The bulk of evolution is seen in Fig.1 to take place at low redshift, at $z \lesssim 1.5$ for Sa, Sb, and at $z \lesssim 1$ for Sc, Sd galaxies, in agreement with observations on damped Lyα systems (see e.g. Pettini and Fall, *this volume*), and completely at variance with what is observed for MgII- and CIV-(halo-)systems.

• The weak redshift evolution of abundances of damped Lyα absorbers is consistent with the long SF timescales for (proto-) galactic disks.
The reason for the much stronger redshift evolution of halo abundances as traced by narrow MgII- or CIV-systems is the comparatively short characteristic timescale for SF of 1 Gyr for halos (see Fritze – v. Alvensleben *et al.* 1991 for details).

• The scatter of heavy element abundances in damped Lyα systems at fixed redshift is consistent with the spread of SF timescales from Sa to Sd disks.
Fig. 1 shows that an Sd galaxy at $z \sim 3$ has a metallicity as low as $1/100\ Z_\odot$, an elliptical at $z \sim 2$ would have $Z \sim Z_\odot$ with a gas content comparable to

that of a present day Sb, consistent with $\langle[Zn/H]\rangle \sim -1 \pm 1$ for damped Lyα
absorbers in this redshift range (cf. Pettini, *this volume*).

- If damped Lyα absorbers are indeed the high redshift progenitors of spiral
  galaxies, our models also predict their spectrophotometric appearance for
  direct comparison with optical identifications.

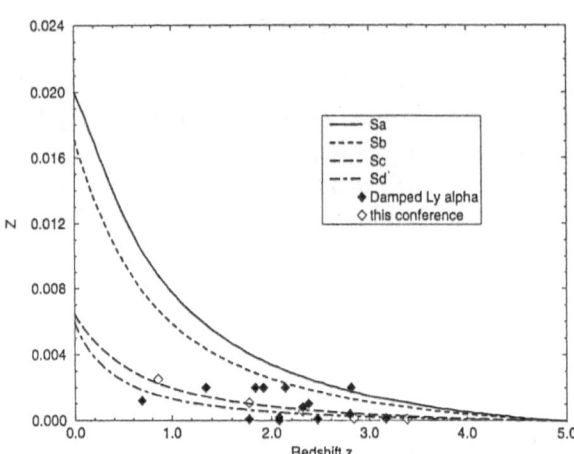

Fig. 1. Redshift evolution of the global metallicity Z for spiral galaxies of various
Hubble types together with metallicities derived from curve-of-growth analysis of
damped Lyα absorbers.

Acknowledgements. This work was partly supported by the Verbundforschung
Astronomie under grant WE-010 R 900-40.

References

Anders, E., Grevesse, N., 1989, Geochimica et Cosmochimica Acta, 53, 197
Fritze – v. Alvensleben, U., 1989, PhD Thesis, University of Göttingen
Fritze – v. Alvensleben, U., Krüger, H., Fricke, K. J., Loose, H.-H., 1989, A&A,
    224, L1
Fritze – v. Alvensleben, U., Krüger, H., Fricke, K. J., 1991, A&A, 246, L59
Kennicutt, R. C., 1983, ApJ, 272, 54
Kennicutt, R. C., 1992, ApJ, 388, 310
Matteucci, F., 1990, in "Chemistry in Space", Kluwer Dordrecht
Sandage, A., 1986, A&A, 161, 89
Schiano, A. V. R., Wolfe, A. M., Chang, C. A., 1990, ApJ, 365, 439
White, R. L., Kinney, A. L., Becker, R. H., 1993, ApJ, 407, 456

Abundances and Ionization in Multiple-Component Absorbers

Richard F. Green[1], Donald York[2], Keliang Huang[3], Jill Bechtold[4], Dan Welty[2], Matt Carlson[2], Pushpa Khare[5], and Varsha Kulkarni[2,6]

[1] National Optical Astronomy Observatories, Tucson, AZ 85726-6732 USA
[2] University of Chicago, 5640 S. Ellis Ave., Chicago, IL 60637 USA
[3] Nanjing Normal University, Nanjing, China
[4] Steward Observatory, University of Arizona, Tucson, AZ 85721 USA
[5] Utkal University, 751004 Bhubaneswar, India
[6] Space Telescope Science Institute, 3700 San Martin Dr., Baltimore, MD, USA

Abstract. The damped Lyman α absorber at z=1.7765 in the quasar MC3 1331+170 is resolved into multiple velocity components, corresponding to an H I region, H II regions consistent with photoionization by hot stars, and higher ionization components that dominate at higher relative velocities. The column densities derived from the N I triplet suggest that [O/H] is depleted by -2.1 dex relative to solar values. Comparison of the predicted vs. observed column densities of S II then shows that only $\sim 1/6$ of the gas on the sightline is in the neutral component producing the damped Lyman α line. In that case, the Zn II / H I ratio provides only a weak upper limit to [Fe/H] for the system. The (S II / Zn II) value implies a strong Fe-peak enhancement relative to α-product elements, compared to Galactic halo stars at [Fe/H] = -2.2. The consequence is that the relative rates of supernovae Type I and Type II may depend on local conditions.

1 Introduction

This paper presents an analysis of some "antique" (i.e., KPNO 4-meter) high-dispersion data to determine absolute abundances in a damped Lyman α absorber. The analysis complements that of the Keck data presented by Wolfe and his collaborators (these proceedings) by going into the ultraviolet down to about 3140 Å. The paper (as noted in its title) was originally intended to be more inclusive. It is now limited to the damped Lyman α absorption system at z=1.7765 in the quasar MC3 1331+170. That spectrum has a velocity resolution of ~ 18 km/s, and a typical $S/N \sim 20$, and is described by Kulkarni et al. (1995).

The goal of abundance determination in quasar absorbers is to trace the star formation histories and supernova rates as a function of environment. The contexts of abundance analyses are therefore the solar and solar neighborhood interstellar medium abundances, along with the fossil record of Galactic halo stars.

The additional context for analyzing gas phase abundances is the depletion of refractory elements onto dust grains. The depletion depends most strongly on condensation temperature, as well as chemical properties. Even in metal-poor gas, the abundances of Fe, Al, and Ca can be reduced by more than a factor of 10. The elements S (of intermediate mass) and Zn (Fe-peak) are nearly undepleted, and can be used as proxies for O and Fe, respectively. N is relatively undepleted, but O can be reduced by more than a factor of 2. In halo stars, the N/O ratio reaches a plateau value of about -0.6 dex for the most metal-deficient objects.

2   Current Picture of Halo Abundances

The current picture of halo stellar abundance patterns has been summarized, for example, by Wheeler, Sneden and Truran (1989). They find that in the most metal-poor stars, the ratio to O of Fe and intermediate-mass elements with even numbers of protons, such as Mg, Si, S, and Ca, are enhanced by about a factor of 3 relative to solar for $[Fe/H] < 1$. The ratio to Fe of other even-proton iron-peak elements, such as Cr and Zn, is essentially independent of metallicity. Strong odd-even effects are seen in both intermediate-mass and iron-peak elements. Odd-proton elements are underabundant (compared to solar ratios) for the lowest metallicities, increasing toward solar values. Examples include Na, Al, P at intermediate mass, and Mn, Co, and Cu near the Fe peak.

The interpretation is that the enrichment patterns are characteristic of the yields of various nucleosynthetic sources:

· Very massive $(30 - 70 M_{\odot})$ stars evolve in $< 4 \times 10^6$ years, and return a very oxygen-rich mixture, with virtually no C or Fe.

· Massive stars $(> 10 M_{\odot})$ evolve in $\leq 2 \times 10^7$ years, returning O and even-proton nuclei from Ne-Ca, and r-process heavy elements in SN Type II.

· Intermediate-mass progenitors $(1 - 10 M_{\odot})$ evolve in 10^8 to 10^9 years, and return C, N and s-process elements (He shell burning).

· Supernovae of Type Ia produce Fe-peak elements along with more Si, S, and Ca. They serve as a time-delay fuse for Fe-peak enrichment.

There is a strong initial metallicity effect in the production of odd-proton nuclei. The most abundant nuclei are self-conjugate (e.g., $^{12}C, ^{16}O, ^{28}Si$), so odd-proton and neutron-excess even-proton production depends on trace abundances of $N > Z$ elements. That trace is very low in extreme metal-deficient stars, producing a pronounced odd-even effect in calculated yields.

3 The Case of MC3 1331+170 - Light Elements

The damped Lyman α system in MC3 1331+170 shows a complex velocity structure, with multiple widely spread components in the higher ions like Si III and C IV. Those ions are not co-extensive in velocity space with the clouds that produce the neutral and first ion absorption, so physical models do not need to produce all ionization states simultaneously in the same clouds. The neutrals and first ions are found in a pair of strong absorbers associated in velocity with the damped Lyman α system, as determined from the velocity of the 21 cm line observed in absorption.

The analysis of abundances is based on the detection of important unsaturated diagnostics. Particularly important is the N I triplet at 1199.5, 1200.2, and 1200.7 Å. These lines are in the Lyman α forest, but show ratios consistent with being nearly unsaturated. The N I is a good tracer of neutral gas, with $NI/NII \sim 40$ in H I gas in the local ISM. In this sytem, $log(NI) = 14.5\pm0.1$, giving $log(NI/HI) = -6.64^{+0.1}_{-0.4}$. The depletion relative to the solar ratio is therefore -2.7.

It is interesting to note the apparent abundance of O I. In this case, the O I line at 1302 Å tends to be quite saturated. The derived column density is $log(OI) = 15.3^{+0.2}_{-0.3}$. The lower limit to the abundance ratio with respect to solar O/H is therefore a depletion of no more than -2.7 dex. The apparent metallicity of gas in this absorber associated directly with H I is low.

Halo star abundances and nucleosynthetic models suggest that S tracks O. From measurements of the local ISM, S appears to be nearly undepleted. Determination of its abundance is therefore likely to be more representative of the relative abundances of α-product elements than a measurement of O itself. S II is the dominant ionization state in both H I and H II regions. Typical values for the local ISM show S I / S II < .002, and S III / S II = 0.05 - 0.1.

We can, in principle, then use the observed abundance of O I to predict the amount of S II expected from the H I region on this sightline. In practice, the one available O I line is saturated. On the other hand, halo stars show an asymptotic limit in the ratio of N/O with decreasing metallicity. We can use the more reliably determined N I column density to correct the O I for saturation, then predict the S II column. If we use a value of log N/O = -0.6 for extreme metal-poor stars, then the depletion in O/H relative to solar values is -2.1. From that, we predict a column density for S II in the H I region of log (S II) = 14.3. The value directly observed is log (S II) = 15.1 ± 0.1.

The conclusion is that most of the S II absorption in this sightline arises in an H II region, with $N(H^+) \sim 6N(H^0)$. The total column density of H is therefore not accurately determined from the damped Lyman α profile.

4   Iron-Peak Elements

It is reasonable to assume a similar composition for H I region and H II region gas in the same velocity components. This result has direct consequences for the determination of absolute abundances of Fe peak elements. Pettini and his collaborators have exploited the virtues of Zn II as a proxy for Fe peak elements. Zn tracks Fe in halo star abundances. Zn is relatively undepleted in the local ISM, and Zn II is the dominant ionization state in H I regions, as well as under many H II region conditions.

For this absorber, Pettini et al. (1994) found:
· $N(ZnII) = (2.4 \pm 1.2) \times 10^{12} \rightarrow depletion = -1.4^{+0.2}_{-0.3}$ and
· $N(CrII) = (5 \pm 1) \times 10^{12} \rightarrow depletion = -2.2 \pm 0.1$.
From this paper, we find:
· $N(FeII) = (2.6 \pm 0.6) \times 10^{14} \rightarrow depletion = -2.2^{+0.1}_{-0.2}$.
The conclusion is that the Cr and Fe abundances are consistent, and show a depletion by ~ 0.8 dex relative to Zn.

[S/Zn] = +0.1, which represents the "true" value of [O/Fe]. (This result depends only very weakly on the distribution of material between the H I and H II regions along the line of sight.) A typical value for metal-poor halo stars is +0.5. This absorber therefore shows a significantly higher contribution of Fe-peak relative to α-product elements than measured in the Galactic halo. The implication is that the enrichment pattern produced by the relative rates of Type I to Type II supernovae may depend on local conditions.

We calculate [Fe/H] = -2.2 for this absorber, typical of metal-poor Galactic halo stars. The Zn II / H I ratio provides only a weak upper limit to the iron-peak abundance because of the substantial fraction of ionized hydrogen inferred to be present in this absorption component.

References

Kulkarni, V.P., Huang, K.-L., Green, R.F., Bechtold, J., Welty, D.E., & York, D.G. 1995, MNRAS, submitted.
Pettini, M., Smith, L.J., Hunstead, R.W. and King, D.L, 1994, ApJ, 426, 79.
Wheeler, J.C., Sneden, C. and Truran, J.W. Jr. 1989, ARA&A, 27, 279.

Element Abundances at High Redshifts: The N/O Ratio at Low Metallicity

Keith Lipman[1], Max Pettini[2] and Richard W. Hunstead[3]

[1] Institute of Astronomy, Madingley Road, Cambridge, CB3 0HA, UK
[2] Royal Greenwich Observatory, Madingley Road, Cambridge, CB3 0EZ, UK
[3] School of Physics, University of Sydney, NSW 2006, Australia

Abstract. Our knowledge of galactic chemical evolution is currently limited to observations of Milky Way stars and H II regions of nearby galaxies. Damped Lyman α systems offer a new approach for tracking the evolution of normal galaxies from early epochs to the present day. Here we report the first measurements of nitrogen abundances in galaxies with less than 1/100 of solar metallicity, a range unexplored by previous observations.

1 Introduction

High resolution spectroscopy of absorbers along QSO sightlines allows us to study element abundances in high redshift galaxies with sufficient accuracy to test models of chemical evolution. The damped Lyman α population, thought to represent the progenitors of present day galaxies (Wolfe, this volume), is particularly well suited to such studies: the large neutral hydrogen column densities not only make accessible absorption lines of weak transitions, but also ensure that the gas is optically thick, largely removing the need for ionization corrections. It is of particular interest that many damped Lyman α systems are very metal-poor, offering us a view of the earliest stages of chemical enrichment in galaxies.

Traditionally, galactic chemical evolution studies have relied on observations of Galactic stars and of bright H II regions in nearby galaxies. Both classes of object have revealed much about the process of metal enrichment; however, each has its limitations. The usefulness of the H II regions is restricted by their abundances; few are known with $Z < 1/10\ Z_{\odot}$, and none with $Z \lesssim 1/50\ Z_{\odot}$. Furthermore, local pollution from evolved stars such as Wolf-Rayets can confuse the interpretation, especially at low metallicities. While old stellar populations do extend to very low abundances, observations of some elements – such as nitrogen – are particularly difficult in metal-poor stars. A further point is that existing abundance measurements in both stars and H II regions refer exclusively to nearby galaxies; at present we have no way of observing these objects in the more distant universe.

High resolution spectroscopy of damped Lyman α systems therefore has much to offer. We have started a program to investigate the abundance of nitrogen in metal-poor damped Lyman α systems, extending the studies of

H II regions and stars to lower metallicities than previously attainable. This approach promises to cast light on the nucleosynthetic origin of N, which is not yet fully understood.

2   Observations

We have analyzed two damped Lyman α systems so far in this project, both chosen for their low metallicities. The $z_{\mathrm{abs}} = 2.53788$ system towards Q2344+124 ($z_{\mathrm{em}} = 2.779$) was observed with the William Herschel telescope on La Palma, while the $z_{\mathrm{abs}} = 2.27936$ absorber towards Q2348−147 ($z_{\mathrm{em}} = 2.940$) was observed at the Anglo-Australian telescope. In both cases we used echelle spectrographs and photon-counting detectors to achieve a resolution of $R \simeq 45\,000$, corresponding to $FWHM \simeq 7$ km s^{-1}. A full description of the observations is given by Pettini, Lipman and Hunstead (1995) and Lipman and Pettini (1995).

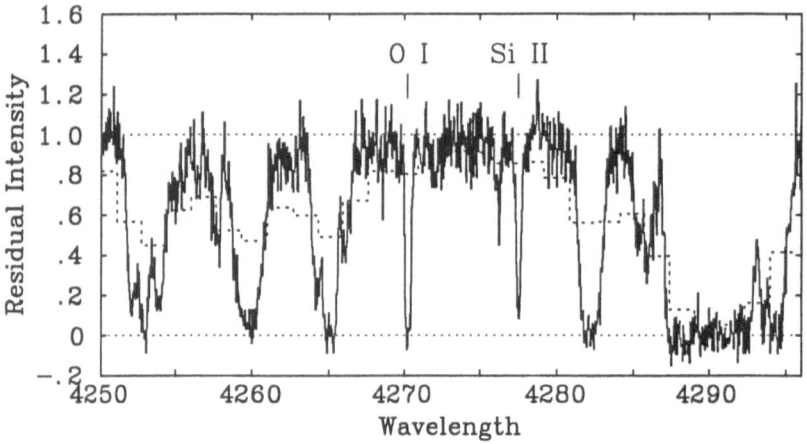

Fig. 1. Portion of the AAT IPCS spectrum of Q2348−147 encompassing the O I λ1302 and Si II λ1304 transitions in the $z_{\mathrm{abs}} = 2.27936$ damped Lyman α system. The resolving power is $R \simeq 45\,000$; the rest-frame equivalent widths of the O I and Si II lines are 146 mÅ and 102 mÅ respectively. The dashed line shows how the same data would appear if recorded with the coarser $R = 1300$ of the HST observations by Reimers et al. (1992). Narrow lines would be lost in the noise and only strong, saturated features – quite unsuitable for abundance measurements – would be discerned.

3   Element Abundances

A portion of the spectrum of Q2348−147 is shown in Figure 1. The velocity structure of the $z_{abs} = 2.27936$ damped system is particularly simple, apparently consisting of a single absorption component with Doppler parameter $b \simeq 10$ km s^{-1}. Column densities N of species which are the major ion stages in H I regions were deduced by fitting theoretical absorption profiles to the metal lines in the manner described by Pettini, Lipman and Hunstead (1995). Errors were estimated by exploring the $b - N$ parameter space to determine the range of values of N compatible with the shape of the observed profiles and their equivalent widths (to within 1σ).

Dividing by the neutral hydrogen column density $N(\text{H I}) = (3.7 \pm 0.7) \times 10^{20}$ cm^{-2}, implied by the damping wings of the Lyman α line, then leads to the element abundances shown in Figure 2.

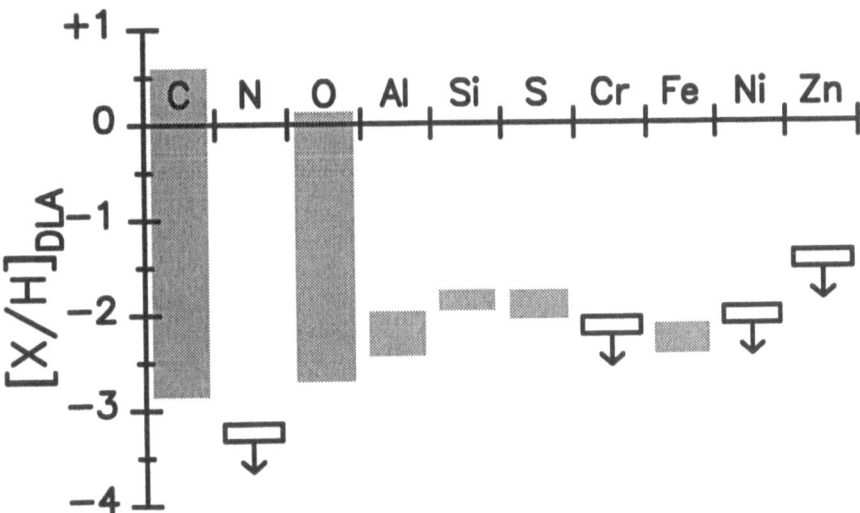

Fig. 2. Element abundances in the $z_{abs} = 2.27936$ damped Lyman α system towards Q2348−147. Values are plotted on a logarithmic scale relative to solar abundances (Anders and Grevesse 1989). Grey vertical bars indicate positive detections, the length of the bar reflecting the allowed range of values. Open boxes show upper limits; the upper and lower edges of the boxes represent 3σ and 2σ limits respectively.

The absorption lines of C II and O I are saturated; in such cases it is not possible to limit the corresponding abundances to better than 3 orders of magnitude, despite the high resolution of our echelle spectra. The problem is exacerbated when dealing with low resolution data, such as those of Reimers et al. (1992), since *only* saturated lines would then be detected (see Figure 1). For this reason, abundance measurements from low resolution observations must be viewed with caution.

Figure 2 shows that most elements in the galaxy giving rise to the $z_{abs} = 2.27936$ damped system are underabundant by about two orders of magnitude relative to the Sun. The fraction of heavy elements present in dust grains is also significantly reduced compared with the local interstellar medium. This is readily realised when we consider that elements such as Fe, Al and S, which locally exhibit widely differing degrees of dust depletions – by up to two orders of magnitude – are found in roughly solar relative proportions in this high-redshift galaxy. A striking feature of Figure 2 is the underabundance of N relative to other elements. This is not a consequence of line misidentification or blending, since the N limit is imposed by a non-detection.

As discussed below, it is useful to compare the N abundance with that of O. To circumvent the large uncertainty in [O/H], we have assumed the O abundance to be equal to that of S in the case of Q2348−147, and that of Si in Q2344+124. These are plausible assumptions since all three elements are created via the α process, and their relative abundances are observed to be roughly constant over the full range of metallicities sampled by H II regions and stellar studies (e.g. Skillman and Kennicutt 1993; Garnett and Kennicutt 1994; McWilliam et al. 1995). Furthermore, we do not expect dust depletions to alter significantly the abundance pattern in either absorption system.

4   The Nitrogen to Oxygen Ratio

Why N/O? Perhaps originally this ratio was studied for no better reason than that both elements are easily observable in H II region emission-line spectra. However, the choice does serve a useful purpose. The nucleosynthetic origin of O is reasonably well understood, while that of N is not; the relative abundances of these two elements can be used to test current ideas on the mechanisms responsible for the formation of N.

Primary production of an element involves its formation from seed nuclei which are produced *in situ* within the star. Primary production does not require the presence of pre-existing nuclei and is therefore independent of the initial metallicity of the star. O is a good example of a primary element, and its production is dominated by supernova explosions of massive stars. The major producers of primary N are thought to be stars of intermediate mass during the third dredge-up phase (Renzini and Voli 1981).

Secondary N, on the other hand, is synthesised from pre-existing seed nuclei. Such a process is thought to be possible in most stars, except those of low mass in which the core temperature is too low to initiate the CNO cycle. If N is produced in this manner, its yield is expected to be proportional to the initial metallicity of the star.

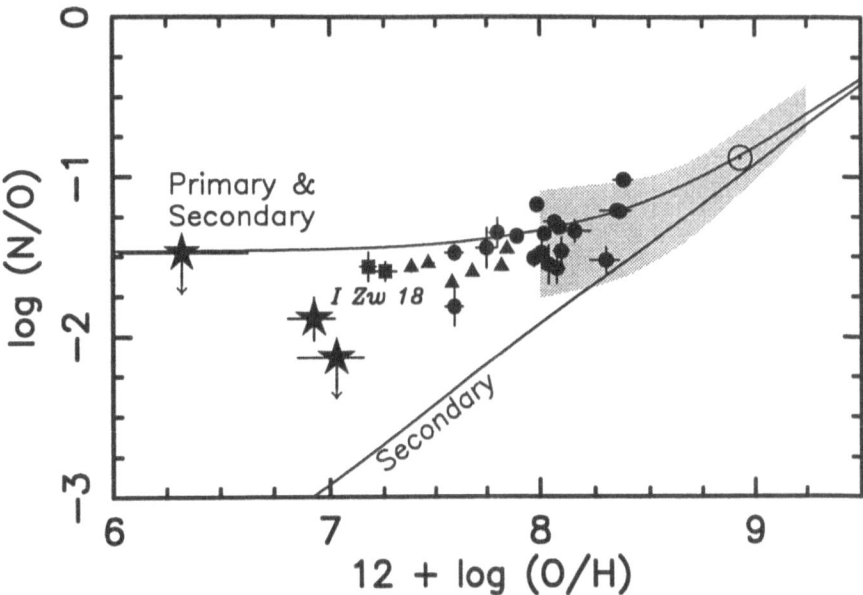

Fig. 3. The N/O ratio for three damped Lyman α systems (star symbols) are compared with values measured in H II regions in spiral galaxies (shaded area) and dwarf star-forming galaxies (filled symbols). H II region data are from the extensive compilation by Vila-Costas and Edmunds (1993); the dwarf galaxies selected are those whose spectra Pagel et al. (1992 – circles) and Izotov et al. (1994 – triangles) considered to be free of contamination by Wolf-Rayet features. The two squares refer to the two H II regions in I Zw 18 observed by Skillman and Kennicutt (1993). The damped Lyman α systems are, from the left, HS 1946+7658 ($z_{abs} = 2.8443$; Fan and Tytler 1994), Q2344+124 ($z_{abs} = 2.53788$; Lipman and Pettini 1995) and Q2348−147 ($z_{abs} = 2.27936$; this paper). Symbols with downward-pointing arrows indicate upper limits to [N/O]. The Sun symbol shows the solar system abundances (Anders and Grevesse 1989). Also shown are the predictions for a purely secondary and a primary + secondary origin of N, reproduced from Vila-Costas and Edmunds (1993).

Figure 3 shows [N/O] as a function of [O/H] for a variety of observations. Also shown are the expectations for secondary only, and combined primary and secondary origin of N (Vila-Costas and Edmunds 1993). Primary production dominates at low values of [O/H], but secondary nitrogen dominates at higher abundances. Our measurements of [N/O] in the damped systems lie in a distinct region of the plot, within the bounds predicted for primary and secondary production, but at lower metallicities than those sampled up to now.

There are several effects to consider when interpreting Figure 3. While H II regions are rich in recent stellar and supernova ejecta, the QSO sightlines probe neutral gas, not necessarily close to regions of active star formation. Therefore, the evolution of the N/O ratio in the population of galaxies traced by the damped systems depends on the rate at which freshly synthesized elements mix with the more generally distributed interstellar medium. Furthermore, the position of each point not only depends on the production mechanism of N, but also on the previous star formation history. For example, were we to observe an H II region soon after a starburst, we would expect that O, produced from short-lived massive stars, would be overabundant relative to N which, being produced mostly in stars of intermediate mass is released back into the ISM over longer timescales (e.g. Edmunds and Pagel 1978). If star formation takes place in isolated starbursts separated by quiescent periods, this delayed-release scenario predicts that the scatter in [N/O] should be greater at lower metallicities (Garnett 1990; Olofsson 1995). Our survey will be able to establish if such a scatter exists among high-redshift damped Lyman α galaxies.

In conclusion, even the preliminary observations presented here provide a clear demonstration of the high potential of damped Lyman α systems, not only for understanding the nucleosynthetic origin of N, but also for studying many other aspects of galactic chemical evolution. We can expect an exciting future in this field now that we have entered the era of 10m-class telescopes.

References

Anders, E., Grevesse, N. (1989). Geochim, Cosmochim. Acta, 53, 197

Fan, X., Tytler, D. (1994). ApJS, 94, 17

Edmunds, M.G., Pagel, B.E.J. (1978). MNRAS, 185, 77P

Garnett, D.R. (1990). ApJ, 363, 142

Garnett, D.R., Kennicutt R.C. (1994). ApJ, 426, 123

Izotov Y.I., Thuan T.X., Lipovetsky V.A. (1994). ApJ, 435, 647

Lipman, K., Pettini, M. (1995). In preparation

McWilliam et al. (1995). ApJ, submitted

Olofsson, K. (1995). A&A, 293, 652

Pagel, B.E.J., Simonson, E.A., Terlevich, R.J., Edmunds, M.G. (1992). MNRAS, 255, 325

Pettini, M., Lipman, K., Hunstead, R.W. (1995). ApJ, in press

Renzini, A., Voli, M. (1981). A&A, 94, 175

Reimers D., Vogel, S., Hagen, H.J., Engels, D., Groote,D., Wamsteker, W., Clavel, J., Rosa, M.R. (1992). Nature, 360, 561

Skillman E.D., Kennicutt R.C. (1993). ApJ, 411, 655

Vila-Costas, M.B., Edmunds, M.G. (1993). MNRAS, 265, 199

Abundances in the Lyman–Alpha Clouds

Sandra Savaglio[1,2] and John Webb[2]

[1] Dipartimento di Fisica, Università della Calabria, I–87036 Arcavata di Rende, Cosenza, Italy
[2] School of Physics, University of New South Wales, P. O. Box 1, Kensington, NSW 2033, Australia

Abstract. We have re–examined the chemical composition of Lyα clouds using the composite–cloud technique, in which each Lyα line in a spectrum is shifted to its rest frame wavelength and all rest frame spectra are co–added to form an 'averaged' Lyα cloud spectrum. We illustrate how various estimates of the spectrum and red-shift evolution of the background ionizing UV flux lead to very different predictions for the relative strengths of heavy element lines in the forest clouds. We also show how the potential abundance limits depend on the various observational quantities and how different procedures may be required for OVI and CIV. Preliminary results are presented from an analysis of two high redshift QSO echelle spectra.

1 Photoionization Models of Non–Primordial Lyα Clouds

The determination of the heavy element abundances in the Lyman forest clouds is an important test of their origin. The observational limitation is that, for an individual cloud, the expected heavy element column densities are low and, generally below typical detection limits. This difficulty prompt-ed Norris *et al.* (1983) to devise a technique in which individual Lyα lines are shifted to their rest frame wavelengths and all rest frame spectra are co-added. The most prominent heavy element transitions can then be searched for in this high signal–to–noise averaged Lyman forest cloud spectrum, yield-ing abundance measurements or upper limits.

However, the results derived in this way rely sensitively on knowing the UV flux impinging on the clouds to a reasonable accuracy. Since there is now substantially more information available concerning the redshift evolution and spectral shape of the UV background ionizing radiation field incident compared to earlier studies, we have re–examined the topic. In particular we have explored which species may yield the best constraints on heavy element abundances in the Lyman forest clouds.

Two quite different models for the redshift dependence of the intensity of the UV flux at the Lyman limit J_{912} have been adopted in this study, repre-senting a plausible range of possibilities. In case (a) we have used an estimate of J_{912} based on the integrated UV flux from observed QSOs, modified by the expected opacity of H and He associated with all QSO absorption systems

(Madau 1992). In case (*b*) we have combined several measurements of J_{912} derived from the proximity effect over a wide range in redshifts (Fig. 1). In both cases we have adopted the spectral shape of the UV background, J_ν, given by Madau (1992).

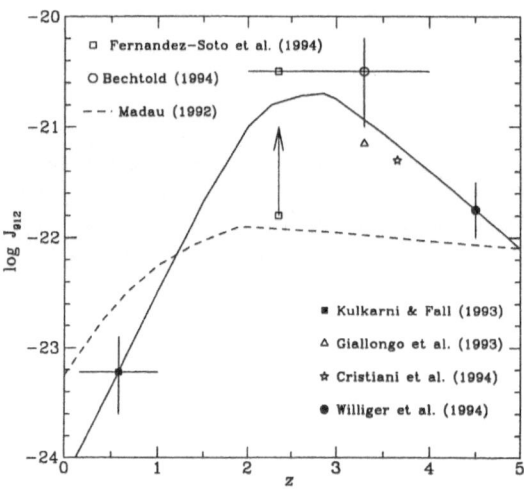

Fig. 1. The two UV flux models used in the photoionization calculations. The solid line is our approximation to the observed proximity effect estimates.

The photoionization code CLOUDY (Ferland 1991) was used to compute ionic column densities for an assumed heavy element abundance of [M/H] = −2 for both UV backgrounds illustrated in Fig. 1. The cloud HI column density used was $\log N(\text{HI}) = 14$, the total hydrogen volumetric density was $\log n_H = -4$ and plane parallel slabs were assumed, illuminated on one side by the UV background. Scaled solar abundances are adopted. Results are shown in Fig. 2. For both UV models, OVI, CIV and NV are all prominent for $z > 2$, but weak at lower z due to the lower UV background. In Fig. 2(*a*), at $z > 2$, the UV flux and ion column densities remain almost constant. In Fig. 2(*b*), the UV redshift dependence of the background radiation is substantially greater and for $z > 1.2$, the ion column densities change rapidly. At redshift $z \sim 3$, CIV and NV become more highly ionized while OVI, which has the higher ionization potential, is less affected by the rapidly evolving UV background flux.

An interesting feature of Fig. 2 is the large variation between the various column densities, particularly for the higher UV background model. For UV background estimates based on the proximity effect, OVI has the highest

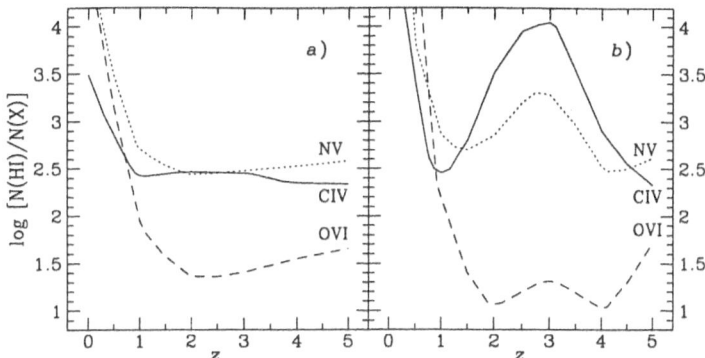

Fig. 2. Heavy element abundances (X) relative to HI as function of redshift for the two different models for J_{912}: a) from Madau (1992); b) from the proximity effect measurements.

column density, being more than 500 times greater than that of CIV at $z \sim 3$. OVI however does of course suffer from confusion with the general forest distribution and thus it is not immediately obvious that OVI can provide the best constraints.

2   Potential Observational Constraints

Here we describe some preliminary results from a limited exploration of parameter–space using simulations of QSO spectra with various input cha-racteristics. Synthetic QSO spectra were generated, using Lyα cloud heavy element column densities based on the lower UV background radiation field model (i.e. the dashed line of Fig. 1). Here we have taken [M/H] $= -2.5$, i.e. lower abundances than those used in deriving Fig. 2. The QSO emission redshift is taken as 3.5, to consider Lyα lines in the redshift range $2.8 - 3.5$.

   In the initial set of simulations (top panel of Fig. 3), we used a spectral resolution and signal–to–noise ratio typical of echelle spectrographs on 4m telescopes (FWHM = 10 km s^{-1}) and S/N = 10 per pixel. In generating the top set of composite spectra in Fig. 3 we computed 10 'observed' QSO spectra and varied the equivalent width (EW) selection threshold from 0.5 Å to 0.1 Å. On the left side of Fig. 3 the region with the Lyβ and the OVI doublet is shown and on the right hand side the CIV doublet. The results suggest that if a low selection threshold is chosen (0.1 Å) any OVI signal is diluted beyond detection (bottom curve). On the other hand, if the threshold

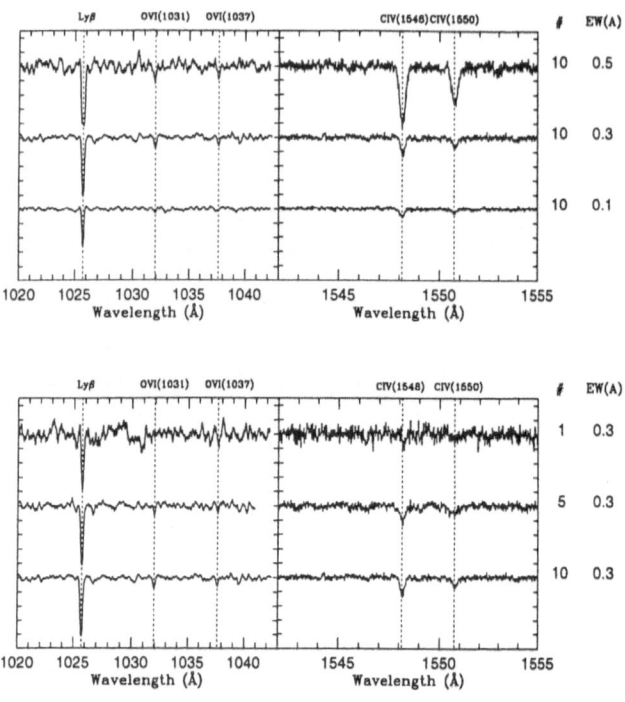

Fig. 3. Composite Lyα forest OVI and CIV doublet spectra derived from simulated QSO spectra. The resolution and S/N are fixed to 10 km s^{-1} and 10 respectively. The right hand column indicates the number of spectra and the EW threshold used in each case.

is rather high (0.5 Å), the number of lines is small and confusion noise due to line blending dominates (top curve). The optimal value of the selection threshold for OVI constraints appears to be close to 0.3 Å. For CIV, where line blending with forest lines is not relevant, the situation is rather different and a selection threshold of EW $\geq$ 0.5 Å gives the strongest detection.

In the second set of simulations (bottom panel of Fig. 3) the EW limit is fixed to 0.3 Å and the number of spectra was changed. The results suggest that CIV could easily be detected with just 5 spectra, whilst OVI needs at least 10, for the assumed spectral characteristics.

Similar numerical experiments were carried out varying the S/N ratio and spectral resolution. The results indicate that, as expected, the detectability of CIV is sensitive to S/N and thus 10m class telescopes are desirable if CIV is to be successfully used, whilst for the confusion–limited OVI the effect of increasing S/N is less marked. Whilst CIV can be detected even in relatively low resolution data, the detection of OVI requires high resolution to overcome line confusion.

In the last example (Fig. 4) we have simulated spectra at different redshifts for constant abundances of −2.5 and a signal–to–noise ratio of 10. 10 spectra were simulated at redshifts of 2.5, 3.5 and 4.5. Fig. 4 suggests that CIV detection is relatively insensitive to redshift, whilst OVI is only detected at $z = 3.5$, presumably due to a more favourable combination of the effects of line confusion, the number of lines per unit redshift, and the redshift dependence of the UV background.

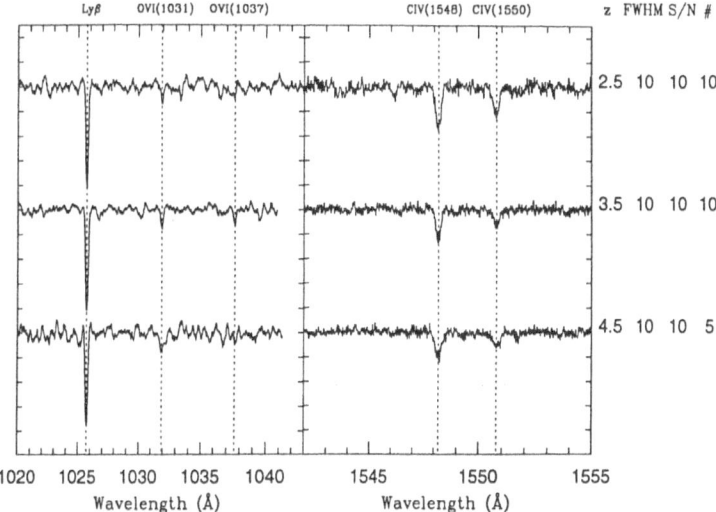

Fig. 4. Composite Lyα forest OVI and CIV doublet spectra derived from simulated QSO spectra with input abundances of [M/H] = −2.5 The right hand column indicates the emission redshift, the spectral resolution, the S/N and the number of spectra.

We emphasize that the results illustrated in Figs. 3 and 4 are based on the *lower* UV background case. Fig. 2 shows that the results would be substantially different if *proximity effect* UV estimates are closer to reality, in which case CIV would be extremely weak and provide no useful constraints. Interesting abundance constraints would then be derived only from OVI.

Apart from one CIV study (Tytler & Fan 1994), all previous work has been at a spectral resolution of about 1 Å or worse. No high spectral resolution study of OVI has been done so far to our knowledge, probably because the line confusion complicates matters and because even for high redshift QSOs, the OVI forest still falls in a relatively blue spectral region, where many instruments/detectors are not efficient. In Fig. 5 we illustrate preliminary results based on just 2 spectra, the $z = 4.1$ Q0000–26 and $z = 4.5$ BRI1033–26. Whilst the Lyβ line is clearly seen, no OVI is detected (the dashed line in Fig. 5). Upper limits on the OVI column density give upper limits on the heavy element abundances of -1.5 and -2 for the two assumed UV flux models (dashed and solid line of Fig. 1 respectively). Further data are clearly required to achieve lower limits or detections.

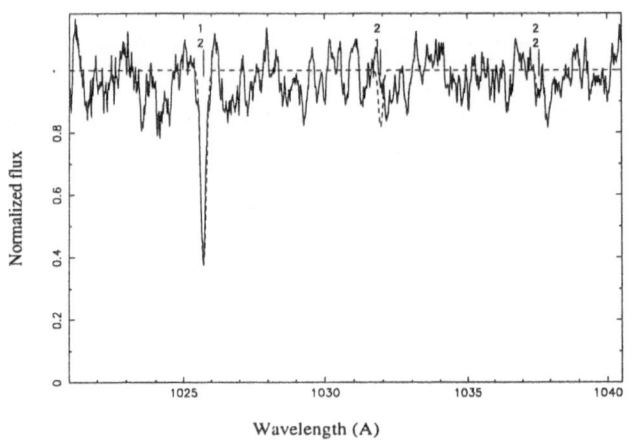

Fig. 5. Composite spectrum of the Lyα forest for the high redshift QSOs Q0000–26 and BRI1033–26 illustrating the Lyβ line and OVI doublet.

References

Bechtold, J., 1994, ApJS, 91, 1.

Cristiani, S., D'Odorico, S., Fontana, A., Giallongo, E., Savaglio, S., 1995, MNRAS, *in press.*

Ferland, G.J., 1991, *OSU Astronomy Dept. Internal Rept.*, 91–01.

Fernandez-Soto, A., Barcons, X., Carballo, R., Webb, J.K., 1995 MNRAS, *submitted.*

Giallongo, E., Cristiani, S., Fontana, A., Trevese, D., 1993, ApJ, 416, 137.

Kulkarni, V.P., Fall, S.M., 1993, ApJL, 413, L63.

Madau, P., 1992, ApJL, 389,L1.

Norris, J., Hartwick, F.D.A., Peterson, B.A., 1983, ApJ, 273, 450.

Tytler, D., Fan, X.-M., 1994, ApJL, 424, L87.

Williger, G.M., Baldwin, J.A., Carswell, R.F., Cooke, A.J., Hazard, C., Irwin, M.J., McMahon, R.G., Storrie-Lombardi, L., 1994, ApJ, 428, 574.

Relative Metal Abundance Patterns in Two Damped Lyman–Alpha Systems

Limin Lu[1], Blair D. Savage[2], Todd M. Tripp[2], and Dave M. Meyer[3]

[1] Dept. of Astronomy, Caltech, Pasadena, CA 91125, USA
[2] Dept. of Astronomy, University of Wisconsin, Madison, WI 53706, USA
[3] Dept. of Phys. & Astron., Northwestern University, Evanston, IL 60208, USA

Abstract. We derive absolute and relative metal abundances for two damped $Ly\alpha$ systems using high resolution, high S/N observations. The results have important implications for understanding the chemical evolution history of galaxies.

Damped $Ly\alpha$ absorption systems in quasar spectra are generally thought to trace young galaxies at high redshifts (Wolfe et al. 1986). These systems are considered well suited for metal abundance studies of high-redshift galaxies because their large H I column densities effectively shield the UV ionizing photons, making ionization corrections minimal. Previous estimates of metal abundances in damped systems have concentrated on the ion species of trace elements (Zn II, Cr II, Ni II) using medium-resolution but high S/N observations (Meyer et al. 1989; Pettini et al. 1990; Meyer & Roth 1990; Meyer & York 1992; Pettini et al. 1994), although there have been a few studies of similar nature carried out at echelle resolution (Rauch et al. 1990; Bergeron & Petitjean 1991; Savaglio et al. 1991; Wolfe et al. 1994). These studies yield the important results that the chemical enrichment levels in damped systems at $z \sim 2$ to 3 are generally low, with typical metallicities 1 to 2 dex below the solar value. Metal abundance determinations of more abundant elements (C, N, O, Mg, S, Si, Fe, etc.) generally require observations with very high spectral resolution in order to resolve the complex velocity structures commonly present in the absorption, and to assess the effects of line saturations. It is important to study the *relative* abundances of different elements in the galaxies at high redshifts because the relative abundance patterns carry important information about the nucleosynthesis history in these galaxies (cf. Wheeler et al. 1989).

We present a study of relative abundance patterns in two damped $Ly\alpha$ systems at $z = 2.8443$ and $z = 1.7382$ in the spectrum of the quasar H-S1946+76 based on high resolution (FWHM=20 km/s), high S/N (40:1 to 80:1 per resolution) observations. Details of the observations are described in a companion paper (Tripp et al., this volume). We use a combination of the profile fitting technique and the apparent optical depth method to derive ion column densities and to assess the effects of line saturation. Only column densities that are considered reliable are used to derive metal abundances. Details of the analysis are presented in an article submitted to the *Astrophysical Journal* by the same authors. We summarize the main conclusions below:

   1. The derived Fe metallicity is ~ 2.5 dex below solar for the $z = 2.8443$ system. This metallicity is consistent with the values for other damped systems at similar redshifts. The exact N(H I) of the $z = 1.7382$ system is not known so we cannot derive the absolute metallicity for this system. The observed lower limits on the Fe/Zn and Cr/Zn ratios indicate that dust depletion effects are very small in the $z = 1.7382$ system.

   2. In both damped systems we find an apparent overabundance of Si relative to Fe compared to their solar ratio by about a factor of 2. We also find $Si/Al=3.0\times(Si/Al)_\odot$ in the $z = 2.8443$ system, and $Mn/Fe=0.32\times(Mn/Fe)_\odot$ in the $z = 1.7382$ system. Photoionization calculations suggest ionization is not likely to produce the observed relative abundance patterns if the intrinsic relative elemental abundances of the gas are solar.

   3. While the overabundances of Si relative to Fe (in both systems) and to Al (in the $z = 2.8443$ system) could be interpreted as due to dust depletion effects, the underabundance of Mn relative to Fe (in the $z = 1.7382$ system) is inconsistent with this interpretation.

   4. The observed abundance ratios among Si, Al, Fe, and Mn in the two damped systems are likely to be intrinsic to the absorbing gas. Similar abundance ratios have been seen in metal-poor Galactic stars, and the ratios are consistent with theories of nucleosynthesis in the early stage of chemical enrichment in galaxies (Wheeler et al. 1989). If this interpretation is correct, it may imply that active star formation in the absorbing galaxies has not proceeded for much longer than the characteristic time scale for producing Type Ia supernova (a couple of billion years) which would bring the Si/Fe ratio close to the solar value. However, in order for this explanation to hold, dust must not exist in enough quantity in these two absorbing galaxies to cause significant Fe and Al depletion. Alternatively, the effects of the dust on heavy element abundances in these damped systems may be different from that found for the Milky Way.

References

Bergeron, J., & Petitjean, P. 1991, A&A, 241, 365

Meyer, D.M., & Roth, K.C. 1990, ApJ, 363, 57

Meyer, D.M., Welty, D.E., & York, D.G. 1989, ApJ, 343, L37

Meyer, D.M., & York, D.G. 1992, ApJ, 399, L121

Pettini, M., Boksenberg, A., & Hunstead R.W. 1990, ApJ, 348, 48

Pettini, M., Smith, L.J., Hunstead, R.W., & King, D.L. 1994, ApJ, 426, 79

Rauch, M. et al. 1990, MNRAS, 242, 698

Savaglio, S., D'Odorico, S., & Moller, P. 1994, A&A, 281, 331

Tripp, T.M., Lu, L., & Savage, B.D. 1994, this volume

Wheeler, J.C., Sneden, C., & Truran, J.W.Jr. 1989, ARA&A, 27, 279

Wolfe, A.M. et al. 1994, ApJ, 435, L101

Wolfe, A.M., Turnshek, D.A., Smith, H.E., & Cohen, R.D. 1986, ApJS, 61, 249

As Metal Poor as It Could Be: the Damped System at $z=3.390$ Towards QSO 0000-2619

G. Vladilo[1], S. D' Odorico[2], P. Molaro[1], and S. Savaglio[3]

[1] Osservatorio Astronomico di Trieste
[2] European Southern Observatory
[3] Dipartimento di Fisica, Universitá della Calabria

Abstract. We present an abundance analysis of the damped Lyα system seen in the spectrum of the QSO 0000-2619 at $z_{abs}=3.390$. From the analysis of high resolution spectra (FWHM $\simeq$ 11 and 14 km s^{-1}) we identify lines of N I, O I, Al II, Si II and Fe II. The gas phase abundances are [Fe/H]$=-2.3$, [Si/H]$=-2.7$, [N/H]$=-2.7$, [O/H]$=-3.1$, and [Al/H]$=-2.8$ relative to the solar value, i.e. at the lowest level found so far in DLAs. The relative abundances show an enhancement of Fe compared to α elements, which is the opposite of what observed in halo stars. Such a behaviour is expected in chemical evolution models when a burst of star formation is followed by a quiescent phase.

The Damped System at $z=3.390$

The observations of QSO 0000-2619 were obtained with the ESO NTT and the EMMI spectrograph at a resolution of 11-14 km s^{-1} (see also Savaglio et al. 1994). Fig. 1 shows some absorption lines of the DLA system at z_{abs} = 3.390. The species shown in the figure are the most common ionization states of N, O, Al, Si and Fe in Galactic H I regions and therefore ionization corrections are expected to be negligible. For these low ionization species the bulk of the absorption is at $z_{abs}=3.3901$, with evidence of two components at 3.3896 and 3.3905.

The abundances were estimated by adopting the value log N(H I) = 21.40 $\pm$ 0.1 dex for the total H column density of the DLA system (Savaglio et al. 1994). The metallicities relative to solar abundances are [Fe/H]$=-2.3$, [Si/H]$=-2.7$, [N/H]$=-2.7$, [Al/H]$=-2.8$, and [O/H]$=-3.1$. From comparison between typical dust-to-gas ratios in DLA systems (Pettini et al. 1994) and in our Galaxy we estimate that gas-phase abundances of Al, Si and Fe might be at most 1.0 dex below real abundances, while O and N should be unaffected.

The metallicity of the system at $z=3.390$ is among the lowest found so far in DLA systems. It is lower than that found in the survey of Pettini et al. (1994) and comparable with the value estimated by Fan and Tytler (1994) for the DLA at $z_{abs}=2.844$ in HS 1946+7658. As far as the element-to-element abundances are concerned, we find [N/Fe]$\simeq -0.4$ and [Al/Fe] $\simeq-0.5$, similar to Pop II stars (Wheeler et al. 1989). The surprising results in our system are the ratios [Si/Fe]$\simeq-0.4$ and [O/Fe]$\simeq-0.8$, which are opposite to what

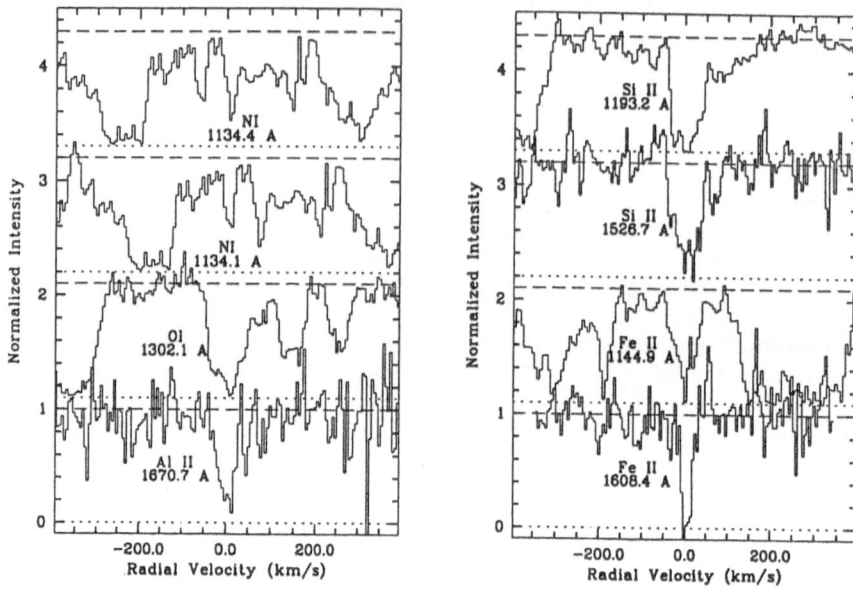

Fig. 1. Portions of the QSO 0000-2619 spectrum including N I, O I, Al II, Si II and Fe II absorptions at $z_{abs} = 3.390$.

is observed in Galactic halo stars where α elements are enhanced. Overabundances of Fe relative to the α elements may be a signature of an early starburst followed by a quiescent phase (Matteucci & Padovani 1993). The passive evolution following the burst allows SN I to contribute significantly to the elemental production. Starburst models are very efficient in producing high metallicities and, for a Salpeter's IMF, solar abundances may be reached in a few 10^8 years after the burst. Since we observe gas with metallicity 2 or 3 orders of magnitude lower than solar, a strong dilution of enriched gas with merging of unprocessed material is required. Alternatively, we might be intercepting an external region of a forming galaxy with primordial material already polluted by a nuclear wind.

References

Fan X.-M., Tytler D., 1994, ApJ Suppl. 94, 17
Matteucci F., Padovani P., 1993, ApJ 419, 485
Pettini M., Smith L.J., Hunstead R.W., King D.L., 1994, ApJ 426, 79
Savaglio S., D'Odorico S., Møller P., 1994, A&A 281, 331
Wheeler J. C., Sneden C., Truran J.W., 1989, ARA&A 27, 279

PART IV

ABSORPTION LINES

FROM GAS

IN OUR GALAXY

Lessons Learned from UV Absorption Lines at $z=0$

Edward B. Jenkins

Princeton University Observatory, Princeton NJ 08544, USA

Abstract. Insights from UV absorption lines that arise from gas in our own galaxy help us to interpret the composition of very distant gas systems that are seen in QSO spectra. For instance, what we know about the depletions of free atoms as they condense onto dust grains nearby can be applied elsewhere, so that we can differentiate between low intrinsic heavy element abundances and the effects of such depletions. Experience from recent, very sensitive observations indicate that some elements created by the r- and s-process neutron capture nucleosynthesis may show absorption features in other systems at about the 1 mÅ level. The abundance of chlorine in neutral form is strongly enhanced by the presence of H_2. Thus a detectable Cl I absorption at 1347 Å should persuade us to search for and study the Lyman and Werner bands of this important molecule. Finally, in the course of vigorous research by many investigators in the field of UV absorption line spectroscopy, we have learned some important lessons on methodology and how to lessen the impact of errors.

1 Introduction

The damped Lyα system at a redshift of approximately 0.0000 has probably been the most intensively studied of all UV absorption line systems. It is the oldest place in the universe that we can observe, it is well evolved chemically, and it's the galactic system that we know best since it serves as our home. It is for these reasons that the $z=0$ metal-line system is a good standard against which to measure the properties of very distant gas systems about which we know very much less. The lessons we can learn from UV absorption lines in our galaxy's interstellar medium are of two types: (1) specific facts about the nature of the matter in space – i.e., its typical composition and physical state and (2) special tricks and pitfalls in the analysis of UV absorption lines. This presentation will concentrate on a few topics that are relevant to contemporary studies of very distant systems.

Much of this discussion will concentrate on what can be learned from a "sweet spot" in the rest wavelength domain that extends between Lyα and about the mid-2000 Å region, above the mess created by the Lyα forest but not so far toward the red that lines from very distant systems are shifted to the remote infrared where observations are more difficult. However there is information conveyed by some lines shortward of Lyα that is so compelling that it's worth the trouble to try and sort things out from the forest (or go to very low z where the forest is not much of a problem).

2 Abundances of Atoms

2.1 General Principles

Much progress has been made in the investigations of atomic abundances in distant systems, as shown by other papers in this volume. Insights from the local ISM are crucial in interpreting the significance of what is seen elsewhere.

The most comprehensively studied features for deriving metal abundances in distant systems, it seems, are the lines of singly-ionized Zn and Cr (Pettini et al. 1990, 1994; Wolfe et al. 1994). These two elements usually have lines that are not badly saturated and at a small wavelength separation from each other. We will see evidence later in this paper that interstellar Zn in our galaxy is not much below its expected abundance and thus has little tendency to condense into solid form. Cr, while not as extensively studied as Zn, seems to exhibit strong depletions, similar to those of Fe (de Boer et al. 1986). Thus, we have a nearly ideal pair for studying the simplest aspects of metal abundances: [Zn/H] may be adopted as a measure of overall metallicity, while [Cr/Zn] tells us how efficient the grain formation process is.

Beyond measures of just the overall level of metallicity, it is important to study various heavy elements to see if their relative abundances mimic the trends seen in old stars in our galaxy, whose ages are traditionally depicted by determinations of [Fe/H] (Wolfe et al. 1994; R. Green, these proceedings). That is, does the pattern of abundances of metal-poor gas systems at high z nearly duplicate that of metal-poor stars in our galaxy? Or is there evidence that the mix of stellar mass ranges, with their different lifetimes and favored modes of element production, differs from that of our galaxy? Other elements that can be studied with relative ease are Si, P, Ni, Fe, and Mn (Meyer et al. 1989; Meyer & Roth 1990; Meyer & York 1992). They have good lines of moderate strength from the singly ionized stage characteristically found in H I regions. The more abundant elements such as C, N, O and often S are more difficult to analyze with much precision: one must either observe very weak transitions or confine the research to gas systems with extraordinarily low metallicity. There are some abundant elements, notably He and Ne, that have no lines from their ground states at wavelengths longward of the Lyman Limit.

The most straightforward task is to study velocity components that are associated with regions that contain fully neutral H. Here, unless the spectrum of the ambient photoionizing radiation is very hard, we can be confident that, at least locally, the dominant ionization stage is the lowest one that has an ionization potential above that of H. For regions that are not optically thick at the Lyman limit, one must have a more detailed understanding of the rates and general configuration of gas subject to collisional ionization (Giroux et al. 1994) and photoionization (Bergeron & Stasińska 1986; Bergeron et al. 1994), and then create a model for the relative ionizations of different species.

For research on the mostly neutral gas, there are some important lessons to be learned from studies of H I regions in our galaxy. First, there is a danger in assuming that *all* of a certain ion that we expect to be most abundant in H I gas does not have an additional contribution from fully ionized gas at a nearly coincident velocity. When we study UV lines in the spectra of stars in our galaxy, every target star has its own H II region, so this is always a potential problem [e.g., λ Sco studied by York (1983)]. For systems with large redshifts, this particular predicament goes away, but nevertheless one probably sees a mixture of an ionized "halo" on top of a contribution from a "disk-like system." To monitor the severity of the possible contribution from photoionized gas, one can observe the strong 1085 Å multiplet of N II, whose absorption features can come only from an H II region. Unfortunately, these features are buried in the Lyα forest. Another possibility, this one in the "sweet spot," is a doublet of Al III near 1860 Å that usually displays lines that are not badly saturated. Singly ionized Al can be ionized to the doubly ionized form by photons with energies in excess of 19 eV, and the ratio of photoionization cross section (Reilman & Manson 1979) to the recombination coefficient (Aldrovandi & Péquignot 1974) is large enough to favor the production of Al III at reasonable temperatures and electron densities.

To do detailed abundance comparisons, we must have a clear understanding of the depletion of atoms as they become a part of the interstellar dust grains. Two tasks are at hand: [1] to comprehend how rapidly one element depletes relative to another and [2] to understand how overall levels of depletion change in our own environment.

2.2 ζ Oph

There seems to be an irresistible tendency to use abundance determinations for the gas in front of ζ Oph as a universal standard for comparison. The reason for this is that the most thorough research on interstellar constituents has been done for the line of sight to this star. Herbig (1968) wrote a comprehensive review of the conditions toward ζ Oph even before we had an opportunity to study this star in the ultraviolet. Over the last quarter century of space observations and more advanced ground-based work, there have been approximately 20 papers per year dealing with interstellar topics that seem to feature this star. The chief reason for the popularity of ζ Oph is that it shows some interesting extremes: its spectrum shows extraordinarily strong molecular features, even in the visible. Of all 109 stars surveyed for H_2 with the *Copernicus* satellite, ζ Oph had the highest fraction of H in molecular form (60%)(Savage et al. 1977), and the depletions in one of the velocity components are more extreme than usual. If ever one wants to look for some new or unusual absorption features, ζ Oph is the first place to go. Thus, it is unfair to say that other galaxies have less extreme depletions than those of our galaxy when the exaggerated conditions in the main cloud in front of ζ Oph are used as a standard for comparison (see *'s in Fig. 1). In fact, depletions in our part of our galaxy vary by large amounts, as we shall discover in the next section.

2.3   Variations in Depletion

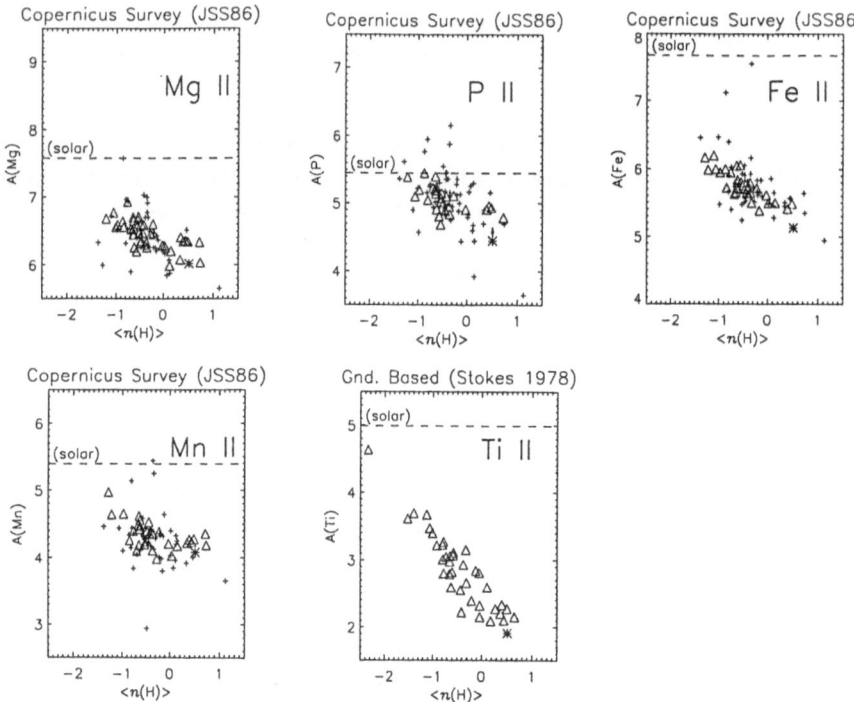

Fig. 1. Logarithmic plots of the gas phase abundances of Mg II, P II, Fe II, Mn II and Ti II relative to that of hydrogen (on a scale normalized to H = 12.0), shown as a function of the average density of hydrogen (atomic plus molecular) along the line of sight (Jenkins et al. 1986). Triangles show determinations where the difference between the top error bar and the bottom one is 0.5 dex or less. Other, less accurate measurements are indicated by small pluses. The asterisk in each panel shows the value for the -15 km s^{-1} component toward ζ Oph (Savage et al. 1992). All of the abundance measurements of Mg II have been adjusted downward by -0.67 dex, in recognition of an improved f value for the 1240 Å multiplet reported by Sofia, et al. (1994). Similarly, the results for Fe II were changed by -0.10 dex to reflect the f value of Shull, Van Steenberg & Seab (1983) for the 1134 Å transition, which was the weakest one used and probably the most influential in determining each column density.

We have discovered that element depletions are strongly correlated with what fraction of the gas is within moderately dense regions, as opposed to that in an intercloud medium of much lower density (Spitzer 1985; Jenkins et al. 1986). A convenient (but approximate) measure of this fraction is the average density of hydrogen gas along the line of sight, $\langle n(\mathrm{H}) \rangle$, equal to the column densities $N(\mathrm{H}) + 2N(\mathrm{H}_2)$ divided by the distance to the star. Figure 1 shows the trend for relative abundances for 4 elements surveyed Jenkins, et al. (1986) using the *Copernicus* satellite. Also shown is behavior of titanium, the only element whose most abundant stage of ionization (Ti II) can be studied from the ground (Wallerstein & Goldsmith 1974; Stokes 1978; Albert 1983).

For each element there is a clear trend in the magnitude of the depletion from the gas phase as a function of the average density. Either the denser regions provide an opportunity for a more complete condensation of atoms onto grains, or they provide an environment where the grains are better protected from disruption by high speed shocks.

What about Zn, an element that is a strong favorite in studies being reported by observers of distant metal-line systems? Unfortunately, lines from this element were in a part of the spectrum that was not covered with much sensitivity by *Copernicus*. Hence we must rely on less accurate data from the IUE satellite (York & Jura 1982; Harris et al. 1983; Harris & Mas Hesse 1986).[1]Van Steenberg & Shull (1988a) performed an analysis of interstellar lines that were recorded in the spectra 261 stars observed with IUE. A strongly filtered subset[2] of their results for Zn, along with a few other elements, is presented in Fig. 2.

Of all the cases shown in Fig. 2, Zn unfortunately seems to show the largest random scatter. There are even some points that indicate that Zn is more abundant than its solar abundance ratio. On average, this element seems to show practically no depletion for low density lines of sight, and possibly there is a mild depletion at high densities.

---

[1] Much newer determinations of the abundances of Zn and Cr from the IUE and HST archives have just been reported in abstract form (Macke et al. 1994; Roth & Blades 1994), and full papers may appear soon.

[2] To reduce the chances for error, Van Steenberg & Shull (1988b) restricted their general statistical conclusions to cases where (1) The target star had a spectral type hotter than B2.5, so that $N(\mathrm{H\,I})$ could be determined from the Lyα profile without interference from stellar Lyα absorption, (2) The weakest line of an element had a central optical depth $\tau_0 < 10$ and (3) the weak lines of S II and Zn II did not have significantly larger W_λ's than the stronger ones, indicating contamination from other features. To be extra conservative, the data shown in *this paper* (Fig. 2) were made to pass an additional test: there was a requirement that the weakest line that had a measured W_λ equal to at least 2 times the uncertainty in W_λ must have had $\tau_0 < 2.0$, assuming the velocity dispersion b and the element's column density are correct. When observations were weeded out according to this last criterion, there was no perceptible change in the general trends (i.e., no bias seemed to be introduced), but there was a reduction in the scatter of the points.

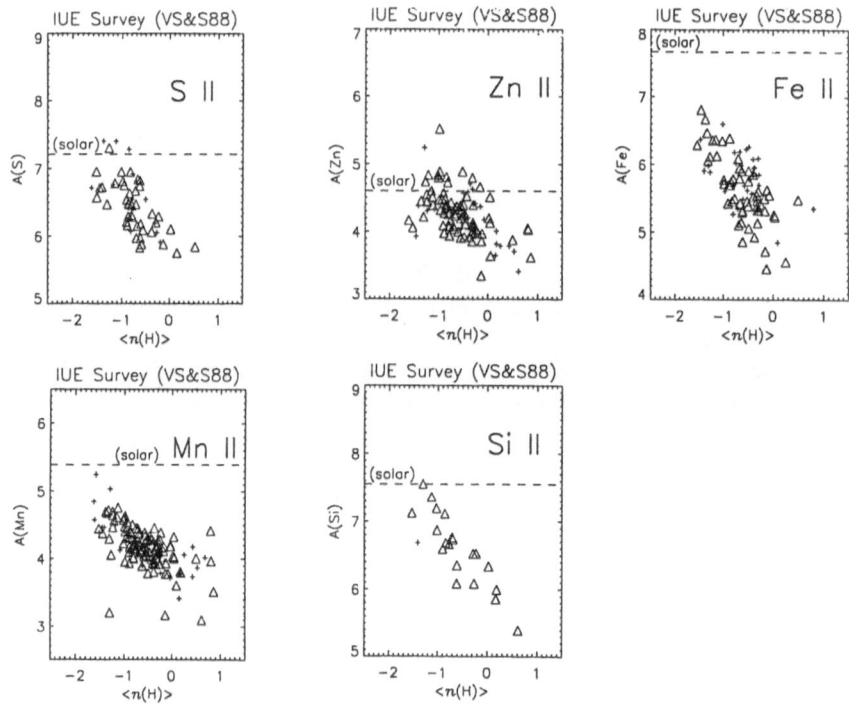

Fig. 2. Gas phase abundances of S II, Zn II, Fe II, Mn II and Si II, using column densities from the IUE survey by Van Steenberg & Shull (1988a) after screening out cases that might be unreliable (see footnote 3 on the previous page). The style of presentation is identical to that of Fig. 1. As with the *Copernicus* results, some revisions in column densities were made in the figure to compensate for changes in f values: Zn was adjusted downward by −0.10 dex, in recognition of improved f values for the 2025 and 2062 Å lines (Morton 1991; Bergeson & Lawler 1993b). Likewise for Mn II: adjustment = −0.085 dex for new f values of the 2576−2606 Å multiplet listed in Morton (1991) and Si II: adjustment = +0.294 dex for an improved f value for the 1808 Å transition (Dufton et al. 1992; Bergeson & Lawler 1993a), which is probably the most influential one because it's the weakest of the three lines used.

2.4   Pitfalls in Interpreting Poorer Quality Data

It is troubling to note that the trends for Fe II and Mn II in Fig. 2 do not agree well with the *Copernicus* data shown earlier [The IUE data seem to show steeper relationships with $\langle n(\mathrm{H})\rangle$]. Also, from HST observations of the $-15\,\mathrm{km\,s^{-1}}$ component of gas in front of ζ Oph [at $\langle n(\mathrm{H})\rangle = 0.51\ \mathrm{cm^{-3}}$] that shows strong depletions for other elements, the abundance of sulfur is approximately solar (Federman et al. 1993). This is in stark contrast with more than a factor of 10 depletion for high density indicated in the IUE dataset. In determining the S abundance to ζ Oph, Federman, et al. (1993) used weak S I lines and then solved for the ionization equilibrium between

S I and S II with conditions indicated by the observed ratio of N(C I) to N(C II) (the ionization potential of S I is similar to that of C I).

The disparities between the IUE results and other, presumably more accurate determinations at large $\langle n(\mathrm{H}) \rangle$ indicate that we should use great caution in accepting abundances from spectra that were recorded at a velocity resolution of only 25 km s^{-1} and a typical signal-to-noise ratio of 15. This is an important lesson that should apply to data of similar quality for extragalactic gas systems.

There are several possible reasons that special precautions in the original analysis and the additional screening done here do not suppress large errors in certain cases. For instance, there is still a danger that a line may *seem* unsaturated when indeed it is not. The filtering of apparently saturated cases may just reject cases where we *think* there is an unacceptably high saturation. Also, Joseph (1989) has pointed out that there is a bias, one that takes some special awareness to avoid, toward measuring too large an equivalent width for strong lines when noise is present. This makes b seem larger, and then this lowers the inferred column density.

Because of limitations in the variety of lines that could be observed, Van Steenberg & Shull had to use curves of growth from some elements to determine the velocity dispersions for others. Often, Fe II had a large range in line strengths and thus this ion (and sometimes Si II) was used as a determinant for the line of sight's b parameter. The use of a highly depleted element to determine the b parameter for a much less depleted one is a hazardous procedure. Van Steenberg & Shull recognized this shortcoming and issued a caution on this point. The problem arises from an effect first discussed by Routly & Spitzer (1952), who found that the ratio of Ca II to Na I generally increased with the displacement of a component's radial velocity from the local standard of rest. This conclusion has been substantiated in more detail by Siluk & Silk (1974) and Vallerga et al. (1993). The effect is also seen for other elements that are usually depleted in the interstellar medium, such as Si II, Fe II (Shull & York 1977) and Ti II (Albert 1983), and it probably results from the destruction of grains when the gas was accelerated in a shock (Spitzer 1976; McKee et al. 1987; Jones et al. 1994; Tielens et al. 1994). It follows that one can imagine that Fe (or Si, or Mn) will be more strongly emphasized in outlying, extra velocity components than for an element that is usually less depleted, such as Zn (or S). As a result, the composite velocity dispersion b for all of the Fe components taken together will be greater than that which applies to the more restricted set of components of Zn. If one then uses the value of b determined from Fe II to interpret a saturated feature of Zn, the resulting abundance for Zn will be too high. This is another principle that must be remembered when analyzing abundances in systems at large redshifts.

2.5   Heavy Elements Created by Neutron Capture Nucleosynthesis

Do we have any chance of seeing elements beyond the Fe peak? If so, we can assess the relative importance of r- and s- process element formation in a distant system and learn something new about the history and character of star formation. Echelle spectrographs on future 8 to 10 meter class telescopes should have the needed sensitivity. Already, we have evidence that one such telescope, the Keck Telescope, can achieve a detection limit well below $1\,\text{mÅ}$ in a reasonable integration time for an absorption system at high redshift (Songaila et al. 1994).

To obtain some idea of the prospects, we must now rely heavily on results for ζ Oph. The value of $\log N(\text{H}_{\text{total}}) = 21.15$ toward this star (Savage et al. 1977) is about comparable to the high end of the damped Lyα systems. The simple equivalent widths will give a measure of what we might see in a distant system, after we multiply them by a factor of about $1/10$ or less to account for the overall lower metallicity of the typical high-z systems. For the more refractory elements, this guide from ζ Oph will be a *conservative* estimate, since depletions elsewhere will probably be less severe.

Table 1. r- and s-Process Elements that could be Sampled

Element	Trans. λ	W_λ for ζ Oph (mÅ)[1]	Percentage Contributions (%r/%s) Synthesis in our Galaxy[2]
Ga	1414	3.8	12 / 88
Ge	1237	9.6	46 / 54
As	1264	0.7	81 / 19
Kr	1236	3.4	61 / 39
Sn	1400	2.7	30 / 70

Notes:
[1] Refs: Ga, Ge, & Kr: Cardelli et al. 1991; As, Sn: Cardelli et al. 1993
    [see also a review by Cardelli (1994)]
[2] From From Käppler, et al. (1989)

The elements listed in Table 1 show a respectable spread in the relative fractions of r and s contributions, although the weak feature of As will probably not be detectable outside our galaxy. If the (%r/%s)'s elsewhere are not too different from here, it is evident that Ge has the best chance of being detected.

3 Molecules

H_2 is the most abundant molecule in the universe, and it is the starting point for a vast network of reactions that involve other molecules (van Dishoeck & Black 1986). An exciting prospect for future investigations of damped $Ly\alpha$ systems is the detection and investigation of absorption by the Lyman and Werner bands of H_2 that show up at rest wavelengths $\lambda \lesssim 1125$ Å. By studying the rotational excitation of H_2, we can determine such important quantities as the molecule's formation rate, the density of radiation that optically pumps and destroys H_2, and the local density and temperature of the H_2-bearing material (Spitzer & Zweibel 1974; Jura 1975a, b; Shull & Beckwith 1982).

While radio emission from CO has been detected at the redshift of an absorption system (Frayer et al. 1994), we have not had much widespread success in detecting the absorption bands of H_2 in the spectra of quasars. A probable detection has been accomplished for only one high-z system, (Foltz et al. 1988), while 4 other systems, 3 of which have $N(\text{H I}) \approx 2 \times 10^{21}\,\text{cm}^{-2}$, have been reported to have strong upper limits for the column density of H_2 or CO [summarized by Levshakov et al. (1992)]. An apparently low probability for intercepting the molecular gas plus the difficulty of working with lines submerged in the $Ly\alpha$ forest are probably the chief reasons for the lack of great progress.

It is useful to draw upon our experience with observations of local gas to see if indicators of gas that is rich in molecules exist at wavelengths longer than $Ly\alpha$. The CO molecule has a band system (the A−X 4th positive system) that has its strongest vibrational members in the wavelength range $1350-1550$ Å. These features are generally not very strong (Morton & Noreau 1994) for typical ratios of CO to H_2 in diffuse interstellar clouds (Federman et al. 1980). Another, perhaps more promising alternative is to find absorption by neutral chlorine. Jura (1974) pointed out that the reactions

$$\text{Cl}^+ + \text{H}_2 \rightarrow \text{H Cl}^+ + \text{H} \quad (R = 10^{-9}\,\text{cm}^3\text{s}^{-1}) \tag{1}$$

followed by

$$\text{H Cl}^+ + \text{e}^- \rightarrow \text{H} + \text{Cl} \quad (R = 10^{-7}\,\text{cm}^3\text{s}^{-1}) \tag{2}$$

are a very efficient way to bring ionized chlorine to a neutral state in the interstellar medium [much more efficient than straight recombination of Cl^+ with free electrons, where $R = 10^{-11}\text{cm}^3\text{s}^{-1}$ at $T \lesssim 100$K (Aldrovandi & Péquignot 1974)]. The effectiveness of this reaction in making Cl I an indicator of H_2 is demonstrated in Fig. 3, taken from the survey of interstellar lines by Bohlin, et al. (1983). In any initial search for molecules in a quasar spectrum, it may be easier to detect the neutral chlorine line at 1347 Å than to try and sort out the H_2 features hidden in the $Ly\alpha$ forest (and for some z's, situated below the transmission cutoff of our atmosphere).

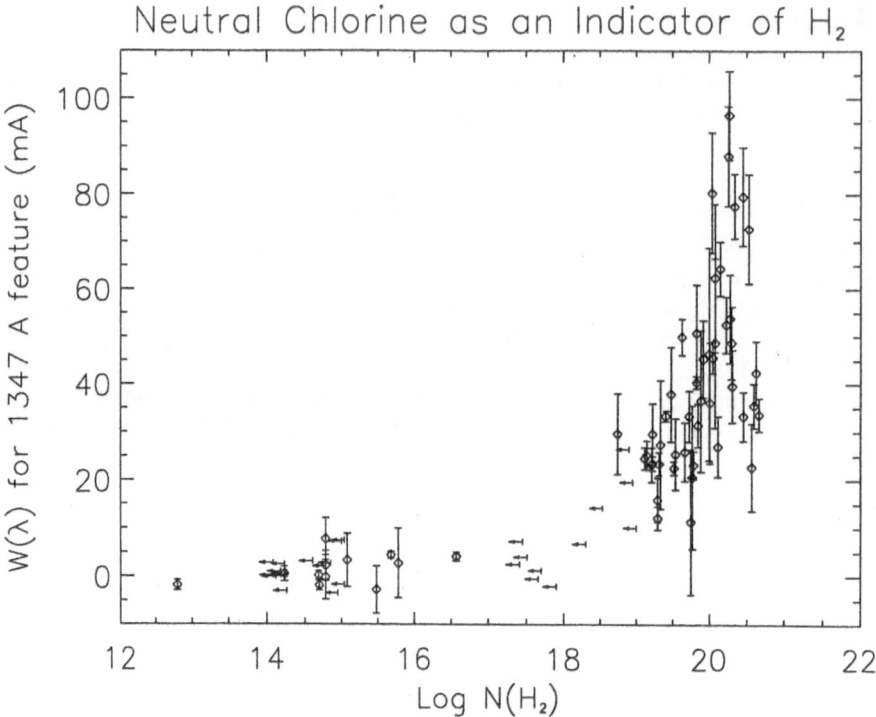

Fig. 3. Equivalent widths of absorption by the Cl I line at 1347 Å *vs* the logarithm of the column density of molecular hydrogen.

4   Concluding Remarks

We have touched upon some selected topics that can be applied to the interpretation of absorption lines that appear in the spectra of quasars. Some of these considerations are covered in more depth in reviews by Spitzer & Jenkins (1975), Cowie & Songaila (1986), and Jenkins (1987). By virtue of its high velocity resolution (3.5 km s^{-1}) and excellent photometric precision, the echelle spectrograph on the GHRS instrument aboard HST is bringing forth much more precise information on specific lines of sight (Savage et al. 1991, 1992; Sofia et al. 1993; Spitzer & Fitzpatrick 1993; Fitzpatrick & Spitzer 1994), but the coverage of many different lines of sight approaching that of the earlier surveys done with IUE and *Copernicus* has yet to be realized.

Acknowledgements. The preparation of this paper was supported by NASA grant NAG5-616. The author is grateful to M. Shull for his useful comments on an early draft of the paper.

References

Albert C.E. (1983): ApJ 272, 509

Aldrovandi S.M.V., Péquignot D. (1974): Rev. Brasil. de Fisica 4, 491

Bergeron J., Stasińska G. (1986): A&A 169, 1

Bergeron J., Petitjean P., Sargent W.L.W., Bahcall J.N., Boksenberg A., Hartig G.F., Jannuzi B.T., Kirhakos S., Savage B.D., Schneider D.P., Turnshek D.A., Weymann R.J., Wolfe A.M. (1994): ApJ 436, 33

Bergeson S.D., Lawler J.E. (1993a): ApJ 414, L137

— (1993b): ApJ 408, 382

Bohlin R.C., Hill J.K., Jenkins E.B., Savage B.D., Snow T.P., Spitzer L., York D.G. (1983): ApJS 51, 277

Cardelli J.A. (1994): Sci 265, 209

Cardelli J.A., Savage B.D., Ebbets D.C. (1991): ApJ 383, L23

Cardelli J.A., Federman S.R., Lambert D.L., Theodosiou C.E. (1993): ApJ 416, L41

Cowie L.L., Songaila A. (1986): ARA&A 24, 499

de Boer K.S., Lenhart H., Van Der Hucht K.A., Kamperman T.M., Kondo Y., Bruhweiler F.C. (1986): A&A 157, 119

Dufton P.L., Keenan F.P., Hibbert A., Ojha P.C., Stafford R.P. (1992): ApJ 387, 414

Federman S.R., Glassgold A.E., Jenkins E.B., Shaya E.J. (1980): ApJ 242, 545

Federman S.R., Sheffer Y., Lambert D.L., Gilliland R.L. (1993): ApJ 413, L51

Fitzpatrick E.L., Spitzer L. (1994): ApJ 427, 232

Foltz C.B., Chaffee F.H., Black J.H. (1988): ApJ 324, 267

Frayer D.T., Brown R.L., Vanden Bout P.A. (1994): Ap. J. (Letters) 433, L5

Giroux M.L., Sutherland R.S., Shull J.M. (1994): ApJ 435, L97

Harris A.W., Mas Hesse J.M. (1986): M.N.R.A.S. 220, 271

Harris A.W., Bromage G.E., Blades J.C. (1983): M.N.R.A.S. 203, 1225

Herbig G.H. (1968): Z. f Ap. 68, 243

Jenkins E.B. (1987): Element Abundances in the Interstellar Atomic Material, in Interstellar Processes D.J. Hollenbach, and H. A. Thronson, eds, (Dordrecht, Reidel), pp. 533-559

Jenkins E.B., Savage B.D., Spitzer L. Jr. (1986): ApJ 301, 355

Jones A.P., Tielens A.G.G.M., Hollenbach D.J., McKee C.F. (1994): ApJ 433, 797

Joseph C.L. (1989): PASP 101, 623

Jura M. (1974): ApJ 190, L33

— (1975a): ApJ 197, 581

— (1975b): ApJ 197, 575

Käppeler F., Beer H., Wisshak K. (1989): Reports on Progress in Physics 52, 945

Levshakov S.A., Chaffee F.H., Foltz C.B., Black J.H. (1992): A&A 262, 385

Macke R.J., Sembach K.R., Steidel C.C. (1994): BAAS 26, 1326

McKee C.F., Hollenbach D.J., Seab C.G., Tielens A.G.G.M. (1987): ApJ 318, 674

Meyer D.M., Roth K.C. (1990): ApJ 363, 57

Meyer D.M., York D.G. (1992): ApJ 399, L121

Meyer D.M., Welty D.E., York D.G. (1989): ApJ 343, L37

Morton D.C. (1991): ApJS 77, 119

Morton D.C., Noreau L. (1994): ApJS 95, 301

Pettini M., Boksenberg A., Hunstead R.W. (1990): ApJ 348, 48

Pettini M., Smith L.J., Hunstead R.W., King D.L. (1994): ApJ 426, 79

Reilman R.F., Manson S.T. (1979): ApJS 40, 815

Roth K.C., Blades J.C. (1994): BAAS 26, 1326

Routly P.M., Spitzer L. Jr. (1952): ApJ 115, 227

Savage B.D., Bohlin R.C., Drake J.F., Budich W. (1977): ApJ 216, 291

Savage B.D., Cardelli J.A., Bruhweiler F.C., Smith A.M., Ebbets D.C., Sembach K.R. (1991): ApJ 377, L53

Savage B.D., Cardelli J.A., Sofia U.J. (1992): ApJ 401, 706

Shull J.M., Beckwith S. (1982): ARA&A 20, 163

Shull J.M., York D.G. (1977): ApJ 211, 803

Shull J.M., Van Steenberg M., Seab C.G. (1983): ApJ 271, 408

Siluk R.S., Silk J. (1974): ApJ 192, 51

Sofia U.J., Savage B.D., Cardelli J.A. (1993): ApJ 413, 251

Sofia U.J., Cardelli J.A., Savage B.D. (1994): ApJ 430, 650

Songaila A., Cowie L.L., Hogan C.J., Rugers M. (1994): Nat 368, 599

Spitzer L. (1976): Comments in Astrophysics 6, 177

Spitzer L., Fitzpatrick E.L. (1993): ApJ 409, 299

Spitzer L., Jenkins E.B. (1975): ARA&A 13, 133

Spitzer L., Zweibel E.G. (1974): ApJ 191, L127

Spitzer L. (1985): ApJ 290, L21

Stokes G.M. (1978): ApJS 36, 115

Tielens A.G.G.M., McKee C.F., Seab C.G., Hollenbach D.J. (1994): ApJ 431, 321

Vallerga J.V., Vedder P.W., Craig N., Welsh B.Y. (1993): ApJ 411, 729

van Dishoeck E.F., Black J.H. (1986): ApJS 62, 109

Van Steenberg M.E., Shull J.M. (1988a): ApJS 67, 225

Van Steenberg M.E., Shull J.M. (1988b): ApJ 330, 942

Wallerstein G., Goldsmith D. (1974): ApJ 187, 237

Wolfe A.M., Fan X.-M., Tytler D., Vogt S.S., Keane M.J., Lanzetta K.M. (1994): ApJ 435, L101

York D.G. (1983): ApJ 264, 172

York D.G., Jura M. (1982): ApJ 254, 88

Probing the Outer Galactic Halo with the GHRS

Blair D. Savage[1], Limin Lu[2], and Kenneth R. Sembach[3]

[1] Dept. of Astronomy, University of Wisconsin, Madison, WI 53706-1582, USA
[2] Dept. of Astronomy, Caltech, Pasadena, CA 91125, USA
[3] Center for Space Research, MIT, Cambridge, MA 02139, USA

Abstract. The Goddard High Resolution Spectrograph (GHRS) on the Hubble Space Telescope (HST) has been used to obtain intermediate resolution (15 to 20 km/s) spectra of selected Milky Way absorption lines through the entire halo toward bright AGNs and QSOs. The objects so far observed include NGC 3783 ($l = 287$, $b = 23$), H 1821+643 ($l = 94$, $b = 27$), Fairall 9 ($l = 295$, $b = -58$), and Mrk 509 ($l = 36$, $b = -30$). This short contribution summarizes the key results.

1 Introduction

In order to use quasar absorption lines to study the evolution of the gaseous matter in the Universe, it is important to have information about the abundances, distribution, and physical processes affecting gas in the halos of galaxies at the current epoch. The gaseous halo of the Milky Way provides an excellent local laboratory for studying the physics of gaseous halos. The current state of understanding of the properties of Milky Way halo gas and the theory of gas in the halo are summarized in several recent reviews (McKee 1993; Savage 1988, 1990, 1995; Spitzer 1990).

Over the last several years we have been pursuing a program to study gas in the outer Galactic halo by obtaining spectra with the Goddard High Resolution Spectrograph (GHRS) of Milky Way absorption lines through the entire halo toward bright AGNs and quasars. The observations were obtained with the GHRS intermediate resolution G160M grating operating in the wavelength regions from 1233 to 1268 Å and from 1521 to 1554 Å. The Milky Way lines seen in these spectral regions include absorption from N V 1238, 1242, Mg II 1239, 1240, S II 1250, 1253, 1259, Si II 1260, 1526, and C IV 1548, 1550. The observations were obtained using the large (2"x2") science aperture. Those obtained after the installation of COSTAR have a resolution of approximately 15 to 20 km/s (FWHM). Those obtained before COSTAR in addition have spectral spread functions with broad wings. The integration times for the observations range from about 1 to 3 hours and the signal to noise in the resulting spectra range from 8 to 20 per diode.

In the following sections we briefly discuss results from our studies relating to the scale height of Galactic C IV, the general prevalence of N V in Milky Way disk and halo, the detection of C IV in the warped outer portions of the

Milky Way, heavy element abundances in Galactic H I high velocity clouds (H I-HVCs) , and the discovery of highly ionized Milky Way C IV-HVCs.

2   The Scale Height of Galactic C IV

Savage et al. (1993a) have used the N|sin$b$| versus |z| technique to estimate the distribution of C IV away from the Galactic plane. In using this technique the column density of C IV perpendicular to the Galactic plane, N(C IV)|sin$b$|, is plotted against the distance away from the Galactic plane, |z|, to each object and a simple model for the distribution is fitted to the data. By combining measurements from the IUE of halo stars from Sembach & Savage (1992) with HST Faint Object Spectrograph data from the quasar absorption line key project, Savage et al. (1993a) estimate an exponential scale height for C IV of $4.9^{+1.8}_{-1.3}$ kpc.

An alternate method for estimating the scale height of interstellar species is to model the asymmetric absorption line profiles toward extragalactic objects which are situated in directions significantly affected by differential Galactic rotation. Such an analysis for the sight line to the Seyfert galaxy NGC 3783 ($l = 287$, $b = +23$) by Lu et al. (1994a) yields a C IV exponential scale height of 3.5 kpc, while for the LMC star HD 36402 ($l = 278$, $b = -33$) a scale height of 4.4 kpc is obtained. In the modeling it is assumed that gas in the halo moves with the same rotational velocity as gas in the underlying Galactic disk. This assumption of co-rotation may be valid for gas in the low halo but will begin to break down for gas at large distances away from the Galactic plane. It is significant that the N|sin$b$| versus |z| technique and the rotational analysis technique for estimating the distribution of the gas away from the Galactic plane are yielding similar results. We note that the Keck Observatory spectra presented at this meeting by Arthur Wolfe suggest the absorption line profiles of metal lines in damped Lyman alpha systems may also be influenced by rotation effects in distant absorbing galaxies.

3   N V in Galactic Halo Gas

The presence of N V in Galactic disk and halo gas is significant because N V is difficult to produce by starlight photoionization since 77 eV is required to photoionize N IV and the likely sources of the ionizing radiation (hot stars) are expected to have strong He$^+$ absorption for E>54 eV. Therefore, among the ions accessible to the HST, N V provides the clearest indicator of hot cooling gas in the interstellar medium. Sembach & Savage (1992) used the IUE to obtain high quality ultraviolet absorption line data toward 12 Galactic stars situated at a variety of distances in the halo and found that N V was generally detected with a column density ratio, N(C IV)/N(N V)= 4.6 ± 2.7. The general presence of N V in Galactic halo gas has also been confirmed toward the extragalactic objects: 3C 273 (Savage et al. 1993b) , NGC 3783 (Lu et al. 1994a), H 1821+643 (Savage et al. 1995), and M 509 (Sembach et al. 1995). The observed values of N(C IV)/N(N V) range from 5.9 ± 1.6 for 3C 273 to 1.7 ± 0.6 for H 1821+643. These and other measurements imply the existence of several types of highly ionized gas in the Milky Way disk and halo (see Savage et al. 1994 and Savage & Sembach 1994).

4   C IV in the Warped Outer Milky Way

The bright quasar H 1821+643 lies in the interesting direction ($l = 94$, $b = +27$) which samples gas in the direction of the upper region of the warped outer Galaxy. Strong Galactic C IV absorption is seen toward this quasar in components centered near -8, -70, and -120 km/s (Savage et al. 1995). The component at -120 km/s which is also seen in 21 cm H I emission suggests absorption and emission from gas at a galactocentric distance of $\sim$25 kpc assuming the gas in the warped outer Galaxy co-rotates with the gas in the underlying disk. Strong C IV absorption at such large Galactocentric distances may be produced by photoionization from the EUV background radiation in the low density gas of the outer Milky Way.

5   Abundances in Galactic H I-HVCs

The Milky Way is surrounded by complexes of high velocity H I with $|v| >$100 km/s (see Wakker 1991). Estimates of the sky covering factor of this gas have doubled twice over the past 10 years. A recent sensitive survey by Murphy et al. (1995) implies a covering factor of $\sim$ 38% not including the gas in the outer Galactic disk or in the Galactic warp. The GHRS observations of NGC 3783 and Fairall 9 have provided important new information about the abundances of heavy elements in high velocity clouds.

Toward NGC 3783, Lu, et al. (1994a) infer a sulfur abundance relative to the Sun of S/H= 0.15 ± 0.05 solar for a HVC at 240 km/s with N(H I) =1.2×10^{20} cm^{-2}. Since S is not normally depleted, this result is probably not affected by the presence of dust in the HVC. This finding favors the idea that the absorption occurs in gas being stripped away by the Galaxy from an extragalactic object such as the Magellanic Clouds or some other member of the Local Group.

Lu et al. (1994b) have directly probed gas in the Magellanic Stream in the direction of Fairall 9. By combining the results for HVC absorption seen at 170 and 210 km/s they infer abundances relative to the Sun of S/H< 0.3 solar and Si/H> 0.2 solar. These abundance limits are consistent with Magellanic Cloud abundances and rule out the Magellanic Stream as being composed of primordial gas. The tidal stripping of gas in the outer regions of galaxies provides an important way of increasing the absorption cross-sectional areas of galaxies.

6   Discovery of Galactic C IV-HVCs

Highly ionized high velocity clouds have been detected by Sembach et al. (1995) in the direction of Mrk 509 ($l = 36$, $b = -30$). Strong C IV absorption is seen extending from -170 to -340 km/s with no associated Si II 1526 absorption. No detectable H I 21 cm emission is seen at these velocities toward Mrk 509. However, there are H I-HVCs with velocities between -270 and -300 km/s approximately 1.5° from the Mrk 509 line of sight. This suggests the C IV-HVCs may occur in the ionized boundaries of the H I-HVCs. Although collisional ionization in a hot gas with T> 10^5 K cannot be ruled out, it appears the measurements are compatible with the idea that the C IV is produced by photoionization by the extragalactic EUV radiation field in the low density boundary of the H I-HVC. It is significant that C IV-HVCs are being found in the Milky Way halo since the observed properties of these clouds closely resemble the high ionization quasar metal line absorption systems.

Acknowledgments . The observations reported here would not have been possible without the efforts of the many people associated with the HST and GHRS projects. We thank them all. Support for this work was provided by NASA through grants NAG 5-1852 and GO-3463.01 from the Space Telescope Science Institute, which is operated by AURA Inc. for NASA under contract NAS5-26555.

References

Lu, L., Savage, B.D., & Sembach, K.R. 1994a, ApJ, 426, 563

Lu, L., Savage, B.D., & Sembach, K.R. 1994b, ApJ, 437, L119

McKee, C. F. 1993, In Back to the Galaxy, eds. S. G. Holt, & F.Verter (New York:AIP), 499

Murphy, E. M., Lockman, F. J., & Savage, B. D. 1995, ApJ, (submitted)

Savage, B.D. 1988, in QSO Absorption Lines: Probing the Universe, eds.J.C. Blades, D. Turnshek, & C. Norman, (Cambridge:Cambridge U. Press), 195

Savage, B.D. 1990, in The Evolution of the Interstellar Medium, ed L. Blitz, (San Francisco:ASP Conf. Series), 33

Savage, B.D. 1995, in Physics of the Interstellar and Intergalactic Medium, eds. A. Ferrara et al.(San Francisco:ASP Conf. Series) , (in press)

Savage, B.D., Lu, L. et al. 1993a, ApJ, 413, 116

Savage, B.D., Lu, L. et al. 1993b, ApJ, 404, 134

Savage, B. D., & Sembach, K.R. 1994, ApJ, 434, 145

Savage, B. D., Sembach, K.R., & Cardelli, J.A. 1994, ApJ, 420, 183

Savage, B. D., Sembach, K.R., & Lu, L. 1995, ApJ, (submitted)

Sembach, K.R., & Savage, B.D. 1992, ApJS, 83, 201

Sembach, K.R., Savage, B. D., Lu, L., & Murphy, E. M. 1995, ApJ, (submitted)

Spitzer, L. 1990, ARA&A, 28, 77

Wakker, B.P., 1991, A&A, 250, 499

Detection of Highly Ionized Galactic HVCs Toward Markarian 509

Kenneth R. Sembach[1], Blair D. Savage[2], Limin Lu[3], and Edward Murphy[4]

[1] MIT, Center for Space Research, Cambridge, MA, USA
[2] Washburn Observatory, University of Wisconsin, Madison, WI, USA
[3] Astronomy Dept., California Institute of Technology, Pasadena, CA, USA
[4] National Radio Astronomy Observatory, Charlottesville, VA, USA

Abstract. We present Goddard High Resolution Spectrograph measurements of the Milky Way halo gas absorption toward Mrk 509 (l = 36, b = -30) in the 1233 – 1268 Å and 1521 – 1558 Å spectral regions at a resolution of 15 km s^{-1}. We detect high velocity C IV absorption between -340 and -170 km s^{-1}(LSR), but find no corresponding Si II 1526 Å or N V 1238 Å absorption. The high velocity C IV absorption has sub-structure that can be modeled with two Gaussian components centered at -227 and -283 km s^{-1}with N(C IV)/N(Si II) > 1.3 and > 5.1 and N(C IV)/N(N V) > 3.0 and > 6.6, respectively. We obtained sensitive observations of the H I 21 cm emission toward Mrk 509 with the NRAO 140 foot telescope. We find no detectable 21 cm emission at the velocities of the C IV–HVCs to a 4σ limit of N(H I) < 4.9x10^{17} cm^{-2}, but mapping of the 21 cm emission in the area around Mrk 509 indicates that H I–HVCs with velocities between -270 and -300 km s^{-1}are located within about 1.5 degrees of the sight line. The C IV–HVC absorption appears to probe the ionized boundaries of the H I–HVCs. Ionization model calculations indicate that it is likely that these C IV–HVCs are photoionized by extragalactic background radiation rather than by collisional ionization within a recombining, hot gas. These C IV high velocity clouds have ionization properties more closely resembling those of the high ionization quasar metal line systems than those of previously studied sight lines through the Milky Way disk and halo.

1 Observations

We have been using the GHRS aboard the Hubble Space Telescope to study the properties of the Milky Way halo through absorption line spectroscopy of highly ionized species such as C IV and N V toward galactic nuclei (Lu et al. 1994, 1995; Savage et al. 1995; Sembach et al. 1995). The aim of this work is to determine the extent, ionization, and kinematics of gas in the Milky Way so that it is possible to better understand the origin and nature of the extended galactic halos favored as the source of the QSO absorption line systems (Bergeron & Boisse 1991; Steidel 1993).

We obtained the GHRS intermediate resolution (G160M) observations of Mrk 509 shown in Figure1. The light of Mrk 509 entered the large (2"x2") science aperture. We used a substep pattern to provide full diode array observations of the off-spectrum backgrounds, which we averaged and subtracted from the on-spectrum data. The on-spectrum exposure times were 3072 and 3584 seconds in the 1240 and 1544 Å observations, respectively.

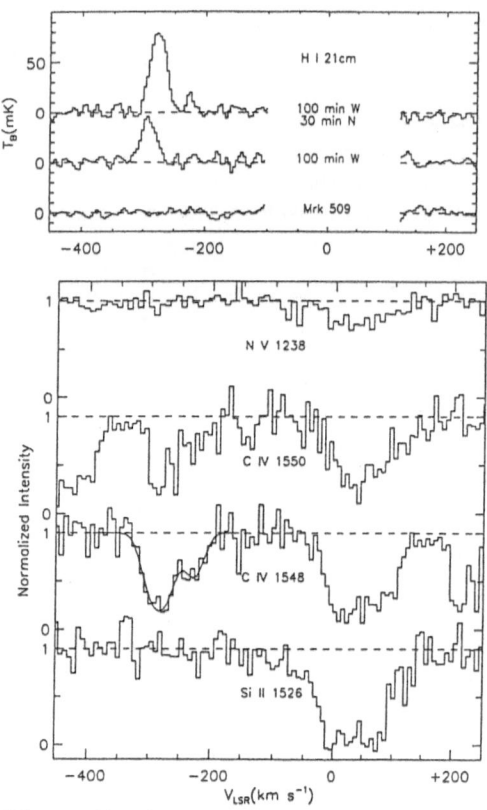

Fig. 1. *Top:* H I 21 cm emission toward Mrk 509 and at two nearby positions obtained with the NRAO 140 foot telescope. *Bottom:* GHRS spectra of the Si II 1526, C IV 1548, 1550, and N V 1238 Å lines toward Mrk 509. Note the strong, high velocity C IV between -340 and -170 km s^{-1}(2-component fit overplotted).

References

Bergeron, J. & Boisse, P. 1991, A&A, 243, 344

Lu, L., Savage, B.D., & Sembach, K.R. 1994, ApJ, 426, 563

Lu, L., Savage, B.D., & Sembach, K.R. 1995, ApJL, in press

Savage, B.D., Sembach, K.R., & Lu, L. 1995, ApJ, in press

Sembach, K.R., Savage, B.D., & Lu, L. 1995, ApJ, 439, 672

Steidel, C.C. 1993, in The Evolution of Galaxies and Their Environment, Proc. of the 3rd Teton Summer Astrophysics Conference, eds. Shull & Thronson (Kluwer), 263

PART V

METAL-RICH SYSTEMS

AT LOW AND INTERMEDIATE

REDSHIFTS

The Subclasses of Metal-Rich Absorbers

J. Bergeron

European Southern Observatory, Karl-Schwarzschild-Str. 2, D–85748 Garching, Germany

Abstract. We have derived the evolution of several physical parameters of metal-rich absorbers from a comparison of the evolution in redshift of the different subclasses of systems. At $z > 2$, absorption from singly-ionized elements always arises from regions of large opacity to UV ionizing photons. This is no longer the case at $z < 1$, and singly-ionized elements are then also found in optically thin regions. Consequently, there is a decrease in the ionization level of absorbers with decreasing redshift. We find higher heavy element abundances at $z < 1$ than at $z > 2$. In the assumption of photoionization by the UV metagalactic field, analysis of the observed column densities for different atoms and ions implies a decrease in the total column density, thus mass, of Mg II clouds between $z = 2$ and 0.5. There is significant clustering of both C IV and Mg II redshifts up to 600 km s^{-1}, with a strong peak at the lowest velocity separations observed of $\sim$30 km s^{-1}. The rest-frame equivalent width of C IV and Mg II scales linearly with the number of subcomponents and is correlated with the total velocity separation between the subcomponents. The evolution of strong Mg II and C IV redshifts thus reflects an evolution in the clustering properties of the absorbers. The high ionization O VI absorption, detected at intermediate and high redshifts, originates from a homogeneous gaseous region of low density in which C II-C III clouds are embedded. The spread in observed ionization states favors photoionization for both phases, and C IV absorption should arise from both regions. Chemical evolution coupled to fragmentation of galactic halos in the early stages of galaxy formation followed by infall of halo gas onto galactic disks can explain most of the observed evolutions.

1 Introduction

Metal-rich absorption line systems, displaced by more than 5000 km s-1 from the quasar emission redshift, trace galactic halos of large extent with radius $R = 75\,L_*^{-0.3}\,h_{50}^{-1}$ kpc, where h_{50} is the Hubble constant in units of 50 km s^{-1} Mpc^{-1} (Bergeron & Boissé 1991; Bergeron, Cristiani & Shaver 1992; Steidel 1993). These identifications have been obtained for absorbers in the redshift range $z = 0.15$–1.1. The metal-rich systems can then be used to study the evolution of galactic halos up to the largest absorption redshifts detected in quasar spectra, more specifically their velocity structure, ionization level, heavy element abundances and star formation activity. These properties can be inferred either from a detailed analysis of individual absorption systems or from a statistical analysis of large homogeneous samples. Elements in different ionization stages are used to build these samples, and those most often detected are H I (Lyα, 21cm line and Lyman limit discontinuity), C IV and Mg II. There is an overlap in H I column density between the metal-rich and metal-poor (Lyα forest) systems. This overlap is more important than previously thought, since some "Lyα forest" systems with N(H I) $\simeq$ $10^{14.5} - 10^{16}$ cm^{-2} show associated C IV absorption (see Tytler et al. in these proceedings). At $z \sim 2$, Lyman limit systems (LLS), N(H I) $> 1.5 \, 10^{17}$ cm^{-2}, are metal-rich. However at higher redshift, $z \gtrsim 3$, a few LLS are metal-poor. The fraction of the latter amongst the LLS is poorly known, since most of the LLS samples are derived from medium resolution observations which do not allow to place meaningful limits on the possible C IV or C II associated absorptions.

At $z \gtrsim 1$, the metagalactic UV radiation field is the dominant source of ionization of the metal-rich absorbers, and elements in lower ionization stages (atoms and singly ionized ions) are only present in regions optically thick to UV ionizing photons (Bergeron & Stasińska 1986; Wolfe 1986). Furthermore the opacity increases with increasing rest-frame equivalent width ratio w_r (Fe II λ2382) / (Mg II λ2796). Consequently, determination of the abundances requires an accurate estimate of N(H I) and observation of a large number of ionization stages. One of the main uncertainties comes from the evaluation of the H I / H ionic ratio, which is a function of the ionization level of the heavy elements and the spectral shape of the ionizing radiation flux. This ionic ratio is better determined when the low ionization ions are the dominant species. The best cases are the damped Lyα systems (N(H I) $> 10^{20}$ cm^{-2}) for which H I / H is very close to unity.

The implications that are derived from a comparison of the evolution of the different subclasses of metal-rich systems at low and high redshifts are outlined in §2. Results derived from the analysis of the velocity structure of C IV and Mg II absorbers are presented in §3. The implications of the presence of a very high ionization phase are discussed in §4. The main results are summarized in §5.

2   Evolution

The evolution of the number density per unit redshift of a given class of absorbers can be written

$$\frac{dN}{dz} = n\sigma \, \frac{d\ell}{dz} \, dz = \frac{c}{H_0} \, n_0\sigma_0 \, \frac{(1+z)^{1+\alpha}}{(1+2q_0z)^{0.5}} \, dz$$

with $n(z)\sigma(z) = n_0\sigma_0(1+z)^\alpha$
where n_0 and σ_0 are the density per unit volume and the mean cross-section of the absorbers at $z = 0$ and ℓ is the distance to the observer. Values of dN/dz are derived from homogeneous samples down to a rest-frame equivalent width limit, $w_{r,min}$, or a given value of N(H I).

Properties of different classes of absorbers can be investigated using ground-based surveys combined with the data base of the Hubble Space Telescope (HST) Quasar Absorption Line Key Project (Bahcall et al. 1993, 1994). The three main samples, C IV, Mg II and LLS, show different evolutions. For the LLS, there is a continuous positive evolution over the redshift interval 0.3-4.1, with $\gamma = 1.50 \pm 0.39$ (Stengler-Larrea et al. 1995). At $0.15 < z < 2.2$, the number density of Mg II systems increases with redshift $\gamma = 0.78 \pm 0.42$ for $w_{r,min} = 0.30$ AA (Steidel & Sargent 1992), whereas at $1.2 < z < 3.7$, there is a negative evolution of the number density of C IV systems with $\gamma = -1.59 \pm 0.79$ for $w_{r,min} = 0.30$ Å (Sargent, Boksenberg & Steidel 1988). Only a crude estimate of the number density of low redshift C IV systems could be made from the small sample obtained with the HST. However, there is preliminary evidence that, as for Mg II systems, the incidence of C IV absorption falls with decreasing redshift at $z < 1.3$ (Bahcall et al. 1993). In addition, the level of evolution for both C IV and Mg II systems is strongly dependent on $w_{r,min}$, the stronger systems evolving more rapidly (Petitjean & Bergeron 1990; Steidel 1990; Steidel & Sargent 1992). The dominant factors governing the evolution are the mass and density of the absorbing gaseous clouds, the heavy element abundances, the ionization level and opacity to the UV ionizing photons and the number of individual clouds per absorber and per line of sight. The dependence of γ on the latter will be discussed in the next section.

The evolution of the mean ionization state of metal-rich systems could be derived from the evolution of the relative number density density of C IV and Mg II systems. However, the samples at lower redshift are yet too small for a statistically meaningful analysis of the evolution of the relative number density of C IV and Mg II absorptions. Alternatively, the evolution of the absorber ionization level can be ascertained from the variation of the equivalent width ratio w_r (C IV $\lambda1549$) / (Mg II $\lambda2796$) $\equiv$ C IV / Mg II for samples with observed C IV and Mg II wavelength ranges. Systems of low ionization, C IV / Mg II ≤ 1, constitute 38% of the HST $\langle z \rangle = 0.53$ sample whereas their fraction comprises only 17% of the $\langle z \rangle = 1.70$ sample (Bergeron et al. 1994). This suggests a lower mean ionization state at smaller redshifts. The strengths of C IV and Mg II absorptions also show different evolutions. While the average strength of Mg II absorption decreases by a factor of four between $z = 1.7$ and 0.5, that of C IV absorption remains roughly constant in this redshift interval.

The ionization structure of the absorbers can be inferred from a comparison between the number density of C IV or Mg II absorption and LLS. At $z \sim 2$, the number densities of Mg II systems down to $w_{r,min}$ (Mg II $\lambda2796$) $= 0.3$ Å and LLS are similar which suggests that Mg II absorption arises in regions optically thick to UV ionizing photons. This is confirmed for individual systems by the observation of a Lyman limit discontinuity for Mg II systems. It is also consistent with the assumption of photoionization by the metagalactic UV radiation field. At $z \simeq 1.7$, Mg II systems account for about

27% of the metal-rich absorbers and nearly all of them have associated C IV absorption. Consequently, the majority of C IV absorbers should be optically thin to UV ionizing photons, as confirmed in several cases by the lack of an associated Lyman limit discontinuity. At lower redshift $z \sim 0.5$, some Mg II and even Fe II absorptions do not show any associated discontinuity at the Lyman edge (Bergeron et al. 1994). This implies a decrease in ionization level of optically thin absorbers between $z = 2$ and 0.5. The ionization level of photoionized regions of small opacities ($\tau_{LL} < 1$) is only a function of $J_{\nu_{LL}}/n_H$ and the spectral shape of the ionizing flux, where $J_{\nu_{LL}}$ is the intensity of the ionizing radiation at the Lyman edge and n_H the gas density. Since, at $z < 2$, $J_{\nu_{LL}}$ varies roughly as $(1 + z)^4$ (see Bergeron et al. 1994 and references therein) the strong evolution in the ionization level of the metal-rich absorbers implies that their gas density does not strongly fall with decreasing redshift. As mentioned above, there is no clear evolution of the number density of Mg II absorbers with $w_{r,min}(\text{Mg II } \lambda 2796) = 0.3$ Å, whereas their heavy element abundances strongly increase between $z = 2$ and 0.5 (Bergeron & Stasińska 1986; Bergeron et al. 1994). This suggests that the total column density, thus mass, of Mg II absorbing clouds decreases with decreasing redshift. This assumption is consistent with studies of low redshift, individual systems: at $z \simeq 0.5$ some Mg II absorbers have total hydrogen column densities no larger than 3×10^{18} cm^{-2} whereas at $z \simeq 2$ $\langle N(H) \rangle \simeq 1 \times 10^{20}$ cm^{-2}.

At $z > 1.5$, the fall in the incidence of weak C IV absorption with redshift, with $\gamma = -1.18 \pm 0.72$ for $w_{r,min}(\text{Mg II } \lambda 2796) = 0.15$ Å, could mainly reflect a decrease with redshift in heavy element abundances. Although C II samples are small, there could also be a negative evolution in redshift of C II absorption, $\gamma = -1.46 \pm 1.73$ for $w_{r,min} = 0.15$ Å (Steidel & Sargent 1992). At high redshift, the number density of Mg II absorbers should then also decrease with redshift, as C II traces similar regions as Mg II, although possibly of smaller opacities to UV ionizing photons. Comparison between the evolution in redshift of C IV (and C II) absorption and LLS is not straightforward since a large fraction of the stronger Lyα forest systems are in fact metal-rich as demonstrated by Tytler et al. (these proceedings). The small intrinsic positive evolution of LLS combined with the negative evolution of C IV and C II absorbers suggest that at higher redshift either the H I column density of metal-rich absorbers increases with redshift and/or the metal abundances become too low for C IV absorption to be detectable in $w_{r,min} = 0.15$ Å surveys. Since $J_{\nu_{LL}}$ is roughly independent of redshift in the interval $2 < z < 3.5$, a strong evolution in the mean opacity of the absorbers would imply higher gas densities at higher redshift. This could be possible at the early stages of galaxy formation. Fragmentation in recently formed galactic halos would result in a decrease of the mass, thus density, of the homogeneous gaseous phase in which the fragments are embedded. All or part of C IV absorption would then originate from this homogeneous phase. At later stages of galactic evolution, this phase would have an even smaller density and be more highly ionized, by collisional or photoionization precesses, and be traced by O VI absorption

3   Velocity Structure

The evolution of the number density of C IV and Mg II absorbers is strongly dependent on the value of $w_{r,min}$ of the sample. In the redshift interval $0.15 < z < 2.2$, the value of γ increases from 0.72 to 1.02 and 2.24 for $w_{r,min} = 0.30$, 0.60 and 1.0 Å respectively (Steidel & Sargent 1992). The incidence of strong Mg II absorption is thus smallest at lower redshifts. For the higher redshifts $1.2 < z < 3.7$ C IV samples, γ decreases from -1.18 to -1.56 and -2.35 for $w_{r,min} = 0.15$, 0.30 and 0.40 Å (Sargent, Boksenberg & Steidel 1988; Steidel 1990). The incidence of C II absorption also falls with redshift (see §2), but the C II samples are too small for investigating the variation of γ(C II) with $w_{r,min}$. The relative smaller number of large equivalent width absorbers at both low and very high redshifts could reflect a very strong evolution of the absorber clustering properties.

There is a significant clustering of both C IV and Mg II redshifts on scales of a few 10^2 km s^{-1} (Sargent, Boksenberg & Steidel 1988; Sargent, Steidel & Boksenberg 1988). For both ions, the two-point correlation function peaks at the smallest velocity separations which could be probed at the spectral resolution of the surveys ($\Delta v = 100$ or 200 km s^{-1}). Higher spectral resolution samples show even stronger clustering at smaller velocity separations, $\Delta v = 30$ km s^{-1}, for both C IV and Mg II redshifts (Petitjean & Bergeron 1990, 1994). Up to $\Delta v \simeq 600$ km s^{-1}, the overall velocity separation between subcomponents is correlated to the total number of subcomponents. This suggests that strong absorptions detected in medium spectral resolution surveys are blends of subcomponents. This trend of increasing rest-frame e-quivalent width with multiplicity of the absorption was already suggested by Wolfe (1986) and York et al. (1986). It has been confirmed by the discovery of a strong correlation between the total equivalent width of Mg II or C IV absorption (individual lines are considered to be subcomponents of a given system if their velocity separation is smaller than 600 km s^{-1}) and the total number of subcomponents. The total rest-frame equivalent width of Mg II or C IV scales linearly with the number of subcomponents. Consequently, large equivalent widths are primarily associated with high multiplicity of the systems and not with large column densities.

Clustering on the smaller velocity scales ($\Delta v \leq 300$ km s^{-1}) could arise from relative motions within galaxy halos, and that on larger scales (up to $\Delta v \simeq 600$ km s^{-1}) from relative motions of galaxy pairs or groups. Modelling of the observed distribution of velocity separations also favors a bimodal distribution. The simple kinematic model used assumes clouds of identical sizes, randomly distributed either within a sphere or a disk and with one or two-component Gaussian velocity distribution (Petitjean & Bergeron 1990, 1994).

At high redshift, the rareness of strong C IV (and C II) systems is consistent with the assumption that we observe on-going fragmentation of the halo gas associated with galaxies. The negative evolution of the number density of metal-rich systems should reflect the chemical evolution of the galaxies. High redshift absorbers should then trace galaxies in early stages of formation. As pointed out by Steidel & Sargent (1992), in situ star formation could be triggered in the halo by cloud-cloud collisions which should then result in a coupling of the observed evolutions of weak and strong C IV absorbers. At $z < 2$, the evolution of strong MgII absorbers reflects the evolution in the number of individual clouds. Infall onto the galactic disk is most likely the dominant process at the origin of the decrease in the number of halo clouds with time.

4   The Very High Ionization Phase

A high ionization phase, traced by O VI absorption, is present at intermediate and high redshift. At $z < 1$, crowding in the Lyα forest is moderate and individual O VI absorption have been detected in the HST Key Project survey. In the redshift interval $0.6 < z < 1$, there are only five C IV systems with observations in the rest-frame wavelength range $\lambda\lambda 1031\text{-}1037$: four exhibit OVI absorption of which two have an associated NV absorption (Bahcall et al. 1993, 1994). In the assumption of photoionization, the observed mean equivalent width ratio $w_r(\text{OVI}\lambda 1031)/w_r(\text{NV}\lambda 1238)$, or its lower limit, implies a much higher ionization state, thus a lower gas density, than for the C III-C II phase (Bergeron & Stasińska 1986; Petitjean, Bergeron & Puget 1992). The estimated column densities of Mg II and Fe II in the systems at $z \simeq 0.6\text{-}1$ imply values of the ionization parameter (ratio of the number density of incident ionizing photons to the gas density) $U \sim (1\text{-}3) \times 10^{-4}$ and those of O VI and N V require $U \sim 10^{-2}$ (Bergeron et al. 1994). Consequently, at $z \simeq 1$ two phases of radically different density (by a factor of ten or more) must be present in galactic halos and photoionization by the UV metagalactic flux could be the dominant process.

The O VI phase has also been detected in several metal-rich absorbers towards the high-redshift bright quasar HS 1700+6416 (Vogel & Reimers 1995) . These HST observations reveal the presence of O VI at $0.722 \leq z \leq 2.579$ in six absorption systems (two are uncertain due to blending) out of ten. Nine of these systems (none at a redshift close to the quasar emission redshift) have observed N V $\lambda\lambda 1238\text{-}1242$ rest-frame wavelength range: three show O VI and N V absorption, O VI only is present in two others and N V only (together with O V) is present in the two highest-redshift systems. For the remaining systems, neither O VI nor N V is detected. In a preliminary analysis, Reimers et al. (1992) have concluded that the column density ratios N(O III):N(O IV):N(O V) are consistent with predictions of single-phase photoionization models. Results of a more detailed analysis (Vogel & Reimers

1995), which do not include singly ionized elements, show for several systems lower values of predicted column densities than observed for doubly ionized elements. This suggests that a two-phase model would better reproduce the observed column density ratios, at least for systems with detected singly-ionized elements. Indeed, the values obtained for all these metal-rich absorbers are close to 10^{-2}, as found by Bergeron et al. (1994) for the high ionization phase of $z \sim 0.6$-1 absorbers. Nearly all the absorption systems detected in HS 1700+6416 show enhanced oxygen and nitrogen abundances relative to carbon and the relative oxygen abundance [O/C] appears to be anti-correlated with the carbon abundance [C/H]. There is no obvious correlation of the abundances and the column density ratios with redshift. Since, at $z < 2$, the intensity of the UV metagalactic flux is expected to decrease with decreasing redshift, roughly as $(1+z)^4$, the lack of clear evolution in redshift of the ionization level of the O VI-N V phase ($U \sim 10^{-2}$ in the redshift interval 0.6-2.5) suggests that the density of the homogeneous high-ionization region should decrease with decreasing redshift. At $z < 0.6$, photoionization by the UV metagalactic flux can lead to a very high ionization phase only in very tenuous gas. Since there is no clear evolution in halos sizes at $0.2 < z < 1$ (Bergeron & Boissé 1991; Steidel 1993), the detection of O VI absorption at very low redshift would imply that either overpressurized gas heated by supernovae occuring in the galactic disk can flow into the halo at very large distances, or the UV radiation field escaping the galactic disk dominates the UV metagalactic field. In both cases, the galaxy should show signs of star formation activity stronger than observed, at least if O VI absorption were detected at very large impact parameters.

Collisional ionization has been favored by Lu & Savage (1993) for the high-redshift O VI phase. Their analysis of a composite spectrum formed of 73 CIV systems at $\langle z \rangle = 2.76$ has revealed the presence of O VI absorption but not of N V. The derived column density ratio N(O VI)/N(N V) $\geq$ 4.4 requires a temperature $T \gtrsim 2.5 \times 10^5$ K for gas in thermal equilibrium. The values found for this column density ratio for the absorption systems studied with the HST varies from 3.2 to > 7.7, except for the two cases at $z \sim 2.5$ with detection of only N V absorption for which N(O VI)/N(N V) < 1. Even if the latter are in fact Lyα absorption lines, both collisional and radiative ionization can lead to the observed N(O VI)/N(N V) ratios. However, collisional ionization would require a multi-phase model to account for the observed N(O IV):N(O V):N(O VI) column density ratios. In the halo of our Galaxy, the N V phase is also thought to be collisionally ionized, with a temperature $T \sim 2 \times 10^5$ K (Sembach & Savage 1992). The observed Al III absorption should trace photoionized gas, and C IV-Si IV absorption should originate from both phases. No single model satisfactorily reproduces all the Galactic observations. In particular, photoionization models fail to produce enough N V and the predicted C IV/Si IV ratio is higher than the observed one (Bregman & Harrington 1986).

In conclusion, photoionization models of galactic halo gas appear to better reproduce all the column density ratios for intermediate and high redshift quasars. The C III/C II clouds are embedded in a more homogeneous phase of very high ionization and both regions contribute to C IV absorption. As demonstrated by Shapiro & Benjamin (1991), ionization parameters as large as $U \gtrsim 10^{-2}$ as required for the O VI phase cannot be achieved in self-ionized galactic fountain flows. However, at low redshift $z < 0.6$ N V and O VI absorption could originate in hot collisionally ionized gas for small impact parameters, since the UV metagalactic flux becomes too low to produce detectable amounts of very high ionization gas.

5   Summary

Metal-rich absorption systems are produced in the extended halos of intervening galaxies (Bergeron & Boissé 1991, Steidel 1993) and the evolution of their properties should then be linked to galaxy formation and evolution. These galactic halos comprise two phases: a diffuse gaseous region in which clouds are embedded. To investigate the evolution of the properties of these two phases, we have analyzed and compared the evolution of different classes of absorbers using data bases build from ground-based and HST observations.

At $0.5 < z < 2$, the ionization level of the absorbers falls with decreasing redshift while their heavy element abundances increase. The diffuse gaseous region is highly ionized and is traced by O VI-N V absorption. A large fraction of the clouds, even some containing singly-ionized elements, are optically thin to UV ionizing photons. In the assumption of photoionization by the diffuse UV metagalactic field, analysis of the various column densities implies a decrease in the total column density, thus mass, of individual clouds. There is also a smaller number of individual clouds at $z = 0.5$ than at $z = 2$, which is reflected by a strong evolution in redshift of the number density of Mg II systems with larger rest-frame equivalent widths. There is only a limited number of C IV absorption detected at intermediate redshift, and most of them at moderate spectral resolution. This prevents studying the clustering properties of C IV redshifts at $z < 1.5$. The O VI phase can be photoionized by the UV metagalactic field, at least at $z > 0.6$. This ionization process is favored since it best explains the N(O IV): N(O V): N(O VI) column density ratios derived for a few systems at $0.7 < z < 2.5$ from HST observations. In this redshift range, the ionization level of the O VI-N V phase does not appear to substantially vary, As the intensity of the UV metagalactic flux strongly decreases with redshift at $z < 2$, this suggests that the gas density of the homogeneous halo phase also falls with decreasing redshift. There is only a small decrease in redshift of the cloud gas density, since the evolution of the strength of the ionizing flux is mainly reflected in the evolution of the cloud ionization level.

At $z > 2$, the number density of metal-rich absorbers strongly decreases with redshift, whereas that of LLS does not show a strong evolution. The fall in the incidence of weak C IV systems at high redshift could mainly reflect a decreases in redshift of the heavy elements abundances, while that much stronger of C IV systems with larger rest-frame equivalent widths points to weaker clustering of the absorbers at higher redshifts on velocity scales $\Delta v \simeq$ (30 to 600) km s^{-1}. The sharp increase in redshift of the relative number of LLS and C IV systems may indicate a rise in the opacity to UV ionizing photons of the metal-rich absorbers. As discovered by Tytler et al. (in these proceedings), some "Lyα forest" systems show in fact very weak, associated C IV absorption, but these systems have low H I column densities, N(H I) $\simeq 10^{14.5} - 10^{16}$ cm^{-2}. If this is also the case for a large fraction of high-redshift LLS, an increase in redshift of the relative number of C II and C IV absorbers should be detectable in larger C IV-C II samples than presently available.

At early stages of galaxy formation, fragmentation of halo gas can explain several observed evolutions. The rareness of strong C IV systems at high redshift, thus the weak clustering on velocity scales of a few 10^2 km s^{-1}, implies a limited number of halo clouds, consistent with on-going fragmentation. Moreover at these early stages of evolution, the relative amount of gas in the fragments would be increasing, thus the gas density and column density of the diffuse phase would decrease with decreasing redshift. Since C IV absorption originates both in clouds and the homogeneous phase, this naturally leads to an increase in redshift of the opacity to UV ionizing photons of diffuse halo gas, which together with chemical evolution could explain the strong evolution in the relative number of LLS and weak C IV systems. If the strength of the UV metagalactic field remains constant or falls at the highest redshifts, $z > 4$, a prediction of this scenario is a decrease of the incidence of O VI absorption at very high redshift. Unfortunately, crowding in the Lyα forest might prevent a clear test on that point.At $z < 2$, infall of halo gas onto the galactic disk results in a decrease of the number of clouds, thus a weakening in the clustering of Mg II redshifts, and of the mass of the diffuse halo component. The evolution of the heavy element abundances between z = 2 and 0.5 implies that chemical evolution is still important at these later stages of galactic evolution, as also indicated by the star formation activity observed in the identified absorbers.

References

Bahcall J.N, Bergeron J., Boksenberg A., Hartig G.F., Jannuzi B.T., Kirkhakos S., Sargent W.L.W., Savage B.D., Schneider D.P., Turnshek D., Weymann R.J., Wolfe A.M., 1993, ApJ Suppl, 87, 1

Bahcall J.N, Bergeron J., Boksenberg A., Hartig G.F., Jannuzi B.T., Kirkhakos S., Sargent W.L.W., Savage B.D., Schneider D.P., Turnshek, D., Weymann R.J., Wolfe A.M., 1994, ApJ Suppl, submitted

Bergeron J., Boissé P., 1991, A&A, 243, 344

Bergeron J., Cristiani S., Shaver P., 1992, A&A, 257, 417

Bergeron J., Petitjean P., Sargent W.L.W., Bahcall J.N, Boksenberg A., Hartig
    G.F., Jannuzi B.T., Kirkhakos S., Savage B.D., Schneider D.P., Turnshek D.,
    Weymann R.J., Wolfe A.M., 1994, ApJ, 436, 33
Bergeron J., Stasińska G., 1986, A&A, 169, 1
Bregman J.N., Harrington J.P., 1986, ApJ, 309, 833
Lu L., Savage B.D., 1993, ApJ, 403, 127
Petitjean P., Bergeron J., 1990, A&A, 231, 309
Petitjean P., Bergeron J., 1994, A&A, 283, 759
Petitjean P., Bergeron J., Puget J.-L., 1992, A&A, 265, 375
Reimers D., Vogel S., Hagen H.J., Engels D., Groote D., Wamstecker W., Clavel
    J., Rosa M.R., 1992, Nature, 360, 561
Sargent W.L.W., Boksenberg A., Steidel C.C., 1988, ApJ Suppl, 68, 539
Sargent W.L.W., Steidel C.C., Boksenberg A., 1988, ApJ, 334, 22
Sembach K.R., Savage B.D., 1992, ApJ Suppl, 83, 147
Shapiro P.R., Benjamin R.A., 1991, PASP, 103, 923
Steidel C.C., 1993, *Proc. of the Third Tetons Summer School on The Environment
    and Evolution of Galaxies*, July 1992, eds. J.M. Shull & H.A. Thronson, Jr.,
    Kluwer Academic Publishers, p. 263
Steidel C.C., 1990, ApJ Suppl, 74, 37
Steidel C.C., Sargent W.L.W., 1992, ApJ Suppl, 80, 1
Stengler-Larrea E.A., Boksenberg A., Steidel C.C., Sargent W.L.W., Bahcall J.N,
    Bergeron J., Hartig G.F., Jannuzi B.T., Kirkhakos S., Savage B.D., Schneider
    D.P., Turnshek D., Weymann R.J., 1995, ApJ, 444, 64
Vogel S., Reimers D., 1995, A&A, 294, 377. Reimers D.,
Wolfe A.M., 1986, *Proc. NRAO Conference on Gaseous Halos of Galaxies*, eds. J.
    Bregman & J. Lockman, NRAO SP, p. 259
York D.G., Dopita M., Green R., Bechtold J., 1986, ApJ, 311, 610

Observational Constraints on the Nature of Low Redshift Lyα Absorbers

Vincent Le Brun[1], Jacqueline Bergeron[2,1], and Patrick Boissé[3]

[1] Institut d'Astrophysique de Paris, CNRS, 98bis Bvd Arago, F-75014 Paris, France
[2] European Southern Observatory, Karl-Schwarzschild-Straße 2, D-85748 Garching, Germany
[3] Ecole Normale Supérieure, 24 rue Lhomond, F-75005 Paris, France

Abstract. We present results of a spectroscopic survey of galaxies around quasars of the HST Key Program. Our sample of Lyα-absorber/galaxy associations, combined with those of published studies, is used to constrain the link between galaxies and Lyα absorbers. There is no strong correlation between the equivalent width of the Lyα absorption and the impact parameter of the galaxy, and the distribution of the velocity difference between the absorber and the galaxy is narrow, FWHM= 240 km s^{-1}. These arguments are compatible with most of the Lyα absorbers tracing the dark matter structures.

1 Presentation of the Survey

The purpose of the survey (Le Brun et al. 1995) is to measure the redshift of the galaxies up to a magnitude $m_r = 22.5$ and an angular impact parameter $\theta = 3'$ from quasars observed by the HST in order to study the origin of low redshift Lyα absorptions. We have obtained 66 galaxy redshifts in three fields around TON 153, 3C 351 and H 1821+64 with MOS-SIS at the CFHT. Among these, 19 are at less than 750 km s^{-1} from a Lyα absorber, and one does not give rise to any Lyα absorption. Combining our results with those from previous studies (Schneider et al. 1992, Morris et al. 1993, Lanzetta et al. 1995), we get a sample of 32 Lyα absorber-galaxy associations and 11 galaxies that do not give rise to any absorption. This sample is used to study various correlations involving the rest equivalent width of the Lyα absorption w_r, the impact parameter D, the velocity difference between the absorber and the galaxy Δv or the magnitude of the galaxy M_r.

2 Results

Lanzetta et al. (1995) found a statistically significant correlation between D and w_r, but their data include a few metal-rich systems and is partially based on low resolution HST data. When excluding these data, and including the 11 cases of galaxies without associated absorption in the rank-correlation calculation, we do not find any significant correlation (Fig. 1). There is a marginally significant correlation (91.3% confidence level) when restricting the sample to $w_r \geq 0.24$Å.

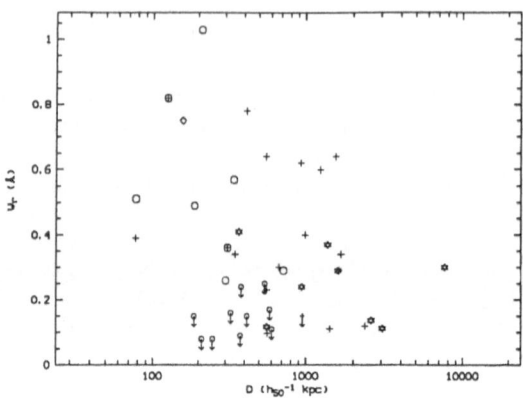

Fig. 1. Impact parameter versus rest-frame equivalent width of the Lyα absorption. Crosses: this work, circles: Lanzetta et al. (1995), stars: Morris et al. (1993), squares: Schneider et al. (1992)

Further, there is no correlation between M_r and D, contrary to intermediate redshift MgII absorbers, for which $R \propto L^{\alpha}$, with $\alpha = 0.2 - 0.3$. The distribution of Δv is narrow, with an half-width at half maximum of 120 km s^{-1}, that is the average value of the velocity dispersion within a large scale structure (Geller, private communication).

All these results indicate two population of Lyα absorbers. A small fraction of them seem to arise in the external part of giant halos around galaxies ($R \simeq 200$ kpc), which may be the outer part of the MgII halos. The majority could be due to gas not associated to any particular galaxy but following the same large scale structures as galaxies. The latter assumption is consistent with recent models (e.g. Cen et al. 1994, Petitjean et al. 1995) where the Lyα clouds trace the dark matter filaments.

References

Cen R., Miralda-Escudé J., Ostriker J.P., Rauch M., 1994, ApJ 437, L9
Lanzetta K.M., Bowen D.V., Tytler D., Webb J.K., 1995, ApJ, in press
Le Brun V., Bergeron J., Boissé P., 1995, submitted to A&A
Morris S.L., Weymann R.J., Dressler A., et al., 1993, ApJ 419, 524
Petitjean P., Mücket J.P., Kates R.E., 1995, A&A 295, L9
Schneider D.P., Bahcall J.N., Gunn J.E., Dressler A., 1992, AJ 103, 1047

The Nature and Evolution of Absorption–Selected Galaxies

Charles C. Steidel

MIT, Physics Department, Cambridge, MA, USA

Abstract. We present results of surveys for high redshift galaxies selected by their having produced detectable Mg II and Lyman limit absorption in the spectra of background QSOs. We discuss the properties of the absorbing galaxies, the connection between galaxy properties and absorption line signatures, and how a combination of QSO absorption line and conventional faint galaxy techniques can be used to study field galaxy evolution to very large redshifts.

1 Introduction

The subject of QSO absorption lines seems to be reaching a remarkable level of maturity, where the statistics of the various classes of systems are well–known over a very large range of redshifts, and efficient high resolution spectrographs on large telescopes are making feasible (and relatively routine) the accumulation of detailed information for individual absorption systems. Nevertheless, the actual *application* of the absorption line studies, as tools in understanding the nature of galaxies (and the intergalactic medium) as a function of time, are in a state of infancy in many respects. It is of course here that the promise of the whole field must be realized: the unique access to physical details of galaxies as a function of time must be exploited and then integrated into the global picture of galaxy formation and evolution.

The aim of this contribution is two–fold: first, in order to understand the overall implications *for galaxies* of the studies of metal line absorption systems, it is crucial to know exactly what types of galaxies represent potential absorbers, how the absorption properties might be related to the galaxy properties, and what parts of those galaxies give rise to various types of absorption signatures. Second, and very much related to the first (but perhaps of more immediate interest to the community interested in galaxy evolution *per se*), what can one infer about the evolution of normal field galaxies using the presence of absorption, rather than flux density in some observed bandpass, as the selection criterion? I will argue that, once the "selection function" of the absorbing galaxies is understood, absorption–selected galaxies can be used to understand the overall evolution of field galaxies to redshifts far beyond where an apparent magnitude limited sample breaks down, and can be used to examine the galaxy luminosity function in a potentially much less biased manner even in the redshift range of "overlap" between the methods.

From the point of view of those of us working in the field of QSO absorption lines, the most basic motivation for the kind of study described below is to provide a "context" in which all of the rich detail that will come out of the present and future absorption line studies can be understood and assimilated by those *outside* of the field. From the point of view of the galaxy evolution community, the methods represent an independent approach to examining the evolution of field galaxies with redshift, and the means to extend field galaxy "redshift surveys" (including, of course, the accompanying detailed physics accessible only through the absorption lines) well beyond $z = 1$. Because many of the results I will discuss have been summarized recently elsewhere (Steidel 1993, Steidel *et al* 1994, 1995, Steidel and Dickinson 1995), I will devote most of the discussion in this article to new results on the gaseous structure of absorbing galaxies and on the dependence of extended gaseous envelopes on galaxy properties.

2 Establishing the "Selection Function"

2.1 Observational Methods and Biases

As discussed in more detail elsewhere (see above references), we use as our absorption selection criterion the Mg II $\lambda\lambda 2796$, 2803 doublet (with rest e-quivalent width $W_0 > 0.3\text{Å}$) because the statistics are well-known for the entire redshift range $0.2 \leq z \leq 2.2$ (Steidel & Sargent 1992) and because it turns out to be a very effective tracer of H I gas that has $N(\text{H I}) \gtrsim 10^{17}$ cm^{-2}; for this reason we usually refer to the sample as being "gas cross-section s-elected" at an effective H I threshold that corresponds to gas with $\tau \gtrsim 1$ in the Lyman continuum (and so the sample could equally well be described as selected by the presence of a Lyman limit system [LLS]). It turns out that the redshift path density dN/dz for such systems is consistent with no evolu-tion in co-moving total cross–section (i.e., the product of the space density of the absorbers and the effective mean cross–section of individual absorbers remains constant) over the whole redshift range observed; this fact becomes very useful in deducing, e.g., luminosity functions and space densities of the galaxies as a function of redshift (see §2.5). The distinct advantages of an absorption–selected galaxy sample (from the point of view of understanding field galaxy evolution) are detailed in Steidel *et al* 1994.

Before summarizing the results of our completed survey for absorption-selected galaxies at $z \leq 1$, it is worth making some general comments on the technique used and possible biases inherent in the method. We have adopted a strategy similar in some respects to the one used in the seminal study of Mg II absorbing galaxies by Bergeron & Boissé (1991) – we obtain deep continuum images (for our survey, they are taken over a very wide wavelength baseline, from the visual to the K–band) of the fields of the QSOs and essentially begin our search for the absorber at the position of the QSO and work our way out in angular separation.

We wish to emphasize that, provided one obtains spectra of the candidate galaxies near the QSO sight line, the search strategy adopted is in no way biased against discovering non–absorbing galaxies; in other words, one is just as likely to identify a putative population of non-absorbing galaxies by searching in fields in which absorption is known to be present as where it is not – if they are there, they will be found. Nevertheless, in our study we have included with our sample of 51 absorber fields (chosen to have Mg II absorption with $W_0 \geq 0.3$ Å and $0.3 \leq z \leq 1.0$) about 25 QSO fields in which no Mg II absorption is present over the same range of redshift. These "control" fields have the obvious advantage that if no objects are seen close to the line of sight, no galaxy spectroscopy is necessary to resolve any confusion about the presence of an "interloper" (i.e., a galaxy close to the line of sight but producing no absorption). For the faint galaxy spectroscopy part of the program, we have made every attempt to obtain spectra of all detected objects within $\sim 10''$ of the QSO, corresponding to $\sim 50h^{-1}$ kpc at the typical redshift of the survey, and often we have obtained spectra of several objects per field at larger angular separations. This search radius was adopted largely from experience; if it had turned out to be necessary to go to larger radii to find a galaxy at the absorption line redshift, then we would have extended the search. What we have found is the nearest galaxy with a redshift matching that of the absorption system, and in several cases we have found evidence for galaxy groups (not surprising, since most present–day galaxies reside in groups). In any individual case, therefore, one cannot *prove* that the identified object is the absorber–it could be a galaxy at the same redshift but larger impact parameter. However, as will be described below, we see very significant correlations of the *identified* galaxy properties with the absorption line properties, and a remarkable absence of interloper galaxies at redshifts different from that of the absorption, so that it is clear beyond almost all reasonable doubt that the adopted strategy has succeeded in finding the actual absorber. Similarly, we cannot go infinitely faint in either the imaging or the spectroscopy, so that the possibility would always exist that the "true" absorber is a hopelessly faint object, perhaps situated directly on top of the QSO (see, e.g., York *et al.* 1986, Yanny & York 1992); however, this too can be rejected in a statistical sense because we have found a galaxy at the right redshift in every case *without* reaching extremely faint levels (although the identified galaxies do in fact represent a very wide range in absolute luminosity), and as will be detailed below, the galaxy luminosities, the distribution of projected impact parameters, and the observed absorption line characteristics form a remarkably coherent relationship.

2.2   The Nature of the Absorbing Galaxies, $z \leq 1$

As stated in §1, many of the results of the survey have been presented elsewhere, and for lack of space in the current contribution, we refer the interested
reader to those references for more details. We concern ourselves here primarily with what types of galaxies are represented in a sample selected by metal
line absorption. It has turned out that the K–band data has been extremely
important to an interpretation of the results of the absorbing galaxy survey;
initially the purpose of the near–IR data was to get an idea about the extent
to which the observed rest–frame B band luminosities constructed from our
optical data[1] were biased by current rates of star formation. It turns out,
however (see §2.3), that it is the K luminosity that is a much more robust
predictor of a galaxy's potential to produce detectable absorption. The wide
color baseline also allows one to classify galaxy "morphological" (really, spectroscopic) types, which are extremely useful in the absence of *HST*–resolution
images for every field.

First, the widely cited conclusion of Bergeron & Boissé (1991) that only
"bright" galaxies are absorbers turns out not to be completely accurate in
view of a much larger sample; while the absorbers have a mean luminosity of
about $0.7L_B^*$ and $0.5L_K^*$ (where these refer to *rest–frame* luminosities), the
actual range of luminosity spans more than a factor of 70, with the faintest
identified absorber having $L_K \approx 0.05L_K^*$. The spectra and the optical/IR
colors of the absorbing galaxies suggest that the sample is drawn from the
full range of galaxy spectroscopic types, from very blue objects with the colors
or present–day late–type spirals and Im galaxies, to objects which have the
colors and spectra of completely unevolved elliptical galaxies (see Figure 1).
There is no evidence for any tendency for the absorbing galaxies to have
particularly high star formation rates, on average. As can be seen from Figure
1, the "average" absorber maintains the rest–frame $B - K$ color of an Sb
spiral for the full range of redshifts observed. This is consistent with a $\sim$
constant rate of star formation for the absorbing galaxy population as a
whole since at least $z \sim 1$. We also see no evidence for luminosity evolution
of the population (in either rest–frame B or rest–frame K) over the same
redshift range (see Steidel *et al.* 1994, Steidel & Dickinson 1995). One of the
most surprising conclusions (see also §2.3 below) is that galaxies of similar K
luminosities but with widely differing current star formation rates appear to
have extended gaseous envelopes that are indistinguishable. The implications
of this fact will be discussed in §4.

---

[1]   It is important for the purposes of obtaining accurate luminosity distributions
to make the appropriate k–corrections to reach rest–frame absolute magnitudes;
for example, simply using an observed R magnitude with no bandpass correction
to obtain M(R) can lead to relative errors of up to 1 magnitude for galaxies in
the redshift range of interest.

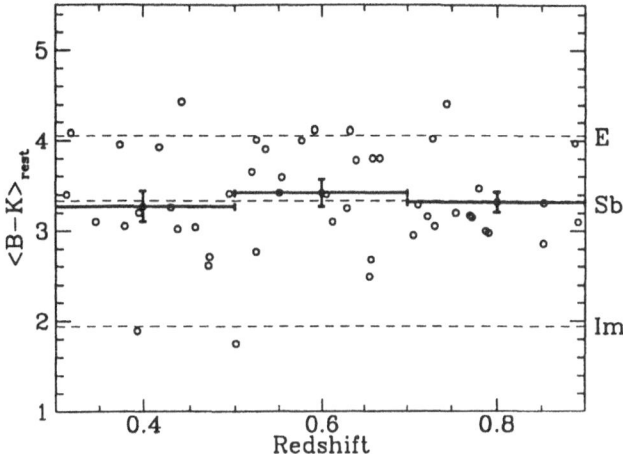

Fig. 1. Plot showing the distribution of rest–frame optical/IR colors of the absorbing galaxies versus redshift. Note that the galaxies appear to represent the full range of spectroscopic types, but have an average color equivalent to a present–day Sb galaxy across the whole redshift range.

2.3 The Geometry of the Extended Gas and the Absorber Selection Function

The advantage of a large sample of absorbing galaxies is that it begins to become feasible to extract significant relationships between the gaseous sizes of the galaxies, the galaxy luminosity, and the absorption line properties. Establishing these statistical trends, together with the determination of the nature of any non–absorbing population, is crucial to inserting the absorbing galaxy population into the context of general field galaxy evolution. For example, it has been somewhat of a "tradition" over the last ~ 20 years to calculate the expected "sizes" of QSO absorbers of various classes based on knowledge of dN/dz and assumptions about the galaxy luminosity function, luminosity/ size scaling relation, and gas–phase geometry. It is now possible to turn the whole problem around, and *infer* (based on direct observations) all of the relations that were always assumed, and directly derive the galaxy luminosity function. If we understand what selection function is in effect for gas cross–section selected galaxies, it is possible to produce a field galaxy luminosity function directly that is almost completely independent of those produced from the apparent magnitude selected surveys of recent years (see, e.g., Lilly 1993, Songaila *et al.* 1994, Colless 1995).

The usual assumption has been that the gaseous sizes of galaxies would obey the Holmberg (1975) relation, $R_{\mathrm{gal}} \propto L^{0.4}$, and the luminosities are drawn from a Schechter (1976) luminosity function. On the basis of a subset of the current sample, we made the claim that the absorbers followed a relation that is closer to $R_{\mathrm{gal}} \propto L^{0.2}$ (Steidel 1993); this exponent, often called β, is a very important number to determine because it controls the extent to which the gas cross-section selection criterion depends upon luminosity, and hence it determines the effective volume probed by an absorption line survey as a function of galaxy luminosity. We have undertaken a new determination of β, using the following method:

We include all galaxies with redshifts (whether they have produced absorption or not) in the survey fields. We then assume spherical "halos" of gas, obeying the relation $R = R^*(L/L^*)^\beta$, searching for the best values of R^* and β, subject to the following criteria. First, we minimize the number of galaxies that fall on the "wrong" side of the $R - L$ curve (i.e., minimize the number of absorbing galaxies falling outside the curve and the non–absorbing galaxies falling inside the curve), and we also insist that the distribution of normalized impact parameters ($D_n = D_i/R_i(L)$, where D_i is the observed impact parameter and $R_i(L)$ is the expected halo extent for a galaxy of the same luminosity as the observed galaxy) is consistent with the expected radial distribution, which should be $n(D_n)dD_n \propto D_n$ for the assumed geometry. This fitting process results in two slightly different relationships, $R(L_K) = 38h^{-1}(L_K/L_K^*)^{0.15}$ and $R(L_B) = 35h^{-1}(L_B/L_B^*)^{0.2}$, with the former having smaller scatter. The halo size–K luminosity relationship is plotted in Figure 2; the most important point to notice is how amazingly well the simple model actually describes the data (and how poorly a model with $\beta = 0.4$ fits the data!). Note that under the best–fit curve there are essentially *no non–absorbing galaxies*. This results in two very important conclusions: first, if a galaxy is brighter than $M_K = -22$ (or about $0.06L_K^*$) and it falls within $R(L_K)$ of a QSO sight line, it will produce detectable Mg II absorption – this means that inside that radius, the covering fraction of the absorbing gas is unity, independent of galaxy spectroscopic/morphological type. Secondly, this same absence of non–absorbers below the curve means that the assumption of spherical halos *must* be close to the truth. For example, if the absorption were all produced by extended thin disks, one would expect, statistically, that 50% of the galaxies within $R(L_K)$ would produce no detectable absorption (see also §2.6 below).

On the other hand, despite the surprising "regularity" of the above results, there is evidence that the existence of extended gas capable of producing absorption depends on environment, at least in extreme cases. Bechtold & Ellingson (1992) have shown that galaxies located in the same clusters as QSOs generally do not produce detectable MgII absorption. However, cluster cores have a very small cross-section on the sky, so it is not surprising that, in a collection of 100 or so random lines of sight, we did not probe any such environments. Again, we already have evidence that many of the absorbers are members of groups, but there is clearly much work to be done in estab-

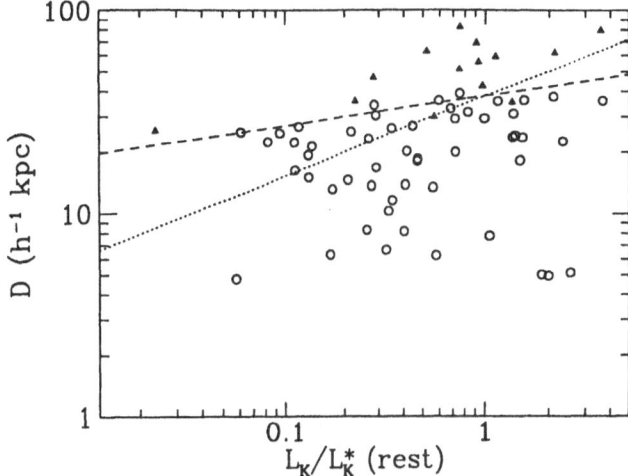

Fig. 2. Plot showing observed impact parameter D versus galaxy near–IR luminosity L_K for the absorbing (open circles) and non–absorbing (filled triangles) galaxies. The dashed curve is the best–fit relation with $\beta = 0.15$ and $R^* = 38h^{-1}$ kpc, as described in the text. The dotted curve has $\beta = 0.4$, which clearly fails badly in describing the data.

lishing the detailed environments of the galaxies and any possible effect on the absorbing cross-sections or halo gas kinematics.

2.4 Dependence of Absorption Line Properties on Galactocentric Distance

Lanzetta and Bowen (1990) have noted a significant inverse correlation between the observed equivalent width of the Mg II absorption lines and the impact parameter between the QSO sight line and the centroid of the absorbing galaxy from the sample of Bergeron & Boissé (1991). Since it is fairly well-established from high–resolution observations that the Mg II equivalent width is proportional to the number of individual velocity components (e.g., Petit-Jean & Bergeron 1990), it is natural to try to obtain a radial "cloud" density distribution function from the observed correlation. Unfortunately, a larger data set reveals that the situation is probably much more complicated; the data for our $z \leq 1$ sample are plotted in Figure 3. Note that most, if not all, of the very high equivalent width systems turn out to be damped Lyman alpha absorbers, with impact parameters smaller than $15h^{-1}$ kpc.[2] Most of

---

[2] We have no direct information on the H I column density for most of the $z < 1$ absorbers; note also that many of the very strong systems are taken from the

the correlation is produced by these systems, where galactic rotation almost certainly enters into the observed kinematics (see Art Wolfe's contribution to these proceedings). Note also that *all* of the known damped systems have $D \leq 15h^{-1}$ kpc, but that several of them have very small Mg II equivalent widths (some of them would not make it into a Mg II selected sample with $W_0 > 0.3$Å). Figure 3 emphasizes the need to explore the detailed relation of the absorption line kinematics to the actual galaxy morphology and sky orientation; this can be done to $z \sim 1$ with HST images in combination with high-resolution spectroscopy (see §2.6).

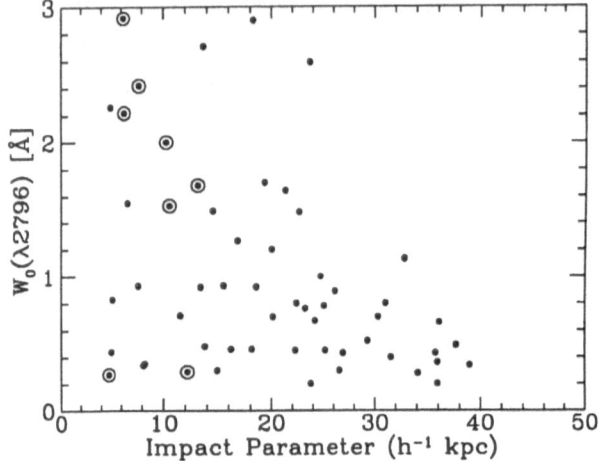

Fig. 3. Plot showing the Mg II equivalent width versus impact parameter. Points surrounded by circles are known damped Lyman α absorbers. The anti-correlation of line strength and impact parameter is significant at the 3.1σ level.

2.5   The Galaxy Luminosity Function

Armed with a very robust empirical determination of the scaling of galaxy gas cross-sections with luminosity, one can then turn the observed distribution of galaxy luminosities into a true luminosity distribution (provided that the absorption systems themselves have been drawn from a statistically homogeneous sample) simply by weighting the histogram by $1/R^2(L)$, which for the K luminosity would be $(L_K/L_K^*)^{-0.3}$. This cross-section weighting is equivalent to dividing by the effective "volume" probed by an absorption line survey

---

literature and are *not* part of our unbiased sample (although we have re-observed all of them as part of our survey). In a sample the size of ours ($\sim$ 60 absorbers) a homogeneous sample would contain only 1 or 2 systems with $W_0 > 2$ Å.

as a function of luminosity, directly analogous to the selection function in an apparent magnitude selected redshift survey. However, note that the volume probed in an absorption-selected survey is much more weakly dependent on luminosity ($L^{0.3}$ rather than $L^{1.5}$!) and therefore one is actually sampling the full range in galaxy luminosities at a given redshift much more uniformly in an absorption–selected sample. The absorbing galaxy K–band luminosity function is is shown in Figure 4; note that it is quite well-represented by a Schechter luminosity function with faint end slope $\alpha \approx -1$, to $M_K \approx -22$. The absolute normalization of the luminosity function is obtained from the same relationship that has traditionally been used to calculate R^* from an assumed luminosity function and scaling relation,

$$\frac{dN}{dz} = \frac{c\sigma n}{H_0}(1+z)(1+2q_0z)^{-0.5}$$

where

$$\sigma n = \pi \int_{L_{min}}^{\infty} \Phi(L)R^2(L)dL.$$

Here dN/dz, $R(L)$, and L_{min} are empirically determined, and Φ^* is calculated by assuming that a Schechter function which fits the observed luminosity distribution applies.

We note that the shape of the rest–frame B luminosity function is quite different (see Steidel et al. 1994, Steidel & Dickinson 1995), resembling a Gaussian rather than a Schechter function. We have attributed this difference to the fact that the population of intrinsically small (faint M_K), star forming "faint blue galaxies" (which contribute significantly to the field galaxy luminosity function even at relatively bright values of M_B) apparently do not possess significant gaseous envelopes. This is further evidence that the existence of the gaseous "halos" is dependent on galaxy *mass*, independent of star formation rate. In fact, we can see from the absorber luminosity function that the luminosity/cross-section scaling relation derived above *must* break down for luminosities smaller than $L \approx 0.05L_K$ – otherwise we would have found either many more very faint absorbers (the relatively weak dependence on luminosity and the knowledge that the luminosity function is rising at faint magnitudes requires this) or more "blank" fields. More work that explores the "transition" region of the luminosity function (at faint intrinsic luminosities), from absorber to non–absorber, would be very important in establishing the nature/origin of the absorbing gas.

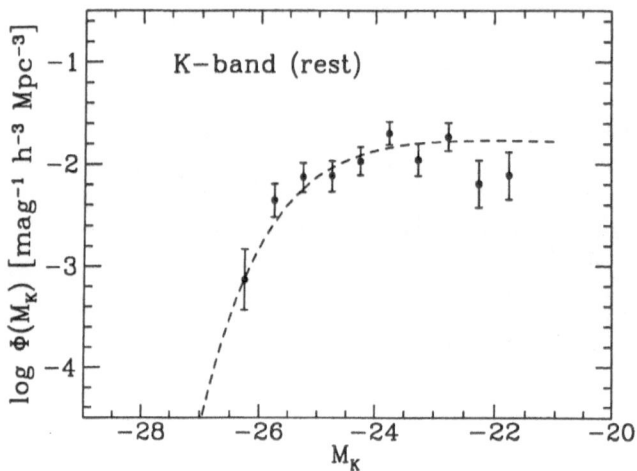

Fig. 4. Plot of the absorbing galaxy K–band luminosity function for the sample with $z \leq 1$ ($\langle z \rangle = 0.65$). A recent determination of the local field galaxy K–band luminosity function (Mobasher *et al.* 1993) is also shown (with arbitrary normalization).

2.6 Some Preliminary HST Imaging Results

We (Steidel, Dickinson, Meyer, & Sembach) have obtained a very deep (24,000 s) image with WFPC-2 on board HST of the field of the QSO 3C 336, which has 5 absorbing galaxies along a single line of sight, allowing us to directly observe the galaxy morphology and orientations with respect to the QSO sight line. First of all, we are very encouraged that the actual observed morphological types match very well the types we had assigned the galaxies on the basis of the spectra and optical/IR colors, in this field ranging from Sd/Im to S0. The $z = 0.892$ absorber is clearly a nearly edge-on mid-type spiral, but it also produces a Lyman α absorption line with $N(H\ I) = 5 \times 10^{19}$ cm^{-2} on the basis of our FOS spectrum. The galaxy has a projected separation from the line of sight of $15h^{-1}$ kpc, but after correction for inclination, if one insists that the absorption arises in the disk, the minimum disk extent would have to be $\sim 70h^{-1}$ kpc, in a galaxy with a luminosity of only $0.3L^*$. Despite intensive spectroscopy, the identification of the $z_{abs} = 0.656$ system, which produces a damped Lyman alpha line with $N(H\ I) = 2 \times 10^{20}$ cm^{-2}, remains ambiguous; the galaxy responsible must be considerably fainter than

$\sim 0.1L^*$. This may serve to caution us against assuming that all high column density gas must be confined to disks of luminous galaxies.[3]

The 5 galaxies producing metal lines along this sight line have impact parameters ranging from $8 - 47h^{-1}$ kpc, and on the basis of our FOS spectra there is a clear trend for the equivalent width ratio W(C IV)/W(Mg II) to increase with galactocentric distance. More data of this kind will clearly be interesting in investigating the actual ionization structure of the halos [see also Bergeron (1995), Lanzetta (1995)].

3 Absorbing Galaxies at Higher z

3.1 $z > 1$

One of the advantages of an absorption–selected sample is that the galaxies can be followed, using identical selection criteria, well beyond $z = 1$, where faint apparent magnitudes for objects of normal luminosity and lack of strong spectroscopic features cause conventional field galaxy surveys to break down. We have recently begun a survey designed to follow the absorption-selected galaxies to $z \sim 1.6$ (see Steidel & Dickinson 1995 for details) in order to establish the field galaxy luminosity function (using techniques described above) at $\langle z \rangle = 1.3$. Insisting on spectroscopic completeness would of course encounter the same difficulties as the field galaxy redshift surveys; however, for the absorption selected sample we make use of very strong predictions based on the $z < 1$ sample (on color, angular separation, apparent magnitude, etc.), together with a priori redshift information so that evolution of the galaxy population producing absorption can be evaluated using deep imaging (one optical passband and one near-IR (K) passband) alone. This allows us to establish the galaxy luminosity function over a significant range of intrinsic luminosity in redshift regimes that are unexplored by the field galaxy redshift surveys. Our preliminary results suggest that no spectacular evolution is occurring even to $z \sim 1.6$ in the field galaxy population, and in fact we have been able to establish that the space density of the normal Hubble sequence galaxies (that appear to dominate the absorption cross-section) remains constant to within $\sim 30\%$ (Steidel & Dickinson 1995), and that there has been at most modest passive evolution of the stellar populations (here again the K–band data are crucial for epoch-to-epoch comparisons) over that entire redshift range. Taken together with the results of the field galaxy redshift surveys, a scenario in which relatively luminous galaxies have been in place since very high redshifts, but where the population of very blue galaxies is undergoing substantial evolution even at relatively modest redshifts, seems to be indicated. A great deal of work is called for in understanding the possibly

---

[3] We also note that the *known* damped Lyman α systems in our $z < 1$ sample show a marginally significant tendency to be fainter and bluer than the Mg II selected sample as a whole. The mean luminosity is only $\sim 0.3L^*$.

intimate connection between the "halo" gas supply of the luminous galaxies (which we see in absorption) and the rapidly evolving "faint blue galaxy" population (see §4).

3.2 $z > 3$

Absorption line selection has clear advantages for "flagging" the sites of known very high redshift objects; since LLSs are now known far beyond $z = 4$, in principle the same selection criteria can be used at these extreme redshifts as at lower redshift to target searches for the absorbing galaxies. At $z > 3$, one can take advantage of a *guaranteed* spectroscopic feature in the spectrum of any object forming stars–a discontinuity at the rest frame Lyman limit of the galaxy, redshifted into the optical portion of the spectrum. One need not make any assumptions about the fraction of ionizing photons escaping a young galaxy, or even the details of the intrinsic spectral energy distribution of the galaxy, as a pronounced discontinuity will be present from consideration of the opacity of the IGM (Lyman α forest blanketing plus optically thick intervening clouds) alone (see Madau 1995). We were able to take advantage of the expected characteristic colors of objects in the redshift range $3 \leq z \leq 3.5$ in a custom photometric system, $U_n G \mathcal{R}$, to identify the object which may be producing the damped Lyman α absorption system at $z = 3.39$ toward Q0000−263 (Steidel & Hamilton 1992), and possibly to identify a group or cluster of objects associated with it (Steidel & Hamilton 1993; Giavalisco *et al.* 1994). Max Pettini and I have been continuing this program in an effort to establish the detectability of other LLS-selected high redshift galaxies in the same redshift range (Steidel & Pettini 1995); in total, we have identified 2 out of 5 of the systems observed (including Q0000−263), and found many $z > 3$ galaxy candidates that may be associated with the absorbers or with the background QSOs. Both putative absorbing galaxies have luminosities only slightly in excess of L^*. The relatively low detection rate compared to the lower redshift surveys may be ascribed to the fact that our current detection limits only allows us to detect galaxies of luminosity $\gtrsim 2L^*$. At present the population of absorbers at $z > 3$ is largely unconstrained (and a luminosity distribution consistent with the lower z absorbers cannot be ruled out), but observations going 1−2 magnitudes deeper (quite possible with 8−10 m telescopes) will be very telling. Also of interest, of course, is the galaxy environment and morphology at such early times. These can also be addressed currently using HST and large ground-based telescopes.

4   What Does It All Mean?

A general result, about which we can all rest assured at this point, is that the QSO metal line systems (at least, those that are optically thick in the Lyman continuum; see Ken Lanzetta's contribution for extensions to lower H I column density regimes) are produced by normal galaxies and therefore in studying the absorption line systems one is justified in claiming that we are learning about the history of such objects. However, the absorbers exhibit (at least to $z \sim 1.5$ or so) a very large range in star formation rate, color, luminosity, and from the limited data that exist, morphological type. This means that when we talk about evolution in metallicity, dust content, kinematics, etc., we are talking about a very wide range of galaxy types and we must keep this in mind – it may well be that the wide range in metallicity observed at $z \gtrsim 2$ (Pettini 1995) reflects very different evolutionary histories for various sub-populations of galaxies.

On the other hand, the one property of the of the galaxies that seems to be most important in determining the size and geometry of the absorbing gas is the galaxy *mass*, and the geometry and sizes inferred are remarkably regular in nature. The picture of spherical halos with a cutoff near $40h^{-1}$ kpc is no doubt a tremendous over-simplification of the true situation; however, again, the data indicate that something is regulating the extended gas distribution to a very large extent. That the star formation rate does not appear to be important to the nature of the halos (I believe) argues very strongly against a model in which the extended gas is produced by outflow from the central regions (at least one which is powered by star formation, as in models of giant "galactic fountains"). Rather, the data seem to point to *inflow* as the dominant source of gas, and the relatively short timescales for the existence of gas in the extended halos (because of cloud–cloud collisions and dissipation) means that the inflow must occur continuously over at least a Hubble time (the time period over which we observe the absorption). What is a self-consistent physical picture that can explain the regularity of the gas at an effective H I column density threshold of $\sim 10^{17}$ cm^{-2}, and make it weakly dependent on mass alone? I suspect that the link is in the pressure (and pressure gradient) of the medium confining the $\sim 10^4$ K clouds, in which case there would be a characteristic radius at which the pressure is sufficient to make the density of the infalling material just optically thick in the Lyman continuum in the presence of the prevailing UV radiation field (whereas the gas distribution will extend, at lower column densities, to much larger galactocentric radii [see Lanzetta 1995]). This pressure, if it is due to some hot medium, is expected to be dependent only on the overall size of the potential well, hence the mass dependence and roughly spherical geometry. Whether or not this picture is correct, there is clearly a great deal of theoretical work to be done in understanding the physics of the absorbing regions.

One final comment: it is not just from the new generation of 8-10m telescopes and high resolution spectrographs that the major advances will come in our field. It is clear that the picture one can assemble from *combining* the information obtained from the QSO spectroscopy and from HST and ground–based imaging and spectroscopy of the same galaxies will allow us to overcome the limitations imposed by photon starvation and actually understand how *normal* galaxies have evolved, from the epoch of their formation to the present.

I would like to thank my collaborators Mark Dickinson, Dave Meyer, Eric Persson, Max Pettini, and Ken Sembach for allowing me to quote the results of joint work prior to publication.

References

Bechtold, J., Ellingson, E. 1992, ApJ, 396, 20

Bergeron, J. 1995, this volume

Bergeron, J., Boissé, P. 1991, A&A, 243, 344

Colless, M. M. 1995, in Wide Field Spectroscopy and the Distant Universe, proc. of the 35th Herstmonceaux Conference, eds. S. Maddox and A. Aragon-Salamanca, in press.

Giavalisco, M., Steidel, C. C., Szalay, A. 1994, ApJL, 425, L5

Holmberg, E. 1975, in Stars and Stellar Systems, 9, Galaxies and the Universe, eds. A. Sandage, M. Sandage, and J. Kristian, (Chicago: University of Chicago Press), p 123.

Lanzetta, K. M. 1995, this volume

Lanzetta, K. M., Bowen, D. V. 1990, ApJ, 357, 321

Lilly, S. J. 1993, ApJ, 411, 501

Madau, P. 1995, ApJ, in press

Mobasher, B., Sharples, R. M., Ellis, R. S. 1993, MNRAS, 263, 560

Petit-Jean, P., Bergeron, J. 1990, A&A, 231, 309

Pettini, M. 1995, this volume

Schechter, P. 1976, ApJ, 293, 97

Songaila, A., Cowie, L. L., Hu, E. M., Gardner, J. P. 1994, ApJS, 94, 461

Steidel, C. C. 1993, in The Environment and Evolution of Galaxies, proc. of the 3rd Teton Astronomy Conference, eds. J. M. Shull and H. A. Thronson, (Dordrecht: Kluwer), p. 263

Steidel, C. C., Dickinson, M. 1995, in Wide Field Spectroscopy and the Distant Universe, proc. of the 35th Herstmonceaux Conference, eds. S. Maddox and A. Aragon-Salamanca, in press.

Steidel, C. C., Pettini, M. 1995, in preparation

Steidel, C. C., Dickinson, M., Persson, S. E. 1995, in preparation

Steidel, C. C., Dickinson, M., Persson, S. E. 1994, ApJL, 437, L75

Steidel, C. C., Hamilton, D. 1992, AJ, 104, 941

Steidel, C. C., Hamilton, D. 1993, AJ, 105, 2017

Steidel, C. C., Sargent, W. L. W. 1992, ApJS, 80, 1

Wolfe, A. M. 1995, this volume

Yanny, B., York, D. G. 1992, ApJ, 391, 569

York, D. G., Dopita, M., Green, R., Bechtold, J. 1986, ApJ, 311, 610

From Metal–Line Absorption Profiles to Halo Kinematics?

Christopher W. Churchill[1], Steven S. Vogt[1], and Charles C. Steidel[2]

[1] University of California, Santa Cruz, CA 95064, USA
[2] Massachusetts Institute of Technology, Cambridge, MA 02139, USA

Abstract. We present work in progress of a detailed study of the kinematic, chemical, and ionization conditions of the halos of intermediate redshift ($0.3 < z < 1.1$) Mg II absorbing galaxies. We aim to incorporate detailed imaging information about the absorbing galaxies themselves with studies of high resolution QSO absorption profiles. We address the questions: "what physical processes are consistent with observed absorption signatures?" and "what parts of galactic halos give rise to what types of characteristic absorption profiles?" Answers to these questions promise to provide a foundation upon which we can interpret both high redshift QSO absorption and Galactic absorption within theories of galactic formation and evolution.

1 Background

Over large look–back times, the imprinted conditions of galactic formation and evolution are available for study in the absorption profiles of QSO spectra. Intensive efforts by several researchers over the last decade have resulted in a basic picture of the nature of metal–line absorbers. Intermediate to low resolution spectroscopic surveys (Young *et al.* 1982, Tytler *et al.* 1987, Sargent *et al.* 1988a, 1988b, Petitjean & Bergeron 1990, Steidel & Sargent 1992, Bergeron *et al.* 1994) have placed metal–line absorbers on a firm statistical foundation, including absorber evolution, clustering, physical size distributions, and a qualitative picture of a typical Mg II absorption selected galaxy (Lanzetta 1992, but see Yanny 1992, Steidel 1992, Lanzetta 1993, Steidel 1993a, 1993b, Bergeron, these proceedings). Ground–based imaging programs of Mg II absorbing galaxies (Bergeron & Boissé 1991, LeBrun *et al.* 1993, Steidel 1993b, Steidel *et al.* 1995) have firmly established the direct association of metal–line absorption with galaxies, provided statistics on the luminosity/color evolution of these galaxies (Steidel *et al.* 1994, Steidel, these proceedings), and paved the road for case by case detailed studies in the drive to discern correlations between galaxy properties and absorption profiles (see for example, Lanzetta & Bowen 1990, 1992, Steidel & Dickinson 1992).

Studies incorporating both high resolution (≈ 6 km s^{-1}) absorption profiles and imaging information of $z < 1$ absorbing galaxies themselves, will ultimately lead to highly accurate measures of the kinematic, chemical, and ionization conditions in the halos of these galaxies. If we are to understand the implications of high–redshift absorption systems for theories of galactic formation and evolution and interpret the wealth of absorption line data available from the Galactic environment, we must first understand what types and what parts of galaxies give rise to intermediate–redshift absorption, where line of sight geometries and galaxy morphologies are accessible.

2   The Program

Our goal is to piece together a detailed picture of the physical conditions in galaxies giving rise to Mg II absorption. We have obtained HIRES (Vogt *et al.* 1994) spectra on the Keck 10–meter telescope for 24 QSO lines of sight, along which Mg II absorption systems are well documented (Steidel & Sargent 1992). These lines of sight include ≈ 10 *identified* (Steidel *et al.* 1995) Mg II absorbing galaxies, which are being further studied. Indeed, we have targeted QSO fields for which infrared and optical ground–based images exist and Hubble Space Telescope images are scheduled for cycles four and five. A comparative study of the various absorption characteristics sampled will allow a broader framework within which to interpret individual galaxies.

3   Kinematical Inferences from the Data

If the motions of ionized gas in halos are systematic, correlations between observed profiles and light paths through absorbing galaxies are implied (see for example, Lanzetta & Bowen 1992). Several models, including galactic fountains (Bregman 1980), chimneys (Norman & Ikeuchi 1989), supershells (MacLow & McCray 1988), cannonballs (Charlton & Salpeter 1989), outflow (Wang 1995), dwarf galaxy accretion (Wang 1993), and general infall (Steidel & Sargent 1992, Steidel, these proceedings) have been proposed. Each is expected to produce characteristic, if not unique, absorption profiles as seen along a given light path through an absorbing galaxy.

In Figure 1 we present a ground–based image of the Q1331+170 line of sight centered on the QSO (which has been removed) and a observed HIRES/Keck Mg II $\lambda 2796$ profile of the edge–on G1 (candidate) absorber at $z = 0.7443$ with impact parameter $D = 19.7h^{-1}$ kpc ($q_o = 0.05$, $h_o = H_o/100$). The redshifts of both G1 and G2 are not spectroscopically confirmed at the time of this writing. A striking aspect of the the Mg II profiles is the quasi-symmetric velocity "splittings". Such signatures are characteristic of radial inflow/outflow (Lanzetta & Bowen 1992). In this case, the regions giving rise to Mg II absorption have line of sight separations of ≈ 30 km s^{-1}.

As a speculative preliminary analysis on this system, we note that the overall profile shape and sub–component columns, velocities, and Doppler b values are *consistent* with multiple expanding supershells (MacLow & McCray 1988) located $R_g \approx 20$ kpc out along the disk and intercepted $z_g \approx$ few kpc above the galaxy plane of G1. A synthetic spectrum of supershells is shown for comparison to the observed profiles. The Mg II would

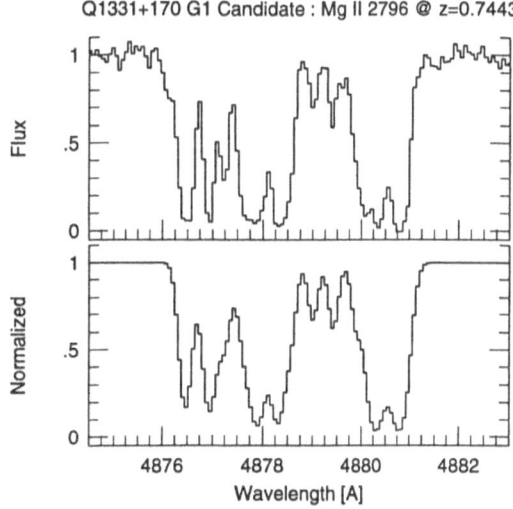

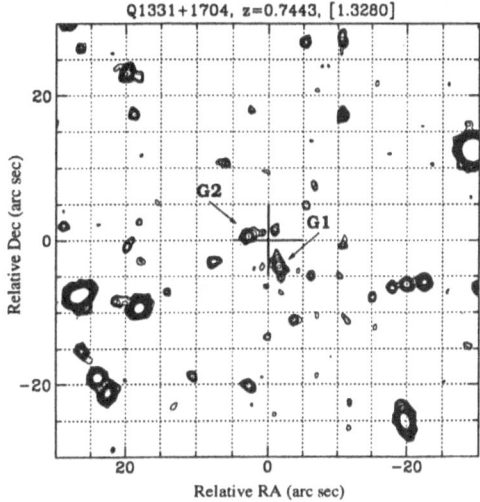

Fig. 1. (a) – (upper panel) HIRES/Keck Mg II profiles of the G1 $z = 0.7443$ candidate absorber to Q1331+170. (lower panel) Simulated spectrum of supershells with parameters discussed in the text. (b) – Ground–based image centered on the QSO (subtracted out).

be confined to thin "onion–layer" shells of $n_H \approx 0.1$ cm^{-3} and $T \approx 10^{4.0-4.5}$ K just inside each cool dense supershell radius, $R = 0.27 \left(L_{38}/n_o t_7^3\right)^{-0.2}$ kpc, where $t_7 = t/10^7$ yr. The ratios of the bubble central densities n_o to the supernovae "luminosities" $L_{38} = L/10^{38}$ are well constrained to be $\approx 10^{-3}$ cm^{-3} erg^{-1} by the 30 km s^{-1} line of sight velocity separations,

$$\Delta v_{\text{los}} = 2\dot{R} \left[1 - (z_g/R)^2\right]^{1/2} \quad \text{km s}^{-1},$$

where $\dot{R} = 16 \left(t_7^2 n_o/L_{38}\right)^{-0.2}$ km s^{-1}. The Δv_{los} remain virtually constant over the observed lifetime of the supershell until "breakout".

Acknowledgments. We are grateful to Ken Lanzetta, Mike Keane, and Jane Charlton for many insightful conversations. This work was partly support- ed by the National Academy of Sciences, through Sigma Xi, The Scientific Research Society, and by the California Space Institute.

References

Bergeron, J., *et al.* (1994), ApJ, 436, 33
Bergeron, J., and Boissé, P. (1991), A&A, 169, 1
Bregman, J. (1980), ApJ, 236, 577
Charlton, J.C., and Salpeter, E. (1989), ApJ, 346, 101
Lanzetta, K.M. (1992), PASP, 104, 835
Lanzetta, K.M. (1993), in *The Environment and Evolution of Galaxies*, eds. J.M. Shull and H.A. Thronson Jr., (Dordrecht : Kluwer Academic Publishers), p237
Lanzetta, K.M., and Bowen, D.V. (1990), ApJ, 357, 321
Lanzetta, K.M., and Bowen, D.V. (1992), ApJ, 391, 48
LeBrun, V., Bergeron, J., Boissé, P., and Christian, C. (1993), A&A, 279, 33
MacLow, M.M, and McCray, R. (1988), ApJ, 324, 776
Norman, C.A., and Ikeuchi, S. (1989), ApJ, 345, 372
Petitjean, P., and Bergeron, J. (1990), A&A, 231, 309
Sargent, W.L.W., Steidel, C.C., and Boksenberg, A. (1988a), ApJ, 334, 22
Sargent, W.L.W., Boksenberg, A., and Steidel, C.C. (1988b), ApJS, 68, 539
Steidel, C.C. (1992), PASP 104, 843
Steidel, C.C. (1993a), in *Galaxy Evolution: The Milky Way Perspective*, PASP Conf. Series 49, ed. S.R. Majewski, (San Francisco : PASP), p227
Steidel, C.C. (1993b), in *The Environment and Evolution of Galaxies*, eds. J.M. Shull and H.A. Thronson Jr., (Dordrecht : Kluwer Academic Publishers), p263
Steidel, C.C, and Dickinson, M. (1992), ApJ, 394, 81
Steidel, C.C., and Sargent, W.L.W. (1992), ApJS, 80, 1
Steidel, C.C., Dickinson, M., and Persson, S.E. (1995), ApJ, in preparation
Tytler, D., *et al.* (1987), ApJS, 64, 667
Vogt, S.S., *et al.*, SPIE, 2198, 362
Yanny, B. (1992), PASP, 104, 840
Young, P., Sargent, W.L.W., and Boksenberg, A. (1982), ApJS, 48, 455
Wang, B. (1993), APJ, 415, 174
Wang, B. (1995), ApJ, submitted

Keck HIRES Spectroscopy: Low-Redshift Mg II and Lyman α Systems

Donna S. Womble

California Institute of Technology, Astronomy 105-24, Pasadena, CA 91125, USA

Abstract. Motivated by the proposed connection between galaxies and the Ly α forest at low redshifts, we report on a sensitive search for Mg II absorption corresponding to Ly α lines detected by the *HST* Key Project at $z \leq 1.3$. Spectra of the bright QSO, PG 1634+706 were obtained using the High Resolution Echelle Spectrometer on the Keck Telescope. We do not detect Mg II absorption corresponding to the majority of Ly α lines although many have similar strength to those which do have associated Mg II lines. The non-detections provide useful limits on the metal abundance in both individual lines and averaged over all Ly α redshifts. We detect an excess of weak Mg II systems which also show corresponding C IV and Lyman-limit absorption when such data are available. In several cases, the extremely narrow line widths are consistent with thermal broadening at temperatures characteristic of Ly α forest lines. From the observed number density, we discuss implications for the size and luminosity of weak-line absorbers.

1 Background

For more than a decade, the consensus has prevailed that high redshift ($z \gtrsim 2$) Ly α forest lines are due to intergalactic clouds (Sargent *et al.* 1980). This hypothesis stems from the lack of detectable metals and velocity clustering in these systems. However, recently it has been suggested that the low redshift ($z \lesssim 1$) Ly α absorption systems arise from the extended gaseous halos of luminous galaxies (Lanzetta *et al.* 1995, also this volume). Similarly, Bahcall *et al.* (1995; also Boksenberg, this volume) found some evidence for velocity clustering in the low-z Ly α systems; they suggest that such line-groupings arise from relatively young clusters of galaxies.

2 Observations

To explore this absorber-galaxy connection from a metallicity viewpoint, we are presently searching for Mg II absorption corresponding to low-z Ly α lines in a subset of objects from the *HST* Quasar Absorption Line Key Project. Spectra of one such object, the bright QSO PG 1634+706 (V=14.9, $z_e = 1.334$), were obtained in June 1994 using the High Resolution Echelle Spectrometer (HIRES) on the Keck Telescope. Using two instrumental setups and a total integration of 2×2000 seconds, these spectra cover the wavelength range 4246–6663 Å with no inter-order gaps. For the Mg II $\lambda\lambda$2796, 2803 doublet, this corresponds to the redshift range, $z = 0.52$ up to the quasar emission redshift. At a velocity resolution of $6\,\mathrm{km\,s^{-1}}$ FWHM and typical signal-to-noise ratio of ≥ 50 to 1 per resolution element, these observations allow us to reach very sensitive limits on the column density of Mg II in foreground gas. These data reach a 5σ rest equivalent width limit, $W_o \leq 15\,\mathrm{mÅ}$ for an unresolved line – over 99% of the sampled redshift path; this corresponds to $\log N(\mathrm{Mg\,II}) \leq 11.6\,\mathrm{cm^{-2}}$ in the optically thin limit. Large portions of the spectrum (in the overlapping regions of both setups) reach an unprecedented limit of $W_o \leq 3\,\mathrm{mÅ}$ (5σ) or $\log N \leq 10.9\,\mathrm{cm^{-2}}$.

The UV spectrum of PG 1634+706 is presented in paper VII of the *HST* Quasar Absorption line Key project (Bahcall *et al.* 1995). This spectrum covers the wavelength range of 2224–3277 Å at a resolution of $230\,\mathrm{km\,s^{-1}}$ FWHM. For Ly α absorption, these data cover only a subset of the redshift range observed with HIRES starting from $z = 0.83$. Bahcall *et al.* (1995) detect numerous Ly α lines and three intervening C IV systems along this sightline. They indicate that in two of these three systems, Ly α lines appear to be clustered about the heavy-element absorption redshifts on velocity groupings of 1900 and $3500\,\mathrm{km\,s^{-1}}$. For most of the sampled redshift path, the *HST* spectrum has a limiting (observed) equivalent width limit of $135\,\mathrm{mÅ}$ (4.5σ) corresponding to $\log N(\mathrm{H\,I}) \lesssim 13.2\,\mathrm{cm^{-2}}$ for a Doppler parameter, $b = 30\,\mathrm{km\,s^{-1}}$ at the lower redshift end.

3 Heavy Element Abundances

Excluding lines associated with C IV systems and probable blends with Galactic absorption, Bahcall *et al.*(1995) detect a total of 43 Ly α lines (4 are "possible" identifications) in the spectrum of PG 1634+706. These Ly α systems have rest equivalent widths ranging from 50 to $770\,\mathrm{mÅ}$. We do not detect any of the corresponding Mg II absorption lines. At the Ly α redshifts, the Mg II rest equivalent width limits range from 3 to $7\,\mathrm{mÅ}$ (5σ).

We used the method, pioneered by Norris, Hartwick, & Peterson (1983), to estimate the typical magnesium abundance of Ly α clouds. In this approach, portions of the spectrum containing the predicted positions of the Mg II absorption lines are shifted to the rest frame of each Ly α line and then added to form a composite spectrum. The resulting spectrum has a significantly higher signal-to-noise ratio from which a more sensitive abundance limit may be derived. However, due to the low velocity resolution of the UV spectrum, this method is prone to errors resulting from inaccuracies in the Ly α redshifts and poorly determined H I column densities. We applied a small zero-point offset to the Ly α redshifts in order to bring the three heavy-element systems (C IV and Mg II) into agreement. With these considerations in mind, we produced a composite spectrum by adding data at the redshifts of 39 Ly α lines with $0.05 \leq W_o \leq 0.77$ Å and a mean redshift, $\bar{z}=1.11$; data for the other 4

Ly α redshifts were ignored due to anomalies in our spectrum associated with poor sky-subtraction, instrumental defects, and nearby telluric features.

No evidence for Mg II absorption is seen in the composite spectrum although the signal-to-noise ratio exceeds 400 to 1 per pixel. The rest equivalent width limit, $W_o(\lambda 2796) \leq 0.65$ mÅ (5σ) corresponds to an optically-thin column density, $\log N(\text{Mg II}) \leq 10.20$ cm^{-2}. The composite Ly α line has an equivalent width $W_o(\lambda 1216) = 0.266$ Å or approximate column density, $\log N(\text{H I}) = 13.99$ cm^{-2} (assuming $b = 30$ km s^{-1}). By adding only a subset of the 10 strongest Ly α lines with $0.40 \leq W_o \leq 0.77$ Å and a mean redshift of $\bar{z} = 1.03$, we reach $W_o(\lambda 2796) \leq 1.25$ mÅ (5σ) or $\log N(\text{Mg II}) \leq 10.48$ cm^{-2} with $W_o(\lambda 1216) = 0.566$ mÅ or $\log N(\text{H I}) = 15.72$ cm^{-2}.

In order to derive metal abundances from these column densities, we must assume a physical model for the Ly α clouds. Chaffee *et al.*(1986) calculated photoionization models for clouds radiated by the integrated flux from QSOs; they evaluate column densities for the most abundant ions as a function of total hydrogen density at three different redshifts. Consistent with these models, we assume the clouds are highly ionized with H/H I= 10^4 and Mg/Mg II= 10^3. These ion ratios are valid for $\log n_{\text{Htot}} = -3.0$ at $z = 1$ however this density may be inconsistent with values derived from high-z limits on the size of Ly α clouds. With these assumptions, we find [Mg/H]≤ -0.2 for our typical Ly α clouds; we have taken Mg/H$\odot = 2.57 \times 10^{-5}$. For just the subset of strong Ly α lines, we get [Mg/H]≤ -1.6. These abundance limits (at $z \simeq 1$) should be compared with those derived in a similar manner for C IV absorption at $z \simeq 2.5$. For Ly α lines of similar strength, Lu (1991) found [C/H]$= -3.2$ whereas Tytler & Fan (1994) derived a limit of [C/H]≤ -2.0 for significantly weaker systems.

4   Velocity Structure, Cloud Temperatures

We do not find any isolated, weak Mg II lines; such lines are seen only as "satellites" to the stronger systems. Here, we define typical "weak" and "strong" lines as having $W_o \sim 10$ mÅ and ~ 35 mÅ, respectively. Based on the photoionization models presented in Steidel & Sargent (1992), some of the "weak" satellite lines appear to have sufficiently low column densities such that they are not optically thick at the Lyman Limit. These Mg II lines are not grouped on velocity scales indicative of galaxy clusters; the observed velocities are consistent with individual galaxy dynamics. In the most extreme system, Mg II components are seen over only 150 km s^{-1}.

In several cases, we detect unresolved, single-component Mg II lines. From these profiles, we derive upper limits to the intervening cloud temperatures assuming the extremely narrow widths are strictly due to thermal broadening. In two specific cases, the implied temperatures are less than 15000 and 35000 K at $z = 0.8182$ and 0.6535, respectively. These are somewhat cooler than typically assumed for high redshift Ly α forest clouds; they imply Ly α line-widths of 16 and 24 km s^{-1} rather than the fiducial value, $b = 30$ km s^{-1}.

5   Number Density of Mg II Systems

Over the sampled redshift path, we detect an excess of Mg II systems a-
long this sightline. Where comparable UV spectra exist, these systems also
show corresponding C IV and Lyman-limit absorption. If we consider this
one sightline to be a mini-survey for Mg II lines with $W_o \geq 0.015\,\text{Å}$, then we
can compare the observed number of systems with the value predicted from
surveys for significantly stronger lines. Integrating (dN/dz) from Steidel &
Sargent (1992) over our redshift path $0.52 \leq z \leq 1.334$, we expect to see a
total of 0.76 ± 0.24 Mg II systems with $W_o \geq 0.300\,\text{Å}$. In fact, we detect one
such system, and a total of five systems with $W_o \geq 0.015\,\text{Å}$. Here, we have
counted all components spread over $\leq 150\,\text{km s}^{-1}$ as a single system.

This excess number of systems may be explained as due to either an
increase in the geometric cross section of galaxies, or an increase in the number of galaxies which produce Mg II absorption. The former explanation implies that absorbers must be a factor of 2.6 times larger in radius. For a
fiducial size, $R(L^*)=35\,h^{-1}$ kpc (Steidel, this volume) for stronger lines with
$W_o \geq 0.300\,\text{Å}$, then the excess implies a radius of $90\,h^{-1}$ kpc for lines with
$W_o \geq 0.015\,\text{Å}$. Alternatively, the excess number of Mg II systems may be
accounted for by a 2 magnitude decrease in the average luminosity of absorbing galaxies. Starting with the average luminosity for galaxies which produce
the stronger Mg II systems, $\langle M_B \rangle = -19.5$ (Steidel, this volume), we need to
integrate a Schechter luminosity function down to $\langle M_B \rangle = -17.5$ in order to
explain the increase in number density for lines with $W_o \geq 0.015\,\text{Å}$. These
last results are very preliminary and we hope to improve on the statistics
with similar data from more sightlines.

Acknowledgments. I wish to thank my collaborator W. L. W. Sargent, and
the *HST* Quasar Absorption Line Key Project Team for permission to discuss
results prior to publication. Support for this work was provided by NASA
through grant number HF-1040.01-92A from STScI, which is operated by
AURA, Inc., under NASA contract NAS5-26555.

References

Chaffee, F.H., Foltz, C.B., Bechtold, J., Weymann, R.J. 1986, ApJ, 301, 116
Bahcall, J.N., Jannuzi, B.T., Schneider, D.P., *et al.* 1992, ApJ, 397, 68
Bahcall, J.N., Bergeron, J., Boksenberg, A., *et al.* 1995, in preparation
Lanzetta, K.M., Bowen, D.V., Tytler, D., Webb, J.K. 1995, ApJ, in press
Lu, L. 1991, ApJ, 379, 99
Morris, S.L., Weymann, R.J., Dressler, A., *et al.* 1993, ApJ, 419, 524
Norris, J., Hartwick, F.D.A., Peterson, B.A. 1983, ApJ, 273, 450
Sargent, W.L.W., Young, P.J., Boksenberg, A., Tytler, D. 1980, ApJS, 42, 41
Steidel, C.C., Sargent, W.L.W. 1992, ApJS, 80, 1
Tytler, D., Fan, X.-M. 1994, ApJ, 424, L87

Mg II Absorption from Low-Redshift Galaxies

David V. Bowen[1], J. Chris Blades[1] and Max Pettini[2]

[1] Space Telescope Science Institute, 3700 San Martin Drive, Baltimore, USA
[2] Royal Greenwich Observatory, Madingley Road, Cambridge, UK

Abstract. We present some of the results of a *Hubble Space Telescope* (*HST*) programme designed to search for Mg II absorption lines in the disks and halos of low-redshift galaxies. We discuss the properties of four galaxies which produce Mg II absorption and show how the environment of a galaxy directly affects the characteristics of the absorption.

1 Introduction

In trying to understand the origin of absorbing gas at high-redshift, observing how *nearby* galaxies absorb is extremely interesting, because the morphology of the galaxy, its environment, and – on a more detailed level – the precise region of its ISM intercepted by a sightline, is most easily discerned at low-redshift. Such galaxies can be studied using the full range of observational techniques currently available, from radio to X-ray wavelengths, data which can then be used to infer the nature and origin of absorbing (and non-absorbing) gas.

Our HST programme uses the Goddard High Resolution Spectrograph (GHRS) to search for Mg II absorption in the spectra of QSOs or supernovae behind or in nearby galaxies, at $\sim 15\,\mathrm{km\,s^{-1}}$ resolution. We describe below four low-redshift Mg II absorbing galaxies and indicate how the high resolution of the data, and information from other species of absorption lines, can be understood with a knowledge of the galaxy and its environment. Full details can be found in Bowen, Blades & Pettini (1995) which also discusses the results of the survey as a whole. Unfortunately, only a few high-redshift Mg II systems have been observed at high spectral resolution (Lanzetta & Bowen 1992 and refs therein); hence information which might be gained from understanding the velocity field of the gas (derived from mapping components comprising a line) is unavailable, even if the absorbing galaxy is identified. Therefore, we have also simulated how our GHRS data would appear if found in a ground-based QSO absorption lines survey, by degrading the resolution of our data to $100\,\mathrm{km\,s^{-1}}$ (FWHM) and signal-to-noise of 20, values typical of recent surveys (e.g. Sargent, Steidel & Boksenberg 1988; Fig. 1). These simulations are included to suggest what types of systems such lines might arise in.

2   Low-Redshift Mg II Absorbers

Q0637−752 and The Large Magellanic Cloud (LMC): Strong Mg II absorption is detected from our Galaxy and from the LMC. Two components can be seen from the LMC, at $v_\odot \approx 255$ and $325\,\mathrm{km\,s^{-1}}$, velocities which correspond to the two H I disks which make up the LMC (e.g. Luks & Rohlfs 1992). The sightline passes 6.7 kpc from the center of the galaxy: this is the furthest that the two H I disks have been traced.

If the absorption were observed from the ground at low resolution, two components would probably be identified and the system classified as a single system. Even if observed at high resolution, the bimodality of the lines might be explained by assuming radial flows of gas from a galaxy. There would be few clues that absorption arose from an irregular starbursting galaxy colliding with a bright spiral (the Milky Way) and a smaller dwarf galaxy (SMC).

Q1219+755 and NGC 4319: The sightline toward Q1219+755 (Mrk 205, $z = 0.07$) passes $3 - 5\,\mathrm{kpc}$ from the center of NGC 4319 ($z = 0.004$). Mg II was found by Bahcall et al. (1993) with the FOS while Mg II and C IV were observed by Bowen & Blades (1993) with the GHRS. The gas is more highly ionized than Milky Way halo gas, and analysis of ROSAT data suggests that NGC 4319 is embedded in hot X-ray emitting gas, possibly arising from the galaxy's interactions with neighbours NGC 4219 and NGC 4386.

If observed at low resolution, the absorption would appear as particularly weak and unremarkable Mg II and C IV lines with an equivalent width ratio of ~ 1. Weak, ionized gas might be expected to arise at the outskirts of a galaxy halo, several tens of kpc from its center. This low-redshift galaxy shows instead that ionized gas can exist close to the center of a galaxy.

Q1543+489 and G1543+4856: G1543+4856 is an edge-on, late-type galaxy $45\,h^{-1}\,\mathrm{kpc}$ from the line of sight at $z = 0.0749$; G1543+4854 is a neighbouring galaxy at $z = 0.0753$, $83\,h^{-1}\,\mathrm{kpc}$ from the QSO sightline. Moderately strong Mg II absorption is detected at a redshift exactly matching that of G1543+4856.

At first sight G1543+4856 seems like a typical Mg II absorber, with absorption arising in the outer regions of the galaxy. However, Ca II is also detected in the optical spectrum of the QSO (Bowen et al. 1991), which indicates a substantial H I column density ($\gtrsim 10^{18-19}\,\mathrm{cm^{-2}}$) along the line of sight. The galaxy is also about half as luminous as that expected for a Mg II absorber with a similar absorbing gas cross-section (see the review by Steidel elsewhere in these proceedings). Although a low luminosity galaxy may remain unrecognized if precisely aligned with the QSO, it seems more likely that the Mg II and Ca II absorption arises in the extended disk of G1543+4856, or from intergalactic debris produced by past interactions between G1543+4856 and G1543+4854. The high-resolution data from the GHRS shows no complex structure which might be expected from an interacting system, but debris so far from either galaxy may have little kinematic structure.

SN1993J and the M81 group: The line of sight towards SN1993J passes through the inner southwest arm of M81, 2.7 kpc from the center of the galaxy. M81 and neighbouring galaxies – M82, NGC 3077, and NGC 2976 – constitute a complex interacting group, with tidally stripped H I linking individual galaxies over an area of 50 x 100 kpc (Yun, Ho & Lo 1994). Mg II, Mg I, and Na I were observed at $4 - 5\,\mathrm{km\,s^{-1}}$ resolution with ECH-B of the GHRS and from the ground (Bowen et al. 1994). Individual absorption components span a velocity range of $-163 < v_\odot < +225\,\mathrm{km\,s^{-1}}$ grouped into three subcomplexes, from the disk of M81, from the Milky Way disk and halo, and from intergalactic tidal debris.

If observed at low spectral resolution, the Mg II lines would be strong with perhaps some suggestion of more than one component. There would be little evidence that the sightline passed through two spiral disks and debris from group interactions, however. Even if the lines were observed at high resolution, it is unclear whether the $390\,\mathrm{km\,s^{-1}}$ span of the components would be interpreted as gas cloud motions within a single galactic halo, or as gas from two galaxies close in velocity space plus tidal debris.

3 Summary

Observing the characteristics of absorption lines – their strength, the complexity of components comprising a line, the ionization state of the absorbing gas – from known, easily studied galactic systems offers an excellent method of 'calibrating' the origin of higher redshift systems. If the low-redshift galaxies studied thus far with HST are in any way representative of high-redshift systems, there may be many dangers inherent in constructing models of generic galactic halos based on statistical properties of QSO absorption line surveys.

References

Bahcall, J. N., Jannuzi, B. T., Schneider, D. P., Hartig, G. F., Jenkins, E. B. 1992, ApJ, 398, 495

Bowen, D. V., Blades, J. C., Pettini, M. 1995, ApJ in press

Bowen, D. V., Blades, J. C. 1993, ApJ, 403, L55

Bowen, D. V., Roth, K. C., Blades, J. C., Meyer, D. M. 1994, ApJ, 420, L71

Bowen, D. V., Pettini, M., Penston, M. V., Blades, J. C. 1991, MNRAS, 249, 145

Lanzetta, K. M., Bowen, D. V. 1992, ApJ, 391, 48

Luks, Th., Rohlfs, K. 1992, A&A, 263, 41

Mathewson, D. S. & Ford, V. L. (1984): Structure and Evolution of the Magellanic Clouds, ed. van den Bergh, S., de Boer, K. S. (Kluwer: Dordrecht), p. 125

Sargent, W. L. W., Boksenberg, A., Steidel, C. C. 1988, ApJS, 68, 539

Yun, M. S., Ho, P. T. P., Lo, K. Y. 1994, Nat, 372, 530

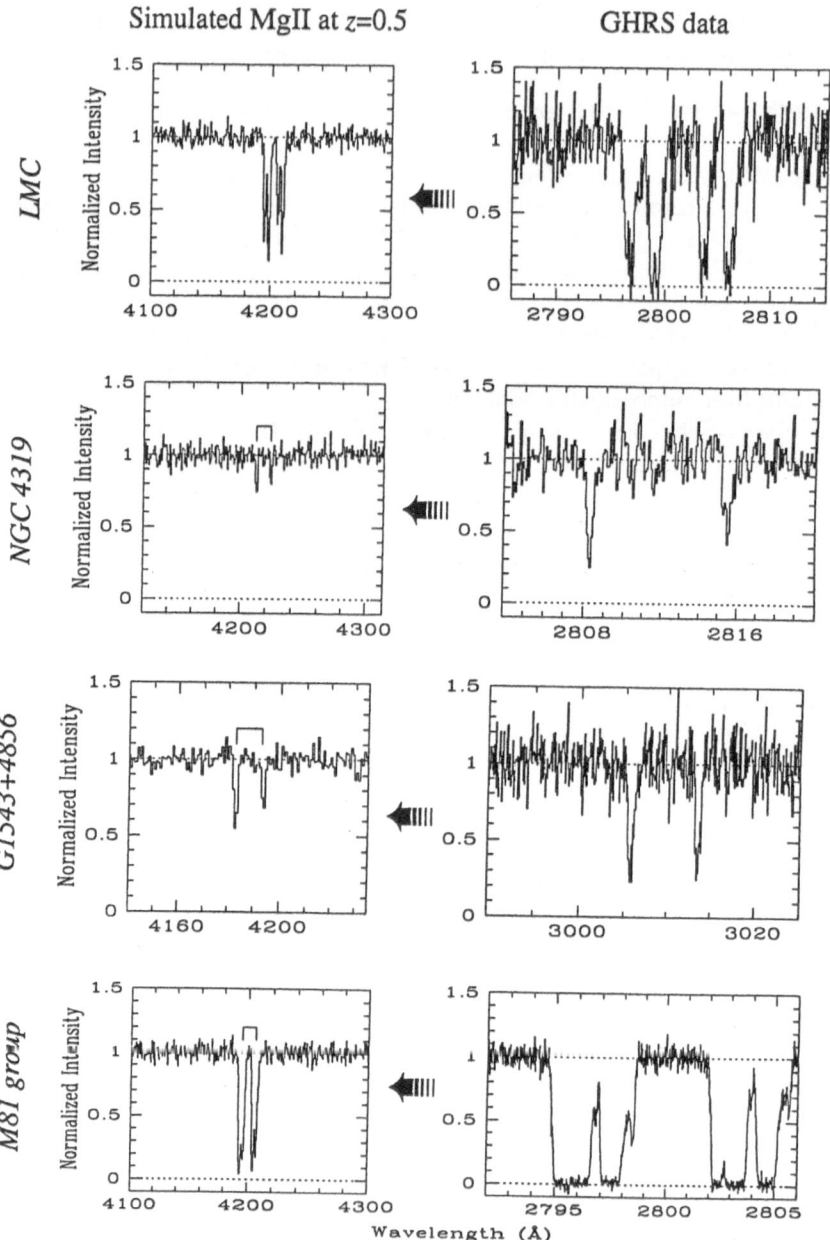

Fig. 1. Spectra from four low-redshift Mg II absorbing galaxies. The right column shows the GHRS data obtained toward QSO sightlines intercepting the galaxies. All data were taken with the G270M of the GHRS, which has a resolution of $\sim 15\,\mathrm{km\,s^{-1}}$ FWHM, except for that of SN 1993J, which was observed with ECH-B ($\sim 3\,\mathrm{km\,s^{-1}}$ FWHM). The left column shows how the Mg II lines would appear if observed at $100\,\mathrm{km\,s^{-1}}$ FWHM and signal-to-noise of 20, i.e., the same as ground-based optical QSO absorption line surveys.

The Sizes of MgII Absorption Systems

Michael J. Drinkwater[1], Rachel L. Webster[2] and Peter A. Thomas[3]

[1] Anglo-Australian Observatory, Coonabarabran, NSW 2357, Australia
[2] School of Physics, University of Melbourne, Parkville, Victoria 3052, Australia
[3] MAPS, University of Sussex, Brighton BN1 9QH, U.K.

Abstract. We present the results of large imaging survey designed to detect the galaxies associated with MgII absorption in a sample of bright QSOs. The redshifts of some of the galaxies have been confirmed spectroscopically and we infer large sizes (up to 30 h^{-1} kpc) for the absorption systems. Our total QSO sample was not selected with any bias towards the presence of MgII absorption, but we find a significant excess of galaxies close to the lines-of-sight to the QSOs (at separations less than about 10 arcsec) in all the sample. We attribute this excess to the magnification by gravitational lensing of QSOs which would otherwise be too faint to be included in our sample.

1 Introduction

The most direct approach to studying the properties of the metal-rich absorbers is to choose systems for which both images and spectroscopic data can be obtained. Ground-based studies are limited to Mg II absorption regions in the redshift range $0.15 < z \lesssim 0.7$, the lower limit given by atmospheric UV-absorption and the upper limit by the limiting magnitude at which galaxies might be imaged by a CCD. One such absorption system is found over the full path to a typical QSO at $z = 2$ (Thomas and Webster 1990).

Bergeron and Boissé (1991; hereafter BB91) have undertaken the most extensive search for galaxies which are at similar redshifts to known low redshift Mg II absorption systems, and are close to the line-of-sight to the QSO. In most cases a galaxy is found near the line-of-sight to the QSO with a similar velocity to the Mg II absorption line. Bechtold and Ellingson (1992; hereafter BE92) have searched for MgII absorption in the spectra of QSOs behind fields containing galaxies of known redshift. Notably they found 3 galaxies closer than $30h^{-1}$kpc to QSO lines-of-sight that did not produce absorption. Both these works estimate very large radii for the galaxy halos which contain the Mg II absorbing gas.

In this paper we take a different approach by looking for galaxies close to the line-of-sight to a sample of bright QSOs that is unbiased with respect to the presence of MgII absorption and then relate the presence of close galaxies to the detection of MgII absorption. We have previously presented the MgII statistics in Drinkwater, Webster and Thomas, (1993, Paper 1) and our further conclusions that there is a bias caused by gravitational lensing in Thomas, Webster and Drinkwater, (1995, Paper 2). Here we summarise those results and present new spectroscopic measurements confirming the conclusions of Paper 1 as well as showing how the lensing bias found in Paper 2 can explain the large absorber radii found by BB91 and BE92.

2 MgII Absorber Sizes

Our QSO sample was based on published samples with intermediate resolution optical spectroscopy, so the presence of low redshift Mg II absorption lines could be determined. We obtained R-band CCD images of 71 bright ($m_V < 18$) QSOs using the Observatoire du Mont Mégantic 1.6m Telescope. The observations are described fully in Paper 1. We searched our data for galaxies close to the line-of-sight to the QSOs, and used a statistical test to identify the galaxies most likely to be associated with the absorption systems. We identified 10 galaxies as being likely absorbers near the 17 QSOs with detected absorption. Four of these are confirmed by BB91. Our recent AAT observations confirm our conclusions that two other close galaxies were not associated with the absorption (16.8″ from QSO 0843+136 at $z_{gal} = 0.372$ c.f. $z_{abs} = 0.6$ and 23.4″ from QSO 1511+103 at $z_{gal} = 1.259$ c.f. $z_{abs} = 0.4$— this 'galaxy' is actually a second QSO). We also found a further example of a close galaxy not associated with any absorption, 5.9″ from QSO 1054-034 with $z_{gal} = 0.372$ and an impact parameter of $20h^{-1}$kpc.

The associated galaxies have a mean luminosity of $0.5\,L_*$, assuming they lie at the absorption redshift. The distribution of impact parameters is shown in Fig. 1. The flat distribution out to large radii ($> 30h^{-1}$kpc) suggests that the absorption systems may not be gravitationally bound to the observed galaxies, but may be part of larger extended systems.

The QSO sample is unbiased with respect to galaxies near the line-of-sight, so we can compare the observed number of absorption systems to that predicted by a simple model with a constant covering factor in Mg II absorbing gas within $30h^{-1}$kpc of each detected galaxy. The model is statistically consistent with a covering factor of unity, but since we have found examples of galaxies with no absorption to the detection limit, the covering factor must be less than unity.

3 Is There a Lensing Bias?

3.1 QSO-Galaxy Association Statistics

We noted that even the QSOs with no detected MgII absorption showed an excess of close galaxies. We restricted our sample further to 64 QSOs with redshifts $z > 1$ (see Paper 2) and found a significant excess of close neighbours at separations less than about 10 arcsec. This is shown in Fig. 2 which plots the distribution of normalised nearest-neighbour distances (defined in

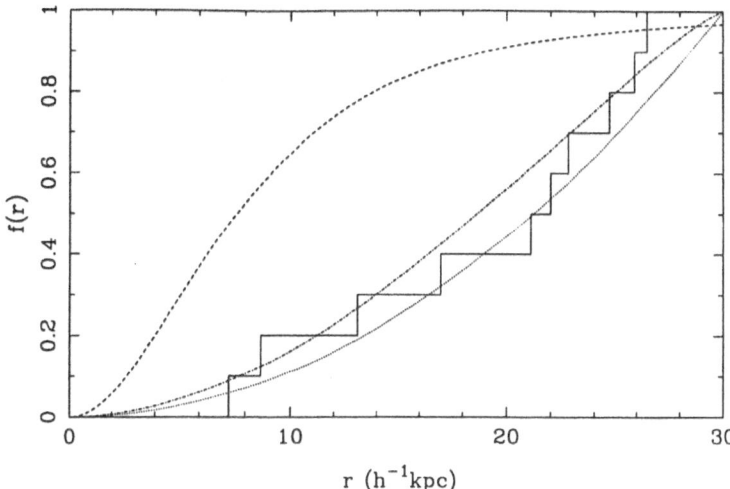

Fig. 1. Cumulative frequency distribution of the impact parameters of the galaxies detected and identified as Mg II absorbers close to the lines of sight to 10 QSOs (solid histogram). The curves show the distributions predicted for three models: a uniform mass distribution (dots), a randomly inclined disc (dash-dot) and a centrally-condensed mass distribution (dashes).

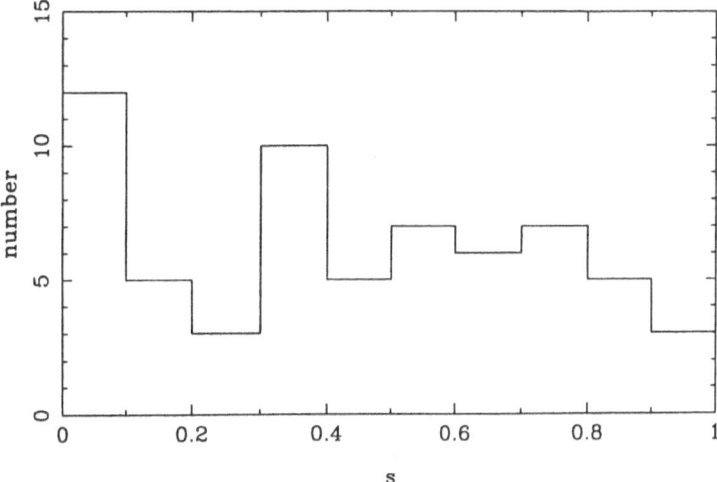

Fig. 2. Distribution of normalised nearest galaxy neighbour statistics s. The expected distribution for randomly distributed galaxies is constant.

Paper 2). There is an excess of galaxies in the smallest bin significant at the 2.4% level. We interpret this as a bias caused by amplification of the QSOs by gravitational lenses associated with the close galaxies.

3.2 Predicted Sizes

The gravitational lensing bias may explain a problem with earlier estimates of the effective absorption radius. The method used by both BB91 and BE92 involves modelling the absorbers as spherical galaxies of radius R_* drawn from a standard luminosity function for field galaxies and comparing the predicted number of absorbers per unit redshift with observed values to solve for the unknown, R_*. This approach can give very large radii, up to 76-$120h^{-1}$kpc when a covering factor of less than unity is used (BE91). Such large radii are not only much larger than the HI in local galaxies, but are also much larger than any impact parameters for the identified absorbing galaxies.

If there is a lensing bias, the true galaxy density near these bright QSOs is higher than for average field galaxies by a factor of between 1.8 and 2.8 (see Table 4 of Paper 2). Since $R_* \propto 1/\sqrt{\sigma}$ where σ is the mean galaxy density, the overdensity of galaxies we detect near the QSOs would reduce the estimate absorber radius by a factor of 0.6 to 0.7 to values still large, but more consistent with observations.

4 Summary

1. Our data are consistent with every MgII absorption system being associated with a $0.1 \leq L/L_* \leq 1.3$ galaxy.
2. The distribution of impact parameters is flat to a maximum value of $b \geq 30h^{-1}$kpc.
3. The covering factor $f < 1$ at impact parameters $b \geq 30h^{-1}$kpc.
4. There are excess galaxies within 10 arcseconds of 30% of all bright QSOs: a bias we attribute to gravitational lensing and which can explain earlier calculations of very large absorber radii R_*.

Acknowledgements. Mont Mégantic Observatory is supported by grants from the Canadian (NSERC) and Québec (FCAR) governments. MJD acknowledges travel support from the Australia/France Cooperation Fund.

References

Bechtold J., Ellingson E., 1992, ApJ, 396, 20 (BE92)
Bergeron J., Boissé P., 1991, A&A, 243, 344 (BB91)
Drinkwater M.J., Webster R.L., Thomas P.A., 1993, AJ, 106, 848 (Paper 1)
Thomas P.A., Webster R.L., 1990, ApJ, 349, 437
Thomas P.A., Webster R.L., Drinkwater M.J., 1995, MN, in press (Paper 2)

Hunting for MgII Absorbers with Fabry-Perot Imaging

Guido Thimm

European Southern Observatory, Karl-Schwarzschild-Strasse 2, D-85748 Garching bei München, Germany

Abstract. The ideal tool to identify faint, extragalactic sources by their line emission is narrow band imaging with a medium resolution (FWHM=1nm) Fabry-Perot filter. We applied this method to search for [OII] 372.7nm emission emitted by MgII-absorbers. Three targets were observed: Q1209+108 (z_{abs}=0.39), Q1248+401 (z_{abs}=0.77) and Q1206+460 (z_{abs}=0.93). The first absorber had been identified by Cristiani (1987) and was re-observed to check the sensitivity of the method. While in the case of Q1248 no emission line galaxy was found, in the field of Q1206 an emission line object at the absorber redshift was identified at a projected distance of $41h^{-1}$kpc (h=H_0/100 km s^{-1}Mpc^{-1}, q_0=0.1)

1 Introduction and Observations

Galaxies at $z \approx 1$ can be discovered by identifying the galaxy causing MgII-absorption lines in the spectrum of a background QSO. However, the number of confirmed MgII-absorbers is still too small (see e.g. Bergeron et al. 1991, Steidel et al. 1994) to investigate if the absorbers exist as isolated objects or whether they are located in groups of irregular galaxies (Yanny & York 1992). To investigate this controversy, a complete and sensitive survey for emission line galaxies in the vicinity of the redshift of an MgII-absorber is necessary.

This is possible by a technique applied by Thimm et al. (1994) to the galaxy cluster around 3C295 at z=0.45. A sequence of Fabry-Perot images covering a certain velocity interval around a certain emission line like [OII] 372.7nm can be used to construct spectra of the galaxies in the field of the QSO. They can be checked for the presence of emission lines. Here, we do not present such sequences but a first result from observations, in which the Fabry-Perot was tuned to cover only the redshifted [OII]-emission as well as the continuum next to the line. All data were obtained with the 3.5m telescope on Calar Alto in Feb. '93.

The absorber close to Q1209+107 was previously identified by Cristiani (1987). It was re-observed to check our sensitivity. Q1248 and Q1206 were observed to search for high-redshift absorbers. For each on- and off-band exposure we integrated 2×1500s. Basic data for our target objects are summarized in Tab. 1.

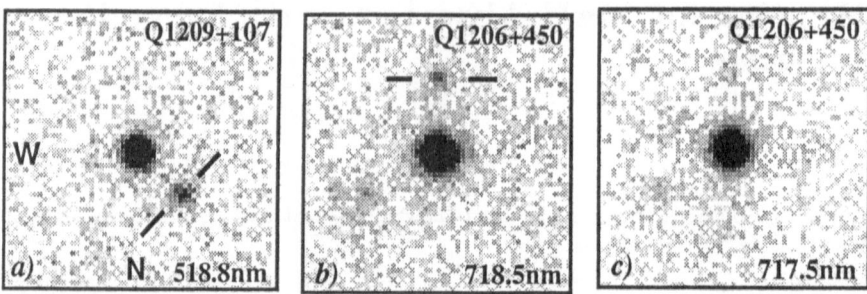

Fig. 1. Fabry-Perot Images of the central $15 \times 15'$ for the two QSOs where MgII-absorbers were identified. Their position is indicated. Fig. 1b) clearly shows the [OII]-emission, while it can not be detected in the off-band image (Fig. 1c).

Results and Discussion

The absorber near Q1209 was clearly resolved (Fig. 1a). We measured an [OII]-flux of $3.8 \pm 0.2 \times 10^{-16}$erg cm^{-2}s^{-1}, consistent with the value of $4.2 \pm 0.2 \times 10^{-16}$erg cm^{-2}s^{-1} obtained by Cristiani (1987). No other emission line galaxy was found in the total field of $3' \times 3'$. Fig. 1b shows the on-band image of Q1206 with the absorber appearing south of QSO. Its [OII]-flux was measured to be $9.2 \pm 1.6 \times 10^{-17}$erg cm^{-2}s^{-1}. The projected distance form the QSO sight-line is $41h^{-1}$kpc In the field of Q1248 we did not detect any emission line feature above our 3σ limit of 5×10^{-17}erg cm^{-2}s^{-1}. We conclude that although the overall exposure time required to determine the absorbers redshift is similar to slit spectroscopy, the latter method cannot avoid to spend telescope time on fore- or background galaxies. Therefore, Fabry-Perot imaging of QSO could be a more efficient way of identifying high-redshift field galaxies.

Table 1. Basic properties of the MgII-absorbers observed[1]

QSO	z_{emi}	z_{abs}	$w_{279.6nm}$/nm	θ /''	$f_{[OII]}/10^{-16}$erg cm^{-2}s^{-1}
Q1209+107	2.19	0.39	0.10	6.9	4.2
Q1248+401	1.03	0.77	0.08	-	-
Q1206+450	1.15	0.93	0.10	8.3	9.2

[1]Redshifts and equiv. widths from Steidel & Sargent, (1987), ApJS, 80, 1

References

Bergeron, J., Boisse, P. (1991), A&A, 243, 344

Bergeron, J., Cristiani S., Shaver, P.A. (1992), A&A, 257, 417

Cristiani, S. (1987), A&A, 175, L1

Thimm, G. et al. (1994), A&A, 285, 785

Yanny, B., York, D. (1992), ApJ, 391, 569

Intrinsic Interstellar Absorption in two Radio Galaxies at z=0.8

Sperello di Serego Alighieri[1], Andrea Cimatti[1] and Sandra Savaglio[2]

[1] Osservatorio Astrofisico di Arcetri, Firenze, Italy
[2] Universitá della Calabria, Cosenza, Italy

Abstract. We have found a strong absorption feature at 2598Å in the spectra of two distant radio galaxies, which we interpret as due to interstellar FeII. We discuss this interpretation and the possibility of obtaining important information on the evolution of the ISM in galaxies, to be compared with the results of the study of intervening and associated absorptions in quasars.

1   Introduction

Being the most distant stellar systems known (Lacy et al. 1994), radio galaxies can provide essential information for the study of the formation and early evolution of galaxies, both for their stellar and interstellar content. So far information on the ISM in distant galaxies has come from bright emission lines by ionized gas, from infrared emission presumably by dust, from a few CO detections from molecular clouds, from the polarization produced by dust and electron scattering, and from the so–called intervening absorptions in quasar spectra. These informations are complementary, in the sense that they sample different phases of the ISM at different locations in the galaxy.

A direct detection of intrinsic interstellar absorption in a distant galaxy has not been obtained yet, because of the very faint continuum of these objects. However such a detection would be important to study the evolution of the ISM — its metallicity and depletion in particular — and to compare the results with those obtained from intervening absorptions in quasar spectra. Radio galaxies are useful in this respect, both because they can be very distant and because they have a relatively higher continuum level in the ultraviolet, where interstellar lines are stronger. Furthermore, if one is prepared to accept that radio galaxies are just quasars viewed from a different angle as foreseen by the AGN Unified Model (Antonucci, 1993), intrinsic interstellar absorptions in radio galaxies can be compared directly with the so–called associated absorptions in quasar spectra, giving information on the ISM around the most luminous AGN.

We discuss here our recent detection (di Serego Alighieri et al. 1994) of a strong ultraviolet absorption in the spectra of two distant radio galaxies.

2   Interstellar Absorption at z=0.8?

During our spectropolarimetric study of distant radio galaxies we have detected a strong absorption feature at 2598Å in the spectra of the radio galaxies 3C 226 (z=0.818) and 3C 277.2 (z=0.766). Because of the needs of spectropolarimetry we have two long exposure spectra, of 45 minutes each, for both galaxies. The absorption is detected in each of the 4 spectra. Since the spectra are all very similar, we show in Figure 1 a section of the summed spectrum in rest–frame wavelengths. The absorption feature has a large rest–frame equivalent width $(8 \pm 2\text{Å})$.

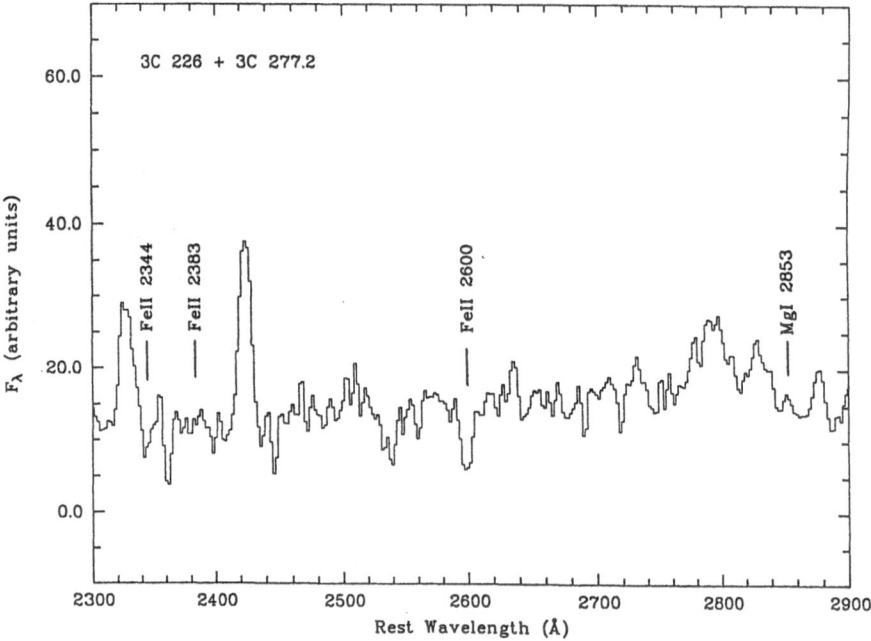

Fig. 1. Summed spectra of 3C 226 and 3C 277.2. The positions of strong ultraviolet interstellar absorptions are marked.

The line is detected at the same rest–frame wavelength in the spectra of the two galaxies, which are at a different redshift and are separated by a very large angle in the sky. Therefore we believe that it is very unlikely that the line is due to an intervening absorber and we conclude that it should be produced by intrinsic absorption in the galaxies.

Could the absorption be of stellar origin? The only stellar absorption at the right wavelength is FeII2600 which is present in F and G type stars (Fanelli et al. 1990). Nevertheless these stars emit a very small fraction of their radiation in the ultraviolet and could hardly explain the ultraviolet to visual colour of the galaxies. Moreover we have evidence from the spectropolarimetry that the dilution of the polarized flux is small in the ultraviolet around the MgII2800 emission line (di Serego Alighieri et al. 1994). Therefore a stellar absorption like FeII2600 in F and G stars could not have such a high equivalent width in the total spectrum, and we exclude that the absorption is of stellar origin.

We tentatively attribute the absorption feature to interstellar FeII in the two galaxies. In support of this interpretation we bring the fact that indeed FeII has a multiplet around 2600Å, whose strongest components, at 2600.2 and 2586.5Å, we cannot resolve, since the resolution of our spectra is about 15Å. FeII2600.2 is the strongest ultraviolet interstellar absorption after the lines of MgII2797+2804 (Kinney et al. 1993), which we cannot detect in our spectra since they are filled by emission.

However there are two problems with the interstellar FeII interpretation. First, in our spectra we do not detect so clearly other strong interstellar ultraviolet lines of FeII and other ions, although we cannot completely exclude their presence, given the S/N ratio of our spectra (see Fig. 1). Second, the rest–frame equivalent width of the absorption feature is much larger than what is normally observed for FeII2586+2600 in the ISM of our galaxy ($\sim$2Å, Kinney et al. 1993). This fact could be due to the different conditions of the ISM in distant radio galaxies, such as a larger column density, different ionization conditions and a lower depletion. Indeed a very strong absorption at 2600Å is present in the IUE spectrum of the supernova SN 1986G, which exploded in the dust lane of the powerful radio galaxy Cen A. Also several components at different velocities, which we see blended in our low resolution spectra, could increase the equivalent width above galactic levels.

The possibility of a lower depletion than the galactic one can be associated with the unification of quasars and radio galaxies. Our polarimetric studies have demonstrated that a large fraction of the ultraviolet continuum from distant radio galaxies is actually scattered and is probably radiation which comes originally from a quasar in the nucleus, hidden from direct view. Therefore this ultraviolet radiation, before and after the scattering, travels through considerable distances inside the strong radiation cone of the quasar, a region where dust can hardly survive, thereby explaining a low depletion.

If radio galaxies are indeed quasars viewed from outside the radiation cones, it would be interesting to compare our results with those on associated absorptions in quasars. The latter seem to have a much higher ionization level: for example CIV is seen, but not FeII (e.g. Savaglio et al. 1994). This could be attributed to the fact that our line of sight to radio galaxies travels through their ISM also outside the radiation cone, where ionization is lower.

3   Outlook

We certainly need to confirm our tentative detection of interstellar absorption in distant radio galaxies. If this is confirmed with spectra taken specially for this purpose, then a further study of these absorptions, in particular with the new 8-10m class telescopes, would allow to obtain important information on the ISM at early cosmological epochs and on its evolution. In particular we could obtain constraints on column densities and metallicity, on the presence of various components at different velocities, on the ultraviolet photoionizing source from the relative abundance of the different ions of the same element, and on the depletion.

We have shown that it is becoming possible to study interstellar absorptions in distant galaxies directly, allowing the study of an essential component of the formation and evolution of galaxies and a useful comparison with the results from intervening and associated absorptions in quasars.

Acknowledgements. We thank Sofia Randich and Roberto Viotti for usefull discussions on the properties of interstellar absorption lines.

References

Antonucci R., 1993, ARAA, 31, 473.

di Serego Alighieri S., Cimatti A. & Fosbury R.A.E., 1994, ApJ, 431, 123.

Fanelli M.N., O'Connell R.W., Burstein D. & Wu C.-C., 1990, ApJ, 364, 272.

Kinney K.L., Bohlin R.C., Calzetti D., Panagia N. & Wyse R.F.G., 1993, ApJSS, 86, 5.

Lacy M., Miley G., Rawlings S., Saunders R., Dickinson M., Garrington S., Maddox S., Pooley G., Steidel C.C., Bremer M.N., Cotter G., van Ojik R., Röttgering H. & Warner P, 1994, MNRAS, 471, 504.

Savaglio S., D'Odorico S. & Møller P., 1994, A&A, 281, 331.

Molecular Absorption in the Millimeter Range

T. Wiklind[1] and F. Combes[2]

[1] Onsala Space Observatory, S–43992 Onsala, Sweden
[2] DEMIRM, Observatoire de Paris
61 Av. de l'Observatoire, F–75014 Paris, France

Abstract. Absorption lines of molecular rotational transitions have been detected at redshifts $z_a = 0.24671$ and $z_a = 0.68466$ towards the two background sources PKS1413+135 and B0218+357. The high sensitivity of absorption line observations, in comparison with emission lines, makes it possible to observe rare molecules such as HCO^+, HCN and HNC. In both systems the line of sight to the background continuum source has a very small impact parameter (a few tenths of an arcsecond) to an intervening galaxy. Although the molecular absorption lines are saturated in their centers, we can derive accurate abundance ratios in the velocity wings.

1 Background

QSO absorption lines can be used to derive properties of the gaseous content of intervening objects. Optical observations mostly probe the hot and tenuous gas in galactic halos. Sometimes the line of sight passes through a galactic disc, producing damped Lyman–α lines, but even in these cases it is a mostly atomic gas component that is observed. The cold and dense molecular gas which is directly involved in the star formation process is not seen in optical spectroscopy, simply because the extinction is high enough to render the background source invisible at optical wavelengths. In a few cases there are background quasars which are not visible at optical wavelengths but are strong radio sources and have an object in the line of sight which is a potential absorber. If the redshift of the intervening object is known, these are good candidates for absorption lines of molecular rotational transitions. We have detected absorption of molecular lines in two such systems.

2 Observations and Analysis

The observations were done mainly with the IRAM 30–m telescope at Pico Veleta in Spain, although the first detection of PKS1413+135 was made with the 15–m SEST telescope in Chile. Spectroscopic observations at millimeter wavelengths have a much higher velocity resolution than what is achieved in the optical regime. At the IRAM telescope we obtained resolutions of $30\,\mathrm{m\,s^{-1}}$ to $40\,\mathrm{m\,s^{-1}}$, which is less than the expected thermal width of the molecular lines when the only heating source is the cosmic background radiation. The continuum level of the background sources were determined both from the chopper–wheel calibrated spectra and using a continuum backend.

The optical depth averaged over the size of the background continuum source can be expressed as $\tau_{\nu_{obs}} = \ln(\frac{T_s}{\Delta T_{mb}})$, where T_s is the continuum level, $\Delta T_{mb} = T_s - |T_{abs}|$, and T_{abs} is the depth of the absorption line measured from the continuum level (cf. Wiklind & Combes 1994, 1995). If f_c is the fraction of the continuum source area covered by molecular gas at the velocity corresponding to the frequency ν, then the true optical depth τ_ν is

$$\tau_\nu = -\ln[1 - \frac{1}{f_c}(1 - \exp(-\tau_{\nu_{obs}}))] \ . \tag{1}$$

If $f_c = 1$ then $\tau_\nu = \tau_{\nu_{obs}}$, if not, we can use this expression to derive a lower limit of the filling factor: $f_c \geq 1 - exp(-\tau_{\nu_{obs}})$.

Assuming 'weak LTE conditions', i.e. the excitation temperature T_x derived from two transitions of the molecule is equal to the rotational temperature T_{rot} characterizing all the levels of the molecule, but T_{rot} is not necessarily equal to T_{kin}, the total column density of a molecular species can be estimated from an absorption line $J \to J + 1$ as

$$N_{tot} = \frac{8\pi}{c^3} \frac{\nu^3}{A_{J+1,J} g_{J+1}} \frac{Q(T_x) \exp(E_J/kT_x)}{1 - \exp(-h\nu/kT_x)} \int \tau_\nu \, dV \ , \tag{2}$$

where $A_{J+1,J}$ is the Einstein A–coefficient, $\int \tau_\nu dV$ the velocity integrated optical depth, $g_J = 2J + 1$ the statistical weight of level J, and $Q(T_x)$ the partition function.

The molecular absorption lines are saturated (see Fig. 1). This means that the covering factor f_c is large. For PKS1413135 we find that at least 90% of the background source is covered by molecular gas of significant column density, for B0218+357 it is >80%. Therefore $\tau_\nu \approx \tau_{\nu_{obs}}$ is a reasonable approximation. The saturated lines also means that the derived column densities are lower limits (see Table 1). By using unsaturated parts of the absorption profiles we can, however, derive quite accurate abundance ratios. Using a 3σ limit on the absorption profile measured both from the continuum level and the zero level, we get abundance ratios which are accurate to within a factor of two.

3   Results and Conclusions

Here we briefly discuss the results achieved so far for the two observed molecular absorption line sources.

3.1 PKS1413+135   This is the best observed molecular line absorption system. We have detected 4 molecules in 7 different transitions: CO(0–1), HCO$^+$(1–2), (2–3), HCN(1–2), (2–3) and HNC(1–2), (2–3). We have also obtained a significant upper limit to the column density of H^{13}CO$^+$(1–2),

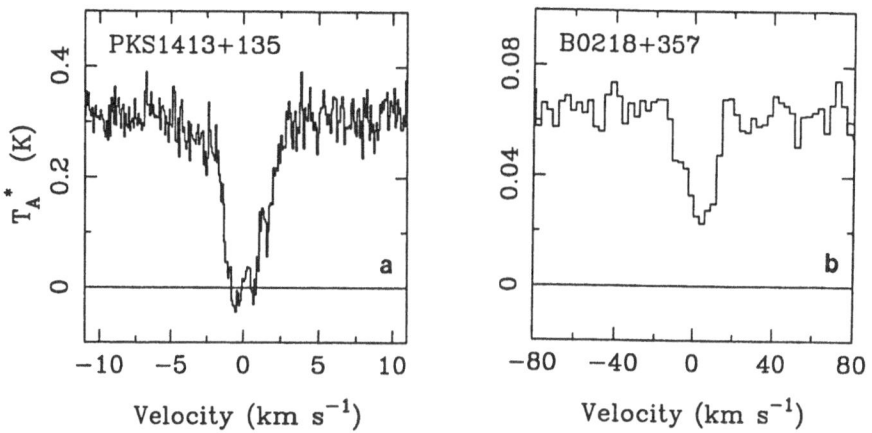

Fig. 1. The $HCO^+(1-2)$ line at a) $z_a = 0.24671$ towards PKS1413+135 and b) at $z_a = 0.68466$ towards B0218+357.

suggesting a $^{12}C/^{13}C$ ratio > 100, as well as a limit to the amount of molecular oxygen of $N_{O_2} < 6 \times 10^{16}\,cm^{-2}$. Molecular oxygen as well as the $J = 1-2$ lines of HCO^+, HCN and HNC are usually not observable with groundbased telescopes, but in this case (and for B0218+357) the lines are redshifted into accessible frequency windows. The width of the absorption profiles is relatively small, $\sim 2\,km\,s^{-1}$. The covering factor $f_c > 0.9$ and the extinction derived from X–ray data (Stocke et al. 1992) of ~ 30 mag suggest that a relatively small and dense molecular cloud is causing the absorption. This means that the size of the background source must be very small. Carilli et al. (1992) detected 21 cm HI absorption with a much larger linewidth, most likely due to a larger size of the background source at lower frequencies.

3.2 B0218+357 This object has been identified as a gravitational lens candidate (O'Dea et al. 1992, Patnaik et al. 1993). It consists of two compact flat–spectrum objects, both within $0\rlap{.}''3$ of the nucleus of the intervening galaxy, and a steep–spectrum ring. 21 cm HI absorption has been observed by Carilli et al. (1993). We detect absorption of CO(1–2), (2–3), $HCO^+(1-2)$ and HCN(1–2). The width of the profiles is $\sim 30\,km\,s^{-1}$, much larger than for PKS1413+135. The large linewidth and the covering factor $f_c > 0.8$ indicate that both compact components could be obscured by molecular gas. Hence, the absorption is caused by relatively diffuse molecular gas of moderate optical depth.

In Table 1 we compare abundance ratios of PKS1413+135 and B0218+357 with those of typical molecular clouds in our own Galaxy. With the present data we can not identify the type of molecular clouds seen at intermediate redshifts.

3.3 Outlook The number of potential molecular line absorption systems are limited by the lack of strong background continuum sources at millimeter wavelengths. There are, however, some 'blank field' objects which are potential candidates. Other molecular absorption line candidates may be found at lower frequencies, were the number of strong background continuum sources are more common. If a substantial number of molecular absorption line systems are found, they can give information about the evolution of metallicity as a function of redshift, the history of star formation activity and, since the molecular gas is an efficient coolant, give upper limits to the temperature of the cosmic background radiation at high redshifts.

Table 1. Molecular line data

	PKS1413+135 $z_a = 0.24671$	B0218+357 $z_a = 0.68466$	TMC–1 South	Orion Hot Core	Sgr B2
$N_{CO}(cm^{-2})$	$> 2.4 \times 10^{16}$	$> 7.0 \times 10^{16}$			
$N_{HCO+}(cm^{-2})$	$> 3.6 \times 10^{13}$	$> 1.2 \times 10^{14}$			
$N_{HCN}(cm^{-2})$	$> 8.5 \times 10^{12}$	$> 1.4 \times 10^{14}$			
$N_{HNC}(cm^{-2})$	$> 5.2 \times 10^{12}$	—			
N_{CO}/N_{HCO+}	770^{+450}_{-200}	660^{+800}_{-230}	13×10^3	$> 1 \times 10^5$	8×10^3
N_{CO}/N_{HCN}	1500^{+980}_{-420}	400^{+550}_{-150}	8×10^3	500	4×10^3
N_{HCN}/N_{HCO+}	$0.4^{+0.4}_{-0.3}$	$1.5^{+0.4}_{-0.3}$	1.7	> 200	2.0
N_{HCN}/N_{HNC}	$1.3^{+0.7}_{-0.3}$	—	1.0	200	0.2
$^{12}C/^{13}C$	> 100	—	65	40	30

References

Carilli, C.L., Perlman, E.S., Stocke, J.T. 1992, ApJ 400, L13
Carilli, C.L., Rupen, M.P., Yanny, B. 1993, ApJ 412, L59
O'Dea, C.P., Baum, S.A., Stanghelli, C., Van Breugel, W., Deustua, S., Smith, E.P. 1992, AJ 104, 1320
Patnaik, A.R., Browne, I.W.A., King, L.J., Muxlow, T.W.B., Walsh, D., Wilkinson, P.N. 1993, MNRAS 261, 435
Stocke, J.T., Wurtz, R., Wang, Q., Elston, R., Jannuzi, B.T. 1992, ApJ 400, L17
Wiklind, T., Combes, F. 1995, A&A in press
Wiklind, T., Combes, F. 1994, A&A 286, L9

CO Emission at $z = 0.437$ Toward 3C 196

David T. Frayer[1] and Robert L. Brown[2]

[1] National Radio Astronomy Observatory and University of Virginia, 520 Edgemont Road, Charlottesville, VA 22903, USA
[2] National Radio Astronomy Observatory, 520 Edgemont Road, Charlottesville, VA 22903, USA

Abstract. We report the detection of CO(1–0) emission during a systematic search at and around the position of the QSO 3C 196 at the redshift of the 21–cm absorption system. The strongest detected CO(1–0) emission was $1'$ south of the line of sight to 3C 196 and represents approximately $10^{11} M_\odot$ of molecular gas. We also detect weaker CO emission at the QSO position, but we are unable to distinguish between the possibilities that this emission arises from the source which is $\sim 1'$ S of the 3C 196 or that the CO emission arises from two or more emission regions spread over ~ 200 kpc.

In an unbiased search for 21–cm absorption toward quasars, Brown and Mitchell (1983) discovered a pronounced absorption line corresponding to redshifted H I at $z = 0.437$ toward the compact, double lobed radio source 3C 196. Subsequently, Foltz et al. (1988) found strong optical metal absorption lines toward 3C 196 at the redshift of the 21–cm absorber. To cover both the optical position and part of the radio lobes, the absorber at $z = 0.437$ must be at least $3''$ in extent, $10h^{-1}$ kpc ($q_o = 1/2$). In order to constrain the total mass and spatial extent of this system, we have searched for redshifted CO(1–0) emission at and around the QSO position.

The CO(1–0) observations were taken using the NRAO 12 Meter Telescope in September 1993 and June 1994. We tuned the dual polarization SIS mixer receivers to a frequency of 80.1870 GHz corresponding to a redshift of $z = 0.43753$. At the observed frequency, the beam of the 12–m is $\theta_{FWHM} = 78''$, and the T_R^* scale is 32 Jy/K assuming that the source is small compared to the beam. We have systematically searched for CO emission in a nine point map, with $1'$ separations, at and directly around the QSO position. During both observing runs, we found strong CO emission $1'$ south of 3C 196 (Figure 1a). Using the standard galactic conversion, the CO detection corresponds to $1.5 \times 10^{11} M_\odot$ of molecular gas. We also have observed weaker CO emission at the QSO position. The seven other positions surrounding the QSO position did not show CO emission (Figure 1b). Both spectra were produced using the same observing and data reduction procedures. We alternated the observations between different positions every 24 minutes so that similar systematic effects would be present at all positions.

In a narrow–band imaging search for redshifted [OII] emission at $z = 0.437$, Nelson and Malkan (1992) found three objects with excess "line" emission. All three of these objects were $\sim 45''$ S of 3C 196, separated by

~ 25″ east–west, and are well within the beam of the CO detection. It is certainly feasible that the CO emission is associated with one or more of these objects. Since these objects appear near the half power in the observations at the central position, the weaker CO emission detected here could be also associated with these same objects. Recent HST observations by Cohen *et al.* (1995) show a spiral galaxy within 3″ of the QSO line of sight, but no redshift is known for the galaxy. If the spiral is responsible for the 21–cm absorption and optical absorption lines, its proximity with the large molecular component of the gas (~ $200h^{-1}$ kpc projected distance) would suggest that the system is a group of galaxies. Otherwise, if the spiral is not at $z = 0.437$, then the CO emission could arise from an extremely large ($\gtrsim 200$ kpc) low surface brightness object which could also produce the 21–cm and optical QSO absorption lines. The list of redshifted CO detections is growing steadily, and these observations provide another example of an early galactic system containing a significant fraction of its mass in the form of metal–enriched molecular gas which could be distributed over a large spatial extent.

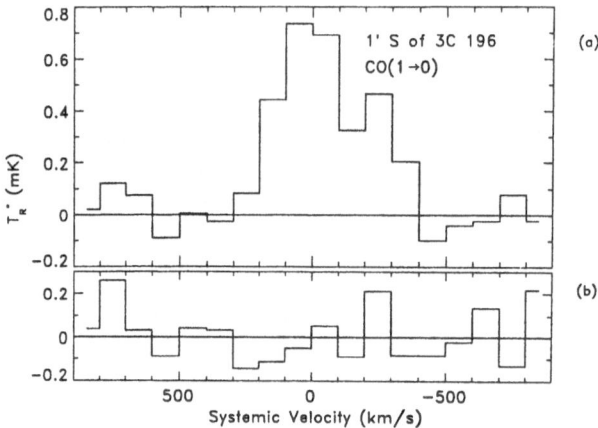

Fig. 1. The upper panel (a) shows the detection of CO(1–0) emission 1′ south of 3C 196 at the redshift of the 21–cm absorption system. The velocity resolution is 100 km/s, and the theoretical rms noise is ~ 0.2 mK. The lower panel (b) shows the non–detection of CO(1–0) emission in the combination of data from all other positions surrounding 3C 196 (1′ N; 1′ E; 1′ W; 1′ N and 1′ E; 1′ N and 1′ W; 1′ S and 1′ E; 1′ S and 1′ W). A zero–order baseline was removed from each spectrum.

References

Brown, R. L., & Mitchell, K.J. 1983, ApJ, 264, 87
Cohen, R. D. *et al.* 1995, in preparation (Poster by M. Burbidge)
Foltz, C. B., Chaffee, Jr., F.H., & Wolfe, A.M. 1988, ApJ, 335, 35
Nelson, B. O., & Malkan, M. A. 1992, ApJS, 82, 447

Automated Search for Absorption Systems in Low Resolution QSO Spectra

Sebastián López and Lutz Wisotzki

Hamburger Sternwarte, Gojenbergsweg 112, 21029 Hamburg, Germany

Abstract. An automated method to search for metal line absorption systems in low and medium resolution QSO spectra is presented. The method is based on pattern recognition in noisy data: it compares rectified QSO spectra with an input absorption line pattern minimizing χ^2 between model and data. The best candidate for a z system is chosen among the minima of the curve $\chi^2 = \chi^2(z)$ and can be subtracted if weaker systems are to be searched for. A modified version of the χ^2 test has been developed in order to compute the probability of finding a system by chance. The method has proved to be successful by detecting 43 metal absorption systems in about 200 low resolution (FWHM$\approx$25Å) QSO spectra of the Hamburg–ESO Survey with $S/N\approx$40–50 and emission redshifts $0.33 \leq z_{em} \leq 3.06$.

1 Data, Model and χ^2 Fitting

A spectrum consists of a MIDAS table with three columns containing wavelength λ_i, flux f_i and an associated error σ_i. In a first step a local continuum is fitted and the flux is normalized. This is done by non linear filtering (Gauss filter with $\kappa\sigma$-clipping). The model m_i consists of an absorption line pattern with p Gaussian–shaped lines centered on $\lambda_0 \cdot (1+z)$ which is shifted over all possible z values. The line widths σ are held fixed (depending on the instrumental profile a suitable value is found) whereas the amplitudes α_j vary. This leads to $p+1$ free parameters to fit. The m_i values are computed according to the formula:

$$m_i \equiv 1 + \sum_{j=1}^{p} \alpha_j \cdot \left(\sum_{k=1}^{n(j)} A_{kj} \cdot \exp\left[\frac{-(\lambda_i - \lambda_{kj}(1+z))^2}{2\sigma^2} \right] \right)$$

with A_{kj}, the initial line amplitudes, chosen to be the corresponding f–values. Input parameters are: a list of absorption lines (rest frame wavelengths, widths and f–values) and an appropiate redshift range. An estimate of the model parameters α_j ist obtained by minimizing $\chi^2 = \sum[(f_i - m_i(\alpha_i, z))/\sigma_i]^2$. For a fixed z the problem is linear and the p α–parameters are computed inverting the matrix corresponding to the normal equations. Once the χ^2 minima have been computed for each possible z, the local minima $\chi^2_{min}(z)$ can be searched for numerically.

2   Detection and Subtraction of Systems

Not all the local minima are associated with real z systems. There also are
"spurious" minima due to possible line combinations that succeed by chance.
In addition, strong systems produce the deepest depressions in χ^2, hiding the
weaker ones. The best way to detect the latter is to *subtract* the absorption
lines at redshift z_{min} corresponding to the first minimum. Nevertheless, such
a subtraction is not always necessary and should be avoided, if possible.

Once a z_{min} has been found, an appropriate confidence level $\Delta\chi^2 = \chi^2 -$
χ^2_{min} is chosen in order to compute an error in redshift.

3   Rejection Parameter

A Rejection Parameter that computes the probability of having found a system by chance is needed. A usual treatment of the χ^2 statistic, however, is
bound to fail, because a description of the data by the absorption lines model
is not complete: for a fixed z other systems may also be present, thus increasing the value of χ^2_{min} artificially. This gives rise to too large (≈ 1) values
of the usual chance probability $P(\nu/2, \chi^2_{min}/2)$ described by the Incomplete
Gamma Function for a chi–square distribution for $\nu = N - p$ degrees of
freedom.

A modified version of the χ^2 test has to be used. The central idea is to
treat the computed χ^2_{min} as belonging to a distribution for $\nu^* = \chi^2_0$ degrees
of freedom, with $\chi^2_0 \equiv \sum[(f_i - 1)/\sigma_i]^2$ being the chi–square value of a model
describing a *null–hypothesis* (spectrum without absorption lines). That is to
say, the statistic is considered to be one, for which a null–hypothesis model
fits "good" to the data. In this way, one is actually *comparing* model with the
hypothesis that the spectrum contains no absorption lines, the new chance
probability $P^*(\nu^*/2, \chi^2_{min}/2)$ being computed for a larger number of degrees
of freedom ν^*.

The absorption lines that do not belong to the z system being tested are
consireded as normally distributed noise. A suitable threshold for P^* can be
computed with the help of numerical simulation and depends strongly on p.

4   Conclusions

1. The method offers objectivity and reproducibility by searching for weak
absorption systems in low resolution QSO spectra. 2. The probabilty of finding a system by chance can be computed. It shows how reliable a system
is, within the noisy data. 3. An application to better resolutions is possible,
provided that the instrumental profile dominates over the intrinsic line shape.

Quantitative Spectroscopy of HS 1700+6416 with the Hubble Space Telescope

Dieter Reimers, Susanne Köhler, and Hans-Jürgen Hagen

Hamburger Sternwarte, Gojenbergsweg 112, 21029 Hamburg, Germany

Abstract. The QSO HS 1700+6416 ($z = 2.72$) has been observed spectroscopically with HST in the range 1160 to 3280 Å at a resolution of $\lambda/\Delta\lambda = 1300$ to 2200. We report the discovery of ~40 new quasar absorption lines from EUV rest wavelengths in 16 redshift systems. In particular, lines from the ions HeI, OIII–OV, NIII, NIV, NeIII–NeVII, SIII–SV have been detected. All 16 metal line systems (MLS) show highly ionized species (OV, OIV, NIV,...). From the number of systems per redshift interval the characteristic radii of the highly ionized gas clouds is estimated as $R \geq 150\ h_{100}^{-1}$ kpc. The wide coverage of ionization stages will permit reliable abundance studies as well as observational constraints for the ionizing radiation field. We find in particular an oxygen overabundance relative to carbon $[O/C] = 0.5...0.9$ similar to that observed in Halo stars. Arguments in favour and against ionization by a metagalactic radiation field due to QSOs are presented.

1 Introduction

High redshift QSO absorption line systems with EUV resonance lines from abundant elements shifted into the UV offer in principle a powerful tool to study quantitatively the early chemical evolution of matter, to investigate the nature of an ionizing metagalactic UV radiation field and to constrain the ionization conditions of the intergalactic medium through the HeI and HeII Gunn-Peterson effects.

However, most high-redshift QSOs are completely obscured at EUV rest wavelengths by the hydrogen Lyman continua of gas clouds at cosmological distances (Møller & Jakobsen 1990, Picard & Jakobsen 1993).

While recently two rather faint QSOs have been detected to be unabsorbed down to HeII 304 and then could be used to perform a HeII Gunn-Peterson test (Jakobsen et al. 1994; Tytler, this volume), the only known high-redshift QSO sufficiently UV bright for quantitative metal absorption line studies is so far HS 1700+6416 ($z = 2.72$, V=16.1 Reimers et al. 1989). Rather spectacular spectra taken with the FOS onboard HST with seven optically thin Lyman Limit Systems (LLS) have opened a new window of extragalactic research (Reimers et al. 1992; Reimers & Vogel 1993; Vogel & Reimers 1993; 1995): Since for O, C and N all relevant ions (OIII to OV, NIII, NIV, CIII, CIV) can be observed simultaneously, relative abundances O/C and N/C could be determined for the first time in several LLSs ($z = 1.8465$ to $z = 2.433$) without having to apply uncertain photoionization corrections.

However, for $\lambda \leq 1700$ Å, the S/N of the FOS spectra was too low for absorption line spectroscopy with the effect that for lower redshift systems ($z = 1.1572, 0.8642, 0.7217$) the coverage of ionization stages for O was incomplete. Therefore in July 1994, after the successful repair of HST, low resolution GHRS spectra have been taken in the wavelength range 1160–1680 Å. While the analysis of the FOS spectra has been completed (Vogel & Reimers 1995), the evaluation of the GHRS data has only recently begun. Therefore, we briefly summarize the results of the quantitative analysis of the FOS spectra and present preliminary results from the GHRS spectra.

2 FOS Spectra

2.1 The Continuum and Emission Lines

The UV energy distribution is characterized by a step structure due to a sequence in redshift of seven optically thin ($\tau \approx 0.5$) LLSs, and the overall wavelength dependence of flux shows the "Lyman valley" due to the cumulative effect of hydrogen continuum and line absorption of "intergalactic" hydrogen as predicted by Møller & Jakobsen (1990) and Picard & Jakobsen (1993). If "deabsorbed", the intrinsic QSO flux distribution is rather flat in the EUV ($f_\lambda \sim \lambda^{-0.7}$ for $\lambda_{\rm rest} \geq 540$ Å, $f_\lambda \sim \lambda^{-1}$ for $330 \leq \lambda_{\rm rest} \leq 540$ Å). The only unambiguously detected EUV quasar emission line is NeV 568.4 Å. In addition, we seem to have detected the far red wing of the HeII 304 Å emission line at $\lambda \leq 1180$ Å (Fig. 3).

2.2 QSO Absorption Line Identification

Superimposed on the continuum step structure are several hundred absorption lines. Most of them can be identified with EUV resonance lines of abundant ions in 15 redshift systems.

Prominent new quasar absorption lines which have not been seen in any cosmic object except the Sun are OV 629.7, OIV 787.7, 608.4, 554.3, 553.3, OIII 702.3, NIV 765.1, NIII 685/685.5 Å as well as HeI 584.33 and 537.03 Å. CIII 977 is also always prominent. The complete list of identifications of metal lines in 15 redshift systems is given by Vogel & Reimers (1995).

Unidentified lines are called Lyα forest lines although one should keep in mind that the EUV metal line lists used for identification might be incomplete. Such tentative Lyα forest lines in FOS spectra (1700 to 3280 Å) of HS 1700+6416, i.e. in the redshift range $0.4 \leq z \leq 1.7$, can be compared with the known distribution from optical data ($z \geq 1.7$) and with the Lyα forest line distribution from the HST absorption line key project (Bahcall et al. 1993). As can be seen from Fig. 2 the Lyα forest line distribution in the redshift range $1 \leq z \leq 1.7$ joins smoothly the optical data (e.g. Murdoch et al. 1986) and the HST data for $z \leq 1$. This indicates that for 2500 Å to 3280 Å, our heavy element absorption line identifications are complete. For

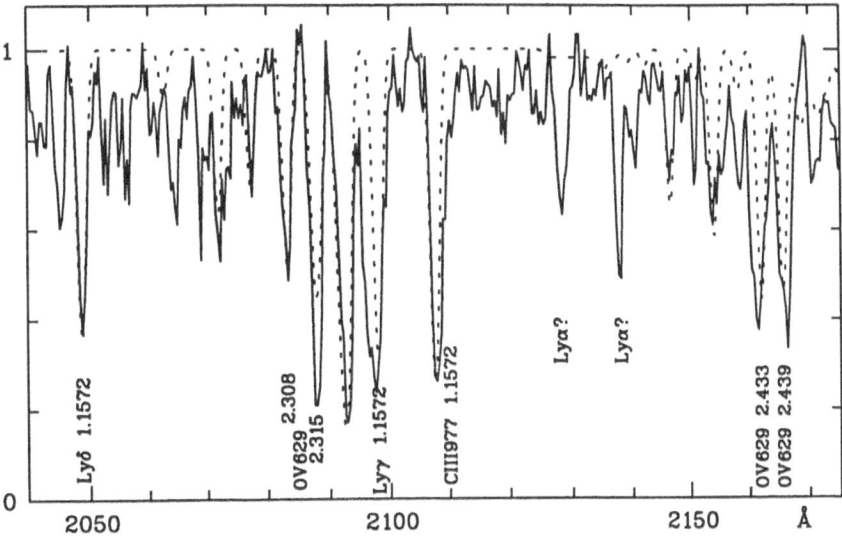

Fig. 1. Observed and calculated (dashed line) part of the FOS spectrum (from Vogel & Reimers 1995)

$0.5 \leq z \leq 1.1$, the tentative Lyα forest lines appear too abundant compared to the HST key programme data. This can have several reasons: The S/N of the FOS spectrum of HS 1700 in this range is rather low, in particular below 2000 Å, and line blending is heavy at the FOS resolution (but this is also the case at longer wavelengths). And finally, there may be indeed additional EUV heavy element lines unknown to us. The extensive EUV resonance line list by Verner et al. (1994) has turned out to be nearly identical with ours and has led to no new identifications in the spectrum of HS 1700+6416. We believe that both higher S/N data and higher spectral resolution are necessary to improve the identifications.

2.3 Abundances and Ionizing Radiation

In the following, we briefly summarize the methods and results of the quantitative analysis of the FOS spectra (Vogel & Reimers 1995). The useful FOS spectral range 1700–3280 Å combined with optical spectra covered a wide range of ionization stages (OIII–OVI, CIII, CIV, NIII, NIV) in particular for the redshift system $z = 1.8465$, 2.1678, 2.433. This allowed to determine both relative CNO abundances (model independent) and absolute CNO abundances through an empirical determination of the shape of the ionizing radiation field (Vogel & Reimers 1993). In a first step, hydrogen column densities N(HI) are determined from the continuum jumps of the 7 LLSs.

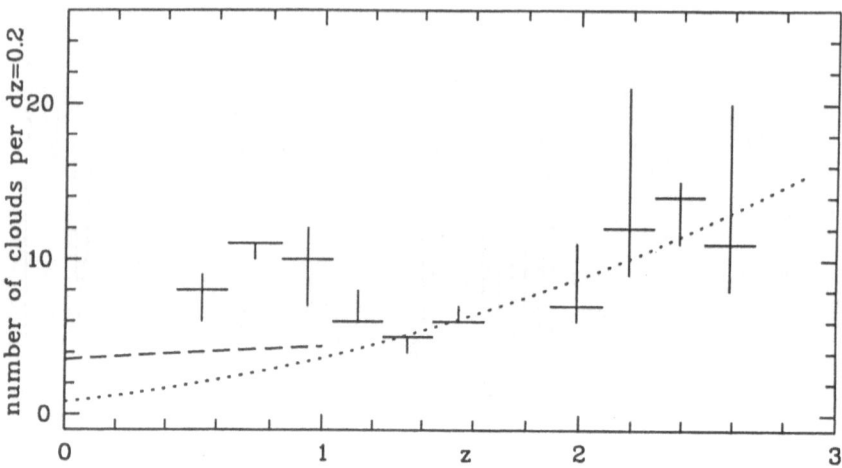

Fig. 2. Observed number of Lyα systems with $W_{rest} \geq 0.32$ Å in FOS spectra (Vogel 1994). The prediction according to the result of Murdoch et al. (1986) is plotted as dots. Short dashes represent the results of Bahcall et al. (1993).

Subsequently, the gas velocity dispersion parameter b is determined by spectral synthesis of the complete Lyman series: b values range between 25 and 42 $km\,s^{-1}$ and are relatively accurate since N(HI) is known with an accuracy of $\pm\,10\%$. It is then assumed that this velocity dispersion is mainly nonthermal, and that the same velocity dispersion applies to heavy element ions. Clearly, this may introduce severe errors. It is known from high-resolution data, that often the lines split into several subcomponents, i.e. at the FOS resolution we may be averaging over several subclouds. However, errors in the relative CNO ion column densities arise only, if subcomponents are saturated and/or different ions are found in different clouds. For the ions mainly used in the FOS study (OIII–OV, CIII, CIV, NIII, NIV) this is improbable. Therefore, we believe, that the relative ion column densities (e.g. N(OV)/N(CIII)) are not grossly in error, while the absolute heavy element abundances may be incorrect since hydrogen lines and high ionization species lines might be formed in different locations.

Under the assumption of photoionization, the large range of ionization potentials of observed ions (OIII to OVI, ...), with some lying even above the HeII ionization potential, can be used to constrain the spectral shape of the ionizing radiation field. It may be possible then to discern between alternative origins (AGN or hot stars in starburst galaxies, metagalactic or local). An optimal fit to the observed ion column densities in the LLSs $z = 1.8465$, 2.1678, 2.433 is obtained with $\alpha = -0.6^{+0.5}_{-0.9}$ ($f_\nu \sim \nu^\alpha$).

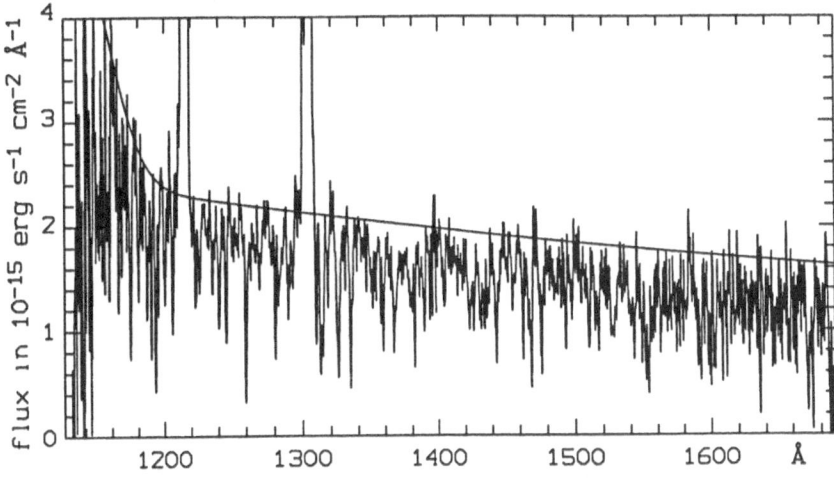

Fig. 3. Spectrum of HS 1700 taken with the GHRS of the HST

All three systems seem to be consistent with one, probably metagalactic, background radiation field. The LLSs $z = 1.8465$ and 2.1678 exclude a break in the ionizing radiation at the HeII 228 Å ionization edge by more than a factor of 5. This seems to exclude young starforming galaxies as the dominant contributors and favours AGN-type sources as being responsible for the ionizing background since models with radiation fields dominated by hot stars typically show large breaks at the 228 Å edge (Miralda-Escudé & Ostriker 1990; Madau 1991, 1992). However, even in AGN dominated radiation fields the opacity of the intergalactic medium causes characteristic steps and softens the unattenuated metagalactic spectrum. One problem with an AGN dominated background is that in order to fulfill both the HI and the HeII Gunn-Peterson effect it has to be rather soft (Madau & Meiksin 1994) which is in conflict with the requirements for metal ionization in LLSs of HS 1700+6416. A second problem is that with a given shape ($\alpha = -0.6$) of the background the size constraints of LLS clouds (see below, ≥ 150 kpc) in combination with the required ionization parameters (log U$= -2.25 \pm 0.1$) lead to an intensity of the background radiation field at the Lyman limit of $J(\nu_{\rm LL}) \approx 10^{-23}$ erg s^{-1} Hz^{-1} cm^{-2} sr^{-1} while the proximity effect demands $3 \; 10^{-21}$ erg s^{-1} Hz^{-1} cm^{-2} sr^{-1} at $z = 2$ (Bechthold 1994). So either the LLS clouds are highly clumped, or the ionization of LLS clouds is due to local sources, an alternative suggested by Friaça & Viegas (this volume), or the proximity effect is not real.

Using the empirical constraints for the assumed metagalactic ionizing radiation and the velocity parameter b derived from the LLSs, a spectrum synthesis of the metal line spectra of 15 redshift systems simultaneously is

Table 1. New lines in GHRS spectra of HS 1700

CII	904.0	0.7217		
	687.05		0.8642	1.1572
CIII	386.2	2.1678	2.315	2.433b
OII	376.75		2.315	2.433
	430.1	2.1678		2.433
	392.0		(2.315)	2.433
OIII	507.4	1.725		
	374.01	2.308	2.579	2.433
NIII	374.2	2.1678	2.315	2.433
NeIII	488.1/489.5	2.1678b	1.8465	
NeIV	543		1.8465	
NeV	568.4	1.1572	1.8465b	
	357.95	2.315		
	480.41	2.315	1.8465b	
NeVI	399/401	2.1678	2.315	2.433:
	558.59	2.1678b	1.8465	2.433
NeVII	465.22	2.1678	1.8465	2.315
SIII	677.75	0.7217	0.8642	1.1572
	698.73	0.7217		1.1572
	724.29		0.8642	1.1572
SIV	657.34		0.8642b	1.1572
	744/748	0.7217	0.8642	1.1572
SV	786.5		0.8642	

b: blended

then performed. Thus obtained heavy element abundances relative to hydrogen show the following characteristics: In several high redshift systems, in particular $z = 1.8465, 2.1678, 2.433$, we find [C/H]$\approx -2$ and [O/C]$\approx 0.5...0.9$, and the relative oxygen overabundances [O/C] seem to increase with decreasing carbon abundances [C/H]. The oxygen-overabundance relative to carbon looks similar to what has been found in extremely metal deficient Pop. II stars in the Halo of our own galaxy and can be explained by SN II ejecta from massive stars. However, the oxygen overabundance in these clouds seems to be accompanied by a similar nitrogen-overabundance relative to carbon which is not observed in Halo stars and which is also not seen in metal deficient emission line dwarf galaxies (Izotov et al. 1994).

On the contrary, in BCGs N is found to be typically overdeficient relative to O ($\log N/O \approx -1.5$) in agreement with theoretical predictions according to which N is a secondary element produced on a longer timescale than oxygen. There are, however, exceptions like FBS 0125-0265 with [O/H]$\approx$[N/H]≈ -1.4 (Izotov et al. 1994). On the other hand, the apparent nitrogen overabundance in the HS 1700+6416 absorption line systems is based on typically only one line per ion (NIV 765, NIII 685) and thus sensitive to line blending at the FOS resolution, since there is barely any unblended line in the spectrum. The insufficient spectral resolution may also affect the absolute O and C abundances

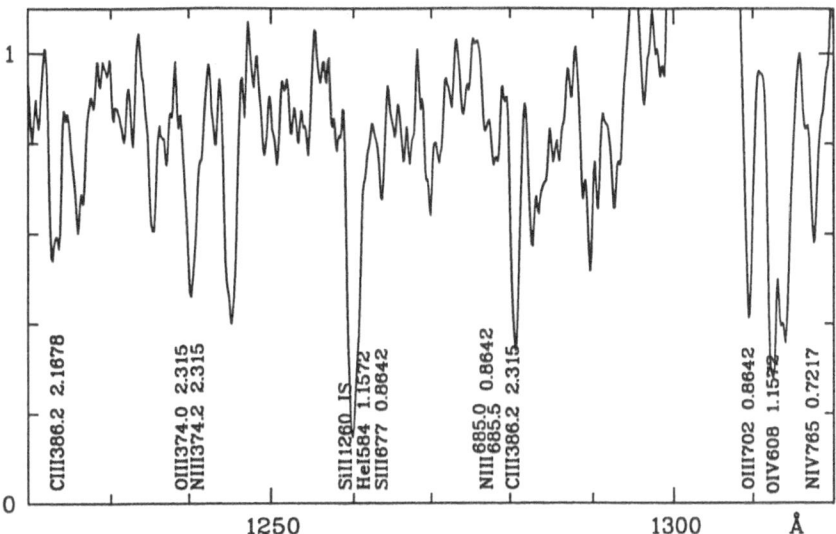

Fig. 4. Part of the HS 1700 spectrum taken with the GHRS with line identifications

since we do not know whether – as assumed – the highly ionized species (OV, CIV,...) are formed in the same physical regions where the Lyman limit steps in the spectrum are formed. At high spectral resolution QSO absorption lines often split into several subcomponents, and both unrecognized saturation effects and a varying distribution of different ions into the subcomponents may lead to spurious results on element abundances. We suspect, e.g., that the discrepancy between the observed neutral hydrogen to helium ratios of ∼20 to 70 on one hand and theoretical values of ∼150 obtained with the same ionizing radiation field that fits the high ionization lines on the other hand (Reimers & Vogel 1993) has its origin in such effects.

3   GHRS-Spectra

3.1   Observations

The wavelength range 1160 to 1680 Å has been observed in July 1994 with the repaired blue side of the GHRS at a resolution of 0.7 Å. The S/N is ∼15 at 1300 Å, but less than 10 for $\lambda \geq 1500$ Å. The wavelength accuracy is better than 0.2 Å. Surprisingly, the absorption line density is still extremely high although the Lyman forest contribution is small in this wavelength range (e.g. Bahcall et al. 1993).

Apparently all these lines are EUV resonance lines with rest wavelengths as short as ∼340 Å for the $z = 2.433$ system. This high heavy-element

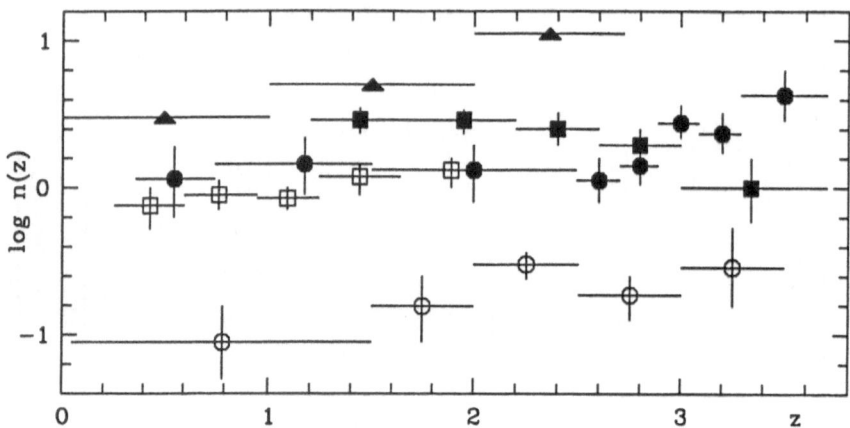

Fig. 5. Number density n(z) versus redshift z for absorption line systems selected from OV (this paper, filled triangle); CIV (filled squares), Lyman limit systems (filled circles), MgII (open squares), and damped Lyα (open circles) (Lanzetta 1993)

absorption line density has to be taken into account when the lack of (or weak) signal for $\lambda_{rest} \leq 304$ Å in QSOs is interpreted in terms of a HeII Gunn-Peterson effect (Jakobsen et al. 1994, Tytler, this volume).

3.2   Line Identification

In addition to the strong absorption lines of OV, OIV, OIII, NIV, NIII with rest wavelengths $\lambda \geq 500$ Å known already from the FOS spectra in high redshift systems, we easily identify the same lines in all lower redshift systems ($z = 0.7217, 0.8645, 1.1572$). One further MLS ($z = 0.5527$) has been detected on the basis of Lyα to Lyϵ, CIII, NIV and OIV 787 Å. In addition we discovered ~30 further new lines not observed before in QSO spectra. For line identification we have used the compilation of Verner et al. (1994). The list of newly observed lines is given in Table 1. Noticeable are lines from five different Neon ions (NeIII–NeVII) as well as SIII, SIV and SV. The Ne lines are particularly strong in the $z = 1.8465$ system, which is consistent with the expectation if Ne follows O (SN II material), since the 1.8465 system has the highest observed O column density.

Important for quantitative work are CIII 386 Å which is a factor of ~ 10 weaker than CIII 977 Å (always saturated?) and NIII 374 Å in addition to NIII 685 Å and 989 Å. We notice that OV 629 Å and NIV 765 Å are also prominent in low redshift systems. The ubiquitous presence of OV, OIV, NIV etc. in all 16 MLSs allows us to estimate the size of the absorbing objects from the observed number density n(z) versus redshift z along the line of sight of HS 1700+6416. By comparison with the CIV, MgII and damped Lyα n(z) versus z relations (Lanzetta 1993) where the MgII and damped Lyα cloud radii can be estimated independently (e.g. Bergeron & Boissé 1991; Steidel 1993; Smette et al. 1995), we find that n(z)$_{\rm OV}$ $\approx$ 3 n(z)$_{\rm CIV}$ – apparently oxygen is a more sensitive tracer of highly ionized halo gas of galaxies due to the higher oxygen abundance –, and we derive characteristic radii of R $\geq$ 150 h$_{100}^{-1}$ kpc. If the absorbing clouds were highly clumped which would be an explanation for the ionization problem mentioned in section 2.3., the smaller covering factor would lead to even larger cloud sizes.

3.3 Further Analysis

The combination of FOS spectra, GHRS spectra and improved optical spectra offers the possibility to tackle several scientific problems:

A fairly complete coverage of ionization stages (OIII–OVI, NIII, NIV, CIII, CIV) is achieved now for high redshift systems (2.1678, 2.315, 2.433) as well as for low redshift systems (0.7217, 0.8642, 1.1572). If the ionization is due to a universal metagalactic background radiation field due to QSOs (with z between 2 and 3), we expect an evolution of the ionizing background according to $\sim (1 + z)^{3+\alpha}$ (α = spectral index $\approx$ −0.6). The ionization parameter is expected then to decrease by a factor of $\sim$10 between the highest and lowest redshift systems with the consequence that the OV/OIII ratio would decrease by a factor $\sim$50! Preliminary results show that this is not the case.

The GHRS spectra will allow to determine the Ne abundance in the $z =$ 1.8465, 2.1678 and 2.315 systems and the S abundance for $z = 0.7217, 0.8642$ and 1.1572. Both Ne and S are expected to follow the oxygen abundance in case of nucleosynthesis in SN II.

The slightly higher spectral resolution, less contamination with Lyα forest lines, and additional lines (e.g. NIII 374 Å) will enable us to get more accurate nitrogen abundances.

Optical spectra of CIV lines with high S/N and spectral resolution will enable us to resolve the lines into subcomponents and to recognize possible saturation effects. Such spectra have been taken by D. Tytler with the Keck 10 m telescope.

4 Future Development

With the available HST data on HS 1700+6416 it will be possible to obtain abundances of C, N, O, Ne, S, Si in both high- and low-redshift systems. We shall also be able to improve the constraints on the nature and cosmic evolution of the ionizing radiation. A precondition, however, for an improved spectral analysis are high-resolution, high S/N optical data in order to be able to perform a decomposition of the unresolved EUV lines into subcomponents via information from CIV lines. Clearly, the study of the chemical evolution of early galaxies via EUV quasar absorption lines can evolve only with more objects (lines of sight) like HS 1700+6416. Recently, 3 new objects with slightly lower redshifts ($z = 2.13, 2.19, 2.40$) have been discovered by the Hamburg Quasar Survey which are all slightly brighter in the UV than HS 1700+6416 (Reimers et al. 1995) and thus ideal targets for HST. Finally, spectroscopy of EUV quasar absorption lines will develop into a flowering branch of extragalactic research when a new high resolution spectrograph onboard HST will come into operation.

Acknowledgements. This work has been supported by the Verbundforschung of the BMFT under grant No. 50 OR 90 058. The Deutsche Forschungsgemeinschaft supports the HQS under Re 353/22-1 to 3.

References

Bahcall, J.N., Bergeron, J., Boksenberg, A. et al., 1993, ApJS 87,1

Bechthold, J., 1994, ApJS 91,1

Bergeron,J., Boissé,P., 1991, A&A 243, 344

Izotov,Yu.I., Lipovetsky,V.A., Guseva,N.G. et al., 1994, in *Dwarf Galaxies*, ESO Conf. Proceed. 49, p.455

Jakobsen P., Boksenberg A., Deharveng J.M. et al., 1994, Nat 370, 35

Lanzetta,K., 1993, in *The Environment and Evolution of Galaxies* (J.M.Shull, H.A. Thronson eds.) Kluwer Acad. Publ., Dordrecht, p.237

Madau, P., 1991, ApJ 376, L33

Madau, P., 1992, ApJ 389, L1

Madau, P., Meiksin, A., 1994, ApJ 433, L53

Miralda-Escudé, J., Ostriker, J.P., 1990, ApJ 350, 1

Møller, P., Jakobsen, P., 1990, A&A 228, 299

Murdoch, H.S., Hunstead, R.W., Pettini, M., Blades, J.C., 1986, ApJ 309, 19

Picard, A., Jakobsen, P., 1993, A&A 276, 331

Reimers, D., Clavel, J., Groote, D. et al., 1989, A&A 218, 71

Reimers, D., Vogel, S., Hagen, H.-J. et al., 1992, Nat 360, 561

Reimers, D., Vogel, S., 1993, A&A 276, L13

Reimers, D., Rodriguez-Pascual, P., Hagen, H.-J., Wisotzki, L., 1995, A&A 293, L21

Smette, A., Robertson, J.G., Shaver, P.A., Reimers, D., Wisotzki, L., Köhler, Th., 1995, A&A in press

Steidel, C.C., 1993, in *The Environment and Evolution of Galaxies*, l.c. p.263

Verner, D.A., Barthel, P.D., Tytler, D., 1994, A&AS 108, 287

Vogel, S., 1994, Ph.D Thesis, University of Hamburg

Vogel, S., Reimers, D., 1993, A&A 274, L5

Vogel, S., Reimers, D., 1995, A&A in press

Reanalysis of the LLS in QSO HS 1700+6416

R. Riediger[1] and P. Petitjean[2]

[1] Astrophysikalisches Institut Potsdam, An der Sternwarte 16, D-14482 Potsdam, Germany
[2] Institut d'Astrophysique de Paris - CNRS, 98bis Boulevard Arago, F-75014 Paris, France

Abstract. We take advantage of the unique information provided by the recent HST observations of three Lyman limit systems toward QSO HS 1700+6416 by Reimers et al. (1992) to constrain photoionization models of the absorbing clouds. Constant density models are ruled out. However, the observed column densities (and in particular the large $N(\mathrm{HeI})/N(\mathrm{HI})$ column density ratio) can be reproduced with a model in which the medium is considered inhomogeneous with two phases of different densities. The oxygen and nitrogen abundances are found to be about 0.1 solar, the carbon abundance may be slightly smaller. The presence of a break of a factor ten in the ionizing spectrum at the HeII edge is not ruled out by these observations. It is clear that the high ionization phase could be thermally ionized.

1 Constant Density Models

HST spectroscopic observations of QSO HS 1700+6416 are unique by the number and variety of the observed absorption lines from HI, HeI, CII to CIV, NIII to NV, and OIII to OVI, some of the latter ions being observed for the first time in a QSO spectrum. The data have been discussed by Vogel & Reimers (1993, 1994) in the framework of constant density photo-ionized clouds. They concluded that (i) the best solution for the ionizing spectrum is a power law of index $\alpha = -0.6$, (ii) a break in the ionizing spectrum at the HeII edge (54.4 eV) by more than a factor of five is excluded, and (iii) oxygen and nitrogen are enhanced relatively to carbon compared to solar with $[\mathrm{C/H}] = -2.3, -1.9$; $[\mathrm{O/C}] = 0.9, 0.8$; $[\mathrm{N/C}] = 0.8, 0.8$ in the $z_{\mathrm{abs}} = 2.1678$ and $z_{\mathrm{abs}} = 2.433$ respectively. Reimers & Vogel (1993) discussed the $N(\mathrm{HeI})/N(\mathrm{HI})$ ratio and found that photo-ionization models give values five times smaller than what is observed with no explanation.

However a break in the ionizing spectrum at 54.4 eV would help explain a high value of the $N(\mathrm{HeI})/N(\mathrm{HI})$ ratio. This would lower somewhat the ionization rate of HeII into HeIII whereas the corresponding photons have small influence on the ionization level of hydrogen.

We have reanalysed the data taking into account all the constraints at the same time. We are led to different conclusions than previously found.

2   Models with Inhomogeneity

When observed at high spectral resolution, metal lines split into several components spanning typically a few hundreds km s^{-1} (e.g. Petitjean & Bergeron
1994). The medium is inhomogeneous and several clouds of different ionization stages may be present along the line of sight. The HST data are of low
resolution thus it is unclear whether absorptions from low and high ionization species, such as OIII and OVI, are produced in the same region. It is
probable that they are not. To mimic this we have modified the code Nebula
(Péquignot et al. 1978, Petitjean et al. 1990) to model the ionization state
in an inhomogeneous photo-ionized cloud consisting of two phases of different densities and a smooth transition zone. The diffuse ionizing flux must be
described in the detail since it can be as strong as the incident flux when
the optical depth is of the order of unity and thus a careful description is
of importance when modeling Lyman limit systems of small optical depth as
those observed in QSO HS 1700+6416.

Assuming a typical ionizing flux of intensity at the Lyman limit $F_o = 5 \cdot 10^{-22}$
erg s^{-1} cm^{-2} Hz^{-1} sr^{-1} and of spectral shape a power law $F_\nu \propto \nu^{-\alpha}$ with
$\alpha = 0.6$ and a break of a factor ten at 54.4 eV, we find a typical solution for a
cloud of total radial dimension 110 kpc and density $n_H = 4 \cdot 10^{-4}$ cm^{-3} with
an inhomogeneity in the centre of density 60 times larger and dimension about
1 kpc. Abundances are found to be $0.08 \cdot Z_\odot$. The fraction of low excitation
elements like HI, HeI and CII is highest in the central region of higher density
while highly ionized ions like NV, OV and OVI are only produced in the
extended low density region. All the column densities except N(CIV) are
reproduced within the uncertainties. The observed CIV column density is too
small, by up to a factor of 5, depending on the absorption system, compared
to what can be reproduced by models with solar [C/O] abundance ratio.
However new MMT data (M. Rauch private communication) show that the
published CIV equivalent widths are in error by a factor three in some cases.
In order to confirm that the [C/O] abundance ratio is smaller than the solar
value, high spectral resolution observations of the CIV wavelength range are
needed.

References

Péquignot, D., Aldrovandi, S.M.V., Stasińska, G. (1978), A&A 63, 313
Petitjean, P., Boisson, C., Péquignot, D. (1990), A&A 240, 433
Petitjean, P. & Bergeron, J. (1994), A&A 283, 759
Reimers, D., Vogel, S. (1993), A&A 276, L13
Reimers, D., Vogel, S., Hagen, H.-J., et al. (1992), Nat 360, 561
Vogel, S., Reimers, D. (1993), A&A 274, L5
Vogel, S., Reimers, D. (1994), A&A, in press

Metal Line Absorptions Toward the QSO HS1700+6416

J.L. Sanz[1], P.M. Rodríguez-Pascual[2], A. de la Fuente[3], M.C. Recondo[2], J. Clavel[4], M. Santos-Lleó[5], W. Wamsteker[2]

[1] Depto. de Fisica Moderna, Univ. de Cantabria, 39005 Santander, Spain
[2] ESA IUE Observatory, P.O. Box 50727, 28080 Madrid, Spain
[3] INSA/VILSPA, P.O. Box 50727, 28080 Madrid, Spain
[4] ESA ISO Observatory, P.O. Box 50727, 28080 Madrid, Spain
[5] Observatoire de Paris, RA 173 CNRS, 92135 Meudon, France

Abstract. We have analyzed high resolution ($b \approx 15\,km/s$) and low resolution ($R \approx 3$Å) optical spectra of the QSO HS1700+6416 ($z_{em} = 2.72$) taken with the ISIS spectrograph at the WHT. We are able to identify more than 50 metal lines.

1 The Metal Line Absorbers

Metal line absorption and Lyman edges, for QSO HS 1700+6416, have been reported by Reimers et al.(1989), Sanz et al.(1993) and Reimers et al.(1992) based on observations with the INT and HST, respectively. Our WHT observations allow to identify more than 50 metal lines. These correspond to the different ions of three metals (C, Si and Mg), associated to nine metal line systems (eight C-IV systems and one Mg-II system). We have obtained redshifts, velocity dispersions, column densities and estimated errors for these lines. A brief comment on each MLS is given below :

(i) $z = 2.7444$.- This system with $z > z_{em}$ consists of a possible CIV doublet ($\lambda\lambda 1548, 1550$) in the CIV emission line of the QSO and no other identified metal line. It is an "associated CIV system".

(ii) $z = 2.7102$.- This system possesses a CIV doublet that appears in the CIV emission line region of the QSO and it does not show any other line. Maybe, it is another "associated CIV system".

(iii) $z = 2.5784$.- This CIV-system with the doublet observed in one of our low-resolution also shows other lines identified in the HST data (HeI $\lambda 584, 537$, CIII $\lambda 386$, OV $\lambda 630$, OIV$\lambda 788$) by Reimers *et al* (1992).

(iv) $z = 2.4336$.- This CIV-system is a Lyman-limit first identified by Reimers *et al* (1989). We also identify SiIV $\lambda 1402$ (the other line of the doublet is confused with the noise) and SiIII $\lambda 1206$. It is also observed HeI, OV,IV,II, NIV,III, CIII and SiIII in the HST data.

(v) $z = 2.315$.- This CIV-system is seen as a Lyman-limit in the HST data. With our high resolution data we are able to distinguish a complex structure for the first time. In fact, three components have been identified using low and high-ionization lines: the SiIV doublet is observed in the low-resolution spectrum but the SiIV λ1394 is also observed in the high-resolution one. NV λ1238 appears blended whereas NV λ1242 is detected as a weak line. Regarding the low-ionization lines we have detected three components for SiII λ1190, 1193, 1304 and CIIλ1335. SiII λ1260 and SiIII λ1206 are blended with other lines. In the HST data, it is also observed HeI, OV,IV,III, CIII,II and NV,IV.

(vi) $z = 2.308$.- This CIV-system is the second complex system found in our spectra of HS 1700+6416. Two components are clearly identified through the low -ionization lines of CI λ1158, 1193, 1277, 1329. CIλ1261, 1280 are blended with other lines. The puzzling result is that the SiIV doublet –that usually appears in the CIV-systems– does not appear in our spectrum. Other lines such as HeI, OV, NV,IV, CIII and SIII are also identified in the HST data.

(vii) $z = 2.1681$.- This CIV-system is clearly identified in our low-resolution spectrum but also shows the SiIV doublet in the high-resolution one. The absorption line corresponding to SiIII λ1206 is observed. It is i-dentified with a Lyman-limit system in the HST data, where other lines are detected (HeI, OV,IV, CIII, AlII).

(viii) $z = 1.8465$.- This CIV-system is identified as a Lyman-limit in the HST data. In our high-resolution spectrum we can detect both the CIV doublet and the SiIV one. From their wavelengths we derive $z = 1.8451$. No other lines are seen in our spectra. In the HST data HeI and OV,IV,III are identified.

(ix) $z = 1.1572$.- This is identified as a MgII-system for the first time. We observe the doublet MgII λ2796, 2803 in our low-resolution spectrum . We also detect MgI λ1828 in our high-resolution spectrum whereas MgI λ2026 is blended with another line. Other elements (OV,IV,III and SIII) are also identified in the HST data.

We have run a photoionization code (CLOUDY) to get information about the physical parameters of the system at $z = 2.315$: a neutral hydrogen density of $log\, n_{HI} \simeq -1.6$, overabundances of N and O, solar abundances of C and Si and abundances below 1/10 solar for the rest of the elements can explain the observed absorption lines if the ionizing background is assumed to be a power-law ($\alpha = -1$) with a flux at the Lyman limit of $log\, J_\nu = -21$.
Acknowledgements. This work was partly supported (J.L.S., M.C.R., M.S.) by the Spanish DGICYT project PB92-0741.

References

Reimers, D., Clavel, J., Groote, D., Engels, D., Hagen, H.-J., Naylor, T., Wamsteker, W., & Hopp, U. 1989, A&A, 218, 71

Reimers, D., Vogel, S., Hagen, H.-J., Engels, D., Groote, D., Wamsteker, W., Clavel, J., & Rosa, M. R. 1992, Nature, 360, 561

Sanz, J. L., Clavel, J., Naylor, T., & Wamsteker, W. 1993, MNRAS, 260, 468

High Signal-to-Noise Echelle Spectroscopy of HS 1946+7658

Todd M. Tripp[1], Limin Lu[1,2], and Blair D. Savage[1]

[1] Dept. of Astronomy, University of Wisconsin, Madison, WI 53706-1582, USA
[2] Dept. of Astronomy, California Institute of Technology, Pasadena, CA 91125, USA

Abstract. We have observed HS 1946+7658 with the echelle spectrograph on the Kitt Peak 4m telescope. The signal-to-noise of the final coadded spectrum ranges from 40:1 to 80:1 per pixel with a resolution of 20 km s^{-1} (FWHM). The heavy element systems that we have identified in the spectrum of this QSO are summarized in Table 1; we detect eleven heavy element systems including two damped Lyman α systems, one associated system (i.e., $z_{abs} \approx z_{em}$), two Mg II systems, and six C IV systems. The results of a detailed analysis of the metal abundances and physical conditions in the two damped Lyα systems are found in Lu et al. (this meeting and 1995). The complex data set, reduction techniques, and results relevant to the remaining absorption line systems are found in Tripp et al. (1995).

Eight hour-long observations of HS 1946+7858 ($z_{em} = 3.051$, $V = 15.9$) were obtained with the echelle spectrograph on the Kitt Peak 4m telescope in May of 1993. To maximize the signal-to-noise (S/N), the echelle grating tilt and/or cross dispersion encoder were changed slightly between exposures to disperse the spectrum on different regions of the 2048×2048 CCD detector and thus reduce residual fixed pattern noise not adequately removed by the flat field. We also improved the S/N by using "optimal" spectrum extraction based on the weighted slit procedure outlined by Horne (1986).

We have used Voigt profile fitting to estimate the redshifts, column densities, and Doppler parameters of the metal lines (using the software described by Lanzetta & Bowen 1992). We have also used the apparent column density method (see Savage & Sembach 1991). The very high quality of the data and the application of careful analysis techniques permits us to infer important information about abundances and physical conditions in the two damped Lyα systems (Lu et al. 1995) and the strong associated system (Tripp et al. 1995). The observed relative abundance patterns in the damped systems are strikingly similar to those observed in low metallicity Milky Way stars and suggest that we are viewing absorption by galaxies in the early stages of chemical enrichment. The ionization in the associated system is very different from the Galactic halo: $N(\text{C IV})/N(\text{Si IV}) \approx 70$ in one of the components of the associated system while in the Milky Way halo, $< N(\text{C IV})/N(\text{Si IV})> = 3.6 \pm 1.3$ (Sembach & Savage 1992). Preliminary photoionization models indicate that the abundances are close to solar in the associated system.

T.M. Tripp, L. Lu, and B.D. Savage

Table 1. Summary of Heavy Element Systems in the Spectrum of HS 1946+7658

System redshift	Species detected
1.11903	Fe II 2600.2
	Mg II 2796.4,2803.5
1.53448	Mg II 2796.4,2803.5
1.7382[a]	Si II 1808.0
	Al III 1854.7,1862.8
	Cr II 2055.6,2061.6,2065.5[b]
	Fe II 2249.9,2260.8,2344.2,2374.5,2382.8,2386.7,2600.2
	Mn II 2576.9,2594.5,2606.5
	Mg I 2853.0
	Mg II 2796.4,[b] 2803.5
2.39528	C IV 1548.2,1550.8
2.64437	C IV 1548.2,1550.8
2.77732	C IV 1548.2,1550.8
2.8435,2.8443,2.8452[a]	O I 1302.2
	Si II 1260.4,[c] 1304.4,1526.7
	C II 1334.5
	Si IV 1393.8,1402.8
	C IV 1548.2,1550.8
	Fe II 1608.5
	Al II 1670.8
2.8925,2.8932	Si IV 1393.8,1402.8
	C IV 1548.2,1550.8
2.91671	C IV 1548.2,1550.8
3.03841	C IV 1548.2,1550.8
3.0496,3.0504[d]	N V 1238.8,1242.8
	C II 1334.5
	Si II 1190.4,[c] 1193.3,[c] 1260.4
	Si III 1206.5[c]
	Si IV 1393.8,1402.8
	C IV 1548.2,1550.8
	Al II 1670.8
	Al III 1854.7,1862.8

[a]Damped Lyman α system.
[b]Blended line (redward of the Lyman α forest).
[c]This line occurs in the Lyman α forest and thus could be contaminated.
[d]Associated system ($z_{abs} \approx z_{em}$).

References

Horne, K. 1986, PASP, 98, 609
Lanzetta, K.M., & Bowen, D.V. 1992, ApJ, 391, 48
Lu, L., Savage, B. D., Tripp, T. M., & Meyer, D. M. 1995, ApJ, in press
Savage, B. D., & Sembach, K.R. 1991, ApJ, 379, 245
Sembach, K. R., & Savage, B. D. 1992, ApJS, 83, 147
Tripp, T. M., Lu, L., & Savage, B. D. 1995, ApJS, in preparation

Redshift Evolution of the Carbon Abundance in Galaxies

Erik A. Stengler-Larrea

Royal Greenwich Observatory, Madingley Road, Cambridge, CB3 0EZ, UK

Abstract. Recent results on the redshift evolution of Lyman limit systems (LLSs) are combined with measurements of the CIV column density in 73 CIV absorption systems to conclude that an abundance effect is responsible for the observed redshift evolution of the latter.

1 CIV Column Densities and Lyman Limit Systems

The redshift evolution of the CIV column density in metal line absorbers, N(CIV), has been measured directly for the first time by Stengler-Larrea (1994, 1995), fitting Voigt profiles to a large sample of 73 absorption systems. Figure 1 shows the redshift distribution of N(CIV) for the 73 systems and a power law fit to the redshift dependence of the form $\log N(CIV)(z) = 15.70(1+z)^{-0.46}$. This result agrees with the previous indirect estimate made by Steidel (1990), based on equivalent width arguments, who concluded on an increase of N(CIV) by *at least* a factor of 3 over $1.5 < z < 3$.

Under the well-founded assumption that LLSs are not largely neutral, but mostly ionised by the UV metagalactic flux, Steidel (1990) showed that such a redshift evolution in N(CIV) in QSO absorption systems must be followed by an increase of the LLSs number density per unit redshift, $N(z)_{LLS}$, by a factor of 3 over $1.5 < z < 3.0$, if both effects are to be attributed to changes in the ionisation state of the absorbers.

The redshift distribution of LLSs has been subject to a substantial controversy (Sargent, Steidel and Boksenberg 1989, Lanzetta 1991) but it now seems established that is well described by a single power law $\propto (1+z)^{\gamma}$ with $\gamma = 1.5$ as given in Stengler-Larrea et al. (1995) and shown in Figure 2.

2 Redshift Evolution of the Carbon Abundance

The estimated evolution of $N(z)_{LLSs}$ implies an increase by a factor of 1.3, with a 1-σ upper limit of 2. Thus, taking the results of the previous sections together, abundance effects must be responsible for the observed redshift evolution of N(CIV). Ionisation effects may be present to a certain extent, but are definitely not dominant.

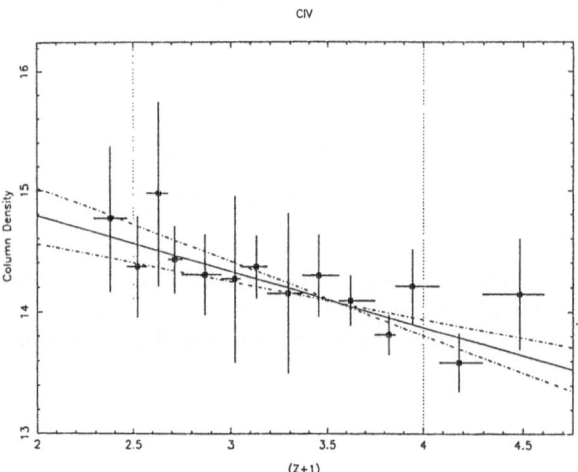

Fig. 1. Distribution in z of N(CIV) as $\log N(CIV) = 15.70(\pm 0.53) - 0.46(\pm 0.15) \times (1 + z)$. The vertical lines mark the range of the predictions of Steidel (1990).

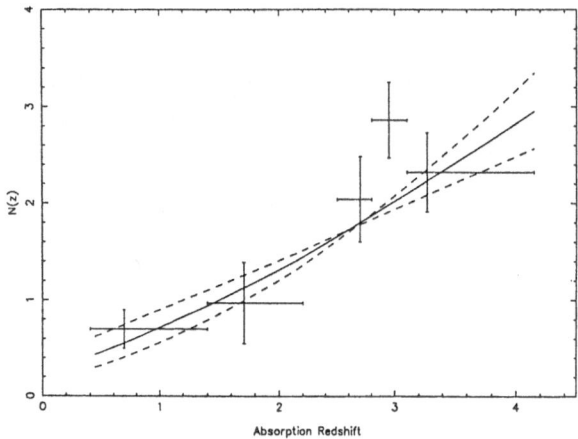

Fig. 2. Redshift evolution of the LLSs number density as $N(z) \propto (1 + z)^{1.5 \pm 0.39}$.

References

Lanzetta, K.M., 1991, ApJ. 375, 1.
Sargent, W.L.W, Steidel, C.C. and Boksenberg, A.,1989, ApJS. 69, 703.
Steidel, C.C., 1990, ApJS, 72, 1.
Stengler-Larrea, E.A., 1994, PhD. thesis, Cambridge.
Stengler-Larrea, E.A. et al., 1995, *to appear in the 1 May issue of ApJ.*
Stengler-Larrea, E.A., 1995, *in preparation.*

Recent Results on QSO Absorption Lines

E. Margaret Burbidge

Center for Astrophysics and Space Sciences and Department of Physics University of California, San Diego, La Jolla, California 92093-0111, U.S.A.

Abstract. Results are shown from four separate studies of QSO absorption lines by the Hubble Space Telescope team at the University of California, San Diego. These results include HST data obtained with the Faint Object Spectrograph (FOS), the High Resolution Spectrograph (HRS), and the new Wide Field/Planetary Camera (WFPCII), as well as data from ground-based observations made at the Keck and Lick Observatories.

Results from an HST study combining Wide Field Camera Images and an FOS spectrum of the radio-loud QSO 3CR 196 (z_e=0.871, V=17) were shown in this poster paper (Cohen et al. 1995, see also Diplas et al. 1994). Subtraction of the point-spread function of the bright QSO from the image taken with the new WFC clearly reveals a barred spiral galaxy ~3 arc seconds SE of the QSO. This galaxy most probably causes the previously observed 21-cm and optical absorption at z=0.437. Radio isophotes measured by A. Oren (private communication) and superposed on the WFC image fit well with this interpretation. The FOS spectrum is complicated by the presence of strong absorption from high ionization lines near the emission redshift (including S IV), but almost certainly indicates the presence of a high column density of H I at z=0.437 in front of the optical quasar, at a location distant from the gas which causes the 21 cm absorption. The large extent of the H I gas is consistent with the presence of a luminous spiral galaxy. The absorption at the emission redshift may not cover the continuum source.

Hamann et al. (1995) have compared recent observations of the high-ionization associated absorption system ($z_a = 2.1340$) in the spectrum of UM 675 ($z_e = 2.147$, $V = 17.1$) with earlier Palomar observations by Sargent et al. (1988). They have found that these absorption lines strengthened by a factor ~ 3 during the 9 year interval between the observations in 1981 and 1990. A study of photoionization models sets limits of < 200 pc for the distance of the absorbing gas from the continuum source, and indicates that the element abundances are roughly solar or higher. Continued monitoring of the spectrum of UM 675 at Lick Observatory since 1990 has shown no further variation in the associated absorption line strengths. The 1981 and post-1990 summed spectra, scaled to the same continuum level, were shown in this poster paper.

Data obtained with the HST/FOS, using gratings G190H and G270H, of the bright QSO PG 1522+101 ($z_e = 1.321$, V = 15.65), which are under study by Diplas et al. (1995), were shown in this poster paper. Of the one hundred and two absorption lines measured at a significance level greater than 4.5σ, most are Ly-α and Ly-β forest lines. There are seventy-four systems with no metal-line absorption, and three with metal-line absorption (three with CIV, one with OVI). The distribution of the Ly-α systems corresponds to a measured dN/dz = 27.3 at a mean redshift of 0.83. Eleven lines are due to absorption from the Galactic Interstellar medium. (Diplas et al. 1993, Diplas et al. 1995).

Barlow and Junkkarinen (1994) are studying the spectra of 3 broad-absorption-line QSOs with the HIRES echelle spectrograph on the Keck 10-m telescope. The profile of the very broad CIV absorption trough centered at ~ 4760 Åin Q 1246-057 ($z_e = 2.236$, $V = 16.73$) had been measured previously with the Lick 3-m telescope, and found to be fairly smooth and, notably, it did not go to zero intensity in the center of the line. It had been conjectured that the profile, when observed with high spectral resolution ($8 km\ s^{-1}$) at the Keck 10-m telescope, might be shown to break up into a large number of narrow lines, as is observed in some $z_a \ll z_e$ absorption systems in QSOs. However, the profile observed at $8 km\ s^{-1}$ showed remarkable similarity to the profile observed at $200 km\ s^{-1}$ resolution at Lick. It did not break up into a large number of narrow optically thick components, and in the central region of the absorption trough it lay above zero intensity by the same amount as determined from the Lick data. The smooth absorption troughs require absorption by thousands of individual clouds in the BAL region. The Lick and Keck profiles of CIV absorption were shown in this poster paper.

Acknowledgements. This work was supported by NASA grant No. NAG5-1630.

References

Barlow, T.A., Junkkarinen, V.T. 1994, B.A.A.S., 26, 1339

Cohen, R.D., Beaver, E.A., Diplas, A., Junkkarinen,. .V.T., Barlow, T.A., Lyons, R.W. 1995, Ap.J., (submitted)

Diplas, A., Cohen, R.D., Barlow, T.A., Beaver, E.A., Junkkarinen, V.T., Lyons, R.W., 1004, B.A.A.S., 36, 050

Diplas, A., Cohen, R.D., Beaver, E.A., Burbidge, E.M., Junkkarinen, V.T. Lyons, R.W., 1993, B.A.A.S., 25, 1307

Diplas, A., Cohen, R.D., Beaver, E.A., Burbidge, E.M., Junkkarinen, V.T., Lyons, R.W., Barlow, T.A., 1995 (in preparation)

Hamann, F.W., Barlow, T.A., Beaver, E.A., Burbidge, E.M., Cohen, R.D., Junkkarinen, V.T., Lyons, R. 1995, Ap.J., in press

Sargent, W.L.W., Boksenberg, A., Steidel, C.C. 1988, Ap.J.S., 68, 539

Fine-Structure Doublets in QSO Spectra and Variability of the Fine-Structure Constant α

S. A. Levshakov[1] and S. D'Odorico[2]

[1] Department of Theoretical Astrophysics, A. F. Ioffe Physico-Technical Institute,
St.Petersburg 194021, Russia
[2] European Southern Observatory, Karl-Schwarzschild-Str. 2, D-85748,
Garching bei München, Germany

Abstract. In Kaluza-Klein theories with extra space dimensions, dependence of the fine-structure constant, $\alpha \equiv e^2/\hbar c \simeq 1/137$, on the volume of the compact space allows one to use fine-structure doublets observed in QSO spectra to probe the cosmological evolution of compact dimensions over time intervals compared with the age of the Universe.

We show, however, that the unresolved velocity structure of absorbing material may affect strongly the shape of the probability density function of the α_z/α ratio (α_z is a value of α at redshift z). This in turn leads to a biased sample mean $\langle \alpha_z/\alpha \rangle$ and a *false* accuracy.

1 Results

At first, note that the sample mean $\langle \Delta\alpha/\alpha \rangle$ has in general a negative bias even in the case of one component symmetrical profiles of the fine-structure doublets because of an asymmetric shape of the probability density function $p_{\alpha_z/\alpha}(x)$, where $p_{\alpha_z/\alpha}(x)\,dx$ is the probability to find the ratio α_z/α between x and $x+dx$ (Levshakov, 1994a). The degree of the asymmetry (see Fig. 1) is a function of a dimensionless parameter ξ/s which characterizes the quality of data involved.

Consider now multicomponent absorption-line profiles which are not completely resolved and are not symmetric in general. Then the subcomponents of different optical depths can affect the value of $\langle \Delta\alpha/\alpha \rangle$ through the line saturation effect. To illustrate this systematic bias, we have calculated a simple model of a blend of two Al III doublets ($\lambda\lambda 1855, 1863$) with different Doppler widths b_i and different τ_i ($i = 1,2$), where τ_i is the central optical depth of the blue component of the Al III doublet. We assume that the radial velocity difference between these subcomponents may vary in the range $0 \leq \Delta v_r \leq \frac{1}{2}$FWHM of an instrumental function. We also assume that b−parameters lie in the range from 4 km s^{-1} (the thermal width for aluminum at $T_{kin} \sim 10^4$ K) to the FWHM. Then, we can calculate maximum possible deviation of $|\Delta\alpha|/\alpha$ from zero using different combinations of $\Delta v_r, \tau_1, b_1,$ and b_2 at a fixed value of τ_2. Obtained dependence of $(|\Delta\alpha|/\alpha)_{max}$

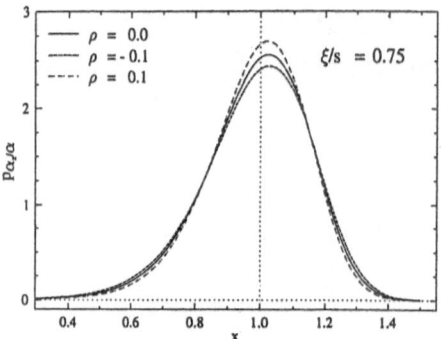

Fig. 1. Probability density function $p_{\alpha_z/\alpha}(x)$ for $\alpha_z/\alpha \equiv 1$. Curvers for different correlation coefficient ρ are shown. For details, see Levshakov (1994b).

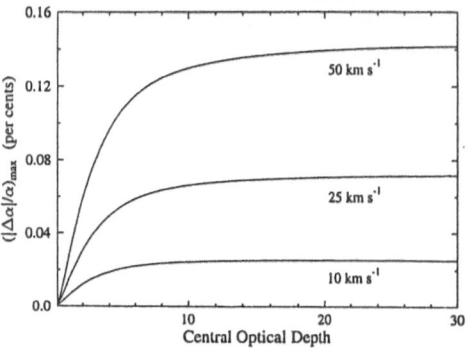

Fig. 2. Maximum deviation $|\Delta\alpha|/\alpha$ from zero for different resolutions FWHM = 50, 25, and 10 km s^{-1}.

upon τ_2 is shown in Fig. 2 for three spectral resolutions : 50, 25, and 10 km s^{-1}.

These calculations show that negligible shifts ($|\Delta\alpha|/\alpha < 0.1$ per cent) can occur in the case of (i) optically thin lines or (ii) at high resolution (FWHM ≤ 10 km s^{-1}).

Our results show that the question on cosmic time variability of α at the level of about 0.1 per cent remains still open. New high resolution observations of QSOs with FWHM ≤ 10 km s^{-1} and $S/N > 30$ are required to reach the realistic value of $\Delta\alpha/\alpha$ at different redshifts.

Acknowledgements. SAL was supported by the ESO C&EE Programme (grant D-06-005) .

References

Levshakov, S. A. 1994a, Astron. Zh., 71, 181.
Levshakov, S. A. 1994b, MNRAS, 269, 339.

Limits on Star Formation Rates in Four Quasar Metal-Line Absorbers

N. Bergvall[1], G. Östlin[1], K.-G. Karlsson[2], E. Örndahl[1], J. Rönnback[3]

[1] Astronomiska observatoriet, Box 515, S-75120 Uppsala, Sweden
[2] Mid Sweden University, S-87188 Härnösand, Sweden
[3] ESO, Karl-Schwarzschild-Strasse 2, D-85748 Garching bei München, FRG

Abstract. In an effort to detect the early phases of star formation in massive galaxies, we have searched for Lyα emission and [OII]λ3727 emission from four quasar metal line absorbers, using narrow-band filters. None of the objects were detected. We discuss the possible effects of dust on the visibility of the objects and present upper limits on the star formation rates, assuming that the effects of dust can be neglected.

1 Observations and Reductions

Using the 2.5-m Nordic Optical Telescope on La Palma, we searched for nebular emission from 4 quasar absorbers with strong CIV or MgII absorption lines, Q1726+344, Q2251+244, Q2343+125 and Q2344+125 (z_{QSO}=2.43, 2.33, 2.52, 2.76). We used narrow band filters (15-20Å) centered on the redshifts of Lyα or [OII]λ3727. On-line integration times ranged from 2 to 8 hours. In the resulting on-off line images we searched for emission from the absorber or associated star forming galaxy. As an extra check of the region close to the quasar luminosity peak, we subtracted the PSF, using nearby stellar images. In none of the resulting images we see extended structure reaching over the 3 σ level that could be due to the absorber. To estimate the upper limits on the line fluxes of the absorbers, we collected statistics of fluxes measured in circular areas with a diameter of approximately 4 arcseconds, sampled in the region around the quasar. The size of such an area corresponds approximately to the projected size of a normal-sized disk galaxy at these redshifts.

To convert these limits to limits in the star-formation rates we firstly assume that the IMF has a Salpeter form and that the mass range is 0.1 - 100 $\mathcal{M}_\odot$. Furthermore we assume that the metallicity of the stars is $\sim$ 5% of the solar and that the gas is about twice as metal rich. We then derive the number of Ly-continuum photons produced by these stars from our spectral evolutionary models. Assuming that Lyα is produced by pure recombination, we then obtain a predicted Lyα luminosity of 2.8×10^{35}W for a star formation rate of 1 $\mathcal{M}_\odot$/yr. From photoionization codes we adopt a typical Lyα /[OII]]λ3727 line ratio of 30.

2 Results and Discussion

One central problem in the analysis is to take proper account of the effects of dust on the emission from the star forming regions. Charlot and Fall (1991) have discussed this issue extensively. Their results indicate that multiple s-cattering and subsequent dust absorption of Lyα is of minor importance for HI column densities below a few times 10^{20}. Judging from the equivalent widths of the Lyα absorption lines, this is a likely situation in the present case, indicating that the attenuation of Lyα is moderate. This prediction however is highly model dependent and it is interesting to compare with observations of nearby galaxies. In a study of nearby metal poor galaxies, Calzetti and Kinney (1992) find that dust extinction is sufficient to explain the observed Hβ /Lyα line ratios and that multiple scattering is probably relatively unimportant. Further support for this comes from s study of nearby starburst galaxies (Valls-Gabaud 1993). These studies and similar studies of low metallicity galaxies by Hartmann et al. (1988) and Terlevich et al. (1993) show that the Lyα /Hβ emission line ratios are attenuated with a factor of 3-8 relative to the dust-free situation. This may be relevant also for the metal line absorbers, in particular when considering that their metallicities are likely to be low. In the table below we present the predicted upper limits of the star formation rates (SFR), assuming $H_0=80$ and $q_0=0.1$, without efforts to correct for the influence of dust. As can be seen, the limits on the total star formation rates, if corrected with a reasonable factor due to the dust, are still not strong enough to rule out that the absorbers have starburst properties.

Table 1. Results

Object	Abs. Line system	Em. Line observed	z	3σ upper limits (W m^{-2} arcsec^{-2})	SFR ($\mathcal{M}_\odot$yr^{-1} arcsec^{-2})
Q 1726+344	CIV	Lyα	2.3	$3 \cdot 10^{-20}$	≤ 0.3
Q 2251+244	MgII	[OII]λ3727	1.1	$4 \cdot 10^{-21}$	≤ 0.2
Q 2343+125	CIV	Lyα	2.4	$3 \cdot 10^{-20}$	≤ 0.3
Q 2344+125	CIV	Lyα	2.4	$7 \cdot 10^{-20}$	≤ 0.6

References

Calzetti, D., Kinney, A.L., 1992, ApJ 399, L39

Charlot, S., Fall, S.M., 1991, ApJ 378, 471

Hartmann, L.W., Huchra, J.P., Geller, M.J., O'Brien, P., Wilson, R., 1988, ApJ 326, 101

Valls-Gabaud, D., 1993, ApJ 419, 7

Terlevich, E., Díaz, A.I., Terlevich, R, García Vargas, M.L., 1993, MNRAS 260, 3

A Group of Galaxies at z=2.38

Paul Francis[1], Bruce Woodgate[2], Steve Warren[3], Palle Møller[4], James Lowenthal[5] and Ted Williams[6]

[1] University of Melbourne, School of Physics, Parkville Vic 3052, Australia
[2] NASA Goddard Space-Flight Center, Code 681, Greenbelt MD 20771, USA
[3] Imperial College, School of Physics , Blackett Laboratory, Prince Consort Road, London SW7 2BZ, UK
[4] STScI, Baltimore, MD 21218, USA
[5] Lick Observatory, Santa Cruz CA 95064, USA
[6] Rutgers University, Physics and Astronomy Department, P. O. Box 849, Piscataway, NJ 08855, USA

Abstract. We have discovered a compact group of ~ 10 large galaxies at redshift $z = 2.38$. Two of the galaxies are strong Lyman-α emitters; all are bright in the near IR. We hypothesise that the group represents a giant elliptical galaxy caught in the act of bottom-up formation.

1 Background

Francis & Hewett (1993) discovered two possible superclusters of high column density Lyman-α absorption-line systems at $z > 2$. They inferred that these superclusters must have overdensities > 30 on scales of ~ 10 co-moving Mpc. If true, the existence of these clusters would pose a threat to models of large-scale structure formation such as CDM, which cannot easily produce such massive structures at such high redshifts (see Francis & Veeraraghavan, these proceedings).

We imaged a small part of one of these putative superclusters, using narrow-band imaging to look for Lyman-α emission (see e.g.. Lowenthal et al. 1991, Møller & Warren 1992, Macchetto et al. 1993), and multicolour optical (Giavalisco, Steidel & Szalay 1994) and near-IR imaging. Only five square arcminutes were imaged (the putative supercluster extends over at least 8 arcmin).

2 Results

Two strongly Lyman-α emitting galaxies were detected in the small field. They show Lyman-α fluxes of 8×10^{-16} and 5×10^{-16}, rest-frame equivalent widths of > 100Å and > 30Å respectively (99% confidence lower limits) and velocity widths of $\sim 700 \mathrm{km s}^{-1}$. Both would easily have been detected in the large blank sky searches for Lyman-α emission carried out by Pritchet &

Hardwick (1990), the fact that they were not tells us that this region of sky has a large (> 30) overdensity of Lyman-α emitting objects (see also Wolfe 1993).

One of the Lyman-α emitting galaxy shows a clearly extended morphology in Lyman-α emission. Half the line flux comes from a marginally resolved nucleus, which is also the source of the near-IR flux. The remainder of the line emission comes from an extended lumpy tail, extending about 8 arcsec from the nucleus.

This Lyman-α emitting galaxy is bright in the near IR ($K = 19.4, B-K = 5.4$, continuum magnitudes). It is surrounded by a group of about 10 similar, very red objects. All show a big spectral break at around 1.2μ, due to the redshifted Balmer break. The galaxy group extends over 20 arcsec (~ 100 kpc), and the galaxies have a velocity dispersion of $\sim 500 \mathrm{kms}^{-1}$, determined from the velocity structure of the absorption line and the Lyman-α emission-line redshifts.

The colours of these red galaxies are well fit by passively evolving stellar populations of mass $\sim 10^{11}$ solar masses and age $\sim 500 \mathrm{Myr}$. The Lyman-α emission may be due to a concealed AGN, or a starburst involving about 1% of the galaxy's mass.

Due to the compact size and high velocity dispersion of this group, we suggest that its members will merge on a timescale of $\sim 10^9$ years, forming a giant elliptical or cD galaxy.

A fuller account of this work has been submitted to ApJ.

References

Francis, P. J. & Hewett, P. C. 1993, AJ, 105, 1633

Francis P. J. & Veeraraghavan, S. 1995, these proceedings

Giavalisco, M., Steidel, C. C., & Szaley, A. S., 1994, ApJ 425. L5

Lowenthal, J. D., Hogan, C. J., Green, et al. 1991, ApJ 377, L73

Machetto, F., Lipari, S. Giavalisco, M. et al. 1993, ApJ, 404, 511

Møller, P. & Warren, S. J. 1992, A& A, 270, 43

Pritchet, C. J. & Hardwick, F. D. A. 1990, ApJ, 355, L11

Wolfe, A. M. 1993, ApJ, 402, 411

A Narrow-line Emission Galaxy at z=2.8 in the Field of Q0831+128

Alfonso Aragón-Salamanca

Institute of Astronomy, Madingley Road, Cambridge CB3 0HA, England

Abstract. In this poster we present the discovery of a narrow-line emission galaxy with $z \simeq 2.8$ found at a projected distance of $\simeq 9''$ from Q0831+128. This galaxy is intrinsically very luminous and compact, with unusual spectral properties.

1   Discovery and Observed Properties

In recent optical spectroscopic observations at the 4.2m William Herschel telescope (La Palma), conducted as part of our on-going campaign to identify the galaxies producing CIV absorption lines in the spectra of high redshift QSOs (Aragón-Salamanca et al. 1994), we have found a rare and potentially very interesting object. It is a relatively narrow line ($\sigma \simeq 600\,\mathrm{km\,s^{-1}}$) compact object with redshift $z = 2.751$ close to the quasar Q0831+128 ($z = 2.734$). The source was identified in a very deep IRCAM K-band image of the QSO field taken at the 3.8m UK Infrared Telescope (Fig. 1). It has $K = 17.9$ and is 8.6$''$ away from the QSO (projected distance: 65 kpc, $H_0 = 50$; $q_0 = 0.5$). The optical spectrum shows strong Lyman-α, NV and CIV emission lines, and a weak red continuum (Fig. 2).

Although intrinsically very luminous ($M_K = -29.6$, i.e. $\sim 60 \times L_*$), the emission line spectrum is very different from that of the quasar. It is reminiscent of the prototypical star-forming Lyman-α radio galaxies. A similar emission line companion was previously found close to a radio-luminous quasar (Steidel et al. 1991), but our object has a Lyman-α line that is quite weak compared with NV and CIV, and it is more luminous. Furthermore, Q0831+128 is an optically-selected QSO, with no reported radio emission. The critical question is to ascertain whether the source is a very luminous star-forming galaxy or an unusual AGN.

References

Aragón-Salamanca, Ellis, R.S., Schwartzenberg, J.-M. & Bergeron, J., 1994, ApJ, 421, 27.
Steidel, C.C., Sargent, W.L.W., Dickinson, M., 1991, AJ, 101, 1187.

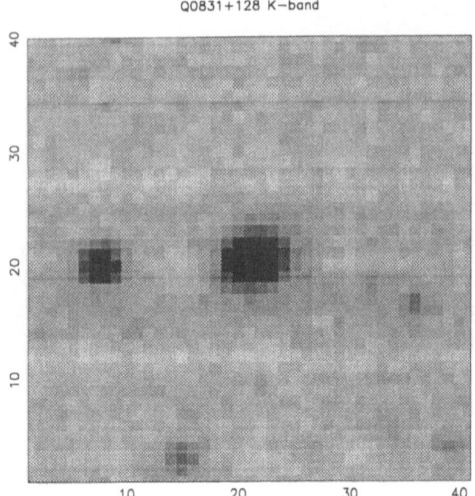

Fig. 1. Near-infrared K-band image of the field near Q0831+128 taken at the 3.8m UKIRT. N is up and E is left. The scale is $0.6''$ pixel^{-1}. The narrow-line emission galaxy is the object $\simeq 9''$ E of the QSO (bright central object).

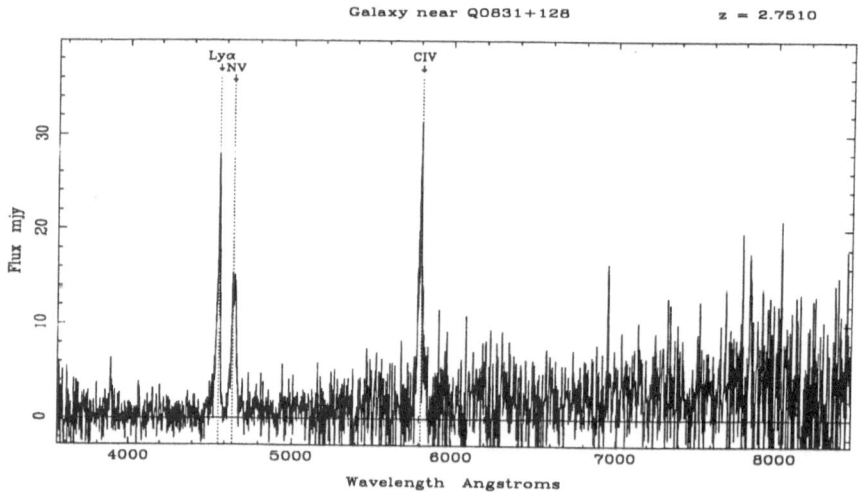

Fig. 2. Optical spectrum of the galaxy from the 4.2m WHT (La Palma). Three relatively narrow emission lines are detected (Lyman-α, NV and CIV).

PART VI

BAL QSOS

AND

QUASAR ENVIRONMENT

Broad Absorption Line Quasars: An Overview

Ray Weymann

Carnegie Observatories, Pasadena CA 91101, USA

Abstract. We review recent dynamical models proposed to explain the high velocity outflows seen in Broad Absorption Line Quasars (BAL QSOs). The arguments leading to the "confinement problem" are reviewed and the way with which these recent models attempt to deal with this problem discussed. We also review the evidence bearing on the anti–correlation between radio–loud QSOs and BAL QSOs and recent suggestions for explaining this anti–correlation.

1   Introduction

It is now fairly clear (in contrast to the situation 15 or 20 years ago) that the vast majority of narrow absorption lines in QSOs which are displaced to the blue by more than a few thousand km/sec with respect to the QSO emission line redshift owe these displacements to the difference in Hubble flow velocity between the cloud in question and the QSO nucleus. The broad absorption QSOs (in which relatively narrow components are sometimes embedded) seem an obvious exception to the "intervening paradigm", so that now we are considering a topic which perhaps ought logically to be considered at a conference on the Broad Emission Line Region of Quasars. For an introduction to the topic of BAL QSOs consult Turnshek (1988) and Weymann et al. (1985).

In this overview I will concentrate primarily on two topics: (1) What are the geometry, kinematics and dynamics of the absorbing material? and (2) What accounts for the remarkable anti-coincidence between broad absorption features and powerful radio emission in QSOs? Other aspects of BAL QSOs, including that of abundance anomalies, are discussed in the contributions by Turnshek and by Wampler elsewhere in this volume. Although most BAL QSOs show only absorption from highly ionized species, a small fraction show low ionization absorption as well; such "Mg II BAL QSOs" may or may not represent a different phenomenon. Wampler discusses elsewhere in this volume an example of the Mg II BAL QSOs. Examples of the variety of absorption line profiles exhibited by the C IV absorption trough may be found in Korista et al. (1993).

Some brief comments should be made about the formal criteria that have been suggested for inclusion of an object in the BAL QSO class (Weymann, Carswell & Smith 1981; Weymann et al. 1991). These criteria were formulated in an attempt to distinguish between clumps of absorption from intervening systems as well as the strong systems frequently seen at the emission line redshifts ("associated absorption") of many QSOs. The essential feature of the criteria is that high velocity ($>$ 5000 km/sec) outflow was required for an object to be formally placed in the BAL QSO category.

2   Geometry, Kinematics & Dynamics

Over the redshift range from about 1.5 to 3.0 the incidence among optically selected QSOs of objects classified as BAL QSOs is close to 10% for absolute magnitudes brighter than about -26. It is also clear that AGNs, at the luminosities of Seyfert 1 nuclei, often generate mild outflows, but never with the very high velocities frequently seen in luminous QSOs (Stocke *et al.* 1994 and references therein). This immediately suggests that radiative acceleration is involved. How is this fraction of $\sim$10% to be interpreted ? Two extreme views are: (1) The BAL region has a covering factor of $\sim$100% (*i.e.*, it is roughly spherically symmetric) in 10% of luminous QSOs. These 10% might be distinguished from their non–BAL counterparts simply by being in an early evolutionary stage lasting 10% of the total QSO lifetime—similar to the notion that the ultra–luminous IR galaxies may evolve into QSOs. (2) The BAL region covers only about 10% of the solid angle seen from the continuum source in all luminous QSOs. It is then simply a matter of orientation which QSOs are detected as BALs and which are not. At least three arguments have been advanced for the 2nd possibility: (i) The resonance line photons which are scattered by the absorbing photons are not easily destroyed, but are redistributed in velocity space. The most detailed analysis of this line of argument is by Hamann, Korista & Morris (1993) and suggests that the covering factors probably do not exceed about 25%. (ii) In most respects the continuum and broad emission line regions of BALs and non–BALs are very similar; it was then argued (*e.g.*, Weymann *et al.* 1991) that this is not likeby to arise if one is dealing with two different types of objects—the 10% in which BAL clouds are present, and the 90% in which they are not. In retrospect, I don't think this latter argument is very compelling if one considers evolutionary scenarios. Indeed, it is a little surprising that more pronounced differences in emission line profiles are not present if orientation does play the dominant role. (iii) Evidence from polarization studies (*e.g.*, Cohen *et al.* 1995) suggests that the material in the BAL clouds is in a disk. I will adopt the "orientation hypothesis" in the rest of the discussion, though one should keep an open mind about the other possibility, especially with regard to the Mg II BAL QSOs, for which there is some evidence that orientation is not the whole story.

It was noted above that radiative acceleration is immediately suggested by the role that luminosity apparently plays in the phenomenon. In addition, the very fact that such a huge amount of the momentum being carried by the photons is intercepted by the gas is an *a priori* argument that radiative acceleration is important. In view of this, I would like to go through the arguments that led me to the view for many years that radiative acceleration, while undoubtedly a factor, was not the dominant agent at work in accelerating the BAL gas to velocities up to 0.1 c. If we: (1) Deduce column densities in various ions from the observed BAL line strengths (which already involves a number of uncertain assumptions) (2) Assume photoionization equilibrium and adopt some assumed shape for the continuum typical of what AGN spectra are thought be like, then the following approximate relations hold (*cf.* Weymann, Turnshek & Christiansen 1985)

$$n_e (D_{pc})^2 / L_{46} \approx 10^9. \tag{1}$$

In addition, from typical deduced column densities, one has, for a single slab of gas of thickness ΔD,

$$n_e \Delta D \approx 10^{20} N_{20} \tag{2}$$

leading to the characteristic path length through a BAL–producing slab of

$$\Delta D \approx 10^{11} N_{20} (D_{pc})^2 / L_{46} \ cm. \tag{3}$$

Evidently, this leads to exceedingly small filling factors for the clouds. Recent echelle observations show that some BAL troughs are smooth and do not break up into individual components having thermal widths (Barlow and Junkkarinen 1994). If one believes that there are really many small clouds whose velocity widths are comparable to the internal sound speed (~ 30 km s^{-1}) then there must actually be many clouds per thermal width spanning the entire velocity range of the troughs—*e.g.*, of order 1000 cloudlets—so that the path length through an individual cloudlet would be only of order $10^8 \ cm$. This extraordinarily small dimension (with a correspondingly small sound crossing time of order 30 sec (!) gives rise to the apparent necessity of confining these clouds by some mechanism (*e.g.*, a warm or hot, moderately low density intercloud medium.) However, unless this medium has extremely low density, even though the clouds experience a substantial radiative force, they would quickly be brought to rest with respect to the confining medium. Thus, as the clouds are seen to be moving at very high velocities, the confining medium would have to be also. But under these circumstances there is insufficient momentum delivered to the cloudlets via scattering of resonance line radiation to accelerate the confining medium as well as the cloudlets. It was this dilemma—the "confinement problem"— which led me to question whether radiative acceleration was the dominant acceleration mechanism operating in BAL QSOs despite its obvious attractiveness. It is then necessary to invoke powerful winds which accelerate the clouds. A detailed investigation of such a coupled cloud, hot wind, and radiation model is discussed elsewhere in this volume by Vilkoviskij *et al.* One must then also propose a way to deposit energy in the confining hot gas. An ingenious suggestion along these lines involving the decay of relativistic neutrons was proposed by Begelman *et al.* (1991). Several groups have recently proposed models for overcoming this confinement problem in rather different ways.

2.1 Cosmic Ray or Magnetically Confined Clouds

Arav, Li, and Begelman (1994) propose resolving the dilemma by a confining medium with extremely low density (*e.g.*, cosmic rays) and/or magnetic fields. They examined the case of confinement by an extremely low density medium which is closely coupled to the cloudlets and show that there are regions of parameter space where scattering of resonance lines provides acceleration comparable to that associated with the pressure gradient in the confining medium. If this is the case, might there be some unique observational signature which would indicate that radiative acceleration is at work? As first noted by Turnshek, one sometimes sees structure in the BAL line profiles with a characteristic velocity of about 6000 km/sec which is the separation between the N V line at $\lambda1240$Å and Lyα, at $\lambda1216$Å. Arav and Begelman (1994) have described a set of circumstances in which the extra radiative acceleration experienced by the clouds as the N V ions reach a velocity approaching ~ 6000 km s^{-1} may provide a detectable signature. Near this velocity the extra flux provided by strong narrow Lyα emission provides an extra "kick" to the clouds. Since large velocity gradients lead to smaller optical depths, a local maximum in the flux in the middle of the absorption profile displaced by about 6000 km s^{-1} may be expected. This signature must be sought in the C IV absorption rather than in the N V/Lyα trough, since in the latter case it is not readily apparent whether such a local maximum is due instead to the partially attenuated underlying Lyα emission. There are in fact profiles which have this behavior. Among the set of circumstances required for this signature to be imprinted upon the line profiles is a rapid drop in continuum flux shortward of Lyα. If this is not the case the modulation produced by the Lyα emission line acting on the N V ion is swamped by a large number of lines further to the UV. See also the discussion by Vilkoviskij elsewhere in this volume.

2.2 Xray–Shielded Flows

A second approach toward models in BAL QSOs dominated by radiative acceleration has been proposed by Murray *et al.* (1994). They suggest that the confinement problem can be avoided altogether by shielding the material responsible for the troughs by very high column densities ($10^{23} - 10^{24}$) of highly ionized gas. This so–called inner "hitchhiking gas" is sufficiently highly ionized that it suffers relatively little radiative acceleration, but with such high column densities it shields the material beyond it from soft X-rays. The shielding provided by this gas prevents material beyond it from becoming so highly ionized that the principle Lithium-like ions responsible for most of the radiative acceleration —*e.g.*, C IV, Si IV, N V, and O VI — are present in the appropriate amounts to drive the outer gas to very high velocities, and in possibly appropriate amounts to provide the observed trough strengths and ratios. The specific model proposed by Murray *et al.* considers both the gas responsible for the high velocity absorption troughs and the inner hitchhiking gas to flow upwards from an accretion disk by a combination of pressure gradients and radiative acceleration normal to the disk. When the gas is exposed to the intense radiation from the innermost region of the central engine, the shielded gas follows a roughly radial trajectory, the inner shielding gas "hitchhiking" behind the wake of the radiatively accelerated material, though it itself is not strongly accelerated. One would expect that the line–locking signature described by Arav and Begelman involving Lyα and N V may apply in the Murray *et al.* model as well.

The idea that the flow in BAL QSOs is primarily along the equatorial plane, skimming above an accretion disk is not a new one. Among other features, it seems to account in a fairly natural way for the high polarization seen in the continuum and the even higher trough polarization seen in some BAL QSOs (Cohen *et al.* 1995). Arav *et al.* point out that if for some reason the material is injected in a non–uniform way as one moves out along a radius vector of the accretion disk, then material along the outer stream lines will be accelerated to lower velocities than material closer in. A line of sight cutting through both flow lines might result in the detached multiple trough structure seen frequently in BAL QSOs. Similarly, Murray *et al.* point out that if one looks sufficiently close to the equatorial plane, the shielding may become so great that low ionization (Mg II) BAL QSOs result. As in the case of the multiple troughs, the material arising from the outer regions of the accretion disk and producing the Mg II material will reach only modest terminal velocities, while the material along the same line of sight will contain higher ionization as well as higher velocity material. This behavior is in fact seen in some Mg II BAL QSOs and is especially pronounced in the remarkable BAL QSO Q0059-27.

2.3   Stellar Contrails

Yet a third class of models involves the shedding of material from stars in a dense stellar cluster surrounding the massive black hole via stellar winds, followed by radiative acceleration. J. Perry and collaborators have proposed models of this sort. Most recently this idea has been taken up by Scoville & Norman (1995). In the Scoville and Norman model, the dominant source of material comes from red giant winds, and the dominant source of radiative acceleration is in fact not via scattering of resonance line radiation but instead radiation pressure on dust grains, which, under plausible circumstances, will be closely coupled to the gas. If the material is shed too close to the massive black hole however, the grains will evaporate. The highest velocity material will be that which originated at the point closest to the nucleus at which the grains can still survive and this leads to an expected scaling law in which the terminal velocities increase as $L_{nuc}^{1/4}$. (In this connection it is interesting to note that in Seyfert nuclei which have luminosities roughly 10^{-4} that of the very luminous QSOs, there appear to be mild outflows with terminal velocities about 1/10 those seen in the luminous BAL QSOs.)

Scoville and Norman then examine the history of the material shed from the star: The locus of the material forms ribbons which are compressed by the radiative forces along the radius vector from the central source, while slowly spreading laterally with the sound speed. The thickness, column densities, particle densities and velocities all seem roughly in accord with the parameters deduced form the observations. Several ribbons may cover portions of the continuum source at any given time and the resultant computed absorption profiles qualitatively resemble BAL QSO profiles.

Thus, all of these models seem to allow for the intuitively appealing dominant role of radiative acceleration. Between the three I must confess to a small bias towards the Scoville–Norman model. Both this model and the Murray *et al.* model account for the origin of the BAL material in a fairly natural way. This bias arises simply because I have concerns about whether cloudlets with such very small dimensions as seem required in the wind models can really survive as distinct entities. What produces such very small sizes and filling factors in the first place and what then prevents them from finally dissipating even further via evaporation ? Further examination of the "micro–surface physics" of such cloudlets, whether magnetic field or cosmic ray confinement is invoked, would be useful.

Regarding the Murray *et al.* disk flow models, there are a number of appealing features to such flows, and one prediction of the model seems to have been recently confirmed: Green *et al.* (1994) have compared the averaged ROSAT properties of BAL QSOs with those of radio quiet non–BAL QSOs and find the BAL QSOs to be strongly deficient in X-rays, relative to the the the non–BAL QSOs, consistent with the strong shielding which is the key to the model. A second prediction of the model (for both the BAL QSO phenomenon and the BELRs of luminous QSOs in general) is that the scale of the emission and absorption regions is very small, even for luminous QSOs: lag times of order 10 days or so between the N V emission and continuum variations are predicted. Unfortunately, for whatever reason, variability on that timescale has rarely, if ever, been seen for *luminous radioquiet* QSOs. The Murray *et al.* model has also not yet been carried to the point where detailed comparisons of line profiles of various ions as a function of velocity can be made with the observations. It is not clear if one can find the appropriate amount of shielding which will not prevent the ionization ratio of, *e.g.,* O III/O IV and C III/C IV from becoming unacceptably high without at the same time precluding higher ionization states like O VI from becoming strong enough. A difficulty in making such comparisons is that the troughs of some ions may be highly saturated while those of other ions may be only mildly saturated. If the scattering material does not cover the continuum source, the various troughs may have roughly the same strengths, even though the column densities may differ by one or two orders of magnitude.

In the Scoville–Norman model, one expects variations on time scales much longer than in the Murray *et al.* model—of order a few years. Whether or not the X-ray results of Green *et al.* can be satisfactorily explained in the Scoville–Norman model remains to be seen.

3   Why are BAL QSOs and Strong Radio Sources Mutually Exclusive?

In this section, I will adhere strictly to the definition of a BAL QSO described above. Regardless of the arbitrariness of this definition, the fact is that since it was proposed in 1981, as far as I am aware, no source with radio flux exceeding 100mJ has been found which has absorption satisfying the BAL criteria. One can test for the statistical significance of this mutual exclusivity in either or both of two ways: (1) One can examine the radio properties of a set of known BAL QSOs and compare them with the radio properties of otherwise similar optically selected, but non–BAL QSOs. The most recent published test of this kind was carried out by Stocke *et al.* (1992) who found that the avoidance of strong radio sources among the BAL QSOs was significant at about the 1% level. (2) One can examine the spectroscopic properties of a sample of radio–loud QSOs (satisfying some set of well–defined optical and radio criteria) with another similar set of radio–quiet QSOs. My colleagues and I have been collecting such a sample for several years (though the full results have still not been published) and the statistics are overwhelming: As noted above, the incidence of BAL QSO among optically selected QSOs is very nearly 10%. By contrast, we have examined about 250 strong (> 100mJ) radio sources with redshift and luminosity distributions similar to the optical sample. In this radio–loud sample, none have been found which produced absorption which qualified the object as a BAL QSO according to the criteria described above. About twenty five should have been accepted, so that, needless to say, the significance of the avoidance of BAL QSOs among radioloud QSOs is highly significant. (About 2×10^{-12}). What accounts for this avoidance? Following is a description of some recent attempts to account for this avoidance, some of which are made in the context of the models described above.

From the description of the equatorial flows for the BAL QSOs described above, one obvious explanation immediately comes to mind: If radio–selected QSOs are preferentially viewed pole–on (or at least well away from their equatorial plane) due to radio beaming, then this would immediately explain the situation. Indeed, for the flat spectrum, compact sources, where the 6 cm flux is thought to be strongly beamed, this seems a perfectly adequate explanation. Strong beaming is generally not thought to characterize the extended and mostly steep spectrum radio sources however. But, among our sample are about 115 which are steep spectrum sources and hence likely to be extended sources, though this must be confirmed by actually examining the radio maps when they exist. There are 86 which have flat spectral indices and are likely to be compact, though again this must be confirmed by examination of radio maps. For the remaining 55 sources, radio spectral indices were not available, though I estimate that of these at least 5 more will be steep spectra sources, simply on the basis of the wavelength at which they were selected. Thus among the ~120 steep spectrum sources one would expect to find about 12 BAL QSOs. Again, one can easily verify that the avoidance of BAL QSOs even among the steep spectrum radio sources alone is highly statistically significant. However, this is only true *if* the sample of steep spectrum radio sources is really an isotropic sample as is commonly assumed to be the case (or at least as isotropic as the optically selected samples).

De Kool (1993) has explicitly questioned this last assumption. He suggests that even the extend, steep spectrum radio sources are expanding at subrelativistic velocities. De Kool considers a (radio) flux-limited sample of sources which are intrinsically isotropically distributed and which have a uniform distribution in expansion velocity with $0.3 < \beta < 0.45$. Further, suppose that 10% of a truly isotropic sample displayed the BAL phenomenon (corresponding to a viewing angle above the plane of the equator of about 6 degrees). De Kool finds that even these subrelativistic velocities produce enough beaming such that a radio flux limited sample having the properties described above would produce BAL QSOs only about 2.5% of the time. In the steep spectrum sample I have described above, the absence of BAL QSOs would then be expected to occur only about 5 % of the time by chance – not yet a strong result. However, it is believed that such high values of β are contra-indicated by studies of the ratio of core/lobe separations of the two components (McCarthy 1995, Perley 1995).

An alternative explanation was suggested by Murray *et al.* They find that the terminal velocity is strongly influenced by the amount of soft X-rays which can penetrate to the region where the radiative acceleration occurs: Other things being equal, a strong flux of X-rays will cause the high velocity low density material to become highly ionized and suppress acceleration to high terminal velocities more readily than a weaker X-ray flux. Since it is known that radio loud objects have, statistically, more intense X-ray flux (for given optical flux) than radio quiet objects, Murray *et al.* suggest that radio loud objects simply never develop high enough terminal velocities that, using the BAL QSO definition we have adopted, they are classified as such. One test of the Murray proposal would be to construct a subsample of extended radio sources whose X-ray properties are similar to a subsample of radio quiet objects, and compare the incidence of BAL QSOs in these two subsamples.

Scoville and Norman suggest that under some conditions, free-free absorption can attenuate the radio flux interior to the stellar contrails (typically at a few parsecs). The contrails would also have to prevent the jets from emerging to large distances, which seems difficult. Shchekinov, elsewhere in this volume, studies the quenching of outflows by hydrodynamical evaporation associated with cloud evaporation and suggests this may be relevant to the radio–source–BAL QSO anticorrelation. In neither case, however, are these mechanisms likely to be relevant if jets emerge pole on while, as suggested by the polarization data, the BAL clouds flow along the equatorial plane. Finally, if one accepts the "unified model" of radio–loud QSOs and Radio Galaxies, it could be argued that the radio–loud BAL QSOs are among those radio sources for which the central source is obscured and the BAL phenomenon thus not observed. However, this begs the question of what the basic difference in the obscuration of the radio loud and radio quiet objects near the equatorial plane is.

As noted above, it seems very plausible that the lack of BAL QSOs among the flat spectrum, compact radio sources is simply an orientation effect (though there could be intrinsic differences along the lines of the Murray *et al.* suggestion as well), so the critical issue is the correct explanation for the avoidance of BAL QSOs among steep spectrum radio sources.

Finally, mention should be made of the curious result found by Francis, Hooper & Impey (1993). These authors found that while known BAL QSOs are not radio *loud* they have, in fact, more radio power than the average optically selected QSO. This result is based upon a small number of sources and it will be very interesting to see if the result holds for a larger sample. The explanation for this fact, if it is borne out, is not clear, though de Kool has suggested an explanation in the context of his proposal.

References

Arav, N., Li, Z-Y., & Begelman, M.C. 1994, ApJ, 432, 62.

Arav, N., & Begelman, M.C. 1994, ApJ, 423, 479.

Barlow, T., & Junkkarinen, V. 1995, in preparation.

Begelman, M., de Kool, M., & Sikora, M. 1991, ApJ, 382, 416.

Cohen, M.H., Ogle, P.M., Tran, H.D., Vermeulen, R.C., Miller, J.S., Goodrich, R.W., & Martel, A. 1995, preprint.

de Kool, M. 1993, MNRAS, 265, L17.

Green, P., Brinkmann,W., Fink, H., & Trümper, J. 1994, private communication from P. Green.

Francis, P.J., Hooper, E.J., & Impey, C. D. 1993, AJ, 106, 417.

Hamann, F., Korista, K.T., & Morris, S.L. 1993, ApJ, 415, 541.

Korista, K.T., Voit, G.M., Morris, S.L., & Weymann, R.J. 1993, ApJ Suppl, 88, 357.

McCarthy, P. 1995, private communication.

Murray, N., Chaing, J., Grossman, S.A., & Voit. G.M. 1994, ApJ, submitted.

Perley, R. 1995, private communication.

Scoville, N., & Norman, C. 1995, submitted to ApJ.

Stocke, J.T., Morris, S.L., Weymann, R.J., & Foltz, C.B. 1992, ApJ, 396, 487.

Stocke, J. T., Shull, J.M., Granados, A.F., & Sachs, E. R. 1994, AJ, 108, 1178.

Turnshek., D.A. 1988 in *QSO Absorption Lines* STScI Symposium Series 2 ed. Blades, J.C., Turnshek, D.A., & Norman, C.A. (Cambridge: Cambridge University Press), p. 17.

Weymann, R.J., Carswell, R.F., & Smith, M.G. 1981, Ann.Rev.Astron.Astrophys., 19, 41.

Weymann, R.J., Morris, S. L., Foltz, C.B., & Hewett, P.C. 1991, ApJ 373, 23.

Weymann, R.J., Turnshek, D.A., & Christiansen, W.A. 1985 in *Astrophysics of Active Galaxies and Quasi-Stellar Object* ed. Miller, J. (Mill Valley: University Science Books), p. 333.

The Covering Factors, Ionization Structure, and Chemical Composition of QSO Broad Absorption Line Region Gas

David A. Turnshek

Department of Physics and Astronomy, University of Pittsburgh

Abstract. Recent HST and ground-based observations have led to better and new constraints on the properties of Broad Absorption Line (BAL) gas in QSOs in terms of its covering factors, ionization structure, and chemical composition. We present and discuss these findings. One of the most important results is that the gas giving rise to BALs has non-uniform ionization and likely has very enhanced metal-to-hydrogen abundances relative to solar composition. In particular, the nitrogen abundance relative to other observable metals is enhanced $\approx$ 3 – 10 times solar values. If normal stellar nucleosynthesis is important, this result may be indicative of a relatively flat initial mass function with the nitrogen overabundance produced by secondary processing in massive stars.

1 Introduction

The phenomenon of Broad Absorption Lines (BALs) in QSOs (see Figure 1 for examples) has been observed for over 25 years now (Lynds 1967; Burbidge 1970), but there is still no absolute consensus on the relationship between BAL QSOs and non-BAL QSOs. Much of the work has led to the belief that BALs are normally observable along a relatively small fraction of the possible sight-lines toward central source QSOs, and therefore the viewing angle determines whether or not BALs are observed. This belief generally results from indirect arguments which require some assumptions. In this paper some of the past results on BAL QSOs are reviewed or referenced and some new results pertaining to covering factor issues are discussed. For the most part, however, this paper deals with recent analyses of the observed BAL profiles themselves; these results have implications for the ionization structure and chemical composition of the BAL region gas.

The organization of this paper is as follows. Observational results (§1.1) and some model concepts (§1.2) are briefly summarized or references to them are given. Note that at this conference Weymann has reviewed some of the models, and so a detailed discussion of models will not be given. Some of the considerations involving viewing angle effects and the various types of BAL region covering factors that are relevant to ongoing investigations are reviewed (§2). Results on the column densities (§3), ionization structure (§4), and chemical composition (§5) of BAL gas are then presented, followed by a discussion of some of the implications (§6) and a short summary (§7).

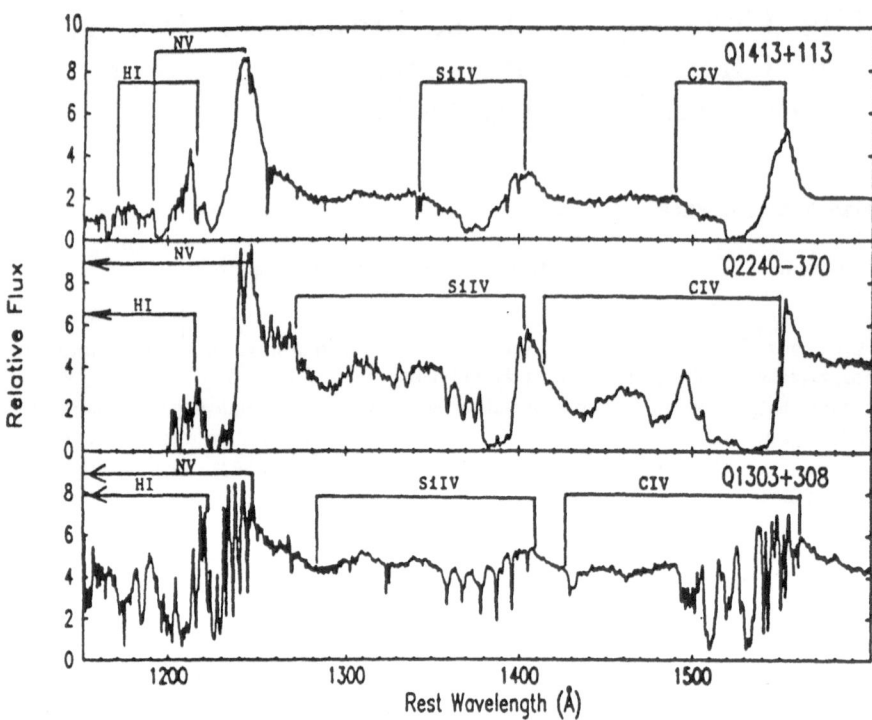

Fig. 1. Some example rest frame spectra of BAL QSOs.

1.1 Review and References to Some Observational Results

As Figure 1 indicates, BAL QSOs are a special class of QSO that show mostly highly ionized gas flowing away from the central source at speeds up to 10,000 to 30,000 km s^{-1} or more. The BALs appear in the spectra of about 10% of all radio quiet QSOs at all redshifts. At the Baltimore QSO Absorption Line Conference, Turnshek (1988) reviewed the topic of BAL QSOs. Since then, there have been a considerable number of new results. New references up until about the end of 1993 are given in Turnshek et al. (1994).

One of the most important pieces of work has been the comprehensive study by Weymann et al. (1991) of the systematics of BAL QSO emission-line and continuum properties relative to those of non-BAL QSOs. A variety of new issues were explored and the study confirmed that, while the distribution of properties of the two classes of objects were remarkably similar and show considerable overlap, the NV, AlIII/CIII], and FeII spectral regions in BAL QSOs show enhancements on average. Stocke et al. (1992) presented an extensive study which confirmed earlier work that, unlike optically selected non-BAL QSOs as a class, 100% of the optically selected BAL QSOs are radio quiet (i.e. with small radio-to-optical luminosities). However, Francis,

Hooper, & Impey (1993) have pointed out that BAL QSOs may tend to be very weak radio sources — clearly not classifiable as radio loud but often with higher radio-to-optical fluxes than "normal" radio quiet non-BAL QSOs. There also continues to be some debate as to whether radio loud QSOs with z_{abs} near z_{em} absorption complexes are related to BAL QSOs. In addition, Hamann, Korista, & Morris (1993) have completed the most careful analysis to date of the resonance line scattering of inner photons by the outflowing BAL region and have compared computed profiles to many different observed profiles. Within the framework of the assumption that scattered photons are not destroyed, these comparisons yield weak constraints on the scattering geometry and constraints on the size of the *global BAL region covering factor*, which is defined as the fraction of the sky covered by BAL region clouds averaged over the entire sky as seen from the central source QSO. The Hamann et al. (1993) analysis indicates that, if scattered photons are not destroyed, the covering factor must *normally* be relatively small, e.g., < 0.2, but there are cases where the covering factor is not tightly constrained to be small. At the same time, Turnshek et al. (1994) completed a thorough study of the unique low-redshift ($z_{em} \approx 0.384$) BAL QSO PG0043+039 and concluded that dust, which may destroy scattered photons, is probably present along the sight-line and that the global BAL region covering factor in this object is probably relatively large. Turnshek et al. discussed further evidence (Weymann et al. 1991; Sprayberry & Foltz 1992; Voit, Weymann, & Korista 1993) which suggests that BAL QSOs with low-ionization BALs (in addition to the normal high-ionization ones) may generally possess dust along their sight-lines. The dust may destroy scattered photons, allowing such objects to have larger global BAL region covering factors than the resonance line scattering constraints normally permit. In addition, it has been known for some time that BAL QSOs are the only radio quiet QSOs which show significant polarization. Recently, spectropolarimetric measurements of BAL QSOs have been presented by Glenn, Schmidt, & Foltz (1994) and Goodrich & Miller (1995). Although the interpretation is not completely clear, the observations indicate that geometry and viewing angle effects are important. The existence of multiple troughs in BAL QSO spectra has also often been cited as evidence for non-spherical BAL region geometries (e.g. Turnshek 1986).

Thus, the studies show that BAL QSOs exhibit very interesting spectral systematics. Their emission-line, continuum, polarization, and radio properties differ from those of non-BAL QSOs — sometimes in subtle ways. Moreover, as Figure 1 suggests, the properties of the BALs and adjacent emission lines exhibit a great deal of variety, ranging from very smooth absorption with strong narrow adjacent emission, to very "broken-up" absorption with broad weak emission. The characteristics of the BAL profiles seem to correlate with those of the corresponding emission lines.

Finally, Korista et al. (1992) has presented important HST-FOS-UV observations of the moderate redshift BAL QSO 0226-1024. This has resulted in reliable measurements of BAL profiles arising from different ions of the same element for a number of different elements for the first time. The analysis of the spectrum of Q0226-1024 forms the basis for much of the results on ionization structure and chemical composition reported on later.

1.2 Model Concepts

The observational results discussed in §1.1 might be interpreted as evidence for very small ($\lesssim 0.01$) global BAL region cover factors in some objects (e.g. radio loud QSOs) and relatively large ($\gtrsim 0.4$) covering factors in other objects (e.g. "low-ionization BAL QSOs"). On the other hand, the anisotropic emission produced by "unified models" may come into play and confuse naive deductions about the size of the global BAL region covering factor in various classes of objects. Much of the work noted above indicates that non-spherical geometries (e.g. disk-like or jet-like geometries) are relevant in BAL QSOs and therefore viewing angle effects are important. However, it is becoming increasingly difficult to envisage a model which can explain all of the systematic differences between BAL and non-BAL QSOs, and the variation in QSO properties, in some type of *single* unified model. Even in models where aspect angle effects dominate, the size of the BAL region covering factor should probably be taken to be variable. With a variable global BAL region covering factor, there are still a number of ways to interpret the large differences in BAL profile types. The variety in profile types may be due to changes in viewing angle, stage of evolution, some intrinsic QSO property, or a combination of these effects.

Some results have been used to constrain BAL region model properties. For example, the data near Lyα/NV indicate that the BAL gas occults the Lyα broad emission-line gas, and this is usually taken as evidence that the BAL region is at a distance > 0.1 pc from the central source, lying outside the broad emission-line region. Also, BAL region electron densities $n_e \gtrsim 10^6$ cm^{-3} are required because otherwise one would expect to see significant *broad* [OIII] emission in BAL QSOs (and QSOs) even for rather small global BAL region covering factors. For this constraint on electron density, photoionization equilibrium calculations then indicate that the BAL region *must lie within* 30 to 500 pc of the central source for the usual range of QSO luminosities. Given these small sizes for the distances between the central source and the outflowing BAL gas, which imply short times for clouds to cross the region compared to QSO lifetimes, there is need for a continuous injection of BAL region material into an outflow. The source of the BAL gas might be stellar atmospheres, supernovae, parts of a torus, the outer extensions of an accretion disk, etc.

Models for BAL QSOs have generally proposed some geometry and location for the BAL region and then tried to understand the acceleration mechanism or some other observed property. Weymann, Turnshek, & Christiansen (1985) discussed a model in which the acceleration of the observed BAL clouds is driven by a thermal wind which is much more highly-ionized and unobserved. The wind is created by evaporating BAL cloudlets and is perhaps 100 or more times more massive than the observed BAL region itself. In this case, the observed BAL cloudlets are formed and accelerated as the wind ablates material off more massive objects. Voit et al. (1993) have applied this idea to a model for low-ionization BAL QSOs and have suggested that these objects may be young QSOs with high BAL region covering factors "in the act of casting off their cocoons of gas and dust" (see also a qualitatively similar suggestion by Hazard et al. 1984). Knerr (1993) has also worked toward developing the thermal wind model. However, observational evidence from absorption-absorption line-locking and emission-absorption line-locking (deduced from the characteristics of the BAL profiles themselves) indicates that radiative acceleration may play some role (Foltz et al. 1987; Turnshek et al. 1988; Korista et al. 1993). Arva & Li (1994), Arva, Li, & Begelman (1994), Arva & Begelman (1994), and Murray et al. (1995) have developed models based on radiative acceleration. If the BAL clouds can be confined and won't evaporate, there would be no need for the massive thermal wind.

Aside from the obvious importance these objects have for the study of QSOs, they may also be important cosmologically. Depending on the details of the correct model, such as the size-scale of the BAL region and especially the properties of any massive thermal wind, the outflows may have an affect on the local environments of QSOs and be a significant source of metals in the young universe (Turnshek 1988, 1994). Below, recent results on covering factors, ionization structure, and chemical composition of the BAL region gas, which will be relevant to future models, is considered.

2 Covering Factors

Here some distinctly different kinds of covering factors which are relevant to BAL QSO investigations are discussed.

2.1 Global and Local Covering Factor

As noted in §1.1, the *global BAL region covering factor* is the fraction of the sky covered by BAL region clouds averaged over the entire sky as seen from the central source QSO. On the other hand, the *local BAL region covering factor* is the fraction of the sky covered by BAL region clouds averaged over special sight-lines (e.g. inside the opening angles of a disk or jet) as seen from the central source QSO. For a large-scale BAL region geometry that is spherically symmetric, the global covering factor would equal the local covering factor. However, as noted earlier, the high incidence of multiple BAL troughs in BAL QSOs and the existence of polarized flux in some BAL QSOs is probably an indication that non-spherical geometries are involved. Here we consider two further pieces of observational evidence which constrain covering factors.

The first piece of evidence involves optical variability. Turnshek (1988) had reported on evidence from the literature which indicated that BAL QSOs are possibly more optically variable than radio quiet non-BAL QSOs. At that time, the strongest evidence came from photographic observations of three out of five BAL QSOs which showed considerable changes in brightness over a baseline of $\approx$ 30 years when compared to results from a sample of over 60 moderate-to-high redshift optically selected non-BAL QSOs. As a follow-up, a QSO monitoring program over a baseline of $\approx$ 6 years with a CCD on the 1-m Swope Telescope at Las Campanas Observatory has been performed to more accurately investigate the optical variability properties. Over the time baseline studied, at most only subtle differences between radio quiet non-BAL QSOs and BAL QSOs are found. Some preliminary results were discussed by Sirola et al. (1992) and a more complete analysis is in preparation. These results places some constraints on covering factor models, aspect angle effects, and the importance of anisotropic continuum emission in QSOs.

Second, Turnshek et al. (1994) has discussed evidence, originally pointed out by Boroson & Meyers (1992), which indicates that the global BAL region covering factor might be very large in QSOs with weak-[OIII] emission (and also strong FeII emission). To investigate this further a sample of weak-[OIII] QSOs has been studied in the UV (mostly with HST) in order to search for BALs (Turnshek et al., in preparation). In a sample of 15 objects observed so far, six have BALs ($\approx$ 30%), which is a relatively large fraction in comparison to what would be expected in a random optically selected sample (but the formal error is still large). Despite this finding, the result indicates that weak-[OIII] alone is not sufficient to predict the presence of BALs. Also, there are examples of strong-[OIII] QSOs which have BALs. However, since the Boroson & Meyers sample was basically selected from the *IRAS* Warm Extragalactic Object sample, there clearly may be other properties that come into play for selecting a sample which will have a very high incidence of BALs. Investigations of this type are relevant to covering factor models and models which incorporate aspect angle effects.

2.2 Continuum Source Covering Factor

The *continuum source covering factor* is the fraction of the observed continuum source covered by BAL region gas along the sight-line. For reliable column density determinations, it is important to know what fraction of the continuum is covered by the absorbing gas. It is possible that, for example, while light from the central continuum source might be completely covered, extended scattered light (e.g. polarized light) might not be covered (e.g. see Goodrich and Miller 1995).

Some constraints on the continuum source covering factor can be derived by considering observations of the gravitationally lensed "Cloverleaf BAL QSO" H1413+1143. HST-FOS observations of the Cloverleaf indicate that, while the BAL profiles along its four sight-lines are definitely not identical, they are very similar. An HST-PC2 image of the Cloverleaf is shown in Figure 2 (see also Sirola et al. 1995). The geometry explored by the four sight-lines and the four FOS spectra are shown in Figures 3 and 4 (see also Turnshek et al. 1995). Given the typical viewing angle differences of ≈ 0.7 arcsec, models of the gravitational lens suggest that the sight-lines intercept the BAL region over a lateral size of $L \approx 10^{16} r_{kpc}$ cm, where r_{kpc} is the distance between the the BAL region and central source in kpc. Observational constraints reviewed by Turnshek (1988) suggest that the interval $10^{-3} < r_{kpc} < 0.5$ is likely. Photoionization equilibrium constraints and the observed column densities suggest that the thickness, x, of a BAL region cloud along a sight-line is $x < 10^{11} r_{kpc}^2$ cm. Therefore, the similar absorption indicates that *average* BAL region column densities *integrated across the continuum source* are relatively uniform (to within $\pm 10\%$) over a lateral extent which is much larger than the thickness of a cloud and approximately the size of the continuum source. Figure 3 trys to depict this by showing a non-uniform absorbing slab. This suggests that there could be some *isolated cases* (when the absorption is very weak) in BAL QSOs where the absorption doesn't completely cover the continuum source. In fact, a few cases have been found where there is reason for concern. These cases usually involve doublet absorption which appears saturated (i.e. nearly equal absorption equivalent widths), but for which the central intensities of the doublet are not near zero. For example, see a SiIV doublet in the spectrum of Q0226-1024 and the discussion of it by Korista et al. 1992. Turnshek, Monier, Wampler, & Hazard (in preparation) have discovered a similar doublet in another high resolution spectrum. These may be representative of the isolated cases where high velocity BAL region gas is not extended enough perpendicular to the sight-line to completely cover the central continuum source. However, while the absorption may not be completely uniform across the continuum source along a sight-line, the Cloverleaf observations clearly show that the average absorption profile is very similar across a region which is about the size of the continuum source. This suggests that the methods used to derive column densities (discussed below) are generally accurate enough to perform an analysis of the ionization structure and chemical composition.

3 Derivation of BAL Region Column Densities

In order to fully understand the ionization structure and chemical composition of the BAL region gas, one must first have the ability to derive the line-of-sight column densities of ions seen in absorption as a function of outflow velocity. The most straightforward procedures for doing this are discussed by Junkkarinen, Burbidge, & Smith (1983), Grillmair & Turnshek (1987), and Korista et al. (1992). The procedures generally assume that the absorbing material completely covers the continuum source (see §2.2), that scattered radiation does not fill in the bottoms of the absorption profiles, and that the

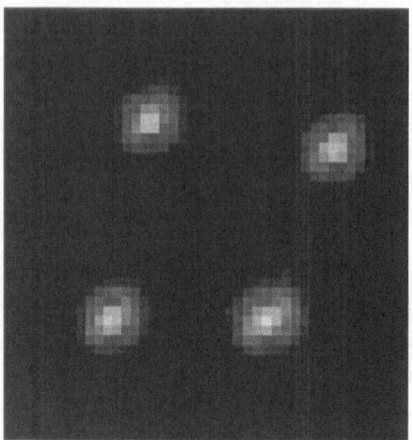

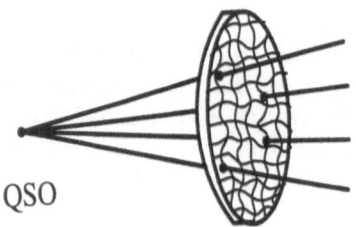

QSO

Fig. 2. HST-PC2 image of the gravitationally lensed Cloverleaf BAL QSO 1413+1143.

Fig. 3. BAL cloud configuration explored by spectroscopy of the Cloverleaf components.

absorption structure is resolved in velocity space, i.e., that absorption in the BAL profile does not get narrower and deeper at higher resolution. Kwan (1990) discusses some of these issues and raises some concerns (see also §2.2). However, precautions can be taken. For example, one can always study BAL profiles in a high enough velocity interval to assure that there is little possibility of resonance line scattered emission filling in the bottoms of the absorption profiles. Moreover, the HST observations of the Cloverleaf BAL QSO suggest that the BAL absorption usually covers the continuum source and is fairly uniform across the continuum source. Therefore, the normal methods used to derive column densities are likely to be generally valid. This view is supported by recent Keck HIRES data on a few BAL QSOs (Burbidge et al., this conference) which show that BAL profiles aren't resolved into many deeper components at very high resolution. For the analysis discussed here, the BAL column densities derived from ground-based and HST-FOS-UV spectroscopy of Q0226-1024 were considered (see Korista et al. 1992).

4 Ionization Structure

Obtaining good data on BAL profiles of different ions of the same element had not been possible until the recent HST-FOS-UV observations of moderate-to-high redshift BAL QSOs. At higher redshift, where important lines might be observed in the optical, intervening Lyman limit absorption often blocks our view of the lines. In the absence of data on different ions of the same element, the ionization structure can not be explored without making assumptions about elemental abundances. While single-zone photoionization models of the BAL region lead to some relevant information, these models have always been viewed as unreliable because of the possibility that elemental abundances are very far from solar ratios. For this study it has been assumed that the main observed metal BALs (e.g. CIV, SiIV, NV) are produced by ions which are near the dominant stages of ionization.

For the BAL region in Q0226-1024, column densities for different ions of the same element are available for carbon (CIII and CIV BAL profiles), nitrogen (NIII, NIV, and NV BAL profiles), and oxygen (OIII, OIV, and OVI BAL profiles). Analysis of the HST-FOS results indicates that a single-zone photoionization model is unable to explain the observed column density ratios for three ions of the same element, indicating that at least a two-zone BAL cloud model should be used. The problem stems from the fact that it is impossible to formulate a single-zone model with, for example, the CIII and SiIV lines forming in the same region as the NV and OVI lines. However, preliminary work indicates that a two-zone model gives fairly good results for N and O, but somewhat poorer results for C. Photoionization calculations using Ferland's code CLOUDY have been performed with the AGN photoionizing continuum proposed by Matthews and Ferland (1987), appropriately enhanced abundances (see below), and an electron density of $n_e \approx 10^8$ cm^{-3}. To obtain agreement with the Q0226-1024 data the inferred preliminary ionization parameters for the two zones are $\log(u_1) \approx -1.2$ and $\log(u_2) \approx -2.7$. There is no significant BAL region HI, HeI, or HeII Lyman continuum opacity included. For clouds in a single region the required ionization parameters suggest a density enhancement of ≈ 30 (but this may be larger since there are only two zones and a two-zone cloud is unphysical). In such a model the absorbing region intersected by a sight-line might consist of a denser inner region surrounded by a less-dense, more highly ionized outer region that has ≈ 50 times the extent of the inner region (see Figure 5). This would be a flattened geometry consistent inferences from the Cloverleaf observations, but along the lines suggested by Williams (1992) for incorporating density fluctuations. Other interpretations of the two-zone geometry are possible. For example, rather than density fluctuations there may be two physically distinct regions with different uniform densities but the same outflow velocities. In either case both zones contribute significantly to the observed column densities — the much larger extent of the lower density region allows it to contribute to the observed column densities despite its much lower density.

Finally, on the ionization structure, it seems somewhat amazing that the observed level of ionization remains roughly constant with outflow velocity. One would not anticipate this since, as the BAL gas gets further away from the central source, it must see an ionizing continuum which becomes increasingly attenuated by far-UV BAL transitions.

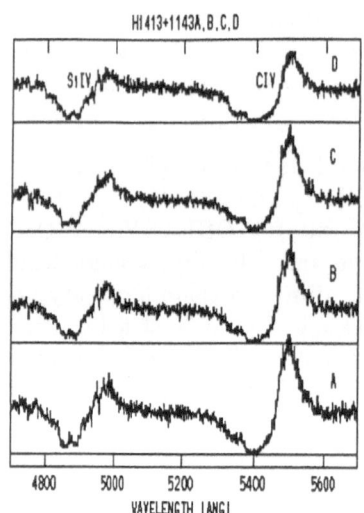

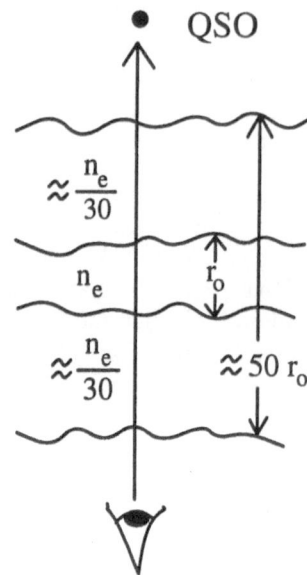

Fig. 4. F_λ spectra of the four components of the Cloverleaf QSO.

Fig. 5. BAL cloud structure inferred from density fluctuation model.

5 Chemical Composition

The two-zone model can explain the ratios of column densities of different ions of the same element for both N and O with the same two ionization parameters. However, the abundances required for an acceptable fit to the metal BALs were very different from solar values. With the Matthews and Ferland (1987) ionizing continuum $[N/O] \approx 3[N/O]_\odot$, $[N/Si] \approx 3[N/Si]_\odot$, and $[N/C] \approx 10[N/C]_\odot$ (but with greater uncertainty) were necessary. Amazingly, $[N/H] \approx 200[N/H]_\odot$ was required to get an acceptable fit to the Lyα BAL.

In addition, there is other evidence for very enhanced abundances in DAL QSOs. There are selected cases of enhancements of S, Ar, and P and/or Fe. For the optical depths inferred for the more common metal BALs, it would be unlikely to observe these rarer elements unless they were enhanced, sometimes in excess of 100 times solar values. Note that for the observed P and Fe transitions, i.e., PV$\lambda\lambda$1117,1128 and FeIIIλ1122, both would produce absorption near a rest frame wavelength of $\approx$ 1122 Å. Therefore, it is difficult to distinguish between a P or Fe enhancement, but some recent spectra indicate that there are cases where the PV doublet separation is seen. Wampler (this conference) has discussed the case of a BAL QSO exhibiting FeII BALs from excited states.

Earlier it was also pointed out that there is now evidence for dust in those BAL QSO spectra which exhibit both high- and low-ionization BALs, however it is not clear if the dust is mixed in with the BAL region (Turnshek et al. 1994). This may be relevant to chemical composition determinations.

Lastly, turning to the broad emission lines in both BAL and non-BAL QSOs, their properties have also been cited as evidence for enhancements. This is relevant if the BAL and broad emission-line regions are physically related or connected. For this work the following should be noted. First, there is an energy budget problem with FeII emission in some QSOs (see Netzer 1990) and part of the solution is likely to rely on large (> 10) enhancements of Fe relative to solar values. BAL QSOs as a class have enhanced FeII emission (Weymann et al. 1991) and the recent study of the BAL QSO PG0043+039 (Turnshek et al. 1994) shows it to be one of the strongest known FeII emitters. Second, based on the NV/CIV and NV/HeII broad emission line ratios in QSO spectra, Hamann & Ferland (1992, 1993) have argued that N is often enhanced relative to solar values. This work is interesting in the context of the results obtained on N enhancements in the BAL region. However, note that Hamann & Ferland did not take into account resonance line scattering in the BAL region which may enhance the NV broad emission line (Turnshek 1984, 19994). The importance of this effect is unclear.

6 Discussion of Some Implications

The results discussed here on ionization structure and chemical composition need to be checked and improved upon in a number of ways. First, additional high-resolution studies of BAL profiles are still needed in order to understand the limitations of column density derivations, especially in the region of the Lyα BAL. Second, theoretical work is needed to explain the nature of the inferred ionization structure which gives rise to multiple ionization zones at the same observed outflow velocities. Third, there is no guarantee that the ionizing continuum follows the one proposed by Matthews & Ferland (1987). BAL QSOs may in fact have different continuum properties as inferred from the fact that they seem to be x-ray quiet. How sensitive the results on ionization structure and chemical composition are to the shape of the ionizing continuum needs to be considered in more detail. It is clear that the deduced metal-to-hydrogen abundance enhancement would decrease if a steeper ionizing continuum was used, because then more hydrogen would become ionized relative to the metals. Fourth, the nature of the photoionization model itself, as applied to BAL column density modeling, needs to be improved. For example, the effects that far-UV BAL profiles and the outflow velocity law have on the shape of the photoionizing continuum and the effects of very enhanced abundances need to be considered more carefully. It is important to understand why the ionization of a BAL trough remains roughly constant with velocity despite the fact that the inner parts of the BAL outflow would absorb a substantial fraction of any ionizing photons from a central source. It would seem that the density of the BAL gas drops just the right amount to balance the loss of ionizing photons and the change in central source distance, keeping the ionization of the gas roughly constant with outflow velocity. Perhaps this is not central source photoionization. Fifth, theoretical work is needed to explain the inferred abundance enhancements. Finally, observations and analysis of far-UV BAL profiles in more QSOs are needed to check the reliability of these results and help pin down any systematics of abundance enhancements.

However, while the shape of the ionizing continuum and its nature may have a substantial effect on the elemental abundances relative to hydrogen, there is less of an effect on the deduced relative enhancements of the metals since the metals have a similar range of ionization potentials. Therefore, despite the limitations in understanding, it seems clear that some mechanism of N enhancement relative to solar values is involved. Hamann & Ferland (1993) have discussed chemical evolution calculations which may be relevant to interpreting an inferred N enhancement. If normal stellar nucleosynthesis is important, their results indicate that a relatively flat initial mass function would be required so that a N overabundance could be produced by secondary processing in massive stars.

7   Summary

The work discussed here suggests the following main points:

1. It is becoming increasingly difficult to envisage a model which can explain all of the systematic differences between BAL and non-BAL QSOs, and the variation in QSO properties, in some type of *single* unified model. The size of the global BAL region covering probably varys in QSOs, ranging up to values considerably larger than 0.2. In some cases the resonance line scattered photons may be destroyed by dust.

2. Evidence is mounting which indicates that the geometry of the BAL region is non-spherical (e.g. disk-like). A non-spherical geometry would produce a situation in which the local BAL region coving factor is larger than the global BAL region covering factor.

3. While absorption from BAL gas appears not to be completely uniform across the continuum source along a sight-line, the average absorption profile is clearly very similar across a region which is at least the size of the continuum source. Thus, with few exceptions the BAL region probably completely covers the unpolarized central continuum source. This suggests that the methods used to derive column densities are generally accurate enough to perform an analysis of the ionization structure and chemical composition.

4. A two-zone photoionization model with ionization parameter fluctuations of ≈ 30 is needed to obtain reasonable agreement with the observed column densities of different ions of the same element. This might be caused by density fluctuations in individual clouds or distinct regions with different densities but similar outflow velocities, so the location of the ionization zones in this model is not currently well-constrained.

5. Compared to solar values, N enhancements relative to H, C, O, and Si are $\approx$ 200, 10 (uncertain), 3, and 3, respectively. The elements of S, Ar, P, and Fe also exhibit large enhancements in some BAL QSOs. Ways to lessen the inferred enhancement relative to hydrogen include adopting a steeper ionizing source or hypothesizing some effect which would cause the neutral hydrogen column density to be underestimated based on analysis of the Lyα BAL profile.

6. The general enhancement of N suggests that, if normal stellar nucleosynthesis is important, a relatively flat initial mass function is involved with the N overabundance being produced by secondary processing in massive stars.

Acknowledgements. I wish to thank some of my more recent collaborators and students for their contributions and helpful discussions on this work: Dr. B. Espey, Dr. C. Hazard, Dr. V. Khersonsky, Dr. O. Lupie, Dr. S. Rao, Dr. J. Wampler, M. Kopko, L. Lee, E. Monier, D. Noll, S. Sherer, and C. Sirola. Much of this work was based on a number of observations made with the NASA/ESA Hubble Space Telescope, operated by AURA under NASA contract NAS 5-26555.

References

Arva, N., & Li, Z. (1994), ApJ, 427, 700

Arva, N., Li, Z., & Begelman, M. (1994), ApJ, 432, 62

Arva, N., & Begelman, M. (1994), ApJ, 434, 479

Burbidge, E.M. (1970), ApJ, 160, L33

Foltz, C., Weymann, R., Morris, S., & Turnshek, D.A. (1987), ApJ, 317, 450

Francis P., Hooper, E., & Impey, C. (1993), AJ, 106, 417

Glenn, J., Schmidt, G., & Foltz, C. (1994), ApJ, 434, L47

Goodrich, R., & Miller, J. (1995), ApJ, submitted

Grillmair, C., Turnshek, D.A. (1987) in *QSO Absorption Lines: Probing the Universe (Poster Papers)* (STScI publication), eds. Blades, Norman, Turnshek,
  1

Junkkarinen, V., Burbidge, E.M., Smith, H.E. (1983), ApJ, 265, 51

Hamann, F., & Ferland, G. (1992), ApJ, 391, L53

Hamann, F., & Ferland, G. (1993), ApJ, 418, 11

Hamann, F., Korista, K., & Morris, S. (1993), ApJ, 415, 541

Hazard, C., Morton, D., Terlevich, R., & McMahon, R. (1984), ApJ, 282, 33

Knerr, J. (1993), PhD Thesis, University of North Carolina

Korista, K., et al. (1992), ApJ, 401, 529

Korista, K., Voit, G.M., Morris, S., & Weymann, R. (1993), ApJS, 88, 357

Kwan, J. (1990), ApJ, 353, 123

Lynds, C.R. (1967), ApJ, 147, 396

Matthews, W., & Ferland, G. (1987), ApJ, 323, 456

Murray, N., Grossman, S., Chiang, J., & Voit, G.M. (1995), ApJ, submitted

Netzer, H. (1990), in *Proceedings of the Saas-Fee Advanced Course 20, Active Galactic Nuclei* (Springer-Verlag), eds. Couvoisier, Mayor, 57

Sirola, C., Turnshek, D.A., Lupie, O., & Espey, B. (1995), ApJ, submitted

Sirola, C., Turnshek, D.A., Monier, E., Sheaffer, S., Weymann, R., Morris, S., Duhalde, O., Krzeminski, W., Kunkel, W., Roth, M. (1992), BAAS, 24, 731

Sprayberry, D., & Foltz, C. (1992), ApJ, 390, 39

Stocke, J., Morris, S., Weymann, R., & Foltz, C. (1992), ApJ, 396, 487

Turnshek, D.A. (1984), ApJ, 278, L87

Turnshek, D.A. (1986), *IAU Symposium 119: Quasars* (Reidel), eds. Swarup, Kapahi, 317

Turnshek, D.A. (1988) in *QSO Absorption Lines: Probing the Universe* (CUP), eds. Blades, Turnshek, Norman, 17

Turnshek, D.A., Foltz, C., Grillmair, C., & Weymann, R. (1988), ApJ, 325, 651

Turnshek, D.A., et al. (1994), ApJ, 428, 93

Turnshek, D.A. (1995), Proceedings of the ESO/EIPC Elba Workshop on *The Light Element Abundances in the Universe*, (Springer-Verlag), ed. Crane, in press

Turnshek, D., Lupie, O., Espey, B., & Sirola, C. (1995), ApJ, submitted

Voit, G.M., Weymann, R., & Korista, K. (1993), ApJ, 413, 95

Weymann, R., Turnshek, D.A., & Christiansen, W. (1985) in *Astrophysics of Active Galaxies and QSOs* (Mill Valley, CA: University Science Books), ed. Miller, 333

Weymann, R., et al. (1991), ApJ, 373, 23

Williams, R.E. (1992), ApJ, 392, 99

The Absorption Spectrum of Nuclear Gas in Q 0059−2735

E.J. Wampler

ESO, Karl-Schwarzschild-Str. 2, D-85748 Garching bei München, Germany

Abstract. Echelle spectra of the $z = 1.58$ broad absorption line (BAL) quasar, Q 0059−2735, show a wealth of absorption lines, mostly from iron-peak elements. Lines from the ground state and from metastable levels up to nearly 4 e.v. are seen. From the absorption line shapes and column densities, strong constraints can be obtained for the absorbing cloud geometry, its chemistry, density, temperature, and distance from the QSO nucleus. Cloud cross sections increase with outflow velocity. Excitation temperatures range from 4 000 K for the lowest velocity, low ionization condensations to 60 000 K for the higher velocity BAL clouds. The low velocity clouds lie between 2 and 10 pc from the central ionizing source and have electron densities that are between 10^6 and $3\,10^8\,\mathrm{cm}^{-3}$. Ionization models suggest that iron in the absorbing clouds is overabundant with respect to carbon, but as the ground-based data do not reach Ly-α, the overall metallicity of these clouds with respect to hydrogen is still unknown.

1   Introduction

Q 0059−2735 is a unique object. First, it is a member of that rare class of quasars that shows a broad absorption line (BAL) flow in both high ionization (C IV) and low ionization (Mg II) lines. About 10% of all quasars are BAL quasars, and about 10% of all BAL quasars belong to the group of low ionization BAL quasars. Second, the spectrum of Q 0059−2735 is unique among the quasars cataloged by Hewitt and Burbidge (1993), as it exhibits many narrow absorption lines of mostly iron-peak elements, that arise from excited levels with substantial population compared to the zero-volt level. The spectrum of Q 0059−2735 is very striking; similar quasars, which would be easily identifiable, have not yet been found.

Q 0059−2735 was discovered by Cyril Hazard in 1986. The early, low resolution spectra, were first described and discussed by Hazard *et al.* (1987). Later, Voit *et al.* (1993) analyzed the spectrum in considerably more detail. Both of these earlier studies used spectra taken with spectral resolutions of $\sim 100\,\mathrm{km\,s^{-1}}$. Voit *et al.* (1993) concluded that the BAL flow is likely composed of a high ionization, high temperature outflow that contains low ionization condensations.

Echelle spectra were obtained over several years using both EMMI on the ESO New Technology Telescope (NTT) and CASPEC on the ESO 3.6-meter Telescope. An extensive analysis of these data is now in press (Wampler,

Chugai & Petitjean; 1995). Presented here is a overview of the results of the Wampler, Chugai & Petitjean (1995) paper, together with a brief discussion of the main results.

2   Discussion

The ESO spectra of the BAL quasar Q 0059−2735 were taken with spectral resolutions of $10 \, \mathrm{km \, s^{-1}}$ in the yellow/red and $20 \, \mathrm{km \, s^{-1}}$ in the ultraviolet/blue. These spectra show that near the quasar emission line redshift, the low velocity portion of a BAL flow overlaps the velocities of 4 distinct mixed ionization clouds that produce narrow, saturated, absorption lines in the quasar spectrum. Many of these narrow absorption lines (NALs) originate from metastable levels that lie several electron-volts above the ground term. The NAL level populations imply excitation temperatures of $\lesssim 10^4 \, \mathrm{K}$. These levels can be populated either by collisional or radiative processes. Despite the presence of strong Mg II absorption, Mg I is not seen. Neither do we see He II λ1640. The absence of absorption from Mg I and He II sets useful constraints on the physical conditions for the absorbing clouds. Also, despite the observed very strong absorption of the continuum radiation by thousands of narrow lines, spectra taken several years apart show that the acceleration of the absorbing gas is $\lesssim 0.03 \, \mathrm{cm \, s^{-2}}$. Models for the absorption line clouds are thus constrained by: (1) the absence of strong Mg I, which sets a limit to the size of the H I region. (2) The existence of cool material. (3) The fact that the Fe II metastable levels are populated. (4) The absence of He II λ1640 line, and (5) the absence of noticeable acceleration of the Fe II narrow lines. These constraints restrict the NAL gas clouds to lie between 2 and 10 parsecs from the central source and to have characteristic gas densities that range between $\sim 10^6 \, \mathrm{cm^{-3}}$ and $\sim 3 \, 10^8 \, \mathrm{cm^{-3}}$.

Many of the NAL absorption features are saturated. This can be seen both from the shape of the lines, which are flat bottomed, and from the relative strength of lines that have a wide range of gf values. However, the saturated lines do not reach zero intensity. Evidently, the NAL absorbing clouds do not completely cover the continuum source. The residual absorption in the bottoms of the saturated narrow lines is $\sim$50% of the continuum level near Mg II but only $\sim$30% of the continuum level near C IV. If the background unabsorbed continuum is a thermal source, then these residual intensities are consistent with a background source temperature that is $\sim$7 000 K.

Cr is depleted relative to Zn by a factor of $\sim$10. This is similar to the depletion seen in metal line absorption of intervening systems that are at very different redshifts from the background quasar redshift (Pettini *et al.* 1994). Pettini *et al.* (1994) attribute the depletion of Cr to the removal of Cr ions from the gas phase by condensation onto dust grains. However, as the amount of flux absorbed by the NALs is quite flat with wavelength, there is little evidence that the background source seen through the absorbing clouds is much reddened. If dust particles do exist in these clouds, they may be larger than those seen in the intergalactic medium of our galaxy.

The BAL absorption troughs are much deeper than the NAL absorption. Thus BAL flow does occult the central continuum region and the high ionization broad emission line (BEL) region. however, the bottoms of the low ionization BAL absorption troughs show that the *low* ionization BAL flow fails to occult the low ionization BEL region. Comparing the BAL absorption line models for Fe II and Fe III shows that the excitation temperatures for the BAL clouds range up to $6\,10^4$ K, thus the BAL cloud temperatures are higher than those found for the the NAL clouds. Four BAL clouds are required to model the BAL structure at low velocity, and an additional 5 are needed to model the broad, wavy absorption that is seen in the higher velocity BAL flow. It is noteworthy that the *same* cloud model, ie. the same b–values, excitation temperatures, relative column densities, and redshifts can be used to model the absorption lines of all ions. Table 1 gives the physical parameters found for the clouds that we have identified in the spectrum of Q 0059—2735.

Table 1. Physical Parameters for the Q 0059—2735 Cloud Models

Cloud No.	Redshift (z)	Velocity ($km\,s^{-1}$)		b-Value ($km\,s^{-1}$)	Exc. Temp. ($10^4\,K$)	Relative Column Density
		NAL				
1	1.58386	+	0.	30	0.4	0.01
2	1.58275		130.	30	0.4	0.0063
3	1.58115		315.	17	0.5	1.0
4	1.58000	+	450.	17	0.7	0.5
		BAL				
5	1.5839	−	5.	80	0.9	0.014
6	1.5828	+	125.	100	0.9	0.02
7	1.5812		310.	80	0.9	0.025
8	1.5801		435.	50	1.1	0.013
9	1.5768		820.	200	3.0	0.0025
10	1.5721		1365.	350	4.0	0.0025
11	1.5652		2140.	250	6.0	0.0005
12	1.5603		2740.	350	5.0	0.00038
13	1.5554	+	3325.	220	5.0	0.00016

The clear implication from the absorption line structure and excitation conditions is that the BAL flow is originating from 4 dense, low ionization clouds. The BAL clouds must be larger than the NAL clouds and, in general, the higher velocity clouds have higher excitation temperatures than the lower velocity clouds. The occulted nuclear region of Q 0059−2735 must be organized into a hierarchical structure with the inner, smaller regions having higher ionization, velocities and temperature than the outer regions. With the exception of the iron-peak elements, which seem to be overabundant with respect to C, Al, and Si, the abundance ratios of the metals in Q 0059−2735 appear to be nearly solar.

Although the physical conditions and implied sizes of the BAL and NAL clouds are not the same, it has impressed us that the models show that both types of clouds have very thin H I layers. The ionized column density is almost exactly equal to the Strömgren column density. We found that the relative thickness of H I layer is approximately of $10^{-2}N_S$. This is an unexpected result. There must be some sort of fine-tuning mechanism for removal of excessive H I material in the Lyman continuum shadow. A possible mechanism is perhaps the existence of a dynamical instability at the H II/H I interface which is connected with the sharp decrease of Lyman continuum flux.

3   Conclusions

All these observed facts can be explained by a paradigm in which low ionization gas condensations lying above an accretion disc are being ablated by the intense radiation field of the quasar nucleus. The condensations and the BAL flow occult different parts of the background emission regions. The origin of the narrow line absorption clouds is uncertain: they could be ablating stellar atmospheres, or stand-off shocks around obstacles in a hot, supersonic flow of low density gas from the central QSO nucleus (Perry & Dyson 1985), or they could be dense gas clouds, ejected by a supernova explosion (Artymowicz *et al.* 1993) or, possibly, by instabilities in the accretion disk.

References

Aldrovandi, S.M.V., Péquignot, D. 1973, A&A 25, 137
Hazard, C., McMahon, R.G., Webb, J.K., Morton, D.C. 1987, ApJ, 323, 263
Hewitt, A., Burbidge, G. 1993, ApJS, 87, 1
Perry,J.J., Dyson, J.E. 1985, MNRAS, 213, 665
Pettini, M. Smith, L.J., Hunstead, R.W., King, D.L. 1994, ApJ, 426, 79
Voit, G.M., Weymann, R.J., Korista, K.T. 1993, ApJ, 413, 95
Wampler, E.J., Chugai, N.N., Petitjean, P., 1995, ApJ, (in press; April 20)

The BAL QSO Theory

E.Y. Vilkoviskij and I.V. Nosov

Fesenkov Astrophysical Institute, National Academy of Sciences, 480068, Almaty, Kazakhstan

Abstract. The main point of the present theory is the joint consideration of the three interacting flows: the cloudy absorbing gas, the hot plasma stream, and the QSO radiation flow. The main conclusions are: i) the line-locking effect is explained, and ii) the comparison of numerical solutions with the observed spectra shows that the BAL QSOs must contain a massive compact stellar cluster which may be the physical reason of the "radio-quietness" of the QSOs.

1 Introduction

The first examples (Lynds 1967, Burbidge 1970) of BAL QSOs were discovered soon after the discovery of the quasars themselves. Despite the fact that these objects have been intensively investigated during the last decades, there is no complete theory of the phenomenon yet.

A step forwards in the understanding of BAL QSOs was published in the work by E. Vilkoviskij (1991).

2 The Model

Considering the main idea of solving together the system of the radiation transfer and of the gas dynamic equations, we consider here the two-phase medium flow (the cold clouds in the hot gas) instead of the continuous cold gas flow. Consequently, we have to solve simultaneously the equation system of the following three interacting flows: cold clouds, hot gas, and radiation.

The equation of motion of the cloud can be written as:

$$m_{cl} \frac{dV}{dt} = F_{lin} + F_{cont} + F_{drag} - F_{gr}, \tag{1}$$

where m_{cl} is the cloud mass, V cloud velocity. The right-hand side contains the forces of the radiation pressure due to the scattering of the radiation by ion lines F_{lin} and due to continuum absorption F_{cont}, the drag-force due to the velocity difference of the cloud and the hot-gas stream F_{drag} (see below) and the gravitation force F_{gr}.

We use the Parker type equation for a radial flow of the hot gas in the form:

$$(1 - \frac{a^2}{v^2})v\frac{dv}{dr} = \frac{4a^2}{r} - \frac{GM(r)}{r^2} + g_{rad} + g_{drag} \, , \qquad (2)$$

where v is the velocity of the hot gas, $a = kT/m$ the sound velocity, r the radius. The right-hand side contains the accelerations due to the gas pressure, gravity, radiation pressure and the drag-force of the clouds.

The notable difference between Equation (2) and the Parker equation lies in the fact that the mass $M(r)$ depends on the distance r. The $M(r)$ is the mass within the r-sphere and it is the sum of the masses of the central object (the massive black hole) and the "distributed mass" of the compact stellar cluster (the stellar kernel). As is shown in our work (Vilkoviskij et al. 1994), Equation (2) usually has three critical points instead of the single critical point of the Parker equation and the position of the transsonic point depends on the relation of the kernel mass to the mass of the central object (the massive black hole), $\xi = M_k/M_{co}$.

3   Results

It can be shown that the outer critical point is more preferable for the BAL QSO models and this means that in this case typically $\xi > 10$. Physically it means that BAL QSOs have a very massive and compact star kernel in their centre, so that the hot gas is confined inside the kernel gravity potential hole, and the hot-gas stream is subsonic into the kernel and becomes supersonic out of the kernel radius, where the broad absorptions are formed.

We use the radiation transfer equation in the form similar to that of Kwan (1990).

Our results permit to suppose an explanation of the absence of strong radio-jets in the radio-quiet quasars: there is too much hot gas inside the massive kernels and it prevents the jets from coming out of the kernel interior.

A more extensive version of the present work will be submitted for publication in Astronomy & Astrophysics.

References

Burbidge, E.M., 1970, ApJ Letters, 160, L33

Kwan, J., 1990, ApJ, 353, 123

Lynds, C.R., 1967, ApJ, 147, 396

Vilkoviskij, E.Y., 1991, Astron. Jur. (Rus), 68, 1150

Vilkoviskij, E.Y., Karpova, O.G., Malkov, E.A., Nosov, I.V., 1994, Izvestiya of NAN
     RK, N4, 40

Evaporation Dominated Flows in BAL Regions of QSOs

Yuri A. Shchekinov

Institute of Physics, Rostov State University, 194 Stachki, 344104 Rostov on Don, Russia

Abstract. Hydrodynamical properties of a two-phase medium with a domination of clouds evaporation have been studied, including analysis of the solution topology of a steady, radial outflow. A gain of mass from evaporating clouds was shown to suppress crucially radial outflows and initiate a convergence of a jet-like flow to a quiet wind. The bulk motion in broad-absorption line regions of QSOs was estimated to be an evaporation-dominated flow. The observational fact that the broad-absorption-line QSOs are exclusively radio-quiet ones is argued to be connected with a suppression of jet outflows by clouds evaporation.

1 Introduction

The broad-absorption line quasars (BAL QSOs) have been recognized recently as representing exclusively the radio-quiet QSO population, and moreover all radio-quiet QSOs are very likely to possess broad-absorption lines (Stocke et al., 1992). A possible explanation is that BAL QSOs are in an intermediate state in which the subrelativistic wind tends to "break through" the surrounding BAL envelope to develop subsequently as a collimated relativistic radio jet (Voit et al., 1993). An alternative view implies that an intense interaction with neighbour gas can destroy a relativistic jet flow. Norman and Miley (1984) mentioned that such an interaction can decelerate and decollimate a relativistic radio jet at kiloparsec scales. It is very likely that on small scales it could work also to violate a jet outflow. Shchekinov (1995) has demonstrated in the framework of a simple hydrodynamical model that when clouds from surroundings pollute the interior of the jet cone it initiates a convergence of a supersonic (presumably relativistic) jet flow to a quiet "breeze"-like one due to clouds evaporation. Sect. 2 summarizes this consideration.

2   Steady, Radial Flows Modified by Evaporation Effects

Inferred from observations of low-ionization BALs (Voit et al., 1993) the ratio of the mass rate of evaporating clouds $\dot{M}_{ev}$ to the mass outflow rate of the confining gas $\dot{M}_{conf}$ can be estimated as $\dot{M}_{ev}/\dot{M}_{conf} \geq 5 - 50$, (Shchekinov, 1995). This demonstrates that the dynamics of BAL regions is dominated by evaporation of clouds gas. One may expect that at equal conditions jet flows must be strongly affected by evaporating clouds getting into the jet cone from the environment.

Let us assume a jet to be at the initial state a monophase flow, corresponding to a dynamical regime which intersects the critical (sound) point being a subsonic outflow at the source and supersonic one at the infinity. If clouds from surroundings intervene in the jet cone, they modify the dynamics of the flow due to an additional mass supplied by evaporation. The equation of motion of a steady, radial outflow can be reduced to the following form (Shchekinov, 1995):

$$\left(M^2 - 1\right) \frac{dM^2}{dr} = F\left(E(\phi, r), \phi, M, r\right),$$

where M is the Mach number, $E(\phi, r)$ is the energy per unit mass, $\phi = \phi(M(r))$ is a function of the solution $M(r)$ and represents the influence of clouds evaporation. In a standard case without evaporation $E = E(r)$ and $F = F(E(r), M, r)$. The dependence of F on the function ϕ was shown to make the critical point $M^2 = 1$ unstable (Shchekinov, 1995). Therefore, in the framework of this consideration a jet outflow is expected to be broken by evaporating clouds when they pollute a jet cone.

As a conclusion, one may speculate that the radio-quietness of BAL QSOs is in an intimate connection with the ability of a quasar's vicinity (i.e., inner accretion disk) to eject a sufficient amount of material which would depress a jet outflow and form a BAL region. Vice versa, the vicinity of radio-loud quasars seems to be inefficient to eject such a material.

Acknowledgements. The author thanks ESO for a travel grant which made possible his participation at the Workshop.

References

Norman, C., Miley, G., 1984, A&A, 141, 85
Shchekinov, Yu.A., 1995, ApJ, to be submitted
Stocke, J.T., Morris, S.L., Weymann, R.J., Foltz, C.B. 1992, ApJ, 396, 487
Voit, G.M., Weymann, R.J., Korista, K.T. 1993, ApJ, 413, 95

Testing Unified X-ray – UV Absorber Models with NGC 5548

Smita Mathur, Martin Elvis, and Belinda Wilkes

Harvard-Smithsonian Center for Astrophysics, Cambridge, MA 02138, USA

1   Introduction

Intrinsic/Associated absorption systems in quasars have been well known but little understood for 20 years. X-ray and UV absorbers were never thought to be one and the same due to their apparently very different properties. The physical properties of the absorber cannot be understood by UV or X-ray studies alone. In fact, the ionization structure of the absorbers was not determined. This situation changed with recent results on 3C351 and 3C212. In 3C351, OVI absorption was observed in HST spectrum (Bahcall et al. 1993) and OVII/OVIII absorption in ROSAT spectrum (Fiore et al. 1993). Through detailed modeling they were found to originate in the same gas. This was the first confirmed X-ray/UV absorber (Mathur et al. 1994). The absorber was found to be highly ionized, outflowing, with high column density, low density and situated outside the BELR. Similar results were obtained for 3C212 (Mathur 1994).

2   Testing The Models with NGC5548

Are the X-ray and UV absorbers one and the same in ALL AGN? Or are the unified XUV absorber models applicable to lobe-dominated radio loud objects only (e.g. 3C351 and 3C212)? The radio-quiet Seyfert galaxy NGC5548 with its extensive echo-mapping data base provides an ideal test case.

The X-ray spectrum of NGC5548 with ASCA (Fabian et al. 1994) shows OVII and OVIII absorption edges with $\tau_{OVII} = 0.26^{+.04}_{-.08}$ and $\tau_{OVIII} = 0.12^{+.07}_{-.03}$. The column density was found to be $N^*_H = 3.8 \times 10^{21}$ cm^{-2}. The HST data (Korista et al. 1994) showed absorption lines of CIV, NV, and Ly$_\alpha$. The photoionization model is presented in figure 1 in which ionization fraction of various ions is plotted as a function of ionization parameter. Filled circles represent the range of values observed with ASCA, filled triangles are HST UV constraints, and vertical lines define the best fit values for the X-ray absorber. A single absorber in NGC5548 reproduces absorption features due to Lyα, NV, CIV, OVII & OVIII. We conclude that the X-ray and UV absorbers are one and the same in NGC5548 (Mathur et al. 1995, in preparation).

The physical properties of the absorber can now be determined through combined X-ray and UV constraints: • The column density is large: $N_H = 3.8 \times 10^{21}$ cm^{-2}. • Ionization parameter is well constrained to 8% (2.4 < U < 2.8). • Density is constrained by the recombination time: $n > 5 \times 10^5$ cm^{-3} $\Rightarrow$ r < 0.8 pc. • Absorber is situated outside the BELR : r > 8 lt. days (BELR) $\Rightarrow$ n< 5×10^9 cm^{-3}. • Thus the density and the radial distance are constrained: 5×10^5 < n < 5×10^9 cm^{-3} and 2×10^{16} < r < 2×10^{18} cm. • The thickness of the absorber is then $\Delta r < 7 \times 10^{15}$ cm. • Outflow velocity $v = 1200 \pm 260$ km s^{-1} is determined by the blueshift of the absorption lines (Shull & Sachs 1993), while the velocity dispersion is small: $b = 40 \pm 5$ km s^{-1} (c.f. FWHM $\sim$ 330 km s^{-1}). • The mass outflow rate is determined to be $\dot{M}_{out} = 0.02$ M$_\odot$ yr^{-1} (for covering factor f=0.1) carrying kinetic luminosity 1.5×10^{40} erg s^{-1} (c.f. Radiative luminosity = 10^{44} erg s^{-1}).

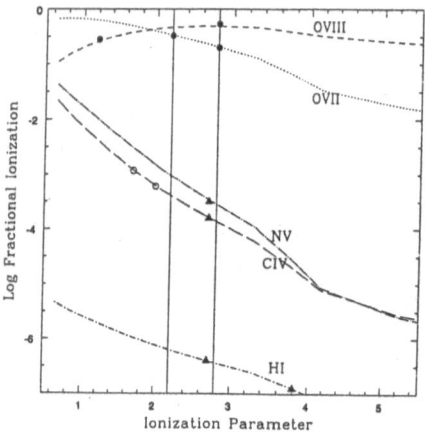

Fig. 1. The photoionization model.

We now believe that all AGN intrinsic absorbers have both X-ray and UV signatures. This association will help us to understand both these and other associated metal line absorption systems, e.g., BAL QSOs.

References

Bahcall J.N. et al. 1993, ApJS, 87, 1
Fiore M. et al. 1993, ApJ, 415, 129
Mathur S. 1994, ApJL, 431, L75
Mathur S. et al. 1994, ApJ, 434, 493
Fabian A. et al. 1994 in *New Horizons in X-ray Astronomy*, Universal Academy Press
Shull J. M., Sachs B. 1993, ApJ, 230, 121

An Associated Absorption System in Front of a Peculiar QSO with $z_{abs} \sim z_{em} \sim 2.7$

Luis E. Campusano[1] and Roger G. Clowes[2]

[1] Department of Astronomy, University of Chile, Casilla 36-D, Santiago, Chile
[2] Centre for Astrophysics, University of Central Lancashire, Preston PR1 2HE, UK

Abstract. We report the discovery of Q1044+0546 ($M_V > -27$), whose optical spectrum is dominated by one very strong (W(rest)$\sim$110 Å) and extraordinarily broad (FWZI(rest)$\sim$270Å) emission line. We argue that this line is Ly-α at z=2.688, imparting a unique character to this object because the CIV λ1549Å emission line which would be around λ5700Å, is weak or absent. We present a preliminary interpretation of this QSO discovered as part of our redshift program in the ESO/SERC field no. 927. The rest-frame Ly-α FWHM and FWZI are $\approx$ 23.000 and $\approx$ 67.000 km s^{-1}, respectively; these values are among the largest found to date. A heavy element absorption line system is present too, with z=2.654, implying a relative velocity with respect to the QSO of $\sim$ 2700 km s^{-1}. There are also several weaker absorption lines in the blueward region of the Ly-α emission, which are presumably from the Ly-α forest. The emission spectrum of Q1044+0546 may be indicative of very special physical conditions and/or rather high metal abundances in its broad line formation region(BLR). It is of great interest to determine if the physical conditions in the clouds of its associated absorption system are also peculiar, because similarities with the BLR would favour the intrinsic hypothesis for the origin of the $z_{abs} \approx z_{em}$ systems (see Petitjean et al. 1994).

1 Introduction

Early in 1994, we completed with the CTIO 4m telescope a redshift survey of $\sim$200 optically selected Quasi-Stellar-Objects(QSO) candidates, in the direction of ESO/SERC field No. 927, for studies of the large scale structure of the universe (Clowes & Campusano 1991). The most unusual object of the set appeared in our last observing run. The most striking property of its spectrum was the presence of one single, strong and extremely broad, emission line dominating the observed λ 3200-7400Å range, and accompanied by several strong narrow absorption lines.

In this paper we argue that the object is a z_{em} =2.688 QSO and identify the narrow absorption lines as a foreground z_{abs} = 2.654 system ($z_{abs} \approx z_{em}$ systems, or 'associated' systems). Apparently very few cases of 'Ly-α only' QSOs have been claimed. Our object is certainly unique in our sample (especially for its emission spectrum but also because of the strong absorption spectrum) of 55 Ly-α QSOs belonging to our homogeneous search in the F927 field.

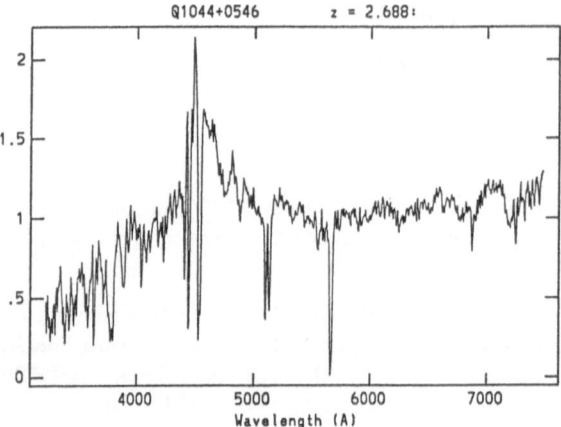

Fig. 1. Smoothed spectrum (f_λ) showing the prominent Ly-α emission.

2   Discussion

Why is the emission spectrum of Q1044+0546 so unusual? Specifically, does the apparent absence of CIVλ1549 imply a low metal abundance? The answer is no, within the frame of photoionization models (Ferland, private communication), given that low abundances generally make UV lines brighter and thus with the cooling being carried out in the UV. A UV spectrum with only Ly-α is produced if the cooling is carried out by IR and optical forbidden lines, which might happen if the density is low and/or the metallicity is high. The fact that UV spectra like the one of Q1044+0546 have been rarely observed then suggests, within the mentioned framework, that the necessary conditions of low density and/or high metallicity are only rarely attained. If high metallicities were to be implied, they will possibly be rather high given the claim by Hamann & Ferland(1993), that the normal BLR metallicities are typically $\sim$ 1 to perhaps even $\geq$ 10 times solar.

A more complete account of this object will be presented elsewhere.

Acknowledgements. LEC has been partly funded by FONDECYT(Chile) grant No. 985/93. The Faculty of Physical and Mathematical Sciences of the U. of Chile partly funded the participation of LEC in this conference.

References

Clowes, R.G. and Campusano, L.E. 1991, MNRAS, 249, 218.
Hamann, F. and Ferland, G. 1993, ApJ, 418, 11.
Petitjean P. et al. 1994, A&A, 291, 29.

X-ray Environment of 3CR Radio Sources

M. Almudena Prieto

Max-Planck Inst. für Extraterrestrische Physik, ROSAT Division, Postfach 1603, D-85740 Garching, Germany

Abstract. The 3CR sample of radio galaxies has been cross-correlated with the ROSAT All-Sky-Survey. 25% of the sources have been detected, the majority being powerful X-ray emitters with a range of luminosities between 10^{44} and 10^{46} erg/sec in the 0.2-2.4 keV PSPC band. A large fraction of the detected sources are lobe-dominated sources (Fanaroff & Riley class II) and are optically classified as QSO/S1 type. Deep PSPC pointing observations reveal most of them unresolved at the resolution of the PSPC ($\sim$25 arcsec at 1 keV), their X-ray size being thus smaller than their typical radio sizes. The lack of extended X-ray emission up to distances comparable to the radio scales is an unexpected result for lobe-dominated sources, in particular, if considering that hot gas halos are thought to be the confining medium in such objects. The absence of such halos may be of relevance for the origin of the absorption lines systems seen in quasars.

1 Brief Description

The 3CR sample of radio galaxies (Laing, Riley and Longair, 1983), has been cross-correlated with the ROSAT/PSPC All-Sky-Survey. The sample contains 173 sources which have been selected with a radio flux at 178MHz $>= 10$ Jy, Dec$>= 10$ deg and $|b| >= 10$ deg. The cross-correlation with the ROSAT survey has resulted in 41 positive detections, the majority being extreme powerful emitters in the radio as in the X-ray domain. The bulk of 0.2-2.4 keV luminosities peaks at about $10^{45.5}$ erg/sec.

We are involved in a project aimed to study the X-ray properties of the 3CR detected sources, and their relationship with the well known optical/radio properties of the 3CR members (Prieto, 1995, in preparation). In this paper, it is presented some of the results derived from a spatial analysis of PSPC images of a selected sample of lobe-dominated sources.

2 Extended X-ray Emission

The PSPC energy band is particularly good to search for soft emission components that in most AGNs are presumably linked to the nuclear central source. In lobe dominated radio sources, an additional extended X-ray component is expected to provide the working surface for, and to some extent confining, the radio structure. For the purposes of this work, we have selected 9 3CR sources with a radio extension well beyond the FWHM of the PSPC

point spread function. The list of the sources including the redshift and size
of the radio structure at 178 MHz is presented in Table 1.

Table 1. 3CR sources spatially analysed in the 0.2-2.4 keV

3CR source	z	size at 178MHz(")
3C 47	0.425	69
3C 215	0.411	60
3C 219	0.174	143
3C 249.1	0.313	53
3C 263	0.652	51
3C334	0.555	58
3C351	0.371	75
3C382	0.058	>120
3C390.3	0.056	189

The radial PSPC light profile for each of those sources has been compared
with the on-axis point spread function (PSF) of the instrument. In all cases,
the observed light profile is found compatible with a point-source distribution
at the resolution of the PSPC. In addition, azimuthal profiles were extracted
for each source to check for asymmetries at any particular direction. All
directions were found to be consisting with an unresolved central component.

3   Conclusions

One of the debated possibilities for the origin of absorption line systems in
quasars is that they could form in giant nebulae or clouds associated with
galaxies. Classical "double radio sources" with radio extensions over several
hundred kpc from their core are plausible candidates for hosting such large gas
halos necessary to confine their radio structure. Though the present sample
is small, it is striking that none of the 3CR double radio sources up to now
analysed show extended X-ray emission up to distances comparable with the
radio scales. The result puts important constrains on the sizes of the potential
gas halos associated to radio galaxies.

References

Laing, R.A, Riley, J.M. & Longair, M.S. 1983, MNRAS 204, 187
Prieto, 1995, in preparation

PART VII

LYMAN-ALPHA SYSTEMS

AT LOW AND HIGH REDSHIFTS

Observations of Lyman-α and Lyman-Limit Systems from the HST Quasar Absorption Line Key Project

Alec Boksenberg

Royal Greenwich Observatory, Madingley Road, Cambridge, CB3 0EZ, UK

Abstract. We determine the evolution of Lyman-α forest lines for redshifts smaller than 1.3 by combining HST spectral observations for 17 quasars, including four new observations of quasars with emission-line redshifts between 1.0 and 1.3. For a power-law fit the density of lines with equivalent widths greater than 0.24 Å is $(dN/dz)_0 = 24.3 \pm 6.6$ per unit redshift, and $\gamma = 0.58 \pm 0.50$. A single power-law fit to this data sample combined with a large-redshift sample from ground-based observations is not as good a fit as a broken power-law. The data are consistent with two populations of Lyman-α lines: one showing significant clumping often in the vicinity (in redshift space) of extensive metal-line systems and evolving relatively slowly, and one being relatively unclustered and evolving more rapidly, with the latter population dominating at large redshifts. The clumps of Lyman-α lines and the inferred rate of evolution of lines at small and intermediate redshifts more nearly resemble the properties of galaxies and of metal-containing absorption line systems than of the familiar large redshift Lyman-α forest lines. For Lyman-limit ($\tau \geq 1.0$) systems we determine the evolution over a wide redshift range from HST observations of 26 quasars combined with high quality data selected from previously available IUE and optical observations, and new optical observations. After taking considerable care to avoid effects which may bias the sample, the data from a total of 169 quasars are well fitted by a single power-law with $(dN/dz)_0 = 0.25^{+0.17}_{-0.10}$ and $\gamma = 1.50 \pm 0.39$, not a broken power-law as has been proposed before.

1   Introduction

The primary goal of the Hubble Space Telescope (HST) Quasar Absorption Line Key Project is to produce a homogeneous, ultraviolet spectroscopic data base for investigating, at small and intermediate redshifts, the properties of the intergalactic medium and of the gaseous content of galaxies and groups of galaxies. When this is combined with ground-based data for larger redshift quasars, the full set can be used to determine how these gaseous systems have evolved with cosmic time. In this paper I describe recent work of the Key Project team in investigations of Lyman-α and Lyman-limit systems (Bahcall et al. 1995; Stengler-Larrea et al. 1995), following the analysis of the first observational results presented in Bahcall et al. (1993, hereafter Paper I).

2   Lyman-α Systems

In Paper I we established some of the characteristics of Lyman-α absorption systems at small redshifts. Here I outline the further observations and analyses of the systems at the intermediate redshifts $0.6 < z < 1.3$ observed in the spectra of the four quasars Ton 153 ($z_{em} = 1.022$), PKS 0122-00 ($z_{em} = 1.070$), PG 1352 + 011 ($z_{em} = 1.121$) and PG 1634 + 706 ($z_{em} = 1.334$), and describe the properties of this sample combined with a sample of 13 smaller redshift objects analysed in Paper I, overall covering the range $0 < z < 1.3$. A full account of this work will appear in Bahcall et al. (1995).

All the new observations, like those in Paper I, were made using the Faint Object Spectrograph (FOS) of the HST. The observing and data calibration procedures basically are described in Schneider et al. (1993, hereafter Paper II). We have used the flat field generated from calibration observations closest in time to each individual quasar observation. Residual errors in the flat fielding may dominate the statistical noise in some portions of the spectra, and can create spurious weak features. We took considerable care to reject such spurious "lines".

Figure 1 shows a portion of the calibrated data obtained for one of the newly-observed quasars, Ton 153. The "continuum", shown dotted, is constructed by fitting cubic splines to a collection of points that represents our best estimate of the underlying continuum and emission line fluxes (and any broad absorption profiles that may be present) against which the narrow absorption lines are measured. The spectra are divided by these continua to produce normalized arrays for our automated analysis program which follows a well-defined procedure for finding and measuring the lines (Paper II); in this, blended features consisting of several closely spaced lines can be separated into individual Gaussian components. This procedure was facilitated for the high line-densities encountered in the new spectra, in contrast to those of the small-redshift quasars in Paper I, by the addition of some line-fitting software options not present in the earlier (Paper II) code. We use an objective procedure for identifying absorption lines, embodied in an evolving software package called $ZSEARCH$, as described in Paper I and in Bahcall et al. (1992). Again, improvements were implemented for the work in this paper, among them an iteration procedure for determining the best-fit redshift for multiple-line systems, and algorithmic tests which result in making specific identifications more robust. A full description of all the new analysis and verification procedures is given in Bahcall et al. (1995).

The spectrum of Ton 153 in Figure 1 shows most of the region we have observed shortward of Lyman-α emission for this object, and serves as a good example of our data set. There is an interesting complex Lyman-α absorption feature, with at least five well-discernible lines over 2029 - 2039 Å, $z = 0.6690 - 0.6773$, which has similar counterparts in some of the other quasars. The Lyman-β absorption corresponding to this complex is present

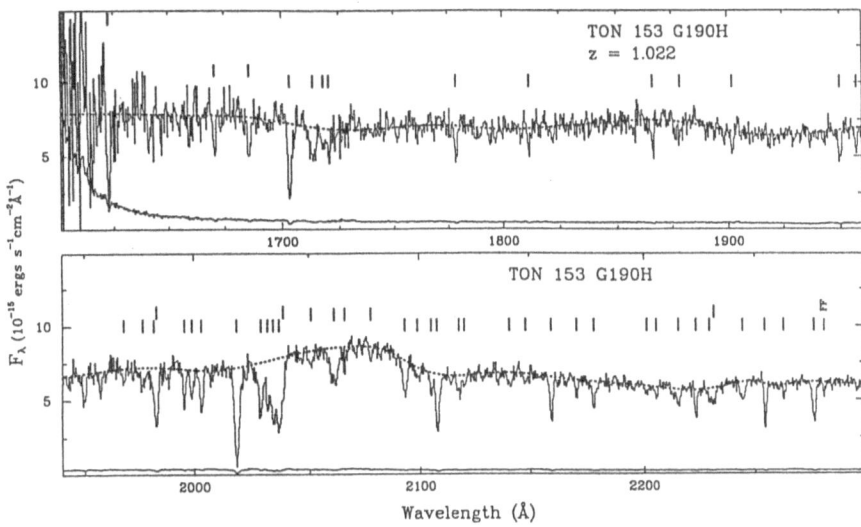

Fig. 1. Ultraviolet spectra of the quasar Ton 153 obtained with the HST FOS. Each panel contains a single spectrum obtained with the G190H grating, the 0.25" × 2.0" slit and the red digicon detector. The short vertical bars indicate the positions of absorption lines in the complete sample. The dotted line is the "continuum" fit: see Paper II. The lower line in each spectrum plot is the $1 - \sigma$ uncertainty in the flux as a function of wavelength. FF indicates a spurious (flat field) feature.

as a broad feature at $\sim$ 1716 Å, $z = 0.6730$. This broad hydrogen absorption complex has a velocity structure extending over $\sim$ 1500 km s^{-1} with individual components having relative velocities amounting to hundreds of km s^{-1}. No metal lines are detected from this complex. However, again as appears in some of the other quasars, there is a neighbouring strong metal-line absorption system, here at $z = 0.6606$ with Lyman-α at 2018.83 Å and Lyman-β at 1703.30 Å and showing corresponding deep Lyman-limit absorption; this system is separated by only $\sim$ 2000 km s^{-1} from the broad hydrogen complex. It is likely that the hydrogen complex and the strong metal-line system are all part of one associated complex of large scale, extending over $\sim$ 3000 km s^{-1}. The $z = 0.6606$ system is dominated by intermediate stages of ionization, with OI, NI, SiIV, CIV and NV all absent, but this is not typical of the sample.

In the four intermediate redshift quasars we identify eight extensive metal-line systems overall, with number of lines ranging between five and fifteen and ionization level of the absorbing ions variously extending from low to high. Amongst these we have detected five associated groups of clustered Lyman-α lines. The apparent velocity spans over the clustered Lyman-α lines with their neighbouring metal-line systems range between 1200 and 3500 km s^{-1}. The average separation of Lyman-α lines is 22 Å for the complete sample of measured Lyman-α lines found in the spectra of the four quasars; by contrast, for the five remarkable clumps of lines there are 6 in 20 Å, 4 in 9 Å, 5 in 12 Å, 4 in 15 Å and 4 in 25 Å, including in each case the Lyman-α line of the extensive metal-line system. Thus the line density of Lyman-α absorptions is greater by nearly an order of magnitude in the vicinity of metal-line systems. While the amount of Lyman-α clumping that is detected depends, among other things, on the spectral resolution, the method and parameters used in selecting the absorption lines and the rules adopted for identifications, and further, it is hard to avoid some degree of *post facto* reasoning in the statistical considerations, nevertheless the overall pattern that is apparent is impressive. We examined the lists of identifications of absorption line features that are reported in Paper I, but it is difficult to compare the results directly with the new data since the way that the line selection software deals with blends has changed since the earlier analyses were done. However, there is no obvious contradiction between the results of the two studies.

We interpret this clumping of Lyman-α absorption systems near extensive metal-line absorption systems as providing evidence that some Lyman-α "forest" lines are associated with galaxies in clusters or superclusters. The metal-line systems may originate either in the halos of individual galaxies within a cluster of galaxies or within an inhomogeneous gaseous medium between the galaxies; the Lyman-α absorptions may occur in the very extended halos of individual galaxies or in primordial gas associated with the cluster that has not yet formed galaxies. That some fraction of the small-redshift Lyman-α absorption lines are clumped previously is apparent from observations of H1821+643 (Bahcall et al. 1992) and 3C 273 (Bahcall et al. 1991; Morris et al. 1991, 1993); and Lanzetta et al. (1994) have also presented evidence for the association of a significant fraction of the small redshift Lyman-α lines with galaxies.

The addition of the Lyman-α systems in the four intermediate redshift quasars to the data from the thirteen smaller redshift objects analysed in Paper I allow us to extend the investigation of the evolution of these absorbers up to a redshift ~ 1.3. We adopt a minimum rest equivalent width for the entire sample, W_{min}, and restrict our analysis to the spectral regions in each quasar for which the $4.5 - \sigma$ detection limit (Paper I) is less than W_{min}. Only lines in regions with $W > W_{min}$ are included in this conservative sample; we refer to this as a "uniform" detection limit sample. While the line densities derived from this procedure advantageously are independent of any assumption about the equivalent width distribution, the sample is of reduced size. We found that the maximum number of lines for such a uniform sample occurs with our data set for $W_{min} \sim 0.24$ Å. We restrict the sample further to systems having velocities more than 3000 km s^{-1} from heavy element systems. In all, this leaves 109 lines in the uniform sample. Using a maximum-likelihood estimator and assuming a distribution of the form $(dN/dz) \propto (1+z)^\gamma$ we find for the line density at zero redshift, $(dN/dz)_0 = 24.3 \pm 6.6$ Lyman-α lines per unit redshift, and $\gamma = 0.58 \pm 0.50$; the uncertainties are approximately $1 - \sigma$

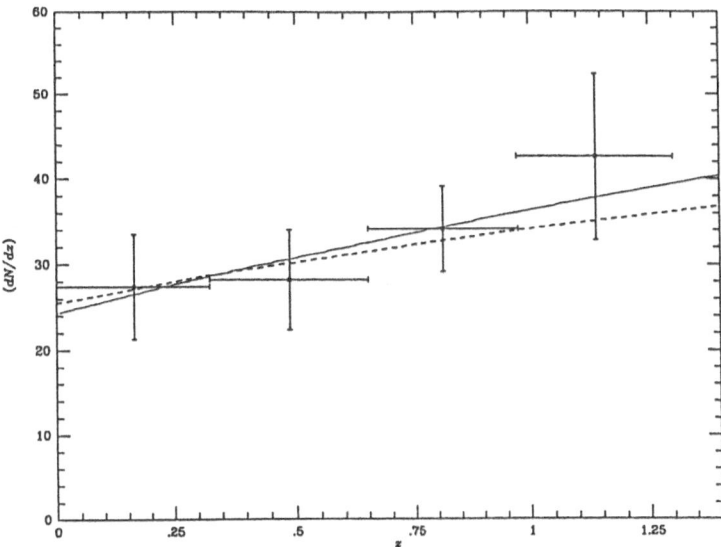

Fig. 2. The evolution of Lyman-α absorption systems over the redshift range from $z = 0.0$ to $z = 1.3$, from the Key Project quasars analysed to date. The solid line is the maximum-likelihood fit to a sample of 109 Lyman-α lines with a uniform detection limit of 0.24 Å rest equivalent width. The best fit yields a value of $\gamma = 0.58 \pm 0.50$ assuming a distribution of the form $(dN/dz) \propto (1 + z)^{\gamma}$. The larger, variable-detection-limit sample of 155 lines yields a value of $\gamma = 0.42 \pm 0.42$. For the uniform sample, we also show the local value of $(dN/dz)_z$ centred at each of four redshift bins of equal increments in Δz between $z = 0.0$ and 1.3. The value of $(dN/dz)_z$ refers to the line density above a rest equivalent width of 0.24 Å. Lines within 3000 km s^{-1} of the emission line redshift of all the quasars have been excluded from the samples analysed in this figure and in the following figure.

values. This distribution is shown as the solid line in Figure 2. Including the set of all lines in our sample with rest equivalent widths > 0.24 Å regardless of detection limit at the wavelength in question increases the sample size by about half. The maximum-likelihood estimation procedure in Paper I yields for this sample of variable detection limit the results $(dN/dz)_0 = 25.5 \pm 5.3$ Lyman-α lines per unit redshift and $\gamma = 0.42 \pm 0.42$. This distribution is shown as the dashed line in Figure 2.

These data further strengthen the conclusions, suggested in Paper I, that the density of Lyman-α lines increases much more slowly with redshift at small redshifts than in the large redshift region observable from the ground, and that a rather abrupt change of slope in the distribution might occur if many of the small redshift systems are of a different character from the more familiar systems observed at large redshifts. To investigate this further we use the large-redshift sample of 905 lines from Lu et al. (1991) and is believed to constitute a uniform detection limit to $W_{min} \sim 0.36$ Å, and a

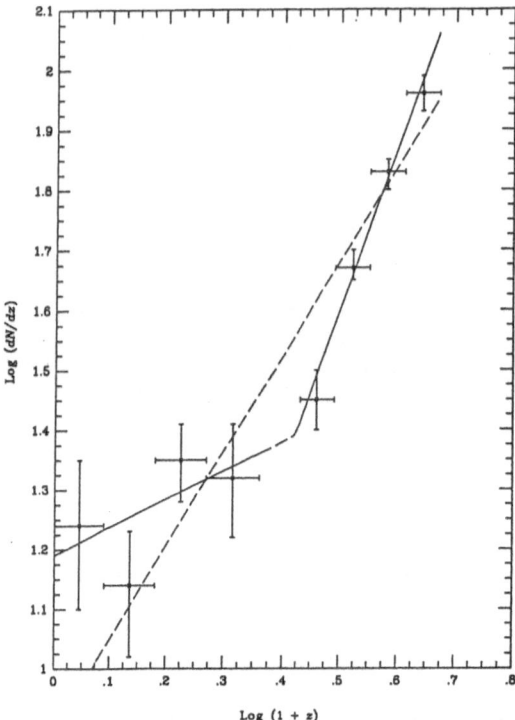

Fig. 3. The evolution of Lyman-α absorption systems using Key Project quasars analysed to date and the large-redshift, ground-based sample discussed by Lu et al. (1991). The Key Project sample used here includes 95 lines and the Lu et al. sample, 905 lines, all with a uniform rest equivalent width of at least 0.36 Å. The separate fits to the HST and to the ground-based data are shown as the two solid lines, with slopes of $\gamma = 0.48 \pm 0.54$ and 2.68 ± 0.27 respectively. The dashed line shows the best fit to a single power law for both the HST and the ground-based data and has a slope of $\gamma = 1.58 \pm 0.13$. We also show local estimates for $\log(dN/dz)_z$ for four redshift bins for the HST data and four bins for the ground based data, with equally-spaced increments of $\log(1 + z)$ between $z = 0.0$ and 1.3 for the HST data and equally-spaced increments between $z = 1.7$ and 3.7 for the ground-based data.

sample of 95 lines from the Key Project which also conform to this detection limit. As before we include in the samples only lines with velocities > 3000 km s^{-1} different from the emission line redshifts. The separate fits to the Key Project and Lu et al. samples give $(dN/dz)_0 = 15.5 \pm 4.2$ and 1.8 ± 0.6 and $\gamma = 0.47 \pm 0.54$ and 2.68 ± 0.27, respectively. These fits are p lotted as the two dashed lines in Figure 3. The interception of these fits occurs at a redshift of 1.65. A single fit to the combined sample, shown as a solid line, yields $\gamma = 1.58 \pm 0.13$; a KS test applied to this fit is acceptable only at the 2.7% level.

The evidence presented for clumping of Lyman-α lines near extensive metal-line systems, together with the results indicating that the evolution of the smaller redshift systems is slower than found at larger redshifts, provides strong support for the suggestion (see, for example, Bahcall et al. 1992; Maloney 1992; Hoffman et al. 1993; Morris et al. 1993) that there are at least two classes of Lyman-α absorbers: 1) the systems, observed most easily at small redshifts, that are associated with galaxies and which, as described here, can be clustered around extensive metal-line systems; and 2) the systems that are not strongly clustered and dominate at large redshifts. The type 1 systems evolve at a rate similar to Lyman-limit systems and general metal-line systems; type 2 systems evolve more rapidly. It is possible that when the dynamical evolution of Lyman-α absorbers is understood and when the variation of the ambient ionising flux with redshift is determined, a proportion of type 2 absorbers will be found to evolve into type 1 absorbers.

3   Lyman-Limit Systems

Following Tytler's (1982) pioneering work, the redshift evolution of Lyman-limit systems has been studied by various workers (e.g. Bechtold et al. 1984; Lanzetta 1988, 1991; Sargent, Steidel and Boksenberg 1989). These estimates, sometimes based on a variety of data inhomogeneously compiled from the literature, have given contradictory results.

In particular, while Sargent, Steidel and Boksenberg (1989) using an homogeneous data set found no significant evolution in the properties of the absorbing objects over the range $0.67 \leq z_{LLS} \leq 3.58$, Lanzetta (1991) using data from the literature deduced the striking result that beyond $z \sim 2.5$ evolution is extremely rapid. However, as described in detail in Stengler-Larrea et al. (1995), we have found that the appearance of the fit is very sensitive to the specific binning of the data. Using a similar data compilation from the literature we obtain a distribution as shown in Figure 4 together with its maximum-likelihood estimated curve, where there seems no hint of a broken power law, whereas using bins similar to those of Lanzetta (1991) the same data seem to fit poorly to the estimated curve and indeed suggest an underlying broken power law distribution as claimed by him (Figure 5).

Because of the various source material this sample compiled from the literature lacks consistency in the quality selection criteria. We therefore decided to re-examine stringently the available spectra and exclude poor quality and severely biased data, in an attempt to homogenize the quality of our data base. We included only those objects having spectra or portions of spectra with appropriate wavelength coverage which we estimated to be of a quality good enough for the reliable detection of Lyman-limit systems with $\tau \geq 1$. For the IUE data we selected from the spectral compilation in the IUE Uniform Low Dispersion Archive (Courvoisier and Paltani 1992) and for the rest we selected from the spectra published in the original papers (Tytler 1982;

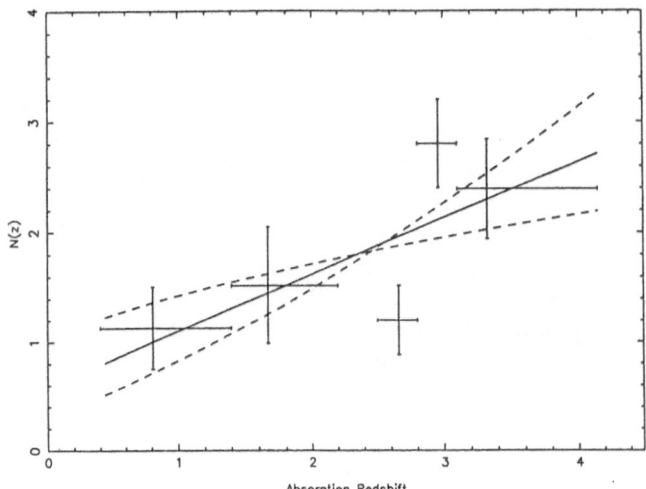

Fig. 4. Redshift evolution of the Lyman-limit systems ($\tau \geq 1$) from a literature data compilation, using 154 objects in total. In this and both the following figures we exclude objects having a Lyman-limit system with apparent "ejection" velocity ≤ 5000 km s^{-1}. The full line represents the maximum-likelihood power law fit to the data with $(dN/dz)_0 = 0.60$ and $\gamma = 0.95$. The dashed lines represent the $1 - \sigma$ errors on these estimates.

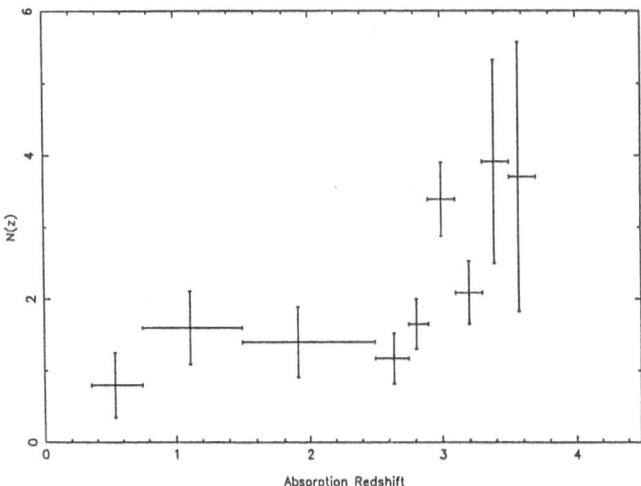

Fig. 5. Redshift evolution of the Lyman-limit systems from the same data sample as in the previous figure, but with binning similar to Lanzetta (1991). The same power law fit as in Figure 1 applies here. Note the scale change relative to Figures 4 and 6.

Lanzetta 1988, 1991; Sargent Steidel and Boksenberg 1989) . To these we added 26 HST objects covering the low redshift region (Bahcall et al. 1993, 1995; Impey et al. 1995). Finally, we added an additional sample of 41 objects with new optical observations from Steidel and Sargent (1995), of which 14 replaced those listed by Lanzetta (1988;1991). The total of 186 objects makes up our revised and extended database. In our analysis we have employed only redshifts for which the velocity relative to the quasar is greater than 5000 km s^{-1}. This reduces our sample to 169 objects.

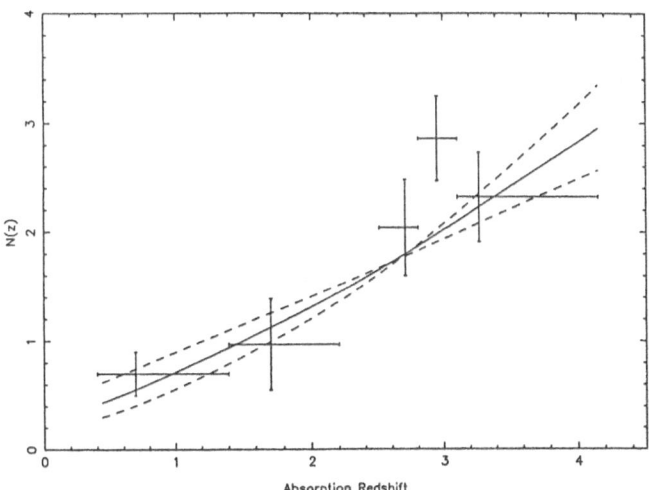

Fig. 6. Redshift evolution of Lyman-limit systems ($\tau \geq 1$) as estimated from our revised and extended sample, using 169 objects in total including 26 objects observed with the HST. The full line represents the maximum-likelihood power law fit to the data with $(dN/dz)_0 = 0.25$ and $\gamma = 1.50$. The dashed lines represent the $1 - \sigma$ errors on these estimates.

The unrevised IUE data, as used by Sargent, Steidel and Boksenberg (1989) and Lanzetta (1991), give at the redshifts covered by the HST sample ($0.4 \leq z_{LLS} \leq 1.4$) $< (dN/dz)_z >= 1.13 \pm 0.38$ (Figure 4), somewhat higher than the value obtained by combining our revised IUE data and the HST data, $< (dN/dz)_z >= 0.70 \pm 0.20$ (Figure 6). The HST data alone give $< (dN/dz)_z >= 0.94 \pm 0.33$ and the revised IUE sample gives $< (dN/dz)_z >= 0.47 \pm 0.23$. The influence of the higher number density at low redshifts obtained with the unrevised IUE sample has the effect of flattening the overall evolutionary behaviour at small redshifts. We note that this is contributory to Lanzetta's (1991) claim that a broken power law is a better description of the evolutionary behaviour. In addition, we observe that in Figures 4 and 5 the region near $z_{LLS} = 2.5$ appears significantly depleted in number of Lyman-limit systems. The apparent underdensity occurs at the inevitably poor-quality short-wavelength end of the optical spectra ($z_{LLS} = 2.5$ at 3200 Å). We paid special attention to the region near $z_{LLS} = 2.5$ when compiling our revised sample. After our removal of poor quality regions from the available spectra a smaller fraction remain with an acceptable quality in the range near the atmospheric cut off.

With our revisions the apparent deficit loses its significance. With our revised database of 169 objects over the whole redshift range $0.32 \leq z \leq 4.11$, we obtain a single power law with $(dN/dz)_0 = 0.25^{+0.17}_{-0.10}$ and $\gamma = 1.50 \pm 0.39$ as shown in Figure 6. This is consistent within errors with a comoving density of absorbers in a Friedmann universe with $q_0 = 0$, but indicates significant evolution if $q_0 = 1/2$. These properties are not greatly different from those found by Steidel and Sargent (1992) for a population of MgII absorbers, over the common redshift interval extending to $z \sim 2$.

References

Bahcall J.N., et al. 1993, ApJS, 87, 1 (Paper I)

Bahcall J.N., et al. 1995, ApJS (submitted)

Bahcall J.N., Jannuzi B.T., Schneider D.P., Hartig G.F., Green R.F. 1992, ApJ, 397, 68

Bahcall J.N., Jannuzi B.T., Schneider D.P., Hartig G.F., Bohlin R., Junkkarinen V. 1991, ApJ, 377, L5

Bechtold J., Green R.F., Weymann R.J., Schmidt M., Estabrook F.B, Sherman R.D., Wahlquist H.D., Heckman T.M., 1984, ApJ, 281, 76

Courvoisier T.J.-L., Paltani, S., eds. 1992, IUE-ULDA Access Guide No. 4A (ESA SP 1153)

Impey C.D., Petry C.E., Malkan M.A., Webb J.K., 1995, in preparation

Lanzetta K.M., 1988, ApJ, 332, 96

Lanzetta K.M., 1991, ApJ, 375, 1

Lanzetta K.M., Bowen D.V., Tytler D., Webb J.K. 1995, ApJ, 442, 538

Lu L., Wolfe A.M., Turnshek D.A. 1991, ApJ, 367, 19

Maloney P. 1992, ApJ, 398, L89

Hoffman G.L., Lu N.Y., Salpeter E.E., Farhat B., Lamphier C., Roose, T. 1993, AJ, 106, 39

Morris S.L., Weymann R.J., Savage B.D., Gilliland R.L. 1991, ApJ, 377, L21

Morris S.L., et al. 1993, ApJ, 419, 524

Sargent W.L.W., Steidel C.C., Boksenberg A., 1989, ApJS, 69, 703

Schneider D.P., et al. 1993, ApJS, 87, 45 (Paper II)

Steidel C.C., Sargent W.L.W., 1995, in preparation

Steidel C.C., Sargent W.L.W., 1992, ApJS, 80, 1

Stengler-Larrea et al., 1995, ApJ, 444, 64 (Paper V)

Tytler D. 1982, Nature, 298, 427

The Gaseous Extent of Galaxies and the Origin of Lyα Absorption Systems

Kenneth M. Lanzetta[1], John K. Webb[2], and Xavier Barcons[3]

[1] SUNY at Stony Brook, Stony Brook, NY 11794–2100, USA
[2] University of New South Wales, P.O. Box 1, Kensington, NSW, Australia
[3] Instituto de Física de Cantabria, Facultad de Ciencias, 39005 Santander, Spain

Abstract. We report current results of an imaging and spectroscopic survey of faint galaxies in fields of *HST* spectroscopic target QSOs. The primary objectives of the survey are to determine the gaseous extent of galaxies and to determine the origin of Lyα absorption systems. The primary results of the survey are that most galaxies are surrounded by extended gaseous envelopes of $\approx 160 \ h^{-1}$ kpc radius and of unit covering factor and that many or most Lyα absorption systems with absorption equivalent widths satisfying $W > 0.3$ Å arise in extended gaseous envelopes of galaxies. We also report some examples of particular interest, including a group or cluster of Lyα-absorbing galaxies at $z = 0.26$ and three "dwarf spiral" Lyα-absorbing galaxies at $z < 0.1$.

1 Introduction

The Lyα absorption systems detected in the spectra of background QSOs are accessible to ground-based observation only at redshifts $z > 1.6$, which is far beyond the realm of galaxies identified in the deepest galaxy redshift surveys. In contrast, these absorption systems are now accessible to *Hubble Space Telescope* (*HST*) observation at just those redshifts $z \approx 0.1 - 0.7$ that galaxies are routinely identified in deep galaxy redshift surveys. This new possibility of observing galaxies and Lyα absorption systems over a common redshift interval provides an unprecedented opportunity to explore the nature and origin of very low column density gas in the universe.

Over the past several years, we and our collaborators have been conducting an imaging and spectroscopic survey of faint galaxies in fields of *HST* spectroscopic target QSOs. The aim of the survey is to combine new observations of galaxies with published and archival observations of absorption systems (e.g. Bahcall et al. 1993) in order to directly determine the gaseous extent of galaxies and the origin of Lyα absorption systems. Initial results of the survey are reported by Lanzetta et al. (1995) and indicate that at redshifts $z < 1$ most galaxies are surrounded by extended gaseous envelopes of $\approx 160 \ h^{-1}$ kpc extent and of roughly unit covering factor and many or most Lyα absorption systems arise in extended gaseous envelopes of galaxies. These results confirm previous speculation that normal galaxies possess extended gaseous halos or disks (Bahcall & Spitzer 1969; Maloney 1992) but are in apparent conflict with the results of a similar survey reported by Morris et al. (1993), which indicate that most Lyα absorption systems are *not* associated with luminous galaxies.

Here we report current results of the survey, which are based on a sample of galaxies and absorption systems several times larger than the initial sample. The current results of the survey strengthen and support the initial results and suggest how the apparent conflict with the results of the survey of Morris et al. (1993) may be resolved. Further results of the survey are reported by Barcons, Lanzetta, & Webb (1995), and results of other similar efforts are reported by Le Brun et al. and Stocke et al. in this volume. Throughout we adopt a dimensionless Hubble constant $h = H_0/(100 \text{ km s}^{-1} \text{ Mpc}^{-1})$ and a deceleration parameter $q_0 = 0.5$.

2   Imaging and Spectroscopic Survey of Faint Galaxies

The observational goal of the survey is to identify in each field under consideration *all* objects with apparent R-band magnitudes satisfying $R < 21.5$ within angular distances to the QSOs of up to several arcmin. At the redshifts $z \approx 0.35$ typical of absorption systems detected by *HST*, this results in sensitivity to galaxies with luminosities greater than $\approx 0.3L_*$ and with impact parameters to the QSOs less than several hundred h^{-1} kpc.

The task of identifying *all* objects in each field under consideration is observationally demanding but allows great flexibility of method. Specifically, it allows us to identify both galaxies that *do* and galaxies that *do not* give rise to absorption systems as well as absorption systems that *do* and absorption systems that *do not* arise in galaxies. This approach is fundamentally different from that of previous surveys that target galaxies that give rise to known absorption systems and eliminates the possibility of the biases inherent in those surveys. The survey is as yet incomplete to the observational goal, but many fields are more than 50% complete to the observational goal within angular distances to the QSOs of ≈ 2 arcmin.

The results reported here are based on a sample of 213 galaxies in 12 fields. Redshifts of the galaxies range from $z = 0.0580$ to $z = 0.8047$, and impact parameters of the galaxies range from $\rho \approx 0$ h^{-1} kpc to $\rho = 733.1$ h^{-1} kpc. A total of 31 galaxies are coincident in redshift to within 1000 km s^{-1} with absorption systems, and a total of 138 galaxies do not give rise to absorption (either Lyα or C IV or both) to within sensitive upper limits (i.e. to within rest-frame equivalent widths satisfying $W < 0.35$ Å).

3 The Relationship Between Coincident Galaxies and Absorption Systems

The survey identifies a large number of galaxies, many of which are coincident in redshift with Lyα absorption systems and many of which are not. What is the relationship between the coincident galaxies and absorption systems? Various lines of evidence indicate that the coincident galaxies are *responsible for* the corresponding absorption systems and are not present as the result of chance coincidence or merely spatially correlated with the absorption systems. This evidence is summarized below.

First, the galaxies and absorption systems are strongly correlated on velocity scales of up to $\approx 250 - 500$ km s^{-1}. This is illustrated in Figure 1, which shows the velocity pairs distribution of the galaxies and absorption systems in the sample. This result demonstrates that the coincident galaxies and absorption systems are not present as the result of chance coincidence but rather are physically associated.

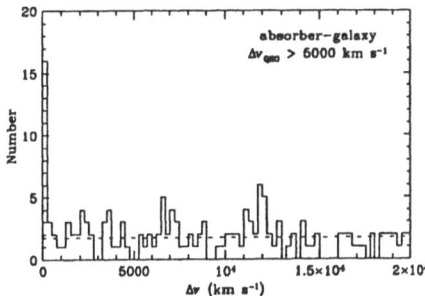

Fig. 1. Velocity pairs distribution of the galaxies and absorption systems in the sample vs. velocity separation Δv.

Second, the coincident galaxies and absorption systems exhibit remarkably small relative velocity differences. This is illustrated in Figure 2, which shows the relative velocity distribution of the coincident galaxies and absorption systems in the sample. The velocity dispersion between the coincident galaxies and absorption systems is 201 km s^{-1}, which includes a substantial contribution from measurement error (typically $\approx 100-150$ km s^{-1}). This result demonstrates that the relative velocities between the coincident galaxies and absorption systems generally span the range < 250 km s^{-1} characteristic of motions within individual galaxies rather than the range $300 - 1000$ km s^{-1} characteristic of motions within groups or clusters of galaxies.

Third—and most important—there is a strong anti-correlation between Lyα absorption equivalent width and impact parameter. This anti-correlation is described in the following section.

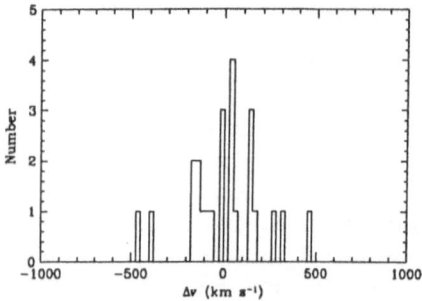

Fig. 2. Relative velocity distribution of coincident galaxies and absorption systems vs. velocity separation Δv.

4   The Gaseous Extent of Galaxies

We first consider the observations from the following point of view: Starting with an unbiased sample of galaxies, what can measurements of the presence or absence of corresponding absorption systems reveal about the gaseous extent of galaxies?

The current observations are illustrated in Figure 3, which shows Lyα and C IV absorption equivalent width versus impact parameter for the galaxies in the sample, excluding those with velocities relative to the QSOs satisfying $v_{QSO} < 5000$ km s^{-1}. (Galaxies in the vicinities of the QSOs are excluded because they exhibit absorption properties different from galaxies far from the QSOs, as is described below.) It is clear that there is a strong anti-correlation between Lyα absorption equivalent width and impact parameter. The anti-correlation is established to a high degree of certainty, even when selected subsamples (for example, galaxies associated with both Lyα and metal ion absorption, or galaxies associated with the strongest Lyα absorption) are arbitrarily removed from consideration. The anti-correlation between Lyα absorption equivalent width and impact parameter strengthens the conclusion that the coincident galaxies are responsible for the corresponding absorption systems and furthermore demonstrates that the absorption systems arise in gaseous structures that are more or less centered on the galaxies.

The observations illustrated in Figure 3 indicate that most galaxies with impact parameters satisfying $\rho < 160 \ h^{-1}$ give rise to Lyα absorption whereas few galaxies with impact parameters satisfying $\rho > 160 \ h^{-1}$ kpc give rise to detectable Lyα absorption. Hence we conclude that most galaxies are surrounded by extended gaseous envelopes of $\approx 160 \ h^{-1}$ kpc radius and of roughly unit covering factor. The observations illustrated in Figure 3 also indicate that some but not all galaxies with impact parameters satisfying $\rho < 65 \ h^{-1}$ kpc give rise to C IV absorption whereas no galaxies with impact parameters satisfying $\rho > 65 \ h^{-1}$ kpc give rise to detectable C IV absorption. Hence we conclude that galaxies are surrounded by extended C IV envelopes of $\approx 65 \ h^{-1}$ kpc radius but that either the fraction of galaxies that possess

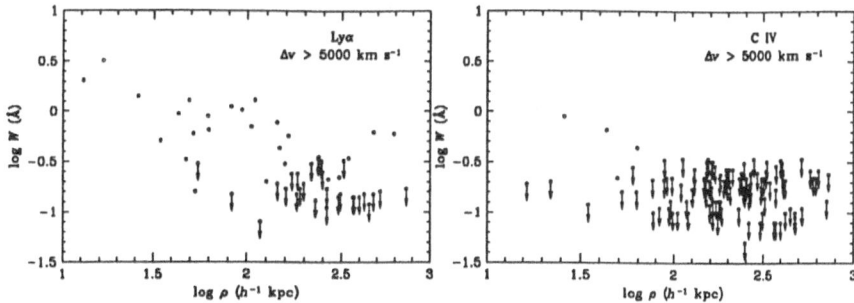

Fig. 3. Logarithm of Lyα (left) and C IV (right) absorption line equivalent width W vs. logarithm of impact parameter ρ.

C IV envelopes or the covering factor of C IV envelopes around galaxies is significantly less than unity.

In principle, the observations illustrated in Figure 3 can be used to determine the geometrical distribution of gas around galaxies. And indeed there is a tentative indication that highly inclined galaxies produce weaker Lyα absorption (at a given impact parameter) than galaxies with low inclination angles, although this result is not statistically significant. If confirmed by future observations, this result would indicate that gas is distributed around galaxies in flattened disks rather than in spherical halos.

We note that galaxies in the vicinities of the QSOs do *not* exhibit an anti-correlation between Lyα absorption equivalent width and impact parameter and do *not* give rise to Lyα absorption with the same frequency or strength as galaxies far from the QSOs. We interpret this as direct evidence of a galaxy "proximity effect," although it is not yet clear whether this proximity effect can be attributed to the radiation fields of the QSOs or to the harsh environments of the galaxy clusters that surround the QSOs.

5   The Origin of Lyα Absorption Systems

We next consider the observations from the following point of view: Starting with a complete, unbiased sample of Lyα absorption systems, what can measurements of the presence or absence of corresponding galaxies reveal about the origin of Lyα absorption systems?

The current observations are summarized in Table 1, which lists the fraction of Lyα absorption systems that arise in galaxies for (1) stronger absorption systems with Lyα rest-frame equivalent widths satisfying $W > 0.3$ Å, (2) weaker absorption systems with $W < 0.3$ Å, and (3) all absorption systems. It is clear that a significant fraction of the stronger Lyα absorption systems arise in galaxies but that apparently a smaller fraction of the weaker Lyα absorption systems arise in galaxies. Of course, the fractions listed in Table 1 *underestimate* the true fractions of Lyα absorption systems that arise in galaxies because the observations are as yet incomplete to the observational goal of the survey and because the observational goal of the survey allows for luminosity and impact parameter thresholds that depend on redshift, so galaxies at large impact parameters or low luminosities could be missed. Hence we conclude that many or most of the stronger Lyα absorption systems arise in extended gaseous envelopes of galaxies whereas a smaller fraction of the weaker Lyα absorption systems arise in galaxies.

Table 1. Identified Lyα Absorption Systems

Condition	Number of Absorbers	Number of Identified Absorbers	Fraction
$W > 0.3$ Å.............	75	25	0.33
$W < 0.3$ Å.............	53	4	0.08
all	128	29	0.23

This difference between the stronger and the weaker Lyα absorption systems is also illustrated in Figure 4, which shows the velocity pairs distribution of the galaxies and absorption systems in the stronger and weaker subsamples separately. It is clear that the stronger Lyα absorption systems are more strongly correlated with galaxies than are the weaker Lyα absorption systems.

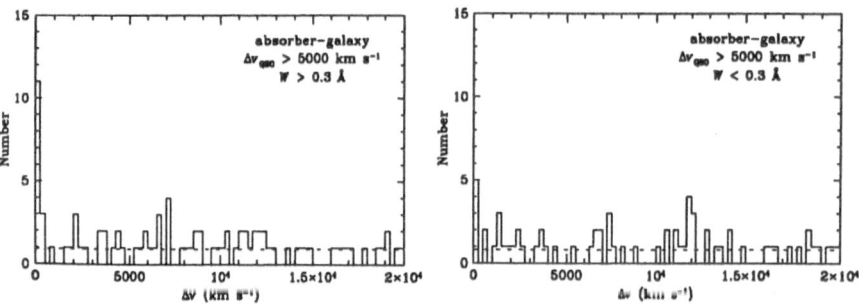

Fig. 4. Velocity pairs distribution of the galaxies and absorption systems in the stronger (left) and weaker (right) subsamples vs. velocity separation Δv.

We believe that it is this difference between the stronger and weaker absorption systems that resolves the apparent conflict with the results of the survey of Morris et al. (1993). The absorption systems considered in our survey were detected with the *Faint Object Spectrograph* (*FOS*) and generally (although not always) have rest-frame equivalent widths satisfying $W > 0.3$ Å, whereas the absorption systems considered in the survey of Morris et al. (1993) were detected with the Goddard High Resolution Spectrograph (*GHRS*) and generally (although not always) have rest-frame equivalent widths satisfying $W < 0.2$ Å. The apparent conflict between the two surveys can thus be understood if stronger absorption systems generally arise in galaxies and weaker absorption systems generally do not.

6   Some Examples of Particular Interest

Here we report some examples of particular interest, including a group or cluster of Lyα-absorbing galaxies at $z = 0.26$ and three "dwarf spiral" Lyα-absorbing galaxies at $z < 0.1$.

6.1   A Group or Cluster of Lyα-Absorbing Galaxies at $z = 0.26$

One of the objectives of the survey is to determine whether or not Lyα absorption systems can arise in groups or clusters of galaxies by explicitly comparing groups or clusters of galaxies with the presence or absence of Lyα absorption systems. This issue is particularly relevant because the "standard" picture (Sargent et al. 1980) maintains that Lyα absorption systems arise in unclustered, intergalactic clouds. Here we report observations of a group or cluster of Lyα-absorbing galaxies at redshift $z = 0.26$.

We have recently identified five galaxies of a group or cluster of galaxies that appears to give rise to a Lyα absorption system at redshift $z = 0.2641$ toward the QSO 1545+2101. We have also identified a galaxy that appears to give rise to a Lyα absorption system at redshift $z = 0.2526$, just 2700 km s^{-1} blueward of the absorption system at $z = 0.2641$. The galaxies and absorption systems occur near the emission redshift of the QSO, and the absorption systems show Lyα absorption but no corresponding metal ion absorption. Impact parameters of the six galaxies range from 47 h^{-1} kpc to 300 h^{-1} kpc. Figure 5 shows a portion of the *FOS* spectrum of 1545+2101 together with tick marks indicating the expected positions of Lyα at the redshifts of the identified galaxies.

The observations presented in Figure 5 provide an exciting but tentative indication that the Lyα absorption line at $z = 0.2641$ is produced by material sharing the virial motion of the group or cluster of galaxies, rather than by material associated with only one galaxy or by material associated with the QSO itself. The redshift centroid and velocity dispersion of the Lyα absorption line ($\langle z \rangle = 0.26406 \pm 0.00012$ and $\sigma = 416 \pm 31$ km s^{-1}) are in excellent agreement with the redshift centroid and velocity dispersion of the galaxies ($\langle z \rangle = 0.26452 \pm 0.00054$ and $\sigma = 285 \pm 89$ km s^{-1}). Furthermore, the Lyα absorption line at $z = 0.2641$ is clearly resolved by the *FOS* spectrum (in contrast to the Lyα absorption line at $z = 0.2526$ and the Galactic Si II λ1526 absorption line, which are unresolved by the *FOS* spectrum), and the absorption profile shows suggestive hints of structure, which is to be expected if it is produced by the cumulative effect of several overlapping galaxies.

What are the implications of this observation? First, it demonstrates that at least *some* Lyα absorption systems arise in groups or clusters of galaxies.

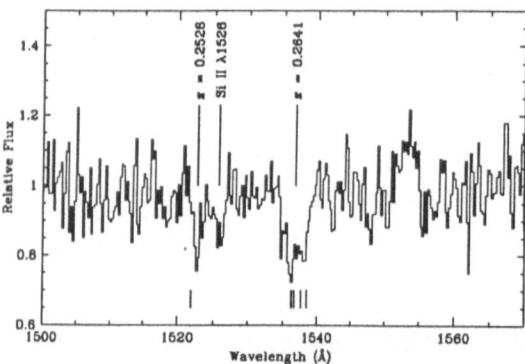

Fig. 5. Portion of the *FOS* spectrum of 1545+2101 (obtained from the *HST* archive) together with tick marks indicating the expected positions of Lyα at the redshifts of the identified galaxies.

This is certainly contrary to the standard picture of Lyα-forest absorption systems as unclustered, intergalactic clouds and indicates that galaxies may be able to maintain significant absorption cross sections even in rich galaxy environments. Second, it suggests that significant clustering of Lyαabsorption systems can be missed due to blending of overlapping absorption lines. This raises the possibility that previous analyses of clustering of Lyα absorption systems may have significantly underestimated the amplitude of the absorption system autocorrelation function on velocity scales of up to several hundred km s^{-1}. We note that a significant fraction of the Lyα absorption systems identified in the survey of Bahcall et al. (1993) have measured velocity dispersions satisfying $\sigma > 300$ km s^{-1}, i.e. velocity dispersions characteristic of motions within groups or clusters of galaxies. If some or all of these absorption systems arise in groups or clusters of galaxies that go unrecognized in the *FOS* spectra, then there may be significant clustering of Lyα absorption systems that is as yet undetected.

6.2 Three "Dwarf Spiral" Lyα-Absorbing Galaxies at $z < 0.1$

Most of the coincident galaxies and absorption systems that have as yet been identified occur at redshifts $z > 0.2$, where galaxies have small angular sizes and faint apparent magnitudes and hence are extremely difficult to study. Hence galaxy and absorption system pairs at redshifts $z < 0.1$ are of particular interest because the spatial and kinematic structure of the galaxies can be resolved by ground-based observations. Here we report observations of three "dwarf spiral" Lyα-absorbing galaxies at redshifts $z < 0.1$.

We have recently obtained accurate ground-based observations of three galaxies that give to Lyα absorption systems at redshifts $z < 0.1$. These observations include accurate optical rotation curves of two of the galaxies and accurate optical images of the three galaxies. Properties of the galaxies are summarized in Table 2, which lists for each galaxy and absorption system pair the galaxy redshift z, Lyα absorption equivalent width W, galaxy impact parameter ρ, galaxy radial scale length r_0, galaxy absolute B-band luminosity M_B, galaxy $B - V$ color, galaxy circular rotation velocity v_c, and galaxy luminosity L. The three galaxies have exponential disk surface brightness profiles, with little or no evidence of central bulges. Details of some of these observations are reported by Barcons, Lanzetta, & Webb (1995).

Table 2. Properties of Three Lyα-Absorbing Galaxies at $z < 0.1$

Property	1545+2101	1704+6048	2135−1446
z	0.0958	0.09163	0.07578
W (Å)	0.18	0.89	0.33
ρ (h^{-1} kpc)	53.4	62.2	47.6
r_0 (h^{-1} kpc)	1.0	1.8	2.9
M_B	−16.0	−17.3	−17.7
$B - V$	0.3	0.6	0.6
v_c (km s^{-1})	...	150	120
L (L_*)	0.04	0.2	0.1

It is clear that the three galaxies are not ordinary, luminous galaxies but rather are underluminous dwarf galaxies. Specifically, the surface brightness profiles, radial scale lengths, $B - V$ colors, circular rotation velocities, and luminosities are consistent with those of extreme late-type "dwarf spiral" galaxies. Thus it appears that the first three Lyα-absorbing galaxies to have been examined in detail are dwarf spiral galaxies.

What are the implications of these observations? First, they suggest that extreme late-type dwarf spiral galaxies contribute significantly to the total cross section for Lyα absorption. If such galaxies are the faded remnants of the "faint blue galaxies" detected in deep imaging surveys, then the steep evolution of the number density of Lyα absorption systems observed at redshifts $z > 1$ may be related to the excess of the number counts of faint blue galaxies above the no-evolution predictions. Second, they demonstrate that at least *some* extreme late-type galaxies possess substantial cross sections at the moderate column densities traced by Lyα absorption. This is somewhat surprising because such galaxies are apparently *not* represented in surveys of galaxies responsible for Mg II absorption systems (Steidel, Dickinson, & Person 1994). It thus appears that some Lyα absorption systems trace a population of galaxies that are not traced by Mg II absorption systems.

7   Conclusions and Prospects for Future Work

We consider the primary results of the survey to be established to a high degree of certainty, namely that at redshifts $z < 1$ most galaxies are surrounded by extended gaseous envelopes of $\approx 160\ h^{-1}$ kpc extent and of roughly unit covering factor and many or most Lyα absorption systems with $W > 0.3$ Å arise in extended gaseous envelopes of galaxies. We believe that the next step is to understand the source of scatter in the relationship between Lyα absorption equivalent width and impact parameter, that is to determine how the presence or absence of gas in the outer parts of galaxies depends on the underlying masses or luminosities, morphological types, past or present star formation rates, and environments of the galaxies. This dependence is crucial to understanding the origin and nature of the extended gaseous envelopes of galaxies. The current results of the survey suggest that this dependence may be complex and that galaxies of a wide range of morphological types (including extreme late-type spiral galaxies) and in a wide range of environments (including groups or clusters of galaxies) may possess extended gaseous envelopes and give rise to Lyα absorption systems. A sample of galaxies and absorption systems significantly larger than the one considered here will be needed to explore these issues.

Finally, we emphasize the possibility of using galaxy and absorption system pairs as probes of galaxy kinematics and dynamics. Galaxy and absorption system pairs are potentially extremely interesting probes of galaxy dynamics because (1) they are identified at projected separations of $\approx 25 - 200\ h^{-1}$ kpc and hence probe spatial scales at which few other dynamical probes have been applied, and (2) they are identified at redshifts as large as $z \approx 0.8$ and hence may provide a unique means of determining the extent and distribution of galaxy dark matter halos at cosmologically significant look-back times.

Acknowledgements. The authors thank the conference organizers for hosting an enjoyable and productive meeting.

References

Bahcall, J. N. et al. 1993, ApJs, 87, 1

Bahcall, J. N. & Spitzer L. 1969, ApJ, 156, L63

Barcons, X., Lanzetta, K. M., & Webb, J. K. 1995, Nature, submitted

Lanzetta, K. M., Bowen, D. V., Tytler, D., & Webb, J. K. 1995, ApJ, in press

Maloney, P. 1992, ApJ, 398, L89

Morris, S. L., Weymann, R. J., Dressler, A., McCarthy, P. J., Smith, B. A., Terrile, R. J., Giovanelli, R., & Irwin, M. 1993, ApJ, 419, 524

Sargent, W. L. W., Young, P. J., Boksenberg, A., & Tytler, D. 1980, ApJS, 42, 41

Steidel, C. C., Dickinson, M., & Persson, S. E. 1994, ApJ, 437, L75

Are Low Redshift Ly-α Absorption Systems Associated with Galaxies ?

M. Rauch[1], R.J. Weymann[1], and S.L. Morris[2]

[1] Observatories of the Carnegie Institution of Washington, 813 Santa Barbara Street, Pasadena, CA 91101, USA
[2] Dominion Astrophysical Observatory, West Saanich Road, Victoria, B.C., V8X 4M6, Canada

Recently the degree of physical association between low redshift Lyα absorption line systems and galaxies has been subject to some debate, and the observational evidence has been interpreted variously as both being in favour of and against the hypothesis that galaxies produce the majority of the low redshift absorption systems (Morris et al. 1993, Mo & Morris 1994, Lanzetta et al. 1994, Stocke et al. 1994).

Summarizing, the "galaxy-hypothesis" (as compared to a mostly intergalactic origin of the absorption) is consistent with the large fraction of high column density systems which appear to occur close to (in redshift space) and within a transverse distance of a couple of 100 kpc from a galaxy. Moreover, there is an anticorrelation between the HI column density of an absorber and the impact parameter to the nearest galaxy neighbour, as expected if there were a physical link. On the other hand, both the velocity autocorrelation of Lyman α absorbers and their cross-correlation with galaxies are weaker than the correlation of galaxies among themselves, and for many low column density systems there is no corresponding galaxy nearby. Obviously, many absorption systems need not arise in galaxies, at least not in those for which these statistical studies are valid.

The possibility remains that *all* absorption at low redshift could be caused by galaxies, but the population giving rise to the bulk of the absorption may consist of low surface brightness (LSB) objects, remaining below the detection threshold of the above-mentioned galaxy surveys. Such LSB galaxies could be drawn from the whole range of the galaxy luminosity function; bright but extended giants like Malin1 (Bothun et al. 1989) could provide the required total absorption probability (product of number density and cross-section) just as well as a population of more numerous but smaller dwarf galaxies. The proximity of the 1012 and 1582 km/s absorbers to the Virgo cluster suggests in particular extended dE/Im "Virgo dwarfs" (Binggeli & Sandage 1984) as possible candidates.

To address this question we have performed a deep broad band (Gunn r) imaging search for low surface brightness (LSB) galaxies in the well-studied field of QSO 3C273. Specifically, we attempted to identify LSB objects responsible for the two lowest redshift Lyman α absorption systems (cz=1012km/s and 1582 km/s) in the line-of-sight (LOS) toward 3C273. We have obtained two mosaic images centered on 3C273, a 15.7'×15.9'image taken with the COSMIC prime focus camera at the Palomar 5m telescope (total exposure time 1800 s), and a larger 53.3'×52.4' mosaic taken at the Las Campanas 1m telescope (total exposure time 3600 s). This larger mosaic provides a radial coverage of $109h_{75}^{-1}$ kpc (cz=1012km/s absorber) or 164 h_{75}^{-1} kpc (cz=1582km/s) away from the LOS to 3C273. To determine the sensitivity for detecting LSB objects we added fake galaxies at random positions and angles and tried to detect them by eye. It turned out that we should have been able to see galaxies down to central surface brightness limits of ∼ 26.4 (26.8) r mag/arcsec^2 in the Las Campanas (Palomar) images. This statement is valid for the size/magnitude range of the known Virgo dwarfs from Binggeli & Sandage (1984), as well as for Malin1 type galaxies, placed at a distance of 16 Mpc (corresponding to cz=1200km/s for h_{75}=1).

Nevertheless, in our real images no galaxy candidates were found. Still - a glance at existing galaxy surveys (cf. in particular Hoffman, Lewis & Salpeter 1995) shows that it is possible to find for each of the two lowest redshift absorbers more than one galaxy with transverse distances several hundreds of kpcs or even Mpcs away from the LOS but remaining within less than 150 km/s of each absorption redshift.

We conclude that at least one of the following must be true to explain these results:
(1) if there are absorbing galaxies within a couple of 100 kpcs from the absorbers they must be of a type even fainter than our detection limit;
(2) the impact parameter of the LOS relative to the absorbing galaxies is even larger than the extent of our images;
(3) in the cases considered here galaxies are not responsible for the absorption.

We suggest our results can be reconciled with previous ones (see above) if the Ly α systems arise in flattened extended large scale structures, with galaxies embedded in the most overdense regions. This picture is consistent with recent structure formation simulations and the large coherence lengths of Ly α absorbers (Miralda-Escudé et al., Petitjean, Mücket & Kates, Dinshaw et al., Rauch & Haehnelt, this workshop). At small galaxy impact parameters, the absorbers are embedded in hot virialized spherical gaseous halos; away from galaxies the absorbing gas is associated with extended, flattened, infalling, unvirialized structures. These may be the sites of the numerous very low column density absorbers.

References

Bothun, G.D., et al., 1987, AJ, 94, 23
Hoffman, G.L., Lewis, B.M., Salpeter E.E., subm. to ApJ, 1995
Mo, H. & Morris, S. L. 1994, MNRAS, 269, 52
Morris, S.L., et al. 1993, ApJ, 419, 524
Sandage, A., and Binggeli, B., 1984, AJ, 89, 919

Gravitational Lenses and Quasar Absorption Lines

Alain Smette

Kapteyn Astronomical Institute, P.O. Box 800, NL-9700 AV, Groningen

Abstract. The properties of gravitational lensing are first summarized. We then consider the relations existing between this phenomenon and the quasar absorption lines. We show how the absorption lines can lend additional support to the lens interpretation of multiply imaged quasars and how they can constrain the physical characteristics of the lensing galaxy. We present preliminary results of a study on the occurrence of associated systems in multiply imaged quasars and new results on the size of absorption systems based on recent spectral studies of the Double Hamburger HE $1104-1805$; in particular, we set a 2 σ lower limit to the diameter of Ly-α clouds of 100 h_{50}^{-1} kpc. We recall that micro-lensing in BAL quasars may help in discriminating between different scenarios for the mechanism(s) responsible for the BAL. Magnification bias due to gravitational lensing may affect the inferred importance of the proximity effect, and, combined with the described 'by-pass' effect, the observed number density and statistical properties of damped Ly-α systems.

1 Introduction

The presence of gaseous matter close to the line of sight to distant quasars gives rise to (at least) two phenomena:

1. Formation of absorption lines in the quasar spectra,

2. Gravitational Lensing.

As these Proceedings describe recent studies of quasar absorption lines, we first only summarize the formation and properties of the gravitational lensing phenomenon. See also the Proceedings of the 31st Liège International Astrophysical Colloquium (1993), hereafter LIAC93, Schneider, Ehlers, & Falco's 'Gravitational Lenses' (1992), or Refsdal & Surdej (1994) for a recent review. The following sections deal with 'What can gravitational lenses bring to the absorption line studies?', and 'What can absorption lines tell us about lenses?'

2   Gravitational Lenses

Figure 1 presents the geometry of a typical single lens system. The lens equation can be written as

$$D_{OS}\theta_I = D_{OS}\theta_S + D_{LS}\alpha(M, D_{OL}, \theta_I),\qquad(1)$$

where I refers to the considered image (we mainly deal with single or double quasar images, so that I = A or B), D_{OL}, D_{OS} and D_{LS} are the angular-diameter distances between the observer and the lens, the observer and the source, and the lens and the source in a Lemaître Universe (i.e., described by the Friedmann-Robertson-Walker metric); θ_I is the angle between the quasar image and the (gravity) center of the lens; θ_S refers to the angle between the actual position of the quasar and the lens i.e., its true position in the absence of gravitational lensing; the deviation angle itself is measured by α. All these angles are supposed to be very small. An additional angle of interest is the so-called Einstein radius, $\theta_E = (D_{LS}/D_{OS})\alpha$ (for an axisymmetric lens) which is easily derived by setting $\theta_S = 0$ in Eq.(1). The corresponding image is a ring, called an Einstein ring.

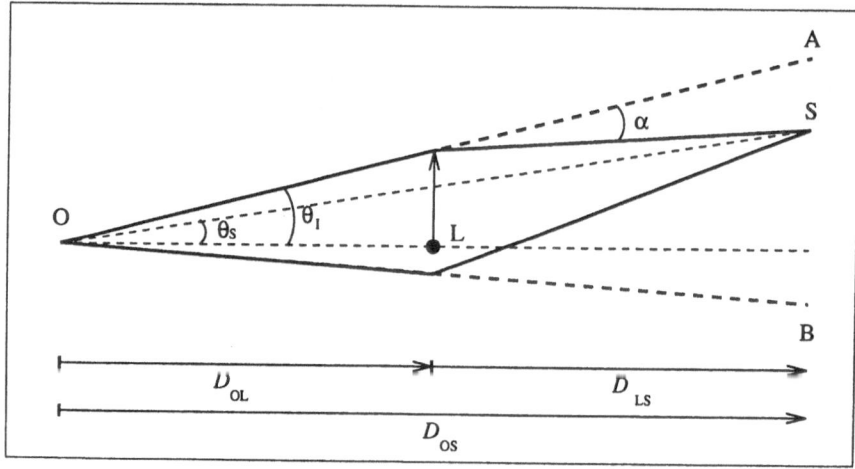

Fig. 1. Geometry of a lensed system

The main properties of gravitational lensing can be summarized as follow:

1. The quasar image is displaced: it does not indicate the real position of the source on the sky. Furthermore, compared to an 'unlensed' situation, the light-path avoids to cross (the center of) extended lenses ('by-pass effect').

2. Formation of multiple images can occur if $\theta_S < \theta_E$, for axisymmetric lenses. For more generic lenses, the source has to lie within a diamond-like curve, called a caustic (cf Schneider, Ehlers, & Falco 1992). If an odd number of images is expected in astrophysical situations (Burke 1981), one can show that at least one must be very faint and situated very close to the center of the lensing galaxy: thus it is not surprising that only 2 and 4 images configurations have been observed so far.

3. In a multiply imaged quasar, we are provided with at least two close but distinct lines-of-sight towards the quasar. The separation $S(z)$ between them can be determined for a given set of cosmological parameters by the observed angular separation between the two images and by the redshifts of both the lens and the quasar (McGill 1990, Smette et al. 1992) For $z > 0.5$, $S(z)$ is rather constant for a pair of distinct QSOs; on the contrary, $S(z)$ decreases to 0 for $z > z_L$ in the case of a multiple QSO images.

4. For the case of a multiply imaged quasar, the different light paths do not have the same length; furthermore, light is 'slowed down' (as seen by a remote observer) in the gravitational well of the galaxy, so that a variation in the observed flux of the source will be detected at different epochs for the different images. Such time-delays are the basic observables to constrain H_0 (and q_0) and the mass of the lensing galaxy (Refsdal 1964, 1966).

5. Image amplification: the total brightness of the image(s) is affected, as its (their) angular area is modified by lensing, while its (their) surface brightness is not. In astrophysical cases, at least one image is amplified (Schneider 1984). The amplification factor depends on θ_S, the relative size of the source and of the Einstein radius projected on the source plane $D_{OS}\theta_E$; it can be very important if the source is very small and (intrinsically) well aligned behind the lens. This amplification is achromatic, as the bending is independent of the wavelength. However, if the source presents wavelength-dependent features, chromatic effects may occur. Micro-lenses can lead to such situations: a micro-lens is a star in an intervening galaxy which lies exactly on the actual light-path towards the quasar; one can show that the $D_{OS}\theta_E$ can be comparable to the region that gives rise to the continuum spectrum of the quasars, while being much smaller than the emission line region. In such a case, the quasar continuum spectrum can be preferentially amplified with respect to the emission lines.

6. Amplification bias: a consequence of the previous point is the amplification (or magnification) bias. This bias is caused by the enhancement of the probability that a quasar of given magnitude is lensed. It is due to the lensing amplification of intrinsically fainter quasars to the observed quasar magnitude, and affects all quasar samples that are magnitude-limited. The amplification bias factor strongly depends on the steepness of the intrinsic quasar luminosity function at the bright end. Fukugita & Turner (1991) have estimated that the probability to find a multiply lensed quasar in a sample of quasars brighter than $m_B = 17$ is 25 times larger than the number expected if each source represented a random line-of-sight.

3 Absorption Lines to Support the Lens Hypothesis of a Multiply-Imaged QSO

Absorption lines can bring evidence or proof for the gravitational lens nature of a candidate, generally in much the same ways as used by Walsh, Carswell & Weymann (1979) to argue for the lens explanation of the 'Old Faithful' Q 0957+561 (more references on the following individual cases can be found in Surdej & Soucail 1993 and Pospieszalska-Surdej, Surdej & Véron 1993):

1. The Broad Absorption Lines in H 1413+117 – the cloverleaf quasar – bring very strong evidence that the 4 images are images of the same quasar (Magain et al. 1988).

2. UM 425 A has recently been identified as another BAL quasar (Michalitsianos & Oliversen 1995); a UV spectrum of the other images (especially B) could therefore bring strong support for the lensing hypothesis, alike the case for H 1413+117.

3. Absorption lines in the combined spectra of Q 1208+1011 A & B reach zero intensity argue for the fact that they are two quasar images (but can't discriminate between the pair and the lens hypotheses) (Magain et al. 1992).

4. Associated systems were found to vary in UM 675 (Hammann et al. 1995, cf also Burbidge, these Proceedings) and 3C 205 (Aldcroft et al. 1994). In the case of UM 675, Hammann et al. showed that they can only arise very close (~ 200 pc) to the quasar continuum source. If all associated systems are produced in small clouds, their presence in the spectra of the images of a gravitationally lensed quasar candidate is a further strong argument for the lensing interpretation. Such an argument has been used for HE 1104-1805 described below (Smette et al. 1994). On the other hand, its absence in one image and not in the other is not sufficient to dismiss the gravitational lens hypothesis, except if such a situation lasts for a longer time than the time-delay expected between the two lensed images. If associated systems are produced by small clouds, differences in the spectra of two images can also result from micro-lensing effects (Surdej, private communication).

5. When plotted as a function of z, absolute or relative differences in the equivalent widths for the CIV and MgII absorption line doublets (depending on the quasar redshift) detected in the spectra of two images of a gravitational lens candidate can bring support for the lensing hypothesis. Small differences are expected if $S(z) \approx 20\ h_{50}^{-1}$ kpc (for $z > 0.5$): no variation with z is expected for a pair of distinct QSOs, while differences should decrease with z in the case of a gravitationally lensed QSO. This behavior has been noted for HE 1104-1805 (Smette et al. 1994, see below).

6. A double quasar, suspected to be gravitationally lensed, can sometimes provide a large sample of Lyα cloud lines for each image. For the purpose of the argument, one assumes the null hypothesis, i.e. that each component is a distinct quasar. A corresponding (lower) limit D_{min} for the size of the Ly-α clouds can then be derived following the procedure described in McGill

(1990) or Smette et al. (1992). D_{min} can then be compared with the separation S between the two members of a known, real pair of distinct quasars for which no common Lyα cloud lines in their spectra has been detected. Obviously, $D_{min} > S$ excludes the null hypothesis, and thus favors the lens interpretation. This argument has also been used for HE 1104-1805 (Smette et al. 1994, see below).

7. On the other hand, the MgII doublet is *not* a good indicator of the lens redshift when the lensing agent is an elliptical galaxy, as it is usually the case. Indeed, this doublet is not detected in the spectra of any accepted lensed quasars for which the lens is an elliptical galaxy whose redshift has been directly obtained, and for which spectra covering the suitable wavelength range, resolution and S/N exist: there is no MgII at the lens redshift $z_l = 0.36$ in the spectra of Q 0957+561 A or B, with a 3σ upper limit of $\simeq 0.3$ Å on the rest equivalent width of the $\lambda2796$ line (Young et al. 1981); nor at $z_l = 0.49$ in the spectra of UM 673 A & B, with 3σ limits of 0.06 Å in A and 0.4 Å in B (Smette et al. 1992); nor at $z_l = 0.29$ in the spectrum of PG 1115+080 A with a 3 σ upper limit of $\simeq 0.1$ Å (Young, Sargent & Boksenberg 1982). Impact parameters range from $\simeq 2$ to $\simeq 36\ h_{50}^{-1}$ kpc. Note that in these cases the CIV doublet, if it is strong enough, should fall below the atmospheric cutoff and waits for HST spectra. However, a MgII doublet due to the lensing galaxy (at $z_l = 0.6847$) appears in the spectrum of the optical counterpart of the Einstein ring B 0218+357 (Browne et al. 1993): in this case, the lens is most probably a spiral galaxy (cf. below for more details on this system, or Wiklind, these Proceedings).

4   Proximity Effect and Gravitational Lensing

The fact that gravitationally lensed quasars are amplified can have the following consequences for the study of the proximity effect:

1. as the selection of quasars for the observations are usually based on their bright observed magnitude, the magnification bias is certainly present: compared to a redshift limited sample, a magnitude-limited one has a larger probability of containing gravitationally amplified quasars. This leads to an overestimation of the quasar flux at the Lyman limit (Schneider, Ehlers, Falco 1992; Bechtold 1994); however, the expected average amplification is probably not very important (10 % ?), except if the sample contains a large number of multiply imaged or micro-lensed quasars: this situation can occur for a sample of very luminous quasars.

2. the amplification probability $p(A)dA$ that a quasar has its flux increased by a factor A due to gravitational lensing is an important function that enters in the theoretical evaluation of the magnification bias. A detailed comparison between the observed importance of the proximity effect in a quasar and the expected one based on its apparent magnitude could theoretically determine the amount of amplification A that affects its apparent

magnitude. A suitably defined sample of quasars could thus provide most welcome observational constraints on the shape of $p(A)dA$. Practically, one has to deal with small numbers of Ly-α lines and thus poor statistics, and on the other hand, with a number of other, not fully understood, phenomena affecting the determination of the importance of the proximity effect (Bechtold, these proceedings).

3. however, it is more likely that such comparison could select quasars that are subject to a large gravitational lensing amplification, as it occurs in multiply imaged or micro-lensed quasars (Schneider, Ehlers, Falco 1992).

5   Associated Systems and Gravitational Lensing

Associated systems (mainly revealed by CIV doublets for which $z_{abs} \approx z_{em}$) are the source of some debates relative to their origin. The Foltz et al. (1986) interpretation that they are mainly associated with radio-loud quasars is not supported by new studies (Barthel 1991). Møller & Jakobsen (1987) suggested that they arise in intrinsically faint quasars ($M_V < -27$).

As multiply imaged quasars sample the underlying population of quasars with usually the same selection effects as those that affect quasar surveys, they can be used to test this hypothesis, as the lensed quasars are intrinsically fainter than estimated from their apparent magnitude.

There are 10 accepted or proposed cases of multiply imaged, non-BAL quasars for which there exist spectra covering the CIV emission line (Surdej & Soucail 1993). Out of the 10, 8 present asssociated ($|z_{abs} - z_{em}| < 0.2$) CIV systems, leading to a sample of 12 systems. This is a factor 6 larger than expected using the number density derived from sample A1 from Sargent, Boksenberg & Steidel (1988); there is apparently no link with the radio-loudness of the quasar. We refer to Smette, Møller & Surdej (1995) for more details.

6   Multiply Imaged Quasars and the Size of Absorption Systems

The property that multiply imaged quasars provide nearby, but distinct lines-of-sight is certainly the one that has been mostly used so far for the QSO absorption line studies. Indeed, close pairs of QSOs are very seldom and often very faint. On the other hand, gravitational lenses offer close lines-of-sight while usually at least one image being quite bright due to lens amplification. Walsh, Carswell & Weymann (1979) and Weymann et al. (1979) were the first to recognize the importance when they obtained spectra of the 'Old faithful' Q 0957+561.

If the extent of MgII and CIV systems confirmed by such observations is in itself very interesting, the use of two close lines-of-sight is crucial to determine important lower limits to the Lyα cloud size. Weymann & Foltz (1983) obtained a first lower-limit of $\sim 0.5 \ h_{50}^{-1}$ kpc, as an absorption line appears in the spectra of PG 1115+080 A & C. The values obtained by Foltz et al. (1984) using the (unfortunately faint) images A & B of the candidate gravitational lens Q 2345+007 were finally interpreted in very diverse ways. As several Lyα lines are common, while at least one is not, they derived a lower limit of about 5-20 $\ h_{50}^{-1}$ kpc for the transverse extent of the Lyα clouds. This was either considered as a typical size or even as a typical scale for clusters of smaller clouds.

With UM 673 A & B, Smette et al (1992) disposed of the first lens for which the geometry is fully known ($z_l = 0.49$, $\theta = 2.2''$) while the quasar redshift is large enough ($z_{em} = 2.727$) that a large portion of the Ly-α forest is accessible from ground-based spectroscopy. The brightness of the two images allowed to obtain 2Å resolution spectra. Except for 2 lines that can be interpreted as a possible MgII doublet, all Lyα lines are common in both spectra, with identical equivalent widths and velocities within the noise. This proves that the 2 lines-of-sight actually cross the same clouds. A firm 2σ statistical lower-limit of 12 h_{50}^{-1} kpc was set for spherical clouds.

With Gordon Robertson, Peter Shaver, Dieter Reimers, Lutz Wisotski and Thomas Köhler (Smette et al. 1994), we recently completed a spectroscopic study of the new gravitationally lensed quasar candidate HE 1104-1805 (the Double Hamburger).

Our 1.2 Å resolution AAT spectra of HE 1104-1805 A & B cover the range 3175 - 7575 Å. The characteristics of this object ($z = 2.31$, $m_B = 16.7$ and 18.6, separation = 3.0 arcsec; $cf.$ Wisotzki et al. 1993) allows good S/N data. Special care was taken to extract the spectra via a 2-steps Gaussian modeling.

We find that the Ly-α absorption lines are strongly correlated in equivalent width, suggesting that the two lines of sight pass through the same clouds in all cases, and there are no significant differences in velocities within $\simeq 10$ km s^{-1} between corresponding pairs of lines. From the 72 Lyα lines with $W_{rest} > 0.085$ Å (and $\lambda > 3395$ Å) detected at 5σ in A and 3σ in B we statistically derive a 2σ lower limit of 100 h_{50}^{-1} kpc for the diameter of spherical Ly-α clouds, assuming that the lens redshift $z_L > 1$, $H_0 = 50 \ h_{50}$ km s^{-1} Mpc^{-1}, $q_0 = 0.5$, and $\Lambda = 0$. Similarly, 2σ lower limits of 100, 60 and 20 h_{50}^{-1} kpc are obtained for the region inside the clouds giving rise to lines with $W_{rest} > 0.17$, 0.32 and 0.60 Å respectively. These values and the strong correlation between Ly-α cloud line equivalent widths are hardly compatible with the predictions given by hydro-dynamical evolutions for the spherical mini-halo model (Meiksin 1994).

For the systems causing the CIV and MgII absorption lines, we find 2σ lower limits of 28 and 22 h_{50}^{-1} kpc respectively for $W_{rest} > 0.3$ Å. Again the equivalent widths of corresponding lines are well correlated, though some lines are not seen in both spectra; the redshift differences between corresponding pairs of lines are small.

A damped Ly-α system is present in one spectrum (A) but not in the other (B), indicating a typical size of the order of the separation of the two lines of sight at that redshift ($z_{abs} = 1.6616$), $\sim 8 - 25h_{50}^{-1}$ kpc depending on the actual geometry of the system. Some high-ionization lines in this system have similar characteristics in both spectra, while all low-ionization lines are much weaker in the spectrum of B. This suggests that, while the region causing the low-ionization lines may be similar in size to the damped Ly-α region, that causing some of the high-ionization lines must be considerably larger.

The Double Hamburger is a peculiar system: the emission lines perfectly agree in redshifts and shapes, but the ratio of the corresponding equivalent widths increases with wavelength. Wisotzki et al. (1993) showed that the difference $f_A - 2.8f_B$ leaves a featureless residual in the range 4000 – 9000 Å, well approximated by a power-law with steeper index than any of the two original spectra. Their interpretation is that a star in an intervening galaxy, located along the line-of-sight towards the A image, is micro-lensing the continuum source of the QSO.

Our spectra extend more to the blue and show that the representation of $f_A - 2.8f_B$ as a power-law cannot be extended below the Ly-α emission line, corresponding to a break in the continuum spectrum of A, supported by IUE observations (Reimers et al. 1994). If micro-lensing is actually at work, two scenarios are possible: the Einstein radius (projected on the source plane) of the micro-lens is very similar to the region giving rise to the quasar continuum region, whose size furthermore depends on wavelength. In such a case, our spectra indicate that the continuum region size is relatively constant between 970 and 1215 Å (in the rest frame) (leading to an amplification factor $\mu \approx 3$) and gradually increase with longer wavelengths (so that the amplification decreases). Another possibility is that the micro-lens greatly ($\mu > 10$) amplifies a new component of the quasar continuum region, with a wavelength-independent size. Such a component does probably never contribute much to the quasar spectra, and thus usually stays unnoticed. The spectrum of this new component is revealed here by the residual $f_A - 2.8f_B$. Such micro-lensing scenarios can obviously occur also for singly-imaged quasars, leading to important consequences: as the continuum spectrum would present a break at Ly-α, power-law based extrapolation of the continuum spectrum redward of Ly-α would lead to an over-estimation of the Gunn-Peterson effect or of the average Ly-α decrement D_A in such particular quasars.

7   Damped Ly-α Systems and Gravitational Lensing

If damped Ly-α systems are produced by neutral hydrogen located in the disk of spiral galaxies, their number density may be affected by several biases due to gravitational lensing. Note that the 3 first biases listed below are more likely to affect low-z systems (say, between 0.2 and 1, the exact value depending on the quasar redshift and (weakly) on the luminosity of the galaxy). The first two are linked the by-pass effect. Assuming that a galaxy can be modeled as a Singular Isothermal Sphere where σ equals the one-dimensional velocity dispersion, it is easy to show that the linear displacement l is equal to the Einstein radius projected on the lens plane and equals $l = (D_{OL} D_{LS}/D_{OS})(4\pi\sigma^2/c^2)$. Furthermore, the Tully-Fisher relation leads to $\sigma = \sigma_*(L/L_*)^{0.34}$ (Fukugita & Turner 1991), so that $l = 2.9\ 10^{-6}\ (D_{OL}\ D_{LS}/D_{OS})\ (L/L_*)^{0.68}$. A typical value is $l \simeq 2\ h_{50}^{-1}$ kpc for an L_* galaxy at $z = 0.25$ in front of a QSO at $z = 1$. Note that l depends on the quasar redshift and on the lens (i.e., damped Lyα system) redshift and luminosity!

1. The line-of-sight avoids the central part of the galaxy (by-pass effect): the inner disk of radius l is then never probed by the line-of-sight towards the quasar (except by faint images, in the case of a multiply imaged quasar): as a consequence, a large number of large column density and molecular systems can simply be missed, possibly affecting the determination of the neutral hydrogen content of the Universe.

2. Lines-of-sight that would have crossed the outer edge of the galaxy can be deviated in a region such that the column density is smaller than the minimal value necessary to enter a given sample, thus affecting the expected number of damped Ly-α systems.

The next two biases are magnification biases: a magnitude-limited sample of quasars is more likely to present a damped Ly-α system than would be expected if each line-of-sight was randomly distributed, either due to:

3. the damped Ly-α system as a whole (in which case it can counter-act the previous bias), or,

4. at higher-z (where the by-pass effect is less important, so that lines-of-sight can 'again' cross the central part of the galaxy) due to micro-lensing by individual stars, that are more concentrated in the very center of the galaxy. Amplification bias due to micro-lensing also depends on the inclination of the galaxy! A detailed account shall be given in Smette, Claeskens & Surdej (1995).

8   Multiply Imaged Quasars and Magnetic Field in Intervening Systems

When linearly polarized radiation passes through a Faraday active plasma, its polarization angle is rotated by an amount $\lambda^2 RM$, where λ is the observed wavelength and the rotation measure, RM, is the integrated value along the light path of the product of the electron number density n_e and line-of-sight magnetic field strength $B_\parallel$. Singly imaged quasars only allow statistical studies of the residual (i.e., Galaxy-corrected) rotation measure (RRM) caused by intervening clouds. Multiply imaged quasars often offer the possibility that two lines-of-sight cross the lensing galaxy with very different impact parameters, and thus offer a unique way of measuring the RRM due to the (lensing) galaxy by making the difference between the RRM of each image (Welter, Perry & Kronberg 1984, Greenfield, Roberts & Burke 1985, Perry, Watson & Kronberg 1993 and references therein). A difference of 100 rad m^{-2} between the RRM of Q 0957+561 A & B has been obtained. There is however some discussion in the literature about its origin as only two systems appear in optical spectra at $z = 1.3913$ and $z = 1.1249$. One expects that the lensing galaxy at $z = 0.36$ produces a damped Ly-α in front of B (if this cD galaxy contains enough neutral hydrogen) and be the main responsible for the differences in RRM due to the differences in impact parameter ($\simeq 6$ and 31 h_{50}^{-1} kpc); however, its computed wavelength $\lambda \simeq 1650$ Å would make it fall blueward of the observed Lyman Limit decrement caused by the $z = 1.3913$ system (Michalitsianos et al. 1993); search for the corresponding 21cm absorption line in front of A and B would be worthwhile in order to estimate the magnetic field in the lensing galaxy.

In fact, all lensed radio-sources, and especially the Einstein rings, can map the variation of the RM across the observed image(s). A nice example is given by Chen & Hewitt (1993) for MG 1131+0456. We note also the value of 870 rad m^{-2} for the difference between the RM of components A & B of 'the smallest Einstein ring' B 0218+357, (Patnaik et al. 1993, O'Dea et al. 1992), for which 21 cm HI (Carilli, Rupen & Yanny 1993) CO, HCO^+ and HCN (Wiklind & Combes 1994, Wiklind, these proceedings) absorption lines at the lens redshift ($z = 0.68466$) have been detected, providing strong constraints on the inclination and position angle of the lensing galaxy and on its interstellar medium.

9 Micro-Lensing and the BAL Region

The cloverleaf H1413+117 (Magain et al. 1988) is a famous BAL QSO, that presents 4 images separated by $\approx 1''$. Micro-lensing of the D image was invoked to explain the observed differences between the NV, CIV, SiIV and AlIII BALs in this image compared to the other ones (Angonin et al. 1990). Hutsemékers (1993) showed that monitoring of the micro-lensing induced variations of the BAL profile can discriminate between the various scenarios invoked for the origin of the BAL.

10 Magnification Bias and Absorption Lines

Additionally to the described effects on the number density and statistical properties of the damped Ly-α lines, the magnification bias may have an important influence on the statistics of other absorption lines. Except for York et al. (1991), who found a hint in this direction, especially for $z \sim 2 - 3$ quasars, most observational studies do not find any correlation between the presence of absorption lines and the quasar magnitude. Let us however quote two additional results.

Nottale (1987) selected 69 quasars in the Asiago catalog, whose redshifts lie in the range $1.6 < z < 2.2$ and that present absorption lines in their spectra. He found a highly significant correlation between their absolute magnitude and the predicted magnification due to gravitational lensing, assuming that these absorption lines arise in rich galaxy clusters (typically 0.5 Mpc from the center of a cluster of $3 \cdot 10^{15} M_\odot$).

In an attempt to find indications for gravitational lensing in samples of narrow metal absorption lines, Borgeest & Mehlert (1994) found a strong dependence of the narrow CIV absorption line density on the quasar emission redshift z_e; most significant is a break at $z_e \simeq 2.8$, in the sense that the number density of CIV systems is smaller in quasars with $z_e > 2.8$ than in those with $z_e < 2.8$. The most plausible explanation seems to be a strong selection effect varying with z_e. They note that for $z_e > 2.8$, most quasars are selected with respect to the detection of the Lyα emission line in objective survey: high redshift quasars that present strong Lyα emission line are also certainly the ones that show less gas in the direction of the lines-of-sight. Their result suggests that the association hypothesis for at least a large fraction of the absorbers is more probable.

Acknowledgements. It is a great pleasure to thank my colleagues Gordon Robertson, Peter Shaver, Dieter Reimers, Lutz Wisotski and Thomas Köhler, Jean-François Claeskens, Palle Møller and Jean Surdej for their collaboration, and the ΦV for their friendship. I am grateful to Jean Surdej and Peter Barthel for their critical reading of the manuscript. I acknowledge financial support under grant no. 781-73-058 from the Netherlands Foundation for Research in Astronomy (ASTRON) which receives its funds from the Netherlands Organisation for Scientific Research (NWO).

References

Aldcroft, T.L., Bechtold, J., Elvis, M.: 1993, ApJS, 93, 1

Angonin, M.-C., Remy, M., Surdej, J., Vanderriest, C.: 1990, A&A, 233, L5

Barthel, P.: 1991, Proceedings of the ESO Mini-Workshop on "Quasar Absorption Lines" held in Garching, ESO Scientific Report no 9, p. 79

Bechtold, J.: 1994, ApJS, 91, 1

Borgeest, U., Mehlert, D.: 1993, A&A, 275, L21

Browne, I.W.A., Patnaik, A.R., Walsh, D., Wilkinson, P.N.: 1993, MNRAS, 263, L32

Burke, W.L.: 1981, ApJ 244, L1

Carilli, C.L., Rupen, M.P., Yanny, B.: 1993, ApJ, 412, L59

Chen, G.H., Hewitt, J.N.: 1993, Aj, 106, 1719

Foltz, C.B., Weymann, R.J., Roeser, H.J., Chaffee, F.H.: 1984, ApJ, 281, L1

Foltz, C.B., Weymann, R.J., Peterson, B.M., Sun, L., Malkan, M.H., Chaffee, F.H.: 1986, ApJ, 307, 504

Fukugita, M., Turner, E.L.: 1991, MNRAS, 253, 99

Greenfield, P.E., Roberts, D.H., Burke, B.F.: 1985, ApJ, 293, 370

Hammann, F., Barlow, T.A., Beaver, E.A., Burbidge, E.M., Cohen, R.D., Junkkari-
    nen, V., Lyons, R.: 1995, ApJ, in press
Hutsemékers, D.: 1993, A&A, 280, 435
Magain, P., Surdej, J., Swings, J.-P., Borgeest, U., Kayser, R., Kühr, H., Refsdal,
    S., Remy, M.: 1988, Nature, 334, 325
Magain, P., Surdej, J., Vanderriest, C., Pirenne, B., Hutsemékers, D.: 1992, A&A,
    253, L13
Michalitsianos, A.G., Oliversen, R.J.: 1995, ApJ 439, 599
Michalitsianos, A.G., Nichols-Bohlin, J., Bruhweiler, F.C., Kazanas, D., Kondo, Y.,
    De La Pena, M., Maran, S.P., Meylan T., Perez, M., Thompson, R.: 1993, ApJ,
    417, L57
McGill, C.: 1990, MNRAS, 242, 544
Meiksin, A.: 1994, ApJ, 431, 109
Møller, P., Jakobsen, P.: 1987, ApJ, 320, L75
Nottale, L.: 1987, Annales de Physique, Col. 2, Suppl.5, 12, 241
O'Dea, C.P., Baum, S.A., Stanghellini, C., Dey, A., Van Breughel, W., Deustua,
    S., Smith, E.: 1992, AJ, 104, 1320
Patnaik, A.R., Browne, I.W.A., King, L.J., Muxlow, T.W.B., Walsh, D., Wilkinson,
    P.N.: 1993, MNRAS, 261, 435
Perry, J.J., Watson, A.M., Kronberg, P.P.: 1993, ApJ, 406, 407
Pospieszalska-Surdej, A., Surdej, J., Véron, P.: 1993, in LIAC93, eds Surdej et al.,
    p. 671
Refsdal, S.: 1964, MNRAS, 128, 307
Refsdal, S.: 1966, MNRAS, 132, 101
Refsdal, S., Surdej, J.: 1994, Reports on Progress in Physics, 56, 117
Reimers, D., Rodriguez-Pascual, M., Hagen, H.-J., Wisotzki, L.: 1994, A&A, 293,
    L21
Sargent, W.L.W, Boksenberg, A., Steidel, C.: 1988, ApJS 68, 359
Schneider, P.: 1984, A&A, 140, 119
Schneider, P., Ehlers, J., Falco, E.E.: 1992, 'Gravitational Lenses', Springer-Verlag
Smette, A., Surdej, J., Shaver, P.A., Foltz, C.B., Chaffee, F.H. et al.: 1992, ApJ,
    389, 39
Smette, A., Robertson, J.G., Shaver, P.A., Reimers, D., Wisotzki, L., Köhler, Th.:
    1994, A&A, submitted
Smette, A., Claeskens, J.F., Surdej, J.: 1995, in prepation
Smette, A., Møller, P., Surdej, J.: 1995, in preparation
Surdej, J., Soucail, G.: 1993, in LIAC93, p. 205
Surdej, J., Refsdal, S.: 1994, Rep. Prog. Phys., 56, 117
Walsh, D., Carswell, R.F., Weymann, R.J.: 1979, Nature, 279, 381
Wiklind, T, Combes, F.: 1994, preprint
Wisotzki, L., Köhler, T., Kayser, R., Reimers, D.: 1993, A&A, 278, L15
Welter, G.L., Perry, J.J., Kronberg, P.P.: 1984, ApJ, 279, 19
Weyman, R.J., Chaffee, F.H.Jr., Davis, M., Carleton, N.P., Walsh, D., Carswell,
    R.F.: 1979, ApJ, 233, L43
Weymann, R.J., Foltz, C.B.: 1983, ApJ, 272, L1
York, D.G., Yanny, B., Crotts, A., Carilli, C., Garrison, E., Matheson, L.: 1991,
    MNRAS, 250, 24
Young, P., Sargent, W.L.W., Boksenberg, A., Oke, J.B.: 1981, ApJ, 249, 415
Young, P., Sargent, W.L.W., Boksenberg, A.: 1982, ApJ, 252, 10

Galaxies Close to the Line of Sight to two z=1 Quasars - Implications for Galactic Halos and Gravitational Lensing

Josef W. Fried

Max-Planck-Institut für Astronomie, Königstuhl 17, D-69117 Heidelberg, Germany

Abstract. R-band images of two z=1 quasars show faint ($m = 21 - 22$) galaxies close ($2 - 3''$ or $\leq 10kpc$) to the line of sight to the quasars. For one of the objects a galaxy with a huge ($35''$ or 153 kpc radius) halo has been claimed to cause an absorption line in the quasar spectrum; however these observations show that the absorption may be due to galaxies much closer to the quasar, therefore not requiring huge halos. Since cases like these may easily be overlooked, not all huge galaxy halos may be real! These galaxies also amplify the quasars by gravitational lensing by factors of 1.2 to 2.

1 Sample and Data Reduction

Direct images of the fields of two radio-loud quasars (PKS 2216-038, $z = 0.901, V = 16.38$ and PKS2356+196, $z = 1.066, V = 18.0$) were obtained at the prime focus of the 3.5 m telescope on Calar Alto, Spain. The data are complete for galaxies down to R=26.

2 Results and Discussion

The images of the two quasars appear non-stellar; image deconvolution with the Lucy-Richardson method clearly shows faint ($m = 21 - 22$) galaxies close to the lines of sight ($2 - 3''$) to the two objects (cf. fig.1). Since these galaxies are almost certainly foreground, the impact parameters are on the order of 10 kpc. Therefore these galaxies will (i) probably cause absorption lines in the spectra of the distant quasars and (ii) act as gravitational lenses and thereby amplify the flux of the quasars.

The quasar PKS2216-038 has also been included in the sample of a study of absorbing galaxies by Le Brun et al(1993). These authors have found a weak absorption line, identified with MgII at z=0.202; they believe that this absorption is caused by a galaxy $35''$ from the line of sight to the quasar, which requires a huge halo of 153 kpc! If,however, the galaxy we discovered closer to the line of sight is the true absorber, then no huge halo is required. Since faint galaxies close to the line of sight to relatively bright quasars are difficult to detect, PKS2216-038 may not be the only case in which the assumption of a huge halo is unfounded.

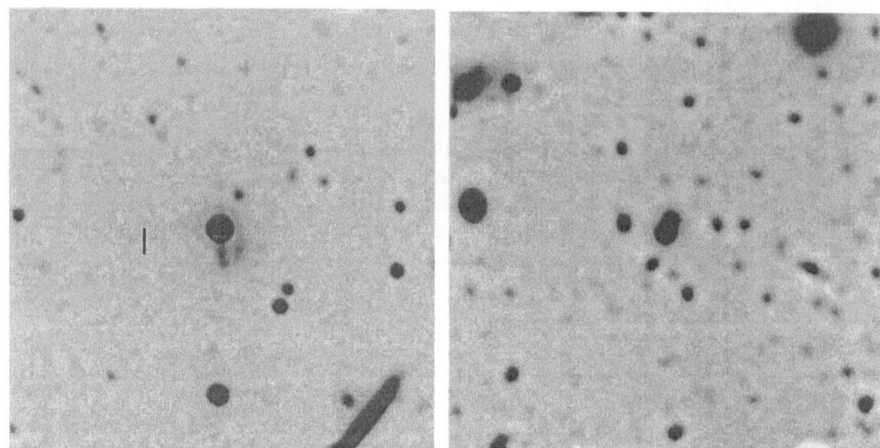

Fig. 1. Fields of the quasars PKS2216-038 (left) and PKS2356+196 (right). The faint galaxies are seen south of PKS2216-038 and north-west of PKS2356+196. The vertical bar left of PKS2216-038 corresponds to 10 kpc length at z=0.2

The galaxies close to the line of sight to the 2 quasars will also amplify the light from the quasars by gravitational lensing. If we assume redshifts of 0.2 and 0.5 (for PKS2356+196) we obtain amplification factors between 1.2 and 2 for velocity dispersions between 150 and 250 $km\,s^{-1}$ using the formula given by Turner et al. (1984), $H_0 = 75\,km\,s^{-1}\,Mpc^{-1}$ and $q_0 = 0.5$. Again it should be noted that cases like the ones presented here are observationally challenging and may easily escape detection in imaging surveys.

A full discussion of these data - including clusters around quasars - will be given elsewhere.

References

Le Brun V., Bergeron J., Boisse P., Christian C. 1993, A&A, 279, 33
Turner E.L., Ostriker J.P., Gott III, J.R. 1984, ApJ, 284, 1

Ionization and Abundances of Intergalactic Gas

David Tytler, Xiao-Ming Fan, Scott Burles, Lance Cottrell, Christopher Davis, David Kirkman, and Lin Zuo

Physics Dept, CASS 0111, UCSD, La Jolla, CA 92093-0111, USA

Abstract. We will briefly present four results – three from the W.M. Keck telescope, and one from the HST – on the abundances and ionization of the Lyman-alpha forest clouds and the intergalactic medium at high redshift. For the first time we have sufficient signal to noise to clearly see the velocity profiles of the Lyα forest lines. Most are Voigt, with little substructure, and no sign of hydrodynamics, and most have velocity dispersions between 22 and 36 km s^{-1} , below the published mean of 34 km/s. There is no correlation between velocity dispersion and column density, but some clouds are apparently very cold. We report the discovery of C IV lines in 60% of high redshift Lyα forest clouds with $logN(HI) > 14.5$. Ionization corrections indicate that about 50% of Lyα clouds have [C/H] > -2.5. If the few Lyα clouds with high $N(\mathrm{HI})$ are the only ones with carbon, then they might all be in the very outer regions of galaxies, especially since they cluster on scales of 1000 km s^{-1} . But if all Lyα clouds have carbon, then most are intergalactic, and the carbon could have three origins: pre-galactic massive stars or very massive objects, stars inside individual intergalactic clouds, and gas ejected from galaxies. We present a low value for the primordial deuterium to hydrogen ratio in one Lyman limit absorber at high redshift, which we believe is the most accurate measurement ever, and we present an HST spectrum of the first QSO to show flux below He II 304, which means that the intergalactic medium (IGM) is highly ionized at $z = 3$ in this direction.

1 Background

These are superb times to be an observational or experimental astronomer. The HST is working, and the new W. M. Keck 10 m telescope is providing some stunning data, especially quasar spectra.

We use quasars as sources of ultraviolet radiation because they are bright and have relatively smooth continua. Along the line of sight to a QSO at $z \simeq 3$ we see over 1000 gas clouds. Most are completely transparent, except at the wavelengths of the UV resonance lines of the ions of the abundant elements. Most clouds are warm, $T \simeq 40,000$ K, so the absorption lines are relatively narrow, but they are so common that they overlap and blend together. Clouds with different redshifts produce lines at different places in the spectrum of a background QSO because observed wavelengths are redshifted from the rest wavelengths: $\lambda_{obs} = (1 + z_{abs})\lambda_{rest}$. The strongest absorption line from most gas clouds is the Lyman-alpha line of H I, which has a rest wavelength of 1215.67 Å. We have now spectra which show over 1000 Lyα lines, each coming from a different gas cloud.

2   Types of Lyman-alpha Lines

We have used the Keck telescope to investigate the properties of the Lyα forest of two QSOs: Q0636+680 (z_{em} = 3.178, V=16.5, 5 hours exposure) and HS1946+7658 (z_{em} = 3.051, V=15.8, 12 hours exposure). Both were chosen to maximize our chance of detecting weak C IV lines coming from the same gas as the Lyα lines. We used the dramatic HIRES echelle spectrograph which gives FWHM =8 km s^{-1} with a 1.14 arcsec slit (Vogt 1994). At this time HIRES efficiency drops rapidly below 4200 Å, while at higher z_{em} the Lyα forest is too crowded and QSOs are fainter, hence both QSOs have $z_{em} \simeq 3$. We covered from below Lyβ emission to beyond C IV emission in a single exposure, with short inter-order gaps at $\lambda > 5200$ Å. Two parts of the Keck spectrum of HS1946+7658 are shown in Figure 1.

We have now tentatively identified six types of Lyα lines in Keck spectra. These are listed in Table 1.

Table 1: Types of Lyα Forest Lines

Name	N(H I) (cm^{-2})	b (km s^{-1})	T (K) (maximum)	N(z) at $z = 2.8$
Narrow	< 13.2	3 – 15	1,000 – 15,000	50
Simple	12.8 – 14.0[a]	22 – 36	30,000 – 80,000	470
Wide	13.0 – 14.0[a]	36 – 60	30,000 - 150,000[c]?	27
Shallow	≤ 13?	60? – 200	150,000 – 2 × 10⁶?	20
Undulating	12.8 ± 0.5	22 – 80	?	9?
Saturated	> 14.5	?[b]	25,000 – 60,000[b]	20

[a] Data are incomplete at $logN(H\ I) < 12.8$ and $b < 40$ km s^{-1} . At $logN$(H I) > 14.0 lines saturate and have uncertain b.
[b] These temperatures are implied by the b values of the thinnest C IV components. Bulk motions are 30 – 100 km s^{-1} .

1. Narrow: These lines are too narrow to be H I at the expected temperatures (Duncan et al 1991). We speculate that they are of three types: (1) not as narrow as data indicate, (2) metal lines, and (3) cool H lines. Observations of Lyβ will be decisive. They can not all be metals, because we would then have too many metal lines: a metal catastrophe. From an identification perspective, many could be C IV lines because we often see systems which show the C IV doublet and H I, but no other lines in optical spectra. We would normally identify C IV by observing both lines of the doublet, but when we see a weak C IV line it is often hard or impossible to tell if the other line is present because it would lie in a region of the spectrum which is strongly absorbed by unrelated H I lines. We are confusion limited, but higher signal to noise certainly helps.

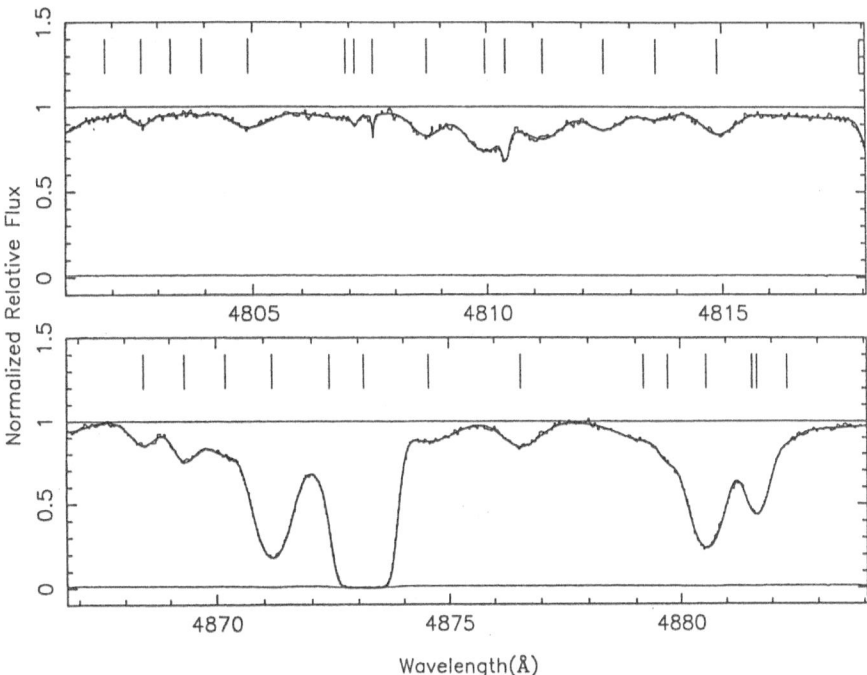

Fig. 1. Portions of the Lyα forest from our 12 hour Keck HIRES spectrum of H-S1946+7658, reduced by David Kirkman. The wavelengths are heliocentric vacuum values. The data are the histogram (SNR = 65 - 105 per 2 km s^{-1} pixel, varying across each order) and the continuous curve is the fit comprised of blended simple Voigt lines. Each line has three free parameters: redshift (shown by the vertical lines), column density and b value. The upper panel shows a region of the spectrum with a continuous blend of weak lines, which covers about 1500 km s^{-1} (4797 – 4821Å). This is an example of *undulating* absorption. The lower panel shows many *simple* lines. Relatively few lines provide an excellent fit to the data. There are two possible *shallow* lines: 4817.9 with b = 191 km s^{-1} and N(H I) = 13.2, and 4881.6 with b = 162 km s^{-1} and $logN$(H I) = 13.1. The line density in this portion of the spectrum is $N(z) \simeq 1100 \pm 300$. The signal to noise is 90 at 4810 Å, 65 at 4850 Å (the end of the order), and about 105 at 4880 Å.

2. Simple: While we see only a few completely isolated lines in each spectrum, there are a few which are very well fit by relatively isolated single Voigt profiles. These lines can be fit by stationary gas at a single temperature. We have not seen the flat bottomed or "w" shaped lines which Miralda-Escudé & Rees (1994a) predicted would be produced by gas falling into dark matter mini-halos. The great majority of lines have $22 \leq b \leq 36$ km s^{-1} .

3. Wide: These lines appear single, but their widths might result from the close blending of two simple lines. Alternatively, some or most of them may be thermally broadened single lines from hot gas.

4. Shallow: These are extremely wide and very shallow (5% deep) lines with no visible sub-structure. We believe they are real, because they are not seen in standard star spectra, but it is difficult to be certain, because they are so shallow and broad. They are hard to recognize except in data of the highest ($>$ 100) signal to noise, and even then we must first fit the ubiquitous simple lines which lie inside and blend with the broad shallow lines. Two are indicated in Figure 2. They could be collisionally ionized hot gas with $T \simeq 10^6$ and $b \simeq 100 - 200$ km s^{-1} , predicted by Verner, Tytler & Barthel (1994). If they are this hot, then their total column densities are huge: 10^{19} to 10^{21} cm^{-2}. The larger the b value, the higher the temperature, and the higher the total column required to produce detectable H I absorption. Lines from gas of these high temperatures are naturally shallow: they are so highly ionized that even the largest total columns of 10^{21} produce only very weak absorption, and the exceptionally large thermal broadening makes them shallow. They might account for a large proportion of the baryons in the universe. They have not been seen before.

5. Undulating: Regions in the spectrum where there is continuous H I absorption by blended shallow lines extending over $500 - 1500$ km s^{-1} . These regions are hard to notice near strong lines, and their extent is not well determined because it is hard to tell if they extend across strong lines. They fill the top of Figure 1. They appear to be regions of space which have lower than average ionization. They might be isolated clouds, with a small volume filling factor, or they could be residual H I Gunn-Peterson absorption from the diffuse IGM (e.g. Miralda-Escudé & Rees 1994b). They are apparently physical associations of lines, rather than chance juxtapositions, because there are many parts of the Lyα forest, at least 5 Å (300 km s^{-1}) in extent, which are completely free of lines of this strength. These undulations have not been seen before.

6. Saturated: These systems have C IV lines, which reveal the gas gas temperature and turbulent bulk motions. They are correlated in velocity over 1000 km s^{-1} and they are the subject of the next section.

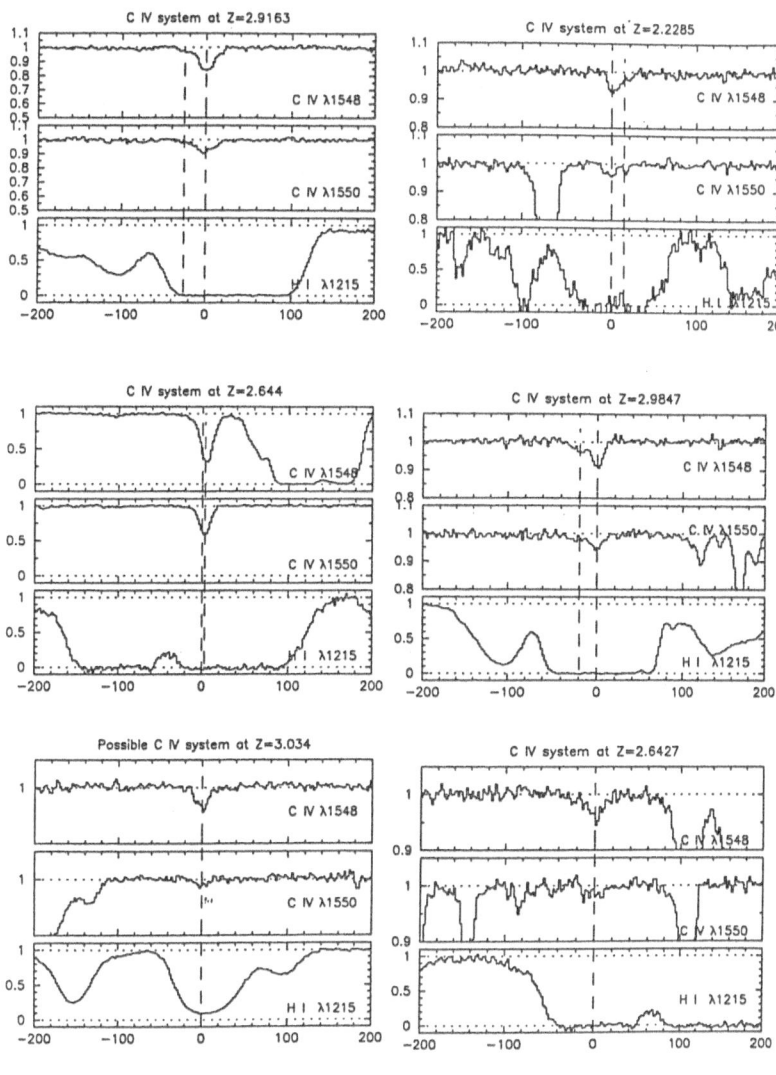

Relative Velocity (km/s)

Fig. 2. Six Lyα forest lines which have associated C IV lines. In each case we show the stronger C IV line at 1548 Å (top), the weaker C IV line at 1550 Å(middle), and the Lyα line (bottom), all on the velocity scale of the apparent center of the Lyα cloud. Since the C IV lines are all weak and on the linear part of the curve of growth, 1548 should have twice the equivalent width of 1550. The carbon lines are often off set from the apparent center of their Lyα line, and there are often several C IV lines separated by about 30km s^{-1} , which shows that the saturated Lyα lines have hidden velocity structure. Line parameters are: ($z = 3.034$, N(C IV)= 12.2, $b = 8.4$) (2.644, 13.3, 7.1) (2.9163, 12.4 and 11.7, 12.4 and 6.5) (2.6427, 12.1, 12.5) (2.2285, 12.1 and 11.9, 5.0 and 7.9).

3   Intergalactic Metals: Carbon in Ly-a Forest Clouds

We searched for C IV lines at the velocities of all Lyα forest lines in the Keck spectra. In Q0636+6801 we saw C IV in 13 Lyα clouds, over a redshift range of 3.11 – 2.52, which corresponds to a line density of $N(z)= 22$. If these Lyα clouds with C IV were all galaxies with the usual scaling and normalization (e.g. Sargent, Boksenberg & Steidel, 1988), then L_* galaxies would be 200 kpc in radius. For HS1946+7658 the signal to noise is significantly higher and we see C IV in more Lyα systems, including those in Figure 2.

We see C IV in all Lyα systems with $log N(\text{H I}) > 15$, and in 60% with $log N(\text{H I}) > 14.5$, but in few with lower $N(\text{H I})$. It appears that we are signal to noise limited, such that all Lyα systems might have similar C IV/H I. Following Tytler & Fan (1994), we have shifted and added the portions of the spectra at the expected positions of C IV lines in the hundreds of Lyα clouds which do not show C IV individually. We see marginal 3.5 σ detections of C IV in Lyα clouds with $14 \leq log N(\text{H I}) \leq 15$. This also supports the impression that we are signal to noise limited, but it is not proof, because proportion of Lyα clouds which show C IV, as a function of $N(\text{H I})$, will depend on the distribution of ionization and abundances, which are unknown *a priori*. Experimental evidence that old data were signal to noise limited can come from data of higher signal, but this does not mean that current spectra are so limited.

We do not know metal abundances accurately because the level of ionization, and especially the shape of the ionizing spectrum (we thank Patrick Petitjean for stressing this) is uncertain. The particular spectrum used by Donahue & Shull (1991) leads to a nearly constant value of C IV/H I over a wide range of likely ionization. We adopt

$$[C/H] = \log \; N(C\ IV)/N(H\ I) \;\; + \;\; 0.377$$

for all clouds. We then use both the measured C IV/H I column densities, and upper limits when C IV is not detected, to determine an abundance distribution. We see a wide range of abundances, from -1 to -2.5 which is the detection limit. The cumulative distribution shows that 50% of Lyα systems have $[C/H] > -2.5$, for the assumed ionization. The next step is to detect O VI lines which should be strong if the ionization is high. We would very much like to measure the O/C abundance ratio, which should be higher than solar, but this will require independent information on the ionizing spectrum, which should come from HST far UV spectra which contain numerous far UV lines of many ions of several elements (Vogel & Reimers 1993).

These results supersede the traditional view of Lyα forest clouds as intergalactic clouds of primordial composition (Sargent et al 1980). We find that Bergeron & Boissé (1984), and Tytler (1987) were to right to speculated that Lyα clouds are related to metal line systems, and we understand why Meyer & York (1987) and Lu (1991) saw C IV in Lyα clouds with high $N(\text{H I})$. It was widely believed that metals were mostly confined to gas clouds with

$logN$(H I) $> 10^{17}$ (e.g. Tytler 1987), but now we have seen that metals are common down to $logN$(H I)= 14.5.

We typically see two components of C IV associated with each saturated Lyα line. Some have one component, and some more, and sometimes C IV is seen in Lyα lines which are not saturated. For the first time we see the velocity sub-structure of Lyα clouds. The C IV components have a component-component velocity dispersion of 30 km s^{-1} (range 10 – 100) when $logN$(H I) < 15. For larger N(H I) the component velocity dispersion is significantly larger: about 100 km s^{-1} , which resembles the dispersion reported by Petitjean & Bergeron (1994) in systems with strong C IV lines. The lower dispersion at lower N(H I), which also applies at low N(C IV), may indicate that these clouds are of lower mass. The C IV lines are sometimes displaced for the centers of their Lyα lines, which is another manifestation of the velocity structure of the Lyα clouds: saturated Lyα lines are often complex blends.

The Lyα clouds which show C IV are clustered on a scale of ≤ 1000 km s^{-1}. This is a property of metal line systems and galaxies, but not of the Lyα forest as a whole. There might still be two separate populations – primordial clouds which do not cluster much, and clouds with C IV which do cluster. But the amplitude of the clustering of the Lyα lines with C IV is lower than that of traditional metal systems, so clustering might decrease smoothly with declining N(H I).

If carbon exists only in those Lyα clouds with high N(H I) in which it has already been seen, then we believe that those absorbers are probably in the outer regions of galaxies. There are four reasons for this deduction: (1) the frequency of occurrence of the C IV lines implies that galaxies are about 200 kpc in radius, which is similar to the radii deduced by Lanzetta et al (1995) for galaxies at low redshift. (2) Giallongo (1991) found that Lyα lines with large equivalent widths, which will often have large N(H I), evolve slowly in redshift, like metal line systems, which are probably galaxies. (3) The Lyα with metals cluster on scales of 1000 km s^{-1} , like metal line systems and galaxies, and (4) metals are made in galaxies.

But if we are signal to noise limited, as we suspect, then most, or all, Lyα clouds may contain metals. There are then at least three options for the origin of the metals. (1) Many clouds, at least those with the largest N(H I), may be gravitationally bound, and they may make their own stars and metals. This idea needs to be explored. (2) The metals can be made in stars in galaxies and distributed widely throughout the intergalactic medium. The clouds would then form during or after galaxies. They could not be primordial perturbations, because it would be difficult to mix in the metals. (3) The metals could be made in population III stars which form throughout the intergalactic medium, before galaxies. While the energy release would not be enough to distort the COBE measurements of the spectrum of the microwave background, David Spergel (private communication) pointed out that the energy release connected with the creation of the carbon is enough to fully ionize all regions of the universe where the carbon is made. If the carbon is confined to Lyα clouds, then they alone would be re-ionized, but if the carbon is wide spread throughout the intergalactic medium, then the stars or very massive objects which made the carbon could have ionized most baryons in the universe.

4 Detection of the IGM: He II Gunn Peterson Effect

Photons which are emitted at wavelengths $\lambda < 304$ Å will redshift up to wavelengths above 304 Å. When their wavelengths are about 304 Å they can be absorbed by He II ions in the intergalactic medium, or in gas clouds. Absorption by smoothly distributed atoms in the IGM is the the Gunn-Peterson test, which is an extremely sensitive to small quantities of gas. We know that the IGM contains very little H I or He I. He II should then be the most promising ion.

Most photons from the far-UV are absorbed as they ionize H I in intervening Lyman limit systems (gas clouds which are optically thick in the Lyman continuum; Tytler 1981; Sargent, Steidel & Boksenberg 1989). The shorter its emission wavelength, the longer a photon spends at wavelengths in the Lyman continuum where $\lambda < 912$ Å, and the lower the probability that the photon will arrive unabsorbed (Moller & Jakobsen 1990; Picard & Jakobsen 1993). Three groups have been searching for the rare QSOs which have detectable far UV flux.

Jakobsen et al (1994) were the first to report a spectrum of a QSO down to 304 Å in the rest frame. They used an objective prism spectrum of Q0302-0019 to determined an optical depth of $\tau > 1.7$ at $\lambda < 304$ Å, which implies that the IGM is of low ionization.

Our group has conducted a systematic search for flux from all high redshift QSOs which might have flux at 304 Å. We used HST FOC UV images. We detected four interesting QSOs, and one, Q1937-69 which has a flux of 2.1×10^{-16} erg cm^{-2}s^{-1} Hz^{-1}, which is 2.2 times the continuum level in Q0302-0019. We obtained a HST spectrum with the FOS G130H grating in 1994. This shows definite flux below 304 Å with $\tau = 1.0 \pm 0.2$, which is apparently different from the Jakobsen et al result. Additional spectra will be obtained to check both results. We are suspicions of the Jakobsen et al spectrum because it shows an extremely strong emission line just to the red of 304 Å. Our spectrum does not show any emission lines in this region, and none are expected from standard photoionization models. For example, we expect that the He II $\lambda 304$ line will have an equivalent width of $0.5 - 1$ times that of He II $\lambda 1640$, which is usually much weaker than 50 Å. However a more interesting possibility is that both results are correct, because the IGM has inhomogeneous ionization which is reasonable (Miralda-Escudé, & Rees 1994b) because the epoch of ionization of He II can be after that of H I.

At this time we do not know whether the we have detected the IGM in He II, because some of the optical depth below 304 Å will be due to the

He II lines of the Lyα forest clouds (Jakobsen et al). Our best estimate of the line optical depth is about 0.5, with the remaining $\tau_{GP} = 0.5$ from the Gunn-Peterson absorption from the distributed IGM He II. It is possible that the Jakobsen et al spectrum overestimated the optical depth, and all of the absorption observed is due to He II Lyα lines. The amount of absorption from the lines depends on the ratio of He II to H I ions, which is set by the shape of the ionizing spectrum (Miralda-Escudé 1993; Jakobsen et al 1993). We believe that this radiation originates in QSOs and is absorbed by intervening gas clouds. We have used our HST images of 82 high redshift QSOs to determine the mean intrinsic spectrum of QSOs from 330 – 1300 Å in the rest frame. We use Monte Carlo simulations of the intervening absorption to determine the intrinsic spectrum which best matches both optical and HST fluxes. The result is that the intrinsic spectral index is $f_\nu \propto \nu^{-1.5}$, with a 67% confidence interval of $\alpha = 1.15$ to 1.65. If $\alpha \geq 1.1$, then the Lyα clouds will have relatively high He II/H I (Giroux, Fardal & Shull, 1995), and the He II lines will be strong, but they saturate, giving a likely upper limit on the line absorption of $\tau_{lines} \simeq 0.5$.

5 The Cosmological Deuterium to Hydrogen Ratio

Our optical spectra of QSO 1937-1009 ($z_{em} = 3.78$, V $\simeq 17.5$) with the W. M. Keck telescope HIRES echelle spectrograph show deuterium in the Lyα and Lyβ lines of an absorption systems at 3.57. The H I Lyman series lines are narrow ($b \simeq 20$ km s^{-1}) and resolved all the way to Lyman-19. They absorb all the flux in their cores, and there is strong Lyman continuum absorption, showing that the H I column density is $log N(H\ I) \geq 17.8$. Weak metal lines in the system, C IV, C II, SI IV and Si II all show the same profiles which are adequately described by two clouds separated by 13 km s^{-1}. A simultaneous fit to all the Lyman series lines, especially the weaker high order lines, indicates a total H I column of about 10^{18}, with 80% in the lower redshift component. The same fit gives D/H $\simeq 2 \times 10^{-5}$ (preliminary, with a systematic error of about 50%), which corresponds to a high cosmological baryon density of $\Omega_b \simeq 0.03\ h_{100}^{-2}$.

Both clouds are moderately ionized, with H I/H $\simeq 0.01$. The metal abundances are about [C/H] $\simeq -3.1$ and [Si/H] $\simeq -2.7$ in the component with the most H I and D I, and [C/H] $\simeq -2.0$ and [Si/H] $\simeq -1.7$ in the other cloud. Both have [C/Si] $\simeq -0.3$, like low abundance halo stars, and their abundances are much too low for stars to have removed significant deuterium.

These results are consistent with D/H of $(1 - 2) \times 10^{-5}$ from the interstellar medium, but not with Songalia et al. (1994) who reported D/H $= 2.6 \times 10^{-4}$, at least ten times more D, in a different QSO absorber. Either their absorption line was H rather than D, or primordial nucleosynthesis was dramatically inhomogeneous.

6 Are Most Baryons in the IGM?

The overall theme of this work is the measurement of the distribution, ionization and abundances of baryons on the largest scale. The relatively high Ω_b deduced from the low D/H value implies that about 80% of baryons are missing. They might be in the IGM, and they might be in the hot Lyα clouds which make the broad and shallow lines. In the next few years we should obtain the data needed to check these possibilities: measurement of the shape and intensity of the spectrum of ionizing radiation, measurement of the He I-I/H I ratio in Lyα clouds, measurement of the He II Gunn-Peterson optical depth and determination of the cosmological density in hot gas clouds.

Acknowledgements. This work is based on observations obtained at the W. M. Keck Observatory, which is a joint facility of the California Institute of Technology and the University of California, and with the NASA/ESA Hubble Space Telescope through the Space Telescope Science Institute, which is operated by AURA, Inc., under NASA contract NAS5-26555. This work was partly supported by NASA grant NAGW-2119, and Space Telescope grants GO-3801.01-91A and GO-5492-01-93A.

References

Bergeron, J., & Boissé, P. 1984, A&A, 133, 374

Duncan, R. T., Vishniac, E. T., & Ostriker, J. P., 1991, ApJL, 368, L1

Giallongo, E. 1991, MNRAS, 251, 541

Giroux, M.L., Fardal, M. & Shull, J.M. 1995 ApJ in press

Jakobsen, P, Boksenberg, A., Deharveng, J.M., Greenfield, P, Jedrzejewski, R. & Paresce, F. 1994, Nature 370, 35

Jakobsen, P. et al 1993 ApJ, 417, 528

Lanzetta, K. M., Bowen, D. V., Tytler, D., & Webb, J. K. 1995, ApJ March

Meyer, D. M., & York, D. G. 1987, ApJ, 315, L5

Miralda-Escudé, J. 1993, MNRAS, 262, 273

Miralda-Escudé, J. & Rees, M. J. 1994a, MNRAS, 260, 617

Miralda-Escudé, J. & Rees, M. J. 1994b, MNRAS, 266, 343

Møller, P. & Jakobsen, P. 1990, AAp, 228, 299

Petitjean, P. & Bergeron, J. 1994, AAp, 283, 759

Picard, A. & Jakobsen, P. 1993, AAp, 276, 331

Sargent, W. L. W., Steidel, C. C. & Boksenberg, A. 1989, ApJS, 69, 703

Sargent, W. L. W., Boksenberg, A., & Steidel, C. C. 1988, ApJS, 68, 539

Sargent, W. L. W., Young, P. J., Boksenberg, A., & Tytler, D. 1980, ApJS, 42, 41

Songalia, A., Cowie, L.L., Hogan, C. & Rugers, M. 1994 Nature, 368, 599

Tytler, D. 1981, Nature, 298, 427

Tytler, D. 1987, ApJ, 321, 49

Tytler, D., & Fan, X.-M. 1994, ApJL, 424, L87

Vogel, S., & Reimers, D. 1993 AAp, 274, L5

Vogt, S. et al. 1994, Instrumentation in Astronomy VIII, ed. D. Crawford, SPIE, 362

The Proximity Effect

Jill Bechtold

Steward Observatory, University of Arizona, Tucson AZ 85721, USA

Abstract. Estimates of the ambient ultraviolet ionizing background from modeling of the observed deficit of Lyα forest lines near quasars (the "proximity effect") are reviewed, including new results from a survey carried out at the Multiple Mirror Telescope. The observed evolution of Lyα forest clouds at z$\sim$ 1.6-4 cannot be reproduced by the evolution in the ultraviolet background implied by the proximity effect measurements, for either the cold-dark-matter minihalo or IGM pressure-confined clouds in their simplest forms. Sources of uncertainty in the ultraviolet background estimates from the proximity effect are discussed.

1 Introduction

The intensity of the ultraviolet ionizing background, and how it evolves with redshift, are relevant to a number of problems discussed throughout this workshop. Rees and Setti (1970) first suggested that quasars are likely to be a significant source of hard UV photons at high redshifts; more recent estimates of the contribution of discrete sources to the ultraviolet background are reviewed by Madau in these proceedings. Photoionization by this radiation background may be important for understanding the evolution of the intergalactic medium (Jakobsen et al. 1993; Jakobsen, these proceedings; and Tytler, these proceedings), the Lyα forest (e.g. Sargent et al. 1980; Ikeuchi and Turner 1991; Charlton, Salpeter and Hogan 1993), the optically thin metal-line systems (Reimers, these proceedings), and the gas in the outermost regions of local galaxies, including the Milky Way (Savage, these proceedings). Comparison of the specific intensity of the background derived from the proximity effect with the estimated contribution from known quasars places limits on the extent to which dust obscuration modifies the observed quasar luminosity function (Fall, these proceedings), and can even be used to place limits on the masses of exotic particles in the early Universe (Sciama, these proceedings). Reviews of the general subject of the UV background are given by Paresce (1990), Bowyer (1991), Henry (1991) and Bechtold (1993).

Weymann, Carswell and Smith (1981) first pointed out that the number density of Lyα forest lines is systematically low near the quasar being used to probe them. They suggested that photoionization by the quasar's UV light might be responsible. Around this time, there was some debate about the value of the evolutionary parameter, γ (see equation 1) for the Lyα forest clouds, with divergent values obtained by different workers with only slightly different data sets (Peterson 1978; Ellis 1978; Sargent et al. 1980; Young,

Sargent and Boksenberg 1982; Carswell et al. 1982; Phillips and Ellis 1983). While part of the problem was that the number of quasars with data was very small, and also that the redshift interval sampled was small, in hindsight it is clear that some of the confusion was caused by neglecting the proximity effect (Murdoch et al. 1986; Tytler 1987). Murdoch et al. (1986) dubbed the deficit the "inverse effect" because the γ they derived for individual quasars, *including* the region near the quasar, was typically smaller than for the sample as a whole.

The focus then shifted to using the proximity effect to estimate the EUV radiation background which is ionizing the clouds. Qualitatively, the extent of the proximity effect indicates roughly how far the ionizing radiation from the quasar is greater than the background. By measuring the brightness of the quasar (often the Lyman limit flux is directly observable), one can therefore estimate the value of the specific intensity of the background, J_ν. Carswell et al. (1987) presented a spectrum of the z=3.78 quasar PKS 2000-330, the highest redshift quasar known at that time, and pointed out that the proximity effect was relatively weak. They found no sign of a decrease in the number of clouds until z=3.74, $\sim$2500 km sec^{-1} from the emission line redshift, z_{em}=3.78. By assuming that the quasar flux at z=3.74 is about equal to the ambient background, they estimated that at the Lyman limit, $J_\nu \approx 3 \times 10^{-21}$ ergs sec^{-1} cm^{-2} Hz^{-1} sr^{-1} (for q_o=0.5). This exceeded the computed background from the integrated contribution of quasars (e.g. Bechtold et al. 1987) by a large factor. However, at that time very little was known about the evolution of the quasar luminosity function at z>3, so these calculated estimates were very uncertain.

A more detailed analysis was then presented by Bajtlik, Duncan and Ostriker (1988, hereafter BDO), who considered a sample of 19 quasars drawn from the literature. They presented clear evidence that the deficit of Lyα forest lines near the quasar was statistically significant, and called this trend the "proximity effect". They fit a simple model for how the number of lines should decline as a function of distance from the quasar, assuming that photoionization by the quasar EUV radiation is the cause of the proximity effect. This model described the data well. They found that log J_ν = 21.0 $\pm$ 0.5 ergs sec^{-1} cm^{-2} Hz^{-1} sr^{-1} for the redshift range of their sample, 1.7 < z < 3.8. Again, this was significantly larger than estimates of the quasar background, particularly at z>3. BDO suggested that there was another source of ionizing radiation, young hot stars in galaxies (Bechtold et al. 1987, Miralda-Escude and Ostriker 1990) or a population of quasars which we don't see today because they are obscured by dust in intervening galaxies (Ostriker and Heisler 1984; Wright 1986; Heisler and Ostriker 1988; Boisse and Bergeron 1988; Najita, Silk and Wachter 1990; Wright 1990; Ostriker, Vogeley and York 1990; Fall and Pei 1989, 1993). Subsequently, many quasars with z>3 were found, and the luminosity function for high redshift quasars became better defined, although the uncertainties are still large (Warren et al. 1987; Schneider, Schmidt and Gunn 1989, 1994; Irwin, McMahon and Hazard

1991; Boyle, Jones and Shanks 1991; Warren, Hewett and Osmer 1991, 1994). The value for the quasar contribution to the background increased (Miralde-Escude and Ostriker 1990; Madau 1992; Madau, these proceedings), and is now much closer to the proximity effect estimates (section 2).

If quasars happen to lie along the line-of-sight they could cause observable voids in the Lyα forest (BDO after a suggestion by Paczynski, 1987) far from the quasar. Although luminous quasars are rare objects, the path length sampled by a typical quasar spectrum is long, and such chance coincidences are common enough to be interesting (see also Moller, these proceedings). Kovner and Rees (1989) discussed possible "clearings" in the Lyα forest from the "two quasar proximity effect" and concluded that the unsuccessful searches for voids in the Lyα forest up to that time were consistent with expectations (Carswell and Rees 1987; Duncan, Ostriker and Bajtlik 1989). In principle, one could use the "two quasar proximity effect" to estimate quasar lifetimes since a population of numerous short-lived ionizing sources would cause fewer large voids than rare, long-lived ones. Also, if the UV light from quasars is strongly beamed, one would expect to see examples of bright quasars along the line of sight with no associated proximity-induced void, or voids in the Lyα forest with no observable quasar to cause them. Of course voids could also be present from real underdensities in the total hydrogen gas, and these would be difficult to distinguish from a proximity-induced void. Dobrzycki and Bechtold (1991ab) found one significant void in the Lyα forest. There is a known quasar nearly but not *quite* at the right redshift, close enough to the line-of-sight, with the right brightness, to cause it. Subsequent searches for a more suitable ionizing quasar have been fruitless. Recently Fernandez-Soto et al. (1995) looked for proximity-induced voids in the spectra of 3 quasars with known quasars along the line-of-sight, but with only 3 objects, they were not able to tell one way or the other whether significant voids are present.

Zuo (1992ab) generalized these ideas and looked for underdense regions in the forest, rather than complete voids. He pointed out that in principle, the nature of the sources of the ionizing background could be deduced from the statistics of underdense regions. A fixed *integrated* background could be made up of rare, bright sources (e.g. luminous quasars) or numerous, fainter ones (e.g. star-forming galaxies): these would cause different signatures in the distribution of Lyα clouds through the proximity effect. Unfortunately, the data available did not allow strong statements to be made.

The BDO sample of 19 quasars was limited, in that it was contained only four quasars with $z_{em} > 3$, the redshifts at which the value of the integrated UV background they derived deviated most significantly from the estimates of the quasar contribution. These four quasars were the most luminous in the sample and had the most marked "proximity" effect. They dominated the statistics, since the expected line density is highest in the highest redshift quasars. Unfortunately, all four were radio-loud. Since radio-loud quasars tend to live in richer cluster environments than radio-quiet ones at lower redshift (Yee and Green 1987; Ellingson, Yee and Green 1991), and tend to have associated metal-line absorption (Foltz et al. 1986, Anderson et al. 1987, Aldcroft, Bechtold and Elvis 1994), one could question whether their proximity effect has anything to do with the photoionization of "normal" Lyα forest clouds (Bechtold 1987). Given the correlations between redshift and luminosity in the BDO sample, the proximity effect could have been the result primarily of redshift, or related to luminosity in some way other than direct photoionization of the clouds.

Subsequently, Lu, Wolfe and Turnshek (1991) and Bechtold (1994) presented larger samples for the redshift range $1.7 < z < 4.1$, for quasars with a wider range of intrinsic luminosities and radio-loudness. They found that the proximity effect is not correlated with radio properties of the quasar, or redshift. The extent of the proximity effect depends on quasar luminosity as predicted by the photoionization picture, but with a large scatter.

Recently, the redshift range over which the proximity effect has been used to measure the UV background has been extended to low redshifts (Kulkarni and Fall 1993, using the HST FOS data of Bahcall et al. 1993) and to z~4.5 with the spectrum of one quasar, BR 1033-0327 (Williger et al. 1994). Results are summarized in section 2.

In order to improve the statistics at z=1.5-2.5, and put better constraints on the evolution of J_ν with redshift, Bechtold et al. (1995) have obtained spectra of about 40 quasars at the Multiple Mirror Telescope (MMT) with the Blue Spectrograph and Loral CCD. The CCD has been optimized for performance in the blue, resulting in a quantum efficiency of ~60% at 3000 Å (Lesser 1994). The spectral resolution is 50 km sec^{-1} FWHM, and the spectra were analyzed in the way described by Bechtold (1994) and Aldcroft, Bechtold & Elvis (1994). This new sample was combined with the data described in Bechtold (1994) and a few z~1 quasars observed with the FOS from the HST archives. Results are described in the next section.

2 dN/dz and the Evolution of the UV Background

We fit the standard parameterization of the evolution of the number of clouds per redshift with rest frame equivalent widths $W_\lambda > 0.32$ Å,

$$dN/dz = A_0(1 + z)^\gamma \qquad (1)$$

using the maximum likelihood method described by Murdoch et al. (1986). Only lines with $(z_{em} - z_{abs}) > 0.15$ are included so that we avoid any influence of the quasar. The new ground-based sample ($1.6 < z < 4.0$) contains 1177 lines, and yields a best fit $\gamma = 1.71 \pm 0.23$ and $A_o = 7.22$. Equation (1) with a single γ is not a good fit, and a significantly better fit is obtained when an extra free parameter is added. While line-blending tends to lower γ (Trevese, Giallongo and Camurani 1992), this sample has higher resolution and signal-to-noise than the compilations of BDO or Lu, Wolfe and Turnshek (1991) and yet gives a lower γ than they report: in our data alone, 37 quasars yield

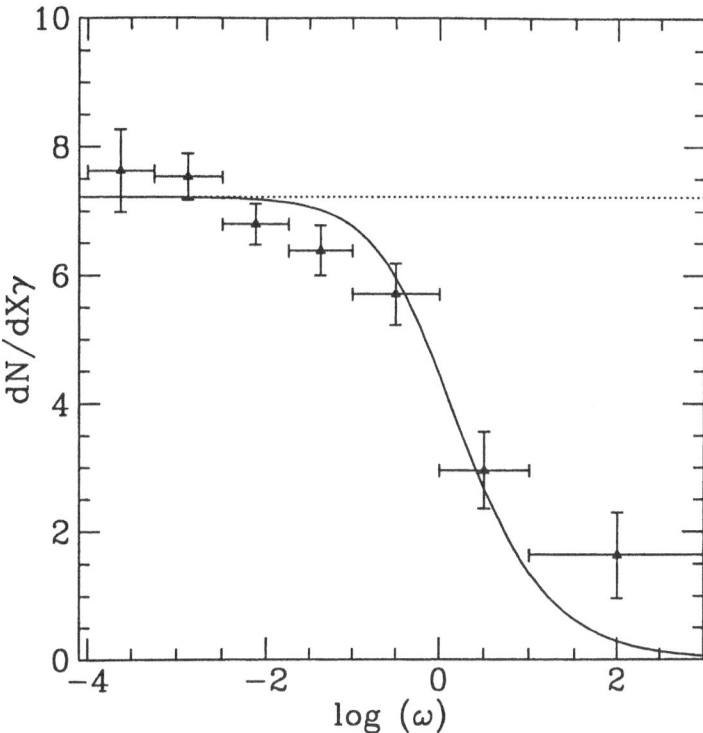

Fig. 1. dN/dX$_\gamma$ vs log ω. The data would fall along the dotted line if there were no proximity effect; the photoionization model is shown by the solid line, assuming J$_\nu$ is constant with redshift.

779 lines and $\gamma=1.51 \pm 0.33$. A discussion of the effects of line-blending in this sample is given by Dobrzycki and Bechtold (1995) and Bechtold et al. (1995). In any event, this parameterization of the line density is a useful empirical description of the cloud properties which is necessary to measure the proximity effect. The best fit for z<1.3 is $\gamma=0.56 \pm 0.61$, based on the FOS Key Project data of Bahcall et al. (1993) and a few quasars drawn from the HST archives. The change in slope in the evolutionary parameter γ suggests a change in the population being probed at z$\approx$ 1.5-2 (see Boksenberg, these proceedings).

In order to estimate J_ν from the proximity effect, we first apply the photoionization model of BDO. In the next section, we discuss the effect of the assumptions of this model. BDO considered the number of lines, X_γ, per co-moving, co-evolving redshift interval, where $X_\gamma = \int (1+z)^\gamma \, dz$, so that equation (1), would become $dN/dX_\gamma = A_o$ if there were no proximity effect. BDO showed that for highly ionized clouds and for a neutral hydrogen column density distribution that is a power law of index $\beta = 1.7$, equation (1) becomes

$$dN/dX_\gamma = A_o(1 + \omega(z))^{-0.7} \tag{2}$$

where $\omega = F^Q/4\pi J_\nu$ and F^Q is the Lyman limit flux of the quasar as seen by each cloud, and J_ν is the specific intensity of the UV background at the Lyman limit at the redshift of the cloud. Thus, a plot of dN/dX_γ versus $\log(\omega)$ shows the line density versus distance from the quasar, scaled according to the ratio of the Lyman limit flux of the quasar at that distance, to J_ν at that redshift. The result for the $4.1 > z > 1.6$ sample is shown in Figure 1. The dotted line shows what one expects for dN/dX_γ if there were no proximity effect (the clouds are closer to the quasar as $\log(\omega)$ increases), and the solid line shows the photoionization model, which is an acceptable fit to the data.

For Figure 1, we have assumed that J_ν is constant with redshift. More generally, we can consider the evolution of J_ν parameterized as $J_\nu = J_{\nu_o} (1+z)^j$ where J_{ν_o} and j are constants. In the sample of Bechtold (1994), j was poorly constrained for the ground-based data since there were very few quasars in the sample with $z \sim 2$: acceptable fits were obtained for $-7 < j < 4$. With the addition of the new sample, the 2σ limits are $-3 < j < 1$, with $\log J_\nu = -20.7$ ergs sec^{-1} cm^{-2} Hz^{-1} sr^{-1} if $j=0$.

A summary of J_ν at the Lyman limit as derived from the proximity effect is shown in Figure 2, which includes the results of Kulkarni and Fall (1993) and Williger et al. (1994). While the uncertainties are still large, broadly speaking, the results are consistent with the predicted contribution from quasars. The general trend that the EUV field is smaller at $z \approx 0$ and $z \approx 4.5$ than at $z \approx 3$ is probably secure. The estimates of the evolution at the low redshift and high redshift ends will improve in the near future, with the completion of the HST Key Project and with observations of more bright quasars at $z > 4$.

3   Relaxing the Simple Assumptions

There are a number of assumptions and limitations of the model used to derive J_ν.

(1) *Quasars emit flux with the same spectrum as J_ν.* In the BDO model, the ratio of the ionization rate caused by the UV background to the rate caused by the quasar UV flux is taken to be equal to the ratio of the flux densities at the Lyman limit. Since the cross section for ionization of H I is a rapidly decreasing function of frequency, this approximation has little effect

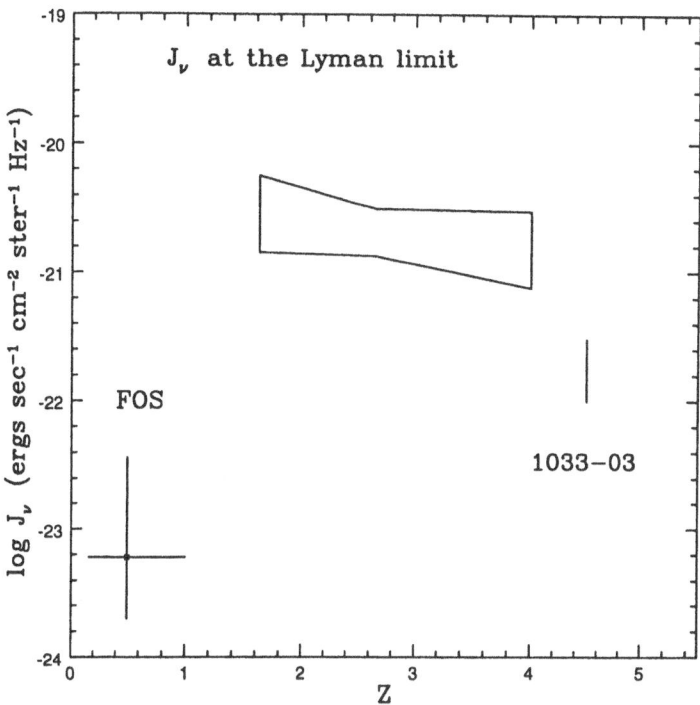

Fig. 2. J_ν at the Lyman limit vs z, as estimated from the proximity effect.

for reasonable spectral energy distributions. Bechtold (1994) estimated that the uncertainty introduced is ~25%.

(2) $\beta=1.7$. If the column density distribution of the clouds is flatter, $\beta=1.3$, then J_ν decreases by a factor of 3 (Chernomordik and Ozernoy 1993; Bechtold 1994). Evidence that $\beta < 1.7$ is given by Tytler, these proceedings.

(3) *The clouds are very ionized.* If the neutral fraction of the clouds is large, then J_ν is overestimated, by potentially a very large factor (see Pettini et al. 1990; their Figure 7; and Bechtold 1994). For example if $N(HI)/N(H)=0.95$, then $\log J_\nu \sim -23$ ergs sec^{-1} cm^{-2} Hz^{-1} sr^{-1}, an order of magnitude lower than the expected sum for observed quasars. Observations of He I absorption may rule out the possibility that the clouds are neutral.

(4) *Quasars may vary* on timescales comparable to the ionization equilibration timescale ($\sim 10^4$ years) for the clouds (see BDO; Bechtold 1994). The ionization of the cloud is determined by the quasar flux over the previous 10^4 years. On average, if the quasar was not systematically brighter or fainter in the past, then the estimate for J_ν should not be affected. However, nothing is known about the variability of luminous quasars on these timescales.

(5) *Gravitational lensing.* Since absorption-line surveys use the brightest quasars known, amplification by lensing may be important (e.g. York et al. 1991; Lanzetta et al. 1991). If lensing has brightened the quasar continuum then the quasars are intrinsically fainter than they appear, and J_ν has been overestimated. The known lenses for which the Lyα forest spectra are available don't seem to have too weak a proximity effect for their apparent luminosity, but the scatter is large (see Bechtold 1994, figure 15). The lack of multiple images for many of the quasars in the sample means that the amplification factor, even if they are lensed, is probably not large (Crampton et al. 1989; Bahcall et al. 1992).

(6) *Reddening by intervening galaxies.* If the quasars are dimmed by dust associated with intervening galaxies, then the quasars are intrinsically brighter than they appear, and J_ν has been overestimated.

(7) *Curve-of-growth effects.* The BDO model assumes the Lyα lines are on the linear part of the curve of growth, which is clearly not true for the stronger lines. If the lines are saturated instead of optically thin, then one has underestimated the factor by which N(H I) has been reduced by the presence of the quasar. This would decrease the estimated J_ν, by a large factor (Dobrzycki and Bechtold 1991b; Chernomordik and Ozernoy 1993).

(8) *The systemic redshifts of the quasars may be systematically underestimated,* since in these high redshift objects the redshifts are based on the broad Lyα and C IV emission lines. These lines may be systematically shifted with respect to the narrow lines and Balmer lines, which for $z=2$-3 quasars are in the near-IR. If the quasar redshift is really higher, the clouds are really farther away from the quasar than assumed, the quasar influence is less, and the real J_ν is lower (Espey 1993; Bechtold 1994). The relative shift may be as high as 1500 km/sec, in which case J_ν is overestimated by a factor of ~ 3.

In summary, most of these factors will *decrease* J_ν relative to the results in Figure 2, by factors of 3 or more. It now seems that we have reached the point where the statistical uncertainty in the proximity effect measurements of J_ν are small compared to these sources of systematic uncertainty, which now limit how well J_ν is determined. Some of the important effects can be easily addressed observationally.

4   Implications for the Lyα Forest

The observed evolution of the number density of the clouds and the evolution of J_ν are related. Qualitatively, in the simplest pressure-confined models, the expansion of the intergalactic medium results in a rapid decrease in pressure

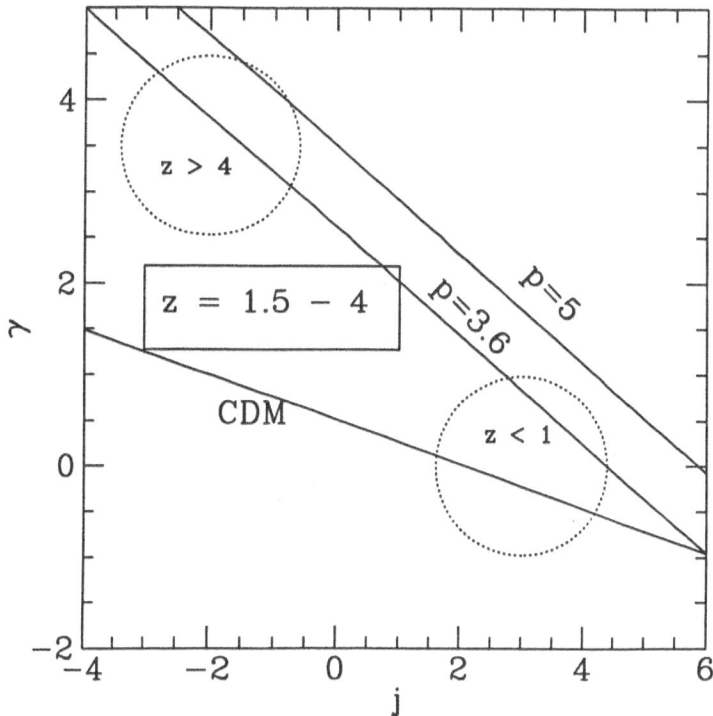

Fig. 3. γ versus j. Predictions for simple CDM confined and pressure confined clouds are shown (see text). The observed constraints for z=1.5 to 4 are shown in the rectangle, and the dotted circles represent schematically the preliminary results for z>4 and z<1.

with z, the clouds expand, become less dense, more ionized (even if J_ν is constant with z), and drop out of the equivalent width limited samples. Hence one expects a rapid evolution of the number of lines, or large γ. For the models where CDM confines the clouds, the evolution is slower, since the density of the clouds is more nearly constant with z. Quantitative predictions for spherical clouds or slabs are given by Ikeuchi and Turner (1991) and Charlton, Salpeter and Hogan (1993), and plotted in Figure 3. Figure 3 shows γ versus j, where $dN/dz \propto (1+z)^{\gamma}$ for the line density evolution and $J_{\nu} \propto (1+z)^{j}$ for the evolution of the UV background. The predictions for CDM confined clouds, and pressure confined clouds, with $P_{IGM} \propto (1+z)^{p}$ are shown, for p=5 (adiabatic IGM evolution) and p=3.6. The observed constraints are shown, for z=1.5 to 4, z>4 and z<1.5 separately.

For z=1.5 to 4, the observed evolution is faster than the CDM models predict, suggesting that mergers, or other effects not accounted for, may be destroying the clouds. The observed evolution is slower than the pressure confined models predict, suggesting that energy input into the IGM may be important at moderate redshift. The z>4 results are schematic, since j is not yet measured for them, but if γ is very steep (see Storrie-Lombardi, these proceedings), the dominance of pressure confinement may be favored. At low redshift, the clouds appear to be closer to the regime where CDM confinement dominates. Note that Rees (in these proceedings) argued convincingly on theoretical grounds that the confinement mechanisms for clouds should have exactly the *opposite* dependence on redshift than suggested here, with pressure confinement only becoming important late, when the intracluster medium is formed. Clearly, understanding the evolution of the ionizing background with redshift will be important for sorting out the origin of, and conditions in, the Lyα forest clouds.

Acknowledgements

It is a pleasure to thank A. Dobrzycki, T. Aldcroft, J. Scott, C. Foltz and M. Lesser for their contributions to the results presented here. Data were obtained with the Multiple Mirror Telescope, a joint facility of the University of Arizona and the Harvard-Smithsonian Astrophysical Observatory. Based in part on observations made with the NASA/ESA Hubble Space Telescope, obtained from the data archive at the Space Telescope Science Institute. STScI is operated by the Association of Universities for Research in Astronomy, Inc. under NASA contract NAS 5-2655. This work was supported by National Science Foundation PYI grant AST-9058510.

References

Aldcroft, T., Bechtold, J. and Elvis, M. 1994, ApJS, 93, 1.
Anderson, S. F., Weymann, R. J., Foltz, C. B. and Chaffee, F. H. 1987, AJ, 94, 278.
Bahcall, J. N. et al. 1992, ApJ, 387, 56.
Bahcall, J. N. et al. 1993, ApJS, 87, 1.
Bajtlik, S., Duncan, R. C. and Ostriker, J. P. 1988, ApJ, 327, 570.

Bechtold, J. 1993, in the *Third Teton School on the Evolution of Galaxies and Their Environment*, ed. J. M. Shull and H. A. Thronson, (Dordrecht:Kluwer), p. 559.

Bechtold, J. 1994, ApJS, 91, 1.

Bechtold, J., Weymann, R. J., Lin, A. and Malkan, M. 1987, ApJ, 315, 180.

Bechtold, J., Dobrzycki, A., Scott, J., Foltz, C. and Lesser, M. 1995, in preparation.

Boisse, P. and Bergeron, J. 1988, AAp, 192, 1

Bowyer, S. 1991, ARAA, 29, 59.

Boyle, B. J., Jones, L. R. and Shanks, T. 1991, MNRAS, 251, 482.

Carswell, R. F., Whelan, J. A. J., Smith, M. G., Boksenberg, A. and Tytler, D. 1982, MNRAS, 198, 91.

Carswell, R. F., Webb, J. K., Baldwin, J. A. and Atwood, B. 1987, ApJ, 319, 709.

Carswell, R. F. and Rees, M. J. 1987, MNRAS, 224, 13P.

Charlton, J. C., Salpeter, E. E., and Hogan, C. J. 1993, ApJ, 402, 493.

Chernomordik, V. V. and Ozernoy, L. M. 1993, ApJ, L5.

Crampton, D., McClure, R. D., Fletcher, J. M., Hutchings, J. B. 1989, AJ, 98, 1188.

Dobrzycki, A. and Bechtold, J. 1991a, ApJ, 377, 69.

Dobrzycki, A. and Bechtold, J. 1991b, in ASP Conf. Ser. 21, *The Space Distribution of Quasars*, ed. D. Crampton (San Francisco:ASP), p. 272.

Dobrzycki, A. and Bechtold, J. 1995, ApJ, submitted.

Duncan, R. C., Ostriker, J. P. and Bajtlik, S. 1989, ApJ, 345, 39.

Ellingson, E., Yee, H. K. C. and Green, R. F. 1991, ApJ 371, 49.

Ellis, R. S. 1978, MNRAS, 185, 613.

Espey, B. R. 1993, ApJ, 411, L59.

Fall, S. M. and Pei, Y. C. 1989, ApJ, 337, 7.

Fall, S. M. and Pei, Y. C. 1993, ApJ, 402, 479.

Fernandez-Soto, A., Barcons, X., Carballo, R., Webb, J. K., 1995, MNRAS, in press.

Foltz, C. B., Weymann, R. J., Peterson, B. M., Sun, L., Malkan, M. A. 1986, ApJ, 307, 504.

Heisler, J. and Ostriker, J. P. 1988, ApJ, 332, 543.

Henry, R. C. 1991, ARAA, 29, 89.

Ikeuchi, S. and Turner, E. L. 1991, ApJ, 381, L1.

Irwin, M., McMahon, R. G. and Hazard, C. 1991, in ASP Conf. Ser. 21, *The Space Distribution of Quasars*, ed. D. Crampton (San Francisco:ASP), p. 117.

Jakobsen, P., Boksenberg, A., Deharveng, J. M., Greenfield, P., Jedrzejewski, R., and Paresce, F. 1994, Nature, 370, 35.

Kovner, I. and Rees, M. J. 1989, ApJ, 345, 52.

Kulkarni, V. P. and Fall, S. M. 1993, ApJ, 413, L63.

Lanzetta, K. M., Wolfe, A. M., Turnshek, D. A., Lu, L., McMahon, R. G., and Hazard, C. 1991, ApJS, 77, 1.

Lesser, M. 1994, Proc. SPIE, 2198, 782.

Madau, P. 1992, ApJ, 389, L1.

Miralde-Escude, J. and Ostriker, J. P. (1990), ApJ, 350, 1.

Murdoch, H. S., Hunstead, R. W., Pettini, M. and Blades, J. C. 1986, ApJ, 309, 19.

Najita, J., Silk, J. and Wachter, K. 1990, ApJ, 348, 383.

Ostriker, J. P. and Heisler, J. 1984, ApJ, 278, 1.

Ostriker, J. P., Vogeley, M. S. and York, D. G. 1990, ApJ, 364, 405.

Paresce, F. 1990, in *The Galactic and Extragalactic Background Radiation*, ed. S. Bowyer, and C. Leinert, p. 307.

Peterson, B. A. 1978, in *IAU Symposium 79, The Large Scale Structure of the Universe*, ed. G. O. Abell and G. Chincarini (Dordrecht: Reidel), p. 349.

Pettini, M., Hunstead, R. W., Smith, L. and Mar, D. P. 1990, MNRAS, 246, 545.

Phillips, S. and Ellis, R. S. 1983, MNRAS, 204, 493.

Rees, M. J. and Setti, G. 1979, AAp, 8, 410.

Sargent, W. L. W., Young, P. J., Boksenberg, A. and Tytler, D. 1980, ApJS, 42, 41.

Schneider, D. P., Schmidt, M. and Gunn, J. E. 1989 AJ, 98, 1507.

Schneider, D. P., Schmidt, M. and Gunn, J. E. 1994 AJ, 107, 1245.

Trevese, D., Giallongo, E. and Camurani, L. 1992, ApJ, 398, 491.

Tytler, D. 1987, ApJ, 321, 69.

Warren, S. et al. 1987, Nature, 325, 131.

Warren, S., Hewett, P. C. and Osmer, P. S. 1991, ApJS, 76, 23.

Warren, S., Hewett, P. C. and Osmer, P. S. 1994, ApJ, 421, 412.

Weymann, R. J., Carswell, R. F. and Smith, M. G. 1981, ARAA, 19, 41.

Williger, G. M., Baldwin, J. A., Carswell, R. F., Cooke, A. J., Hazard, C., Irwin, M. J., McMahon, R. G., Storrie-Lombardi, L. J. 1994, ApJ, 428, 574.

Wright, E. L. 1990, ApJ, 353, 411.

Yee, H. K. C. and Green, R. F. 1989, ApJ, 319, 28.

York, D. G., Yanny, B., Crotts, A., Carilli, C., Garrison, E. and Matheson, L. 1991, MNRAS, 250, 24.

Young, P., Sargent, W. L. W., and Boksenberg, A. 1982, ApJ, 252, 10.

Zuo, L. 1992a, MNRAS, 258, 45.

Zuo, L. 1992b, MNRAS, 258, 36.

The Proximity Effect on the Lyman α Forest Due to a Foreground QSO

A. Fernández-Soto[1], X. Barcons[2], R. Carballo[1] and J.K. Webb[3]

[1] Departamento de Física Moderna, Universidad de Cantabria, Avenida de Los Castros s/n, E-39005 Santander, Spain

[2] Instituto de Física, Consejo Superior de Investigaciones Científicas - Universidad de Cantabria, Avenida de Los Castros s/n, E-39005 Santander, Spain

[3] School of Physics, University of New South Wales, Kensington NSW 2033, Australia

Abstract. We present the results of our study on three close pairs of QSOs. Our results are consistent with the existence of a proximity effect due to the foreground QSO, but due to its weakness we can only reject the absence of such effect at $\sim 1\sigma$ level. By modeling this proximity effect in terms of a simple photoionisation model, we find the best value for the UV ionizing background to be $\sim 10^{-20.5}$ erg cm^{-2} s^{-1} Hz^{-1} srad^{-1} at the Lyman limit, and an absolute lower limit (95% confidence) of $10^{-21.8}$ erg cm^{-2} s^{-1} Hz^{-1} srad^{-1}. This lower limit rejects a number of models for the UV background where it is mostly contributed by QSOs and absorption by Lyman limit systems is taken into account.

1 Introduction

The inverse effect on the QSO Lyman α forest is a well known feature. Several models have been suggested to account for this effect. Some of them point to some kind of local mechanism around the observed QSO (e.g. gravitational infall or photoionisation), while other ones postulate conditions that are totally extrinsic to the QSO itself, as could be the case for small absorbing clouds –unable to totally cover the QSO continuum emitting region when located close to it– or a flattening of the distribution of lines at high redshift. Studying the existence of the inverse effect caused by a QSO close to the line of sight towards a background one in the spectrum of the latter will allow us the distinction between these two families of models.

Table 1. Observational data

| OBJECT | COORDINATES (J2000) | | | | | | $\theta('）$ | z | m_V | WAVELENGTH |
	RA (α)			Decl. (δ)						RANGE (Å)
1055+021	10	57	57	01	54	03		2.73	17.8	
1055+021	10	57	13	01	47	55	12.4	2.29	20.0	3910-4111
1222+228	12	25	27	22	35	13		2.051	15.5	
1222+228	12	25	24	22	31	28	3.8	1.87	19.0	3255-3695
1228+077	12	31	21	07	25	18		2.391	17.6	
1228+077	12	31	08	07	24	39	3.3	1.878	17.5	3360-3845

Three of such pairs have been observed (see Tab. 1), and line lists –with parameters N, b and z for each line– have been obtained by absorption line fitting. We also performed correlation analysis to eliminate lines belonging to metal systems.

2   Results

Following the suggestion by Bajtlik et al. (1988) we have modeled the proximity effect in terms of the enhancement of the UV photoionizing flux (J_ν) by the nearby QSO. Thus, the model will allow us to estimate a value for the ratio of QSO UV flux to background flux –local to the cloud.

The model predicts that the distribution of Lyman α lines in the background QSO will have a 'hole' in the neighbourhood of the foreground one. The relative importance of this clearing will depend –as the *only* free parameter– on J_ν. We can fit this function to our data using Maximum Likelihood methods to obtain the best value for J_ν.

As this method turns out to produce a non-normalizable likelihood function for the values of J_ν, we have also used other statistical tools (mainly Bootstrap and Monte Carlo methods). All of them have led us to the same final values for J_ν.

This procedure finally allows us to marginally detect the Proximity Effect in our sample, but unfortunately the effect is so weak that we cannot reject its absence – there is a probability $\simeq 35\%$ of obtaining a result compatible with ours by chance, without any photoionization effect.

The best value that we obtain with our approach for the UV ionizing background is $\sim 10^{-20.5}$erg cm^{-2} s^{-1} Hz^{-1} srad^{-1} at the Lyman limit, with an absolute lower limit of $10^{-21.8}$erg cm^{-2} s^{-1} Hz^{-1} srad^{-1} (95% confidence). These values are compatible with other determinations of J_ν based on the inverse effect at similar redshifts (Bechtold et al. 1987 and this volume), and we agree with those works in pointing that this flux is stronger than it should be in order to explain all of it with the present models for QSOs (Bechtold et al. 1987, Madau 1992, Meiksin & Madau 1993, Madau, this volume).

More details on this work can be found in Fernández-Soto et al. (1994).

References

Bajtlik, S., Duncan, R.C., Ostriker, J.P. (1988): ApJ, 327, 570
Bechtold, J., Weymann, R.J., Lin, Z., Malkon, M.A. (1987): ApJ, 315, 180
Fernández-Soto, A., Barcons, X., Carballo, R., Webb, J.K. (1994): MNRAS, submitted
Madau, P. (1992): ApJ, 389, L1
Meiksin, A. & Madau, P. (1993): ApJ, 412, 34

Lyα Forest: Results and Prospects

R. F. Carswell

Institute of Astronomy, Madingley Road, Cambridge CB3 0HA, England

Abstract. The distributions of the numbers of Lyα systems with respect to redshift, HI column density and Doppler parameter are examined. While a number of the suggested features may be due to selection effects, it is likely that there is some redshift dependence in the distributions of the other two parameters. At high redshifts there may be a population of relatively shortlived clouds, while those seen at lower redshifts could be a longer-lived component which is often associated with galaxies. The prospects for testing models using line profiles are discussed.

1 Introduction

As with the heavy element systems, studies of the Lyα forest systems are at a turning point because of the existence of HIRES on the Keck telescope, and proposed high resolution capabilities on other very large telescopes. Also considerable strides are being made in identifying their sizes and whereabouts from HST observations (Lanzetta 1995; Lanzetta et al., 1995; Dinshaw et al., 1995a,b).

Most of the statistical work to date has been done using intermediate dispersion spectra on large samples of quasars, particularly with a view to determining the redshift evolution of the numbers of the Lyα systems per unit redshift. This topic has been covered by Bechtold (1995), so I don't propose to repeat it here. Instead I shall outline what one learns from high resolution (full-width-half-maximum < 30 km s^{-1} spectroscopy, and point to some future possibilities in this area. Part of this has been covered by Tytler (1995) in his description of the results of Keck spectroscopy of high redshift quasars.

2 Data

When you are presented with a high resolution spectrum containing many Lyman lines for which, despite your best efforts, corresponding heavy element lines are usually undetectable (but see Tytler, 1995), you can measure, in increasing order of difficulty, the redshift z, Lyman line equivalent widths, HI column density N(HI) and Doppler parameter $b = \sqrt{2}\sigma$ for each component. And that is about all, unless you have really high signal-to-noise ratio (S/N) data so you can examine details of the line profiles. If you have two sightlines close together, then you can obtain highly important size estimates (see Dinshaw et al, 1995), and size scales for variations of the measurable parameters, but otherwise our inferences are based basically on tables of three numbers (z, N(HI) and b) for as large a number of systems as we can get covering as wide a range of redshifts as we can get. To do this requires resolutions of order 10 km s^{-1}, covering a wide wavelength range, so echelle spectroscopy is required. On 4m telescopes this has been a photon-starved subject, with exposures measured in nights rather than hours, and S/N per pixel often < 10. A consequence of the low S/N is likely to be that blends of Lyα systems may not be recognized as such and treated as single lines. Indeed, to avoid over-interpreting the data the strategy adopted is to fit the minimum number of components to a complex that is consistent with the data (see Rauch et al., 1992). Under these circumstances there may be several more systems for which the inferred Doppler parameters are higher, and the HI column densities may be higher, than there is in the underlying population.

3 Distributions

3.1 HI Column Density

The HI column density distribution has been measured using 4m telescopes at a variety of redshifts, and a power law fit with numbers of systems N per unit column density of the form

$$d\mathcal{N} \propto N^{-\beta} dN$$

where $\beta \sim 5/3$ have been obtained. This power law has been fitted for over a small range of column densities, perhaps $13.3 < \log N < 15$ at best.

At the lower end of the redshift range for which the Lyα systems are measurable at optical wavelengths too few quasars have been studied, and so too few Lyα systems measured, to constrain the numbers much above $\log N = 14.5$, and in any case this is the range where the HI column density tends to be uncertain if only Lyα is measured because of line saturation effects. The lower column density limit is set by asking what is the weakest line one would reliably detect at a reasonably large Doppler parameter. Incompleteness effects will become increasingly important at lower column densities. The question here is, what is a "reasonably large Doppler parameter", given that some of the very broad lines are likely to be blends? Basically one want to ensure that sufficiently few lines are missed that the overall distributions will be barely affected, and setting the column density limit to the minimum value which would be detected if the Doppler parameter is twice the median value seems to work reasonably well.

At the highest redshifts the problems in determining the higher HI column densities are eased because of the accessibility of high order Lyman lines to constrain the profile fits, but made difficult by the blending caused by the increased line density. Figure 1 illustrates how bad this can be, and how different the spectra are at redshifts $z \sim 2$ and $z \sim 4$. The further complication introduced at high redshifts by the high line density is that there is a

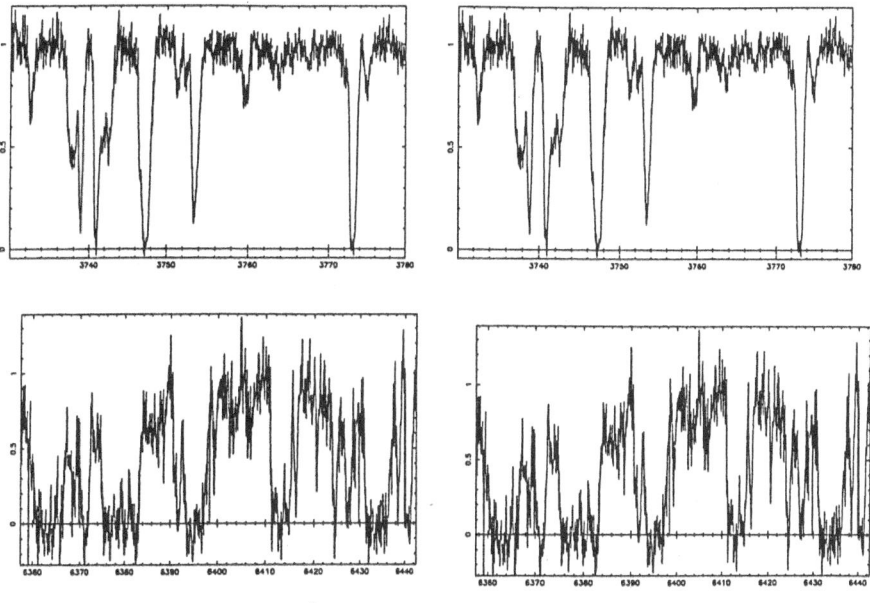

Fig. 1. Echelle spectra of $1100 - 264$ ($z = 2.15$, upper) and $1033 - 03$ ($z = 4.5$, lower) covering the same velocity interval

significant probability that low column density systems will be lost in high column density ones. Williger et al. (1994), in their analysis of the $z = 4.5$ quasar $1033 - 03$, showed that at redshifts $z > 4$ a large fraction of the lines with $\log N < 14$ would be missed in this way. They concluded that, after correction for this effect, the column density distribution function found for redshifts $z \sim 2$ could apply, with only a change in normalization, at $z \sim 4$. Of course, this is not to say that the distributions are necessarily similar – it is more like saying that, after allowing for selection, they are not necessarily different.

Our inability to match the column density distributions over the full range of column densities because of the selection effects at these very different redshifts adds to the uncertainties in inferences about the redshift evolution of the numbers of the Lyα systems. If we are to say anything about the redshift evolution we now either assume that the shape of the distribution remains the same or we look for equivalent width dependence of the redshift evolution. The first approach was used by Morris et al (1991) to compare their 3C273 results with $z \sim 2$ to yield, for the usual assumed form

$$\frac{d\mathcal{N}}{dz} \propto (1 + z)^{\gamma}$$

$\gamma = 0.8 \pm 0.4$. For the range $z \sim 2$ - $z \sim 4$ Williger et al. (1994) found $\gamma = 4.6$. On the other hand, Giallongo et al. (1993) find evidence for HI column density dependence in the number evolution, suggesting that the shape of the distribution does evolve.

3.2 Doppler Parameter

The distribution of Doppler parameters has been discussed in several publications (see e.g. Rauch et al, 1993), with the general result that at redshifts $z \sim 2$ - 3 there is a spread of values about a median of about 30 km s^{-1}. Tytler (1995) finds a value which is a little lower from his excellent Keck spectrum of $0636 + 68$, but at high S/N it is not very surprising that the median b will be a little lower. Presumably several of the lines which earlier workers would have reported as being single are blends. Tytler also reports that the correlation between b and N reported by Pettini et al. (1990) for systems with log$N$ less than about 13 is not seen in his data, but that there seems to be an excess of narrow systems at well below the previous detection limit. To some extent this confirms the conclusion by Rauch et al. that the correlation reported by Pettini et al. arises from uncertainties in the analysis of weak lines and selection effects.

There has been an indication that the b distribution does change with redshift. The Williger et al (1994) analysis of the Lyα forest at $z \sim 4$ towards $1033 - 03$ revealed an excess of systems with Doppler parameters compared with similar analyses at redshifts $z = 2$ - 3. Of course the nature of the Lyα forest spectrum is very different at $z = 4$ than at $z = 2$, so there are concerns that the low b systems may not be real. Most of the systems measured at the highest redshifts are in highly blended systems, and there is an excess of high b systems (figure 2) which presumably is a result of the failure to resolve the components to the level possible at lower redshifts. It is possible that other components of the blends could have artificially low Doppler parameters, simply because their neighbors have been analysed as broad single lines rather than blends. Some simulations, and checks against other spectra, suggest that this effect will not be strong enough to explain the excess seen, but the only way of checking this properly is through the analysis of simulations. This is a big job, and has not yet been done. The best guess at present, though, is that it is likely that there really is a narrower subpopulation at the highest redshifts.

4   Comparison with Models

There have been several suggestions concerning the location of the Lyman forest systems, ranging from pressure-confined systems to associations with normal galaxies. The work described by Lanzetta et al (1995) shows that, at least at low redshifts ($z < 1$), a significant number of the systems which show strong Lyα lines are associated with galaxies. At high redshifts the situation is not at all clear, and it is tempting to associate the narrower systems found by Williger et al. (1994) at $z \sim 4$ with a transient population of (intergalactic?) clouds which disappear as the universe evolves. If this is correct, then the rapid evolution in numbers at redshifts above $z \sim 2$, and the near comoving uniform density below that redshift, suggests that the transient population has largely disappeared by $z \sim 2$. Models involving mini-haloes, or gravitational collapse of small-scale structure, intergalactic medium shocks, etc., then have to satisfy an additional constraint that these mechanisms cease to be effective at low redshifts.

It turns out to be quite difficult to use the line profiles to differentiate between the various possibilities. Even the simple-minded approach, where you might consider that all Lyα systems are at 10^4K and any excess Doppler parameter arises simply from clustering of the systems, is hard to test. To illustrate this we take a fairly clean profile from $1100 - 264$, fit it with the minimum number of components we can get away with at 10^4K, and then ask at what S/N we would see component structure without making this assumption. Table 1 gives the parameters for a line at 3707A, and figure 2 the data with the three - component fit.

Table 1. Lyα at $z = 2.049$ towards $1100 - 264$

z	b km s^{-1}	logN(HI) cm^{-2}
Single component fit:		
2.04925	22.7	13.79
Three components at 10^4K:		
2.04905	12.85	13.02
2.04923	12.85	13.52
2.04942	12.85	13.23

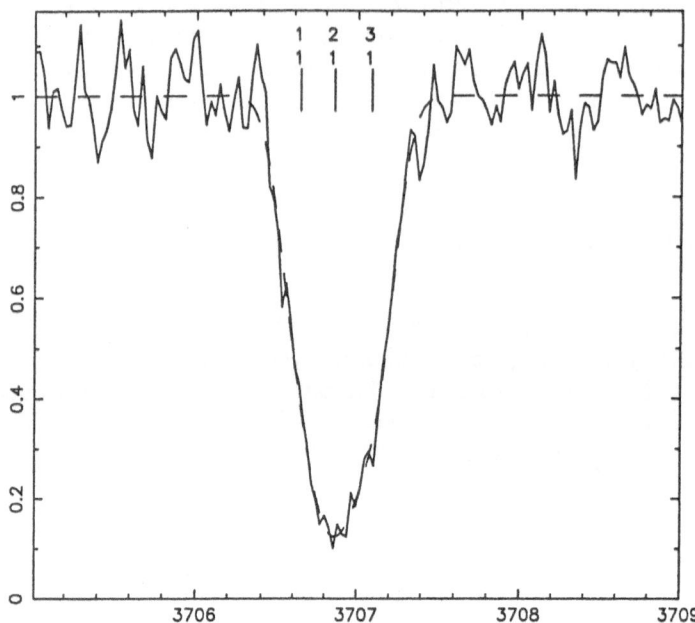

Fig. 2. A Lyα feature in the spectrum of 1100 − 264 (solid line) with three fitted components at 10^4K (dashed line). Tick marks show the positions of each of the components.

Simulations at various S/N show that even if the S/N=50 (per 0.0275A pixel) the single component fit is still acceptable (figure 3), with a χ^2 probability that the fit is a good one of $> 1\%$, and K-S probability $\sim 8\%$. A two-component fit is good, with most of the line as one component with $b = 21$ and logN(HI)=13.78, with a weak $b = 5$ and logN(HI)=12.2 component in the red wing. The fact that this line was chosen because of hints of low level asymmetry in the 4m data, and because it is narrow so few components would be needed, suggests that in the general case even higher S/N will be required to deblend features satisfactorily if the individual components are narrow.

Other cases, based on more realistic models, fare little better. The impressive models involving collapse of small-scale structure by Cen et al (1994), which have quite complex velocity fields, show very close approximations to Voigt profiles when the temperature broadening is folded in. Even uniform expansion requires considerable velocity fields to be detected easily − 20 km s^{-1} across a cloud folded in with thermal broadening shows only hints of the flat bottomed profile which is characteristic of such systems. It may be possible to do better by stacking profiles, or looking at average parameters in some way, but the early indications from Keck data are that Voigt profiles provide a reasonably good fit.

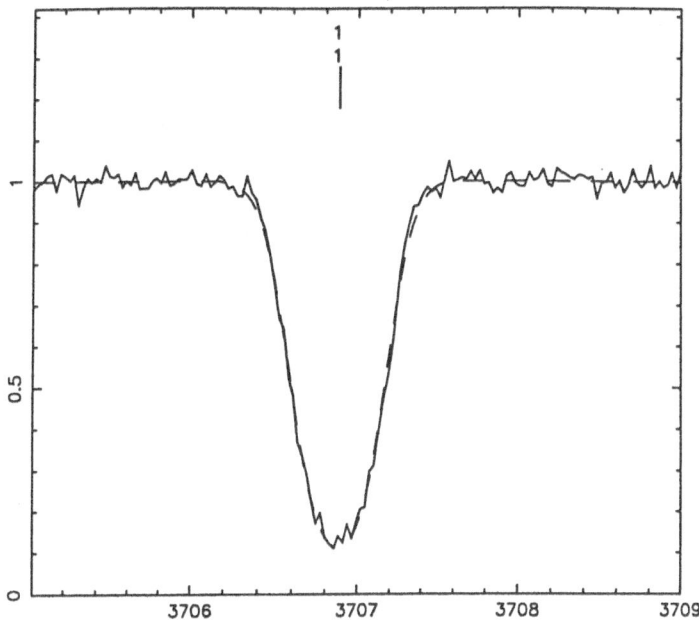

Fig. 3. A single component fit (dashed) to a 3-component line in a simulated spectrum at S/N=50 (solid line)

Note that for programs as demanding as this a good knowledge of the instrument profile as it applies to the object is required. In all fitting procedures Voigt profiles are convolved with the instrument profile are fitted to the data to determine the Voigt profile parameters. The instrument profile determined from the comparison lines will not apply accurately to the object observations unless the seeing profile considerably exceeds the slit width, and this is not generally the case since the observer tries to maximize the S/N by getting a large fraction of the light down the slit. For the S/N=50 line used above overestimating the instrumental profile width by 1/3 reduces the probability that the fit describes the data by about a factor $2\frac{1}{2}$, so an acceptable fit can easily become unacceptable if the wrong instrument profile is used. This will be increasingly true at higher S/N. An easy way around this problem, at the expense of some light loss, is to use a fibre feed to the spectrograph so that the image and comparison profiles are the same as is proposed for the Gemini 8m telescope high resolution optical spectrograph. A less satisfactory alternative is to fit using a range of acceptable instrument profiles and see if departures from Voigt profiles are needed for all.

5   Future Developments

The high S/N spectra from the Keck telescope are already yielding several important new results (see Tytler, 1995). No doubt the distribution functions against b and N(HI) will be refined, and extended to lower column densities over a range of redshifts. Then, with careful analysis, the redshift dependence of the distributions should become evident. Then, by looking at the evolution of features in these distributions it may be possible to trace the average evolution of ionization and velocity field in the clouds with redshift.

One of the results described by Tytler (1995), showing that heavy elements are present in Lyα systems with logN(HI)$>$ 14.5, is potentially very important. Not only does it show that the systems are enriched, and presumably associated with stars in some way, but it also provides a way of testing the nature of the highest redshift systems. Williger et al. (1994) found a large number of systems in the Lyα forest with logN(HI)$>$ 15. If these turn out also to have CIV associated with them then early enrichment of material far from galactic centres is indicated. If at the same time those with low b-values do not show heavy elements, then the idea of a relatively transient narrow-line population would be reinforced. We await the results of the analysis of the Keck data with considerable interest.

References

Bechtold J., 1995, these proceedings.
Cen R., Miralda-Escudé J., Ostriker J.P., Rauch M., 1994, ApJL, 437, L9.
Dinshaw N., Foltz C.B., Impey C.D., Weymann R.J., Morris S.L., 1995a, Nature, in press.
Dinshaw N., Foltz C.B., Impey C.D., Weymann R.J., Morris S.L., 1995b, these proceedings.
Giallongo E., Cristiani S., Fontana A., Trevese D., 1993, ApJ, 416, 137.
Lanzetta K.M., 1995, these proceedings.
Lanzetta K.M., Bowen D.V., Tytler D., Webb J.K., 1995, ApJ, in press.
Morris S.L., Weymann R.J., Savage B.D., Gilliland R.L., 1991, ApJL, 377, L21.
Pettini M., Hunstead R. M., Smith L., Mar D.P., 1990, MNRAS, 246, 545.
Rauch M., Carswell R.F., Chaffee F.H., Foltz C.B., Webb J.K., Weymann R.J., Bechtold J., Green R.F., 1992, ApJ, 390, 387.
Rauch M., Carswell R.F., Webb J.K., Weymann R.J., 1993, MNRAS, 260, 589.
Tytler D., 1995, these proceedings.
Williger G.M., Baldwin J.A., Carswell R.F., Cooke A.J., Hazard C., Irwin M.J., McMahon R.G., Storrie-Lombardi L.J., 1994, ApJ, 428, 574.

Simulation Analysis of the Lyman–Alpha Forest

Adam Dobrzycki[1] and Jill Bechtold[2]

[1] Center for Astrophysics, 60 Garden Street, Cambridge, MA 02138, USA
[2] Steward Observatory, University of Arizona, Tucson, AZ 85721, USA

Abstract. We present a simulation based analysis of moderate resolution spectra of the Ly-α forest regions in 17 quasars. Our goal was to establish the distributions of the physical parameters of the Ly-α forest clouds minimizing the subjective part of the analysis. We find the index of the power law distribution of line column density to be β=1.4±0.1 for N=10^{13}–10^{16} cm^{-2}, somewhat smaller than the value derived in the line list based analyses. We attribute this discrepancy to flattening of the distribution at low N.

Using a simulation based analysis of the Ly-α forest at z≈3, we determine the distributions of the two physical parameters of the Ly-α clouds, the neutral hydrogen column density, N, and Doppler broadening parameter, b. In the traditional approach, this requires fitting a Voigt profile to the observed lines in data with sufficient resolution to resolve the lines. Column densities of the Ly-α lines appear to follow a power law, $d\mathcal{N}/dN \propto N^{-\beta}$, while the Doppler parameter distribution can be approximated with a Gaussian, $d\mathcal{N}/db \propto \exp[-(b - <b>)^2/2\sigma_b^2]$.

The sample consists of 17 Ly-α forest spectra of bright quasars at z≈3, with ~50 km s^{-1} FWHM resolution and S/N>15 (per resolution element). Seven spectra were observed especially for this project, and the remaining ten were taken from Bechtold (1994). All spectra but one were observed with the MMT and the same spectrograph (Blue Channel). The remaining spectrum, which was observed with the Mt.Palomar 5-m Hale telescope with 2D-Frutti Photon Counter, was added to the sample because it had virtually the same resolution and S/N ratio. All spectra were reduced in the same way. Details of the sample can be found in Dobrzycki & Bechtold (1995).

For each set of parameters, and for each quasar in the observed sample, its simulated duplicates were generated. Thus, each set of simulations consisted of 17 "quasar" spectra, having identical z_{em}, binning, continuum, instrumental resolution, and S/N ratios as the actual observed spectra. For each simulation, lists of resolution elements with significant absorption were generated. The distributions of (1) equivalent widths of these elements and (2) separations between them were then compared with the respective distributions of analogical quantities derived from observed spectra.

Fig.1 presents the main result obtained in this study.

• There is good agreement between simulation-based and line-list based derivations of the line normalization. See Fig.1 caption for the definition of the simulation parameter $\mathcal{N}_3$. We got $\mathcal{N}_3$= 45±10. Line-list based studies

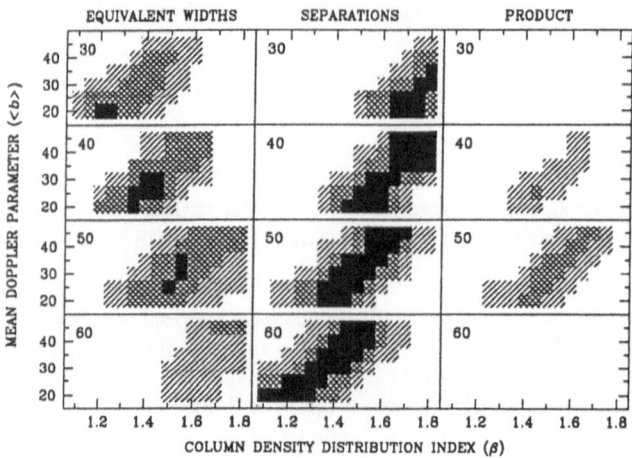

Fig. 1. The points on the simulation grid in which the K-S probability that the two samples were drawn from the same distribution was higher than 0.0003 (single-hatched), 0.05 (cross-hatched), and 0.33 (solid filled). Left panel: distributions of equivalent widths of absorption bins, middle panel: scaled separations between bins, right: the product of the two. The number in the upper left corner is a normalization for each simulation; physical interpretation of this parameter is: the number of strong (>0.36Å) lines in a spectrum of a hypothetical quasar at $z_{em}=3$.

(e.g. Bechtold 1994 and references therein) give $N_3=35$–45 with a typical uncertainty of 15.

- We find $\beta=1.4\pm0.1$, which is quite interesting. It is lower than the values obtained with the analysis of line lists, which give $\beta=1.7$–1.9 (e.g. Petitjean et al. 1993 and references therein), but it is consistent with indirect studies (e.g. Webb et al. 1992; Press & Rybicki 1993). We attribute this difference to flattening of the distribution at $N<10^{13.7}\,\mathrm{cm}^{-2}$, recently confirmed by the observations from the Keck telescope (Tytler 1995).

- Fig.1 yields $<b>=30\pm10\,\mathrm{km\,s}^{-1}$. This is well below the resolution of our sample, which is about twice this value. We can therefore only claim that our result is consistent with previous determinations (Carswell et al. 1991 and references therein).

References

Bechtold, J. 1994, ApJS, 91, 1
Carswell, R.F., et al. 1991, ApJ, 371, 36
Dobrzycki, A., & Bechtold, J. 1995, ApJ, submitted
Petitjean, P., et al. 1993, MNRAS, 262, 499
Press, W.H., & Rybicki, G.B. 1993, ApJ, 418, 585
Tytler, D. 1995, this conference
Webb, J.K., et al. 1992, MNRAS, 255, 319

Evidence for Large, Quiescent Lyman-Alpha Clouds from HST UV Spectroscopy of the Quasar Pair Q0107-025A,B

N. Dinshaw[1], C.B. Foltz[2], C.D. Impey[1], R.J. Weymann[3] and S.L. Morris[4]

[1] Steward Observatory, University of Arizona, Tucson, AZ 85721, USA
[2] Multiple Mirror Telescope Observatory, Tucson, AZ 85721, USA
[3] Carnegie Observatories, 813 Santa Barbara St., Pasadena, CA 91101-1292, USA
[4] Dominion Astrophysical Observatory, Victoria, British Columbia, Canada

Abstract. We present ultraviolet spectra of the Lyα forest of the quasar pair Q0107–025A,B ($z_{em} = 0.956, 0.952$; angular separation 1.44') obtained with the *Hubble Space Telescope*, in which we detect four absorption lines common to both spectra in the redshift range $0.5 \leq z \leq 0.9$, and six lines which are seen in the spectrum of one quasar but not the other. Assuming that these are Lyα lines, the directly-measured lower limit on the characteristic radii of the clouds is between 160 and 180 h_{100}^{-1} kpc (where $h_{100} \equiv H_0/100$ km s^{-1} Mpc^{-1}, $q_0 = 0.5$). The rms velocity difference between the common absorption lines along the two lines of sight is only about 100 km s^{-1}. These direct measurements lead to a picture of absorbing clouds that are both larger in extent and more quiescent than can easily be explained by current theoretical models.

1 Introduction

The absorption lines of Lyα observed in the spectra of high redshift quasars are thought to arise in cosmologically-distributed, intervening "clouds," the origin and physical nature of which are still unknown. Various models have been proposed, including pressure-confined clouds in a hot intergalactic medium (Ikeuchi & Ostriker 1986), relics of primordial density fluctuations associated with the cold dark matter (CDM) scenario for the biased formation of galaxies (Bond et al. 1988), gravitationally-confined clouds in CDM minihalos (Rees 1986), and shocks resulting from explosive galaxy formation (Lake 1988). Reliable information on the densities, ionizations and geometry of the clouds is required to distinguish between these models, which in turn depends on some knowledge of the cloud size.

The best published limits on cloud sizes suggest that they lie in the range from 5 kpc to 1 Mpc (Smette et al. 1992; Shaver & Robertson 1983; Crotts 1989). There are three *new* constraints on Lyα forest absorber sizes, all of which are based on the presence of common absorption lines in the spectra of physical or gravitationally-lensed pairs of quasars. In the first, observations of the lensed pair HE 1104-1805 have been used to infer a 2σ model-dependent lower limit on the cloud radii, at redshift $z \simeq 2$, of 50 h_{100}^{-1} kpc (Smette, these proceedings). Two independent studies of the quasar pair Q1343+2640 A,B separated by 9.5″ suggest cloud radii of at least $\sim 20\,h_{100}^{-1}$ kpc at $z \simeq 2$ (Dinshaw et al. 1994; Bechtold et al. 1994). Here, we present direct evidence for cloud radii larger than 160 to $180\,h_{100}^{-1}$ kpc at $z \simeq 0.7$.

2   Observations

The close pair of quasars Q0107–025A,B were observed on 1994 February 12 with the Faint Object Spectrograph (FOS) on the refurbished *Hubble Space Telescope (HST)*. The G190H grating was used on the red side of the FOS with the 1″ circular aperture yielding a spectral resolution (FWHM) of 1.4 Å, or about 200 km s^{-1}. Fig. 1 shows the flux calibrated *HST* spectra.

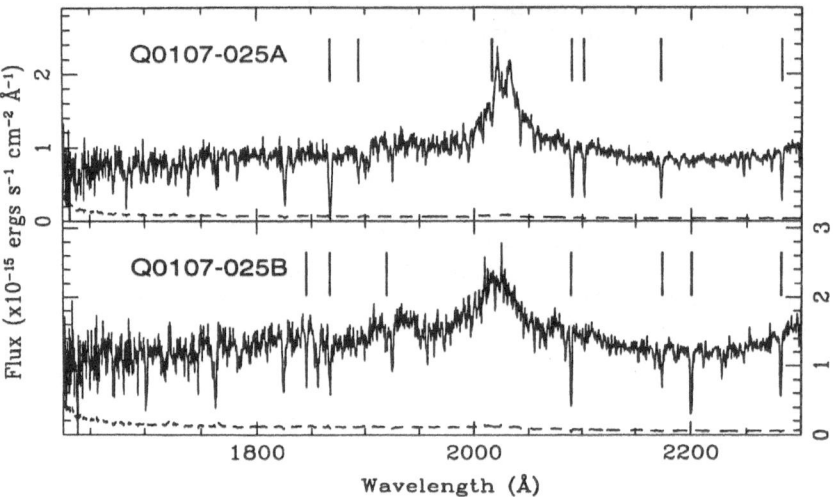

Fig. 1. *HST* FOS spectra of Q0107–025A and B as a function of vacuum, heliocentric wavelength. The dotted line in each panel shows the 1-σ errors. Tickmarks indicate $> 5\sigma$ absorption features with $W_0 \geq 0.33$ Å. Several strong lines which are not marked can be identified as higher order Lyman lines. The emission feature near 2020 Å is Lyman β+OVI $\lambda\lambda$ 1031, 1037.

Absorption lines were fitted by Gaussians unconstrained in width and central wavelength. We considered only those lines detected at the 5σ confidence level that were not identified as Galactic lines, metal lines or higher order Lyman lines. We also restricted our sample to lines with wavelengths longward of 1800 Å, shortward of which the S/N degrades rapidly and the identification of Lyα lines is confused by the presence of higher order Lyman lines. At 1800 Å, the 5σ rest equivalent width limit on an unresolved absorption line is about 0.33 Å in the spectra of both objects so that between 1800 − 2300 Å all lines with $W_0 \geq 0.33$ Å should be detected. These lines are listed in Table 1.

Table 1. Lyα Absorption Lines with $W_0 \geq 0.33$ Å

A				B				A–B
λ(Å)	σ_λ	W_0(Å)	z_{abs}	λ(Å)	σ_λ	W_0(Å)	z_{abs}	Δv(km/s)
		< 0.28		1845.17	0.06	0.44	0.5178	
1867.83	0.04	1.74	0.5364	1867.32	0.10	0.66	0.5360	82 ± 17
1894.34	0.10	0.60	0.5583			< 0.28		
		< 0.22		1919.99	0.33	0.38	0.5794	
2016.20	0.16	0.34	0.6585			< 0.18		
2090.33	0.03	0.94	0.7195	2089.55	0.03	0.79	0.7188	112 ± 4
2102.14	0.03	0.72	0.7292			< 0.16		
2172.52	0.03	0.65	0.7871	2173.58	0.07	0.40	0.7880	-146 ± 7
		< 0.14		2200.36	0.03	0.88	0.8100	
2282.03	0.02	0.67	0.8772	2281.61	0.03	0.58	0.8768	55 ± 5

3 The Size of Lyα Clouds

Table 1 contains four pairs of lines which are seen at very similar wavelengths in each spectrum. In all cases the redshift differences between the two features comprising the pair are small, $\Delta z < 0.0009$. The probability of getting *four* or more pairs of lines out of a total sample of 14 within this redshift difference is $\sim 10^{-7}$. Interpreting these as cases where the lines of sight to the two quasars pierce the same absorber, the lower limit on the transverse *diameter* of an absorber seen along both lines of sight is simply the proper separation of the lines of sight at the redshift of the absorber. For the common pairs observed here, the *model-independent* lower limits on the transverse *radii* range from about $160\,h_{100}^{-1}$ kpc for the lowest redshift pair at $z = 0.536$ to more than $180\,h_{100}^{-1}$ kpc for the pair at $z = 0.877$ (for $q_0 = 0.5$; assuming $q_0 = 0$, these limits become 180 and $225\,h_{100}^{-1}$ kpc, respectively).

If however we adopt the simplest model for the absorbers, *i.e.* identical and spherical clouds, then, using standard maximum likelihood techniques, we may estimate the characteristic size of the clouds. The most probable value and the upper and lower 95% confidence limits for the radius, R, of the clouds are estimated to be $R = 350\,h_{100}^{-1}$ kpc and $270 < R < 860\,h_{100}^{-1}$ kpc for $q_0 = 0.5$, respectively. Here, by "radius" we simply mean the impact parameter which produces column densities corresponding to our equivalent width detection limit. For $q_0 = 0.0$, the radii are larger by about 20%.

In addition to the sizes of the clouds, we may gain some insight into the bulk motions of the absorbing gas over scales equal to the separation of the lines of sight by measuring the difference in the radial velocities along the two lines of sight. The rms velocity difference for the four coincident pairs is 104 km s^{-1} and the limit on the accuracy of these measurements is dominated by random uncertainties arising in the wavelength calibration and line-fitting procedures, conservatively estimated to be about 17 km s^{-1}. Taken together, this suggests small, but statistically significant, velocity differences over scales of several hundred kpc.

4   Discussion

The observations presented here together with those of Q1343+2640A,B (Dinshaw et al., these proceedings) seem to imply an evolution in the cloud radii from $R \simeq 70\,h_{100}^{-1}$ kpc at $z \simeq 2$ to $R \simeq 350\,h_{100}^{-1}$ kpc at $z \simeq 1$. However, (1) these estimates are derived in the context of a simple-minded model; (2) they depend critically on the definition of "common" lines in spectra with low and rapidly-varying SNR; and (3) the upper limits on the sizes are particularly uncertain. Therefore, it may be premature to argue that evolution in the absorber size has been confirmed.

Our observations appear to suggest that the Lyα absorbers are highly correlated in the transverse direction over scales of several hundred kpc. By contrast, the Lyα lines observed with the *HST* Goddard High Resolution Spectrograph at 20 km s^{-1} resolution reveals little evidence for structure, *e.g.* an excess of line pairs with separation $50 - 300$ km s^{-1} (in contrast with ground-based observations which show clustering of the high column density lines [Chernomordik; Cristiani et al., these proceedings]). Though it must be noted that the transverse properties result from studies of the relatively strong lines reported here while the line-of-sight properties are obtained from high spectral resolution studies of weaker lines, taken at face value, the marked contrast between the *transverse* versus *line-of-sight* correlation properties of the absorbers suggests that the Lyα absorbers are coherent structures of low dimensionality, such as sheets or filaments.

If the absorbers are indeed sheet-like or filamentary, our observations cannot in principle provide an upper limit to the size of the Lyα absorbers. However, observations of quasar pairs with different separations can be used to map their topology. For example if the sheets are inhomogeneous or contain holes, then it may be possible to constrain the size of the inhomogeneities.

Acknowledgements. This research was supported by grant GO-5320.01-93A from the Space Telescope Science Institute, AURA Inc.

References

Bechtold, J., Crotts, A.P.S., Duncan, R.C., & Fang, Y. 1994, ApJ, 437, L79.
Bond, J.R., Szalay, A.S., & Silk, J. 1988, ApJ, 324, 627.
Crotts, A.P.S. 1989, ApJ, 336, 550.
Dinshaw, N., Impey, C.D., Foltz, C.B., Weymann, R.J., & Chaffee, F.H. 1994, ApJ, 437, L87.
Ikeuchi, S., & Ostriker, J.P. 1986, ApJ, 301, 522.
Rees, M.J. 1986, MNRAS, 218, 25.
Sargent, W.L.W., Young, P., Boksenberg, M., & Tytler, D. 1980, ApJ, 42, 41
Shaver, P.A., & Robertson, J.G. 1994, ApJ, 268, L57.
Smette, A., Surdej, J., Shaver, P.A., Foltz, C.B., Chaffee, F.H., Weymann, R.J., Williams, R.E., & Magain, P. 1992, ApJ, 389, 39

The Absorption Spectra of Q1107+487 and Q1442+295

Ruth Carballo[1], Xavier Barcons[2], and John K. Webb[3]

[1] Departamento de Física Moderna, Universidad de Cantabria, Santander, Spain
[2] Instituto de Física, CSIC-Universidad de Cantabria, Santander, Spain
[3] School of Physics, University of New South Wales, Kensington, Australia

Abstract. We present the first moderate resolution ($\simeq$ 40–120 km s^{-1}) spectroscopic observations of the bright ($V \leq 16.7$) high-redshift QSOs Q1107+487 (z_{em} = 2.965) and Q1442+295 (z_{em} = 2.669) (Sanduleak & Pesch 1989). The relatively high signal to noise reached in the spectra along with an extensive wavelength coverage of the Lyα and the Lyβ forest allowed us to obtain, through profile fitting, column densities and Doppler parameters of the Lyman clouds towards these QSOs. The spectral coverage of regions longward of the Lyα emission line of the QSOs at the expected wavelengths of CIV $\lambda\lambda$1548,1550 at the redshifts of the Lyα forest allowed us to identify some heavy element absorption systems (hereafter HEASs) towards these QSOs. We have found no Lyman absorption system or HEAS towards these QSOs for which our data allow a deuterium measurement or to provide an interesting upper limit for the D/H ratio. The reason for this is that the Lyman lines with high column density detected towards these QSOs belong to absorption systems showing velocity structure.

1 Observations and Data Analysis

The spectra were obtained on five nights of 1991 January and March using the IPCS detector on the Intermediate Resolution Spectrograph at the 2.5 m INT at the Observatorio del Roque de los Muchachos (La Palma, Spain). For a detailed description of the data reduction, continuum fitting, generation of the line lists and techniques of profile-fitting of the absorption lines, the reader is refered to Carballo et al. (1995).

2 Lyman Forest Absorption Systems

Only the Lyman forest systems with Lyα absorption longward of the QSO Lyβ emission were analised. When data at the corresponding Lyβ with reasonable S/N and spectral resolution were available and the lines were sufficienly unblended there, the Lyα and Lyβ lines were fitted simultaneously, in order to better constrain the N_{HI}, b and z parameters.

We have obtained the profile parameters of 96 Lyman forest lines towards Q1107+487 and 18 Lyman forest lines towards Q1442+295, within the redshift range $2.36 \leq z \leq 2.96$. The line sample towards Q1107+487 is large enough to obtain statistically significant information on the b and N_{HI} parameters of these clouds. The following results were obtained for the Lyα forest towards this QSO.

· At the relatively high S/N of our spectra no correlation between b and N_{HI} is apparent from the data (see Figure 1).

· The column density distribution of the Lyman clouds shows a flattening in the range $10^{13.6}\,\mathrm{cm}^{-2} \leq N_{HI} \leq 10^{14.1}\,\mathrm{cm}^{-2}$, above the estimated completness limit of the sample ($10^{13.5}\,\mathrm{cm}^{-2}$), which we believe to be a real effect.

· The mean Doppler parameter of the Lyman lines is $\langle b \rangle = 31 \pm 2\,\mathrm{km\,s}^{-1}$ with a dispersion $\sigma(b) = 16 \pm 2\,\mathrm{km\,s}^{-1}$.

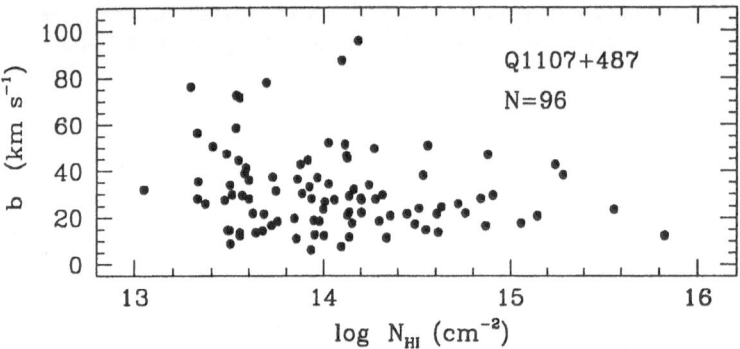

Fig. 1. Doppler parameter b versus $\log N_{HI}$ for the Lyman forest lines towards Q1107+487.

3 Heavy Element Absorption Systems

A rather strong HEAS was found towards each QSO through the detection of the CIV doublet longward of the QSO Lyα emission line. The redshifts of the systems towards Q1107+487 and Q1442+295 are $z = 2.760$ and $z = 2.439$ respectively. Both systems show complex velocity structure, spanning total velocity intervals of $\simeq 270\,\mathrm{km\,s}^{-1}$, with a mean value of the velocity splittings between the different components of $\langle \Delta v \rangle = 161 \pm 23\,\mathrm{km\,s}^{-1}$. The system towards Q1442+295 has detected OI λ1302 and CII λ1334 absorption lines - apart from the CIV doublet - longward of the QSO Lyα emission line. Comparison of the OI, CII and CIV profiles for this system reveals a systematic change of the ionization state with velocity.

References

Carballo R., Barcons X., Webb J.K., 1995, AJ, in press
Sanduleak N., Pesch P., 1989, PASP, 101, 1081

Common Lyα Absorption Toward the Quasar Pair Q1343+2640A,B

N. Dinshaw[1], C.D. Impey[1], C.B. Foltz[2], R.J. Weymann[3] and F.H. Chaffee[2]

[1] Steward Observatory, University of Arizona, Tucson, AZ 85721, USA
[2] Multiple Mirror Telescope Observatory, Tucson, AZ 85721, USA
[3] Carnegie Observatories, 813 Santa Barbara St., Pasadena, CA 91101-1292, USA

Abstract. We present observations of the Lyα forest of the close quasar pair Q1343+2640A,B. We detect 8 absorption lines of Lyα common to both spectra in the redshift range $1.7 < z < 2.1$, and 4 lines which are seen in one spectrum but not the other. At the $9.5''$ separation of the two quasars, this implies a firm lower limit on the characteristic radius of the Lyα clouds of $20\,h_{100}^{-1}$ kpc (where $h_{100} \equiv H_0/100$ km s^{-1} Mpc^{-1}, $q_0 = 0.5$) at a redshift $z \simeq 1.8$.

1 Introduction

The current best limits on the sizes of the Lyα clouds, derived from observations along the two lines of sight to gravitationally-lensed and projected pairs of quasars, lie anywhere in the range from a few kpc to 1 Mpc (Smette et al. 1992, and references therein). In this paper, we present spectra of the Lyα forest of the quasar pair Q1343+2640A,B ($z_{em} = 2.029, 2.031$) which are separated on the sky by 9.5 arcsec, corresponding to a linear separation of $\sim 40\,h_{100}^{-1}$ kpc for $q_0 = 0.5$, and which sample linear scales comparable to the cloud sizes predicted by several theoretical models (Sargent et al. 1980; Ikeuchi & Ostriker 1986; Rees 1986).

2 Observations

Spectra of Q1343+2640A and B were obtained from 1994 March 11 to June 2 UT with the Multiple Mirror Telescope, using the Blue Channel Spectrograph and Loral 3072×1024 CCD. The 832 line mm^{-1} grating, used with a $1''$ slit, yielded a spectral resolution of 1 Å FWHM. Exposures of a He-Ar-Cu calibration lamp were taken before and after each object exposure. Uncertainties in the wavelengths amounted to no more than $\sigma \simeq 0.15$ Å.

The line list, assembled from Lyα lines with observed equivalent widths, $W > 5\sigma_W$ from $z \simeq 1.79$ to the emission redshifts of the quasars, is presented in Dinshaw et al. (1994; Table 1). All the 5σ Lyα lines with a corresponding match within the velocity difference ± 150 km s^{-1} were assumed to be coincident to both spectra. The estimated number of *random* Lyα pairs with

$|\Delta v| < 150$ km s^{-1} is less than ~ 0.2. Under this criterion, eight of the Lyα lines are considered common or coincident and four were designated anticoincident.

3   Estimate of the Lyα Cloud Size

Assuming identical clouds, the firm lower limit on the radius of the absorbers equal to half the angular-diameter distance between the two lines of sight and independent of any assumptions about the variation of the cloud density with impact parameter is $R \simeq 20\,h_{100}^{-1}$ kpc. We note that although we refer to "cloud size" in this paper, we cannot distinguish between coherent entities and correlated but distinct structures, $i.e.$, large ensembles of smaller clouds.

The characteristic size of the Lyα clouds was estimated by means of Monte Carlo simulations, assuming all the clouds are identical, spherical and have constant column density. The simulations are very similar in detail to those reported by Smette et al. (1992), and result in the determination of the ratio, f, of the number of coincidences to the number of coincidences plus anticoincidences which when compared to the observed ratio gives the characteristic cloud size. For 8 coincident and 4 anticoincident lines corresponding to $f = 0.67$, the characteristic radius R was estimated to be $70\,h_{100}^{-1}$ kpc with lower and upper limits of $41 < R < 175\,h_{100}^{-1}$ kpc. (For $q_0 = 0$, these numbers increase by $\sim 50\%$.)

Differences in wavelengths of the common absorption features probe the bulk motion along lines of sight separated by scales of 40 kpc. Existing data from Smette et al. (1992) indicate that the clouds are remarkably quiescent on scales of a few kpc. The rms velocity difference for the 8 coincident lines is 65 km s^{-1}, whereas the total 1σ uncertainty in the wavelengths is estimated to be ~ 24 km s^{-1}, suggesting that the rms velocity differences deviate from zero at the 2.7σ. The significance of this result is not overwhelming, but it indicates real, albeit small, differences in the projected radial velocities of the clouds on scales of $\sim 40\,h_{100}^{-1}$ kpc.

Acknowledgements. This research was supported by the National Science Foundation under grant AST 93-20715.

References

Dinshaw, N., Impey, C.D., Foltz, C.B., Weymann, R.J., & Chaffee, F.H. 1994, ApJ, 437, L87.
Ikeuchi, S., & Ostriker, J.P. 1986, ApJ, 301, 522
Rees, M.J. 1986, MNRAS, 218, 25
Sargent, W.L.W., Young, P., Boksenberg, M., & Tytler, D. 1980, ApJ, 42, 41
Smette, A., Surdej, J., Shaver, P.A., Foltz, C.B., Chaffee, F.H., Weymann, R.J., Williams, R.E., & Magain, P. 1992, ApJ, 389, 39

The Lyα Forest towards HS1700+6416 at High Resolution

P.M. Rodríguez-Pascual[1], A. de la Fuente[2], J.L. Sanz[3], M.C. Recondo[1], J. Clavel[4], M. Santos-Lleó[5], W. Wamsteker[1]

[1] ESA IUE Observatory, P.O. Box 50727, 28080 Madrid, Spain
[2] INSA/VILSPA, P.O. Box 50727, 28080 Madrid, Spain
[3] Depto. Física Moderna, Univ. de Cantabria, 39005 Santander, Spain
[4] ESA ISO Observatory, P.O. Box 50727, 28080 Madrid, Spain
[5] Observatoire de Paris, RA 173 CNRS, 92135 Meudon, France

Abstract. We have analyzed high resolution ($b \approx 15$ km/s) optical spectra of the QSO HS1700+6416 ($z_{em} = 2.72$). The high S/N ratio in the Lyα forest allows a reasonable determination of the neutral hydrogen column density as well as of the velocity dispersion of the clouds. A weak, but significant, trend for stronger lines to be broader suggests that their column densities (N_{HI}) and Doppler parameters (b) are related. A simplified model of very flattened, highly ionized clouds, gravitationally confined by dark matter, gives a good representation of the data, where the observed N_{HI}–b correlation is due to orientation effects.

1 The Lyα Forest and the Distribution of b and N_{HI}

The determination of the distribution of N_{HI} and b parameters of the Lyα absorption lines in the QSO spectra and the presence or absence of a correlation between them is a crucial item for the understanding of the nature of these intergalactic systems. Up to now, the studies reporting a positive detection of such a correlation have been disputed to be driven by SNR and selection criteria effects.

The high resolution spectra of HS1700+6416 taken at the WHT at La Palma have a SNR larger than 10 for the whole Lyα forest. The line selection criteria imposed on the data, assure that our sample is complete at the 90% for lines with $N_{HI} \geq 10^{13.5}$cm^{-2} and $b < 60$ km/s. The probability distribution $p(N_{HI}, b)$ can be computed directly if we assume that N_{HI} and b are independent and that the observed distributions of N_{HI} and b are good representatives of $p(N_{HI})$ and $p(b)$. Since we have determined the errors in these parameters, to obtain their 2D distribution, each point in the (N_{HI},b) plane is substituted by a 2D gaussian of dispersions ΔN_{HI} and Δb. From Fig. 1 is clear that the expected distribution under the assumption of N_{HI} and b independent does not match the actual observed distribution: the broader lines tend to be stronger than the narrow ones, although this does not imply that there are not narrow strong lines.

We have tested another alternative that includes the following assumptions: (i) Very flattened (slabs) clouds. (ii) The probability distribution of neutral hydrogen column density perpendicular to the slab (N_0) is given by a power law of index $\beta = -1$ with a cut-off near $N_0 = 10^{14}\,cm^{-2}$. (iii) There exist bulk motion in the plane of the slab, characterized by a Doppler parameter b_1 such that $b_1^2 = k \cdot N_0$, as well as an isotropic velocity distribution characterized by b_0. (iv) The orientation effects on the clouds are translated into the following observed column density and Doppler parameter $N = N_0/cos\,i$ and $b^2 = b_0^2 + b_1^2 \cdot sin^2 i$, where i is the inclination of the cloud with respect to the line of sight.

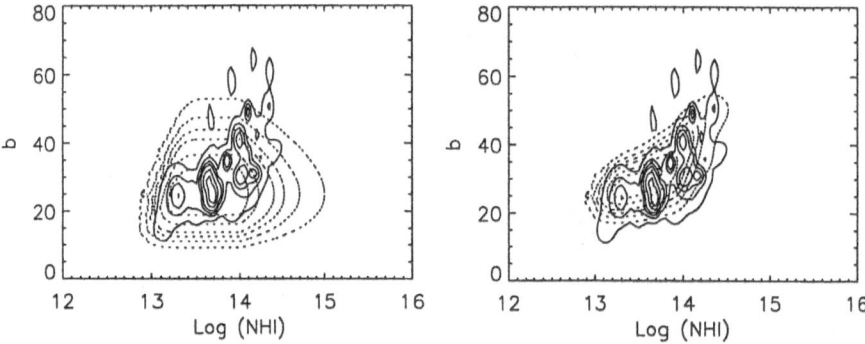

Fig. 1. Comparison of the observed distribution of ($log\,N, b$) (solid lines) with that expected from $p(log\,N)$ and $p(b)$ being independent (dotted lines). **Fig. 2.** As the previous one but for the model of flattened clouds described in the text.

These assumptions imply (Rodríguez-Pascual et al. 1994): (a) A broken power law distribution in N with index -2 for $N >> N_0$. (b) A b distribution strongly peaked at b_0 with a tail toward large Doppler parameters and with $p(b) = 0$ for $b < b_0$. (c) The 2D distribution of ($log\,N, b$) is much closer to the observed one. To state this point, a simulation of 5000 lines according to the previous assumptions is shown in Fig. 2, where we have used the following inputs: $b_0 = 20\,km/s$, $k = 300 \times 10^{-13} cm^{-2} (km/s)^2$ and $p(N_0) \propto N_0^{-1}$ for $12 < log\,N_0 < 14$. Then, this simulated ($log\,N, b$) distribution and the observed one are very much closer -when typical errors in the measurements and the actual detection limits are considered- that in the previous case of b and N_{HI} being independent.

A possible interpretation is that the clouds have a virialized baryonic matter component with a flattened geometry and a dark halo, with a proportion of 5% and 95%, respectively. Moreover, the clouds are highly ionized, i.e. the fraction of neutral hydrogen must be $\simeq 2 \times 10^{-5}$.

A complete report on the results presented in the paper are in Rodríguez-Pascual, P. M., et al. 1994, ApJ, submitted.

High-Resolution Spectroscopy of the Bright Quasar HS 1946+7658

A. de la Fuente[1], Rodríguez-Pascual[2], J.L. Sanz[3], M.C. Recondo[2], J. Clavel[4], M. Santos-Lleó[5], W. Wamsteker[2]

[1] INSA/VILSPA, P.O. Box 50727, 28080 Madrid, Spain
[2] ESA IUE Observatory, P.O. Box 50727, 28080 Madrid, Spain
[3] Depto. de Física Moderna, Univ. de Cantabria, 39005 Santander, Spain
[4] ESA ISO Observatory, P.O. Box 50727, 28080 Madrid, Spain
[5] Observatoire de Paris, URA 173 CNRS, 92195 Meudon, France

Abstract. The optical spectrum of the QSO HS 1946+7658 ($z_{em} = 3.05$) has been observed at high resolution ($b \sim 15$km/s) in the wavelength interval $4115 - 4970$Å. The signal-to-noise ratio (≥ 10) in the Lyα forest allows the determination of the neutral hydrogen column density as well the velocity dispersion of the intervening clouds by fitting Voigt profiles. The trend for stronger lines to be broader suggests that their column densities (N_{HI}) and Doppler parameters (b) are related. It is difficult to explain this observed log $N_{HI} - b$ diagram (specially the deficit of strong lines with low b values) as a selection effect.

1 Observations and Data Reduction

We have made spectroscopic observations of the QSO HS 1746+7658 ($z = 3.05$) with V = 15.85 (Hagen et al. 1992) on 1992 June 27 and 28 at the ISIS/WHT at La Palma. High resolution optical spectra were taken with the holographic grating H2400B and TEK CCD detector providing a dispersion of 0.19Å/pixel. These observations yielded five adjacent sub-spectra covering the interval 4117–4971Å and with an overlap of 30Å. The data reduction was carried out with the MIDAS software package. The continuum in each sub-spectrum has been fitted to a polynomial function though an iterative method allowing σ-clipping. In each step, the σ level is estimated by a gaussian fitting to the positive tail of the distribution of the residuals. The S/N ratio is ≥ 10 for the whole region. The influence of a limited S/N ratio in the Ly-α clouds analysis has been discussed by Rauch et al. (1993).

We restrict our analysis to the spectral range 4117–4808Å which covers almost the whole Ly-α forest but excludes a blue band of QSO Ly-α emission line to avoid the "proximity effect". We have also exclude from this analysis all the absorption lines related to identified Metal Line Systems using the data recently reported by Fan and Tytler (1994) and also our low and high resolution spectra.

2 The Ly-α Sample

We have imposed two constrains to select bands as regions suitable to be fitted: *(i)* At least one pixel must lie below the 4σ level and *(ii)* the number of pixels lying below the 2σ level must be larger than the FWHM. The absorption lines have been fitted to Voigt profiles convolved with the instrumental point spread function. The cloud parameters have been determined thought a minimization using the MINUIT package from the CERN software library.

This sample is complete at the 90% for the lines with log $N_{HI} \geq 13.4$ and $b<50$km/s. The analysis of the distributions of the Ly-α clouds parameters has shown:

(1) Weak (log $N_{HI}<14$) lines are characterized by b parameters of ~28km/s, with a dispersion of ~11km/s, that is marginally consistent with a constant value, given the uncertainties.
(2) There exists a high tail in the b distribution, and it is associated to lines with large column densities.
(3) The distribution of observed neutral hydrogen densities seems to be better described by a broken power law, with the break at log N_{HI}~14 as already noted by Carswell et al. (1987) and Giallongo et al. (1993).
(4) A weak, but noticeable, correlation between b and log N_{HI} is observed in the data. There is a lack of strong lines with low b values. It makes difficult to accept that N_{HI} and b are independent.

This results are in agreement with our similar observations of the QSO HS 1700+6416, and suggest a geometrical model of very flattened, highly ionized clouds, gravitationally confined by dark matter, where the observed $N_{HI}-b$ correlation is due to orientation effects (Rodríguez-Pascual et al. 1994). This simplified model is able to reproduce the observed distribution of neutral hydrogen column densities and Doppler parameters.

Acknowledgements. This work was partly supported (J.L.S., M.C.R., M.S.) by the Spanish DGICYT project PB92-0741.

References

Carswell, R. F., Webb, J. K., Baldwin, J. A., Atwood, B. 1987, ApJ , 319, 709
Fan X.-M. , Tytler, D. 1994, ApJS, 94, 17
Giallongo, E., Cristiani, S., Fontana, A., Trèvese, D. 1993, ApJ , 416, 137
Hagen, H-J., Cordis, L., Engels, D., Groote, D., Haug, U., Heber, U. Köhler, Th., Wisotzki, L., Reimers, D. 1992, A&A, 253, L5 246 ,545
Rauch, M., Carswell, R. F., Webb, J. K., Weymann, R. J. 1993, MNRAS , 260, 589
Rodríguez-Pascual, P. M., de la Fuente, A., Sanz, J. L., Recondo, M. C., Clavel, J., Santos-Lleó, M., Wamsteker, W. 1994, ApJ, submitted

PART VIII

LARGE-SCALE STRUCTURE

The Spatial Correlation of Ly-α Clouds

Stanislaw Bajtlik

Copernicus Astronomical Centre, ul. Bartycka 18, 00-716 Warsaw, Poland

Abstract. The usual expectation is that the spatial correlation of Ly-α clouds should be comparable to the correlation of galaxies. This is based on the assumption that the spatial separation of clouds is much greater then their size. Observations of the lensed QSOs and close pairs indicate this may not be the case. We show there are several factors diminishing the spatial correlation of clouds, namely: non-negligible size to separation ratio, redshift evolution and mixing in the same catalog objects located at a very wide range of redshift. The implication for the observing strategy is discussed.

1 Introduction

Thirty years have passed since the first theoretical predictions (Bahcall and Salpeter 1965, Wagoner 1967) and observational detections of the Ly-α clouds (Burbidge et al. 1966, Stockton and Lynds 1966). We still do not know their physical properties, such as: size, shape, density, ionization level, temperature, confinement mechanism or the formation epoch and scenario. There exists a plethora of models of clouds ranging from pressure confined (Ikeuchi and Ostriker 1986), to clouds in the mini-halo of the dark matter (Rees 1986), to slabs of intergalactic material (Charlton, Hogan, Salpeter 1993). Suggestions are made that Ly-α absorption lines arise in the extended gaseous halos of the most luminous galaxies (Lanzetta et al. 1995). For the review of the proposed models and basic observations of Ly-α forest see: Bajtlik (1993).

One of the most intriguing properties of the population of Ly-α clouds is the apparent lack of the spatial correlations in their distribution. Since the first extensive study of this problem, done by Sargent, Young, Boksenberg and Tytler (1980) (hereafter SYBT) most of the research resulted in no detection of the correlations. Most researchers used a one-dimensional version of the correlation function analysis (Carswell et al. 1984, Rees and Carswell 1987, Lu et al. 1991). Other statistical methods were also used. Crotts (1989) and Ostriker, Bajtlik and Duncan (1988) employed the nearest neighbour statistics, more suitable than the correlation function for detection of completely empty voids (rather then over-dense and under-dense regions). Dobrzycki and Bechtold (1991) also claimed a detection of a single void in the spectrum of a quasar they observed. Webb (1987) have claimed a detection of a positive correlation function on velocity scales between 50 and 200 km/s. However, so small separations are strongly affected by line blending and metal lines contamination. Duncan, Bajtlik and Ostriker (1988) developed a mathematical model for the description of the line blending effect. Other classes of the intergalactic absorbers (metal line systems) do show a strong correlation on a velocity scale of a few hundred km/s (SYBT, Sargent, Boksenberg and Steidel 1988).

Just recently Cristiani et al. (1994) announced a detection of significant clustering with $\xi \simeq 1$ at $\Delta v = 100$ km s^{-1}. Because of a very long controversy and recent observations with the opposite conclusions (Carswell, 1994), we must wait for further confirmation of a possible positive detection of correlations.

Some of the models of the formation of Ly-α clouds view this process as a part of the process of structure formation in the Universe in general (Bond, Szalay and Silk 1988, Miralda 1994). As most of the models predict or postulate cloud masses to be comparable to the galactic masses, we should expect clustering properties of clouds to be comparable to the clustering of galaxies.

In this paper we present the preliminary results of a detailed study indicating we should not expect to detect any significant one-dimensional correlation function. These reasons are: large cloud sizes relative to the mean cloud separations, rapid redshift evolution of the correlation function, wide redshift coverage in Ly-α catalogs, large scale peculiar motions.

2 Correlation Function Analysis

Galaxies and clusters are strongly correlated. Their correlation function is of the form:

$$\xi(r) = (r_0/r)^\gamma, \tag{1}$$

where $\gamma \simeq 1.8$. The correlation scale for galaxies is $r_0 \simeq 5h_{100}^{-1}$ Mpc and for clusters $r_0 \simeq 24h_{100}^{-1}$ Mpc (Peebles 1993). This correlation is equivalent to an excess probability of finding a second object at a distance $r = |\vec{r_{12}}|$ from the first one. A differential probability is:

$$\delta P = n^2 (1 + \xi(r)) \, dV_1 \, dV_2, \tag{2}$$

where n is a comoving mean number density of objects, and dV_1 and dV_2 are volume elements.

There is a crucial difference between 3-D or 2-D catalogs and a 1-D survey which QSO spectra represent. In the case of catalogs of galaxies we correlate positions of objects regardless of their physical sizes. In a one-dimensional catalog however physical sizes of objects – their cross sections – are crucial.

SYBT discussed this problem. The excess differential probability in a one dimensional survey is:

$$dp \doteq \phi_0 \, \sigma \, (1 + \zeta(s)) \, ds, \tag{3}$$

where ϕ_0 is comoving number density of objects, σ their cross section and s separation along the line of sight. SYBT concluded that provided cloud sizes are small compared with their separations the relation holds: $\zeta(s) = \xi(s)$.

The only evidence we have about cloud sizes comes from studies of spectra of lensed QSOs or close pairs of quasars. Foltz et al. (1984) and Bajtlik and Duncan (1991) studied spectra of Q2345+007 A and B (at $z_{em} = 2.14$) and concluded that clouds radii are comparable to $10 - 80$ kpc. Smette et al. (1992) from the analysis of common absorption features in UM 673 A and B (at $z_{em} = 2.727$) obtained a lower limit of 6 kpc for clouds radii. Just recently Dinshaw et al. (1995) obtain much larger limits for the cloud sizes from the spectra of Q0107-025 A and B (at $z_{em} = 0.956$ and 0.952). According to them clouds can be as big as several hundred kpcs. This means that their sizes could be comparable to their mean separations.

The situation is visualized on Fig. 1.

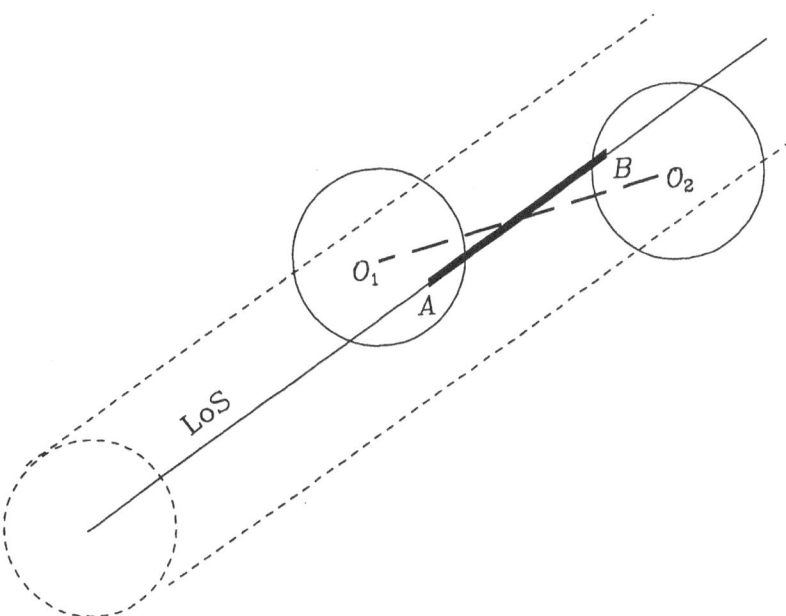

Fig. 1. Clouds of radii R are intercepted by line of sight (LoS). $\overrightarrow{O_1A} = \vec{r_1}$, $\overrightarrow{O_2B} = \vec{r_2}$, $\overrightarrow{O_1O_2} = \vec{r}$, $AB = s$.

Estimation of the correlation function in this case is equivalent to the three dimensional case, with the volume of integration corresponding to the volume of a narrow cylinder (marked with the dashed lines on Fig. 1).

We assume all clouds are of the same size R (this can represent pressure confined clouds in an inhomogeneous IGM). The relation between a true correlation function $\xi(r)$, where r is a physical distance between clouds centers and observed one dimensional function $\zeta(s)$, where s is a projected distance along the line of sight is:

$$\zeta(s) = 2\pi \int_0^R dr_1 \int_0^R dr_2 \int_0^{2\pi} d\phi \, r_1 \, r_2 \, \xi(r), \tag{4}$$

where

$$r = \sqrt{r_1^2 + r_2^2 - 2r_2 r_1 \cos\phi + s^2}. \tag{5}$$

Observations of Ly-α absorption in the optical range can cover a large redshift interval (typically $\Delta z \simeq 0.6$ for a high redshift QSO). The correlation function can vary quite substantially in this range. We can factorize time and space dependence:

$$\xi(r) = (r_0/r)^\gamma D^2(t). \tag{6}$$

For the flat Universe ($\Omega = 1$):

$$D(t) \propto \begin{cases} t^{2/3}, & \text{for } \xi \ll 1; \\ t^2, & \text{for } \xi > 1. \end{cases} \tag{7}$$

This is a very rapid evolution in redshift space. Changing variable from t to z, again for the flat Universe we have:

$$D(z) \propto \begin{cases} (1+z)^{-1}, & \text{for } \xi \ll 1; \\ (1+z)^{-3}, & \text{for } \xi > 1. \end{cases} \tag{8}$$

Taking this into account, from equations (4) and (6) we get:

$$\zeta(s) = 2\pi r_0^\gamma D^2(t) \int_0^R dr_1 \int_0^R dr_2 \int_0^{2\pi} d\phi r_1 r_2 r^{-\gamma/2}. \tag{9}$$

For a specific epoch we can calculate the numerical factor by which a "true" correlation function ξ is smoothed out by the fact that a given physical separation can correspond to a range of projected line of sight separations (and vice versa):

$$\zeta(s) = f(u)\xi(s). \tag{8}$$

Figure 2 presents this numerical factor f(u) as a function of the ratio $u = s/R$.

3   Conclusions

For clouds of sizes comparable to their separations the amplitude of the correlation function can be diminished by up to 50 percent. As the expected amplitude of the correlation function should not be very big (it is less then 1 for metal systems) this may be an important effect and should be taken into account in the analysis of the data.

Rapid evolution of the correlation function (see $D(t)$ above) makes searches for correlations in optically (i.e. $z > 1.6$) collected data hopeless. Contributing to the washing out of the correlations of Ly-α lines is a standard procedure of combining in the data sample lines from very different regions in redshift space. The more strongly correlated signal from lower z part of

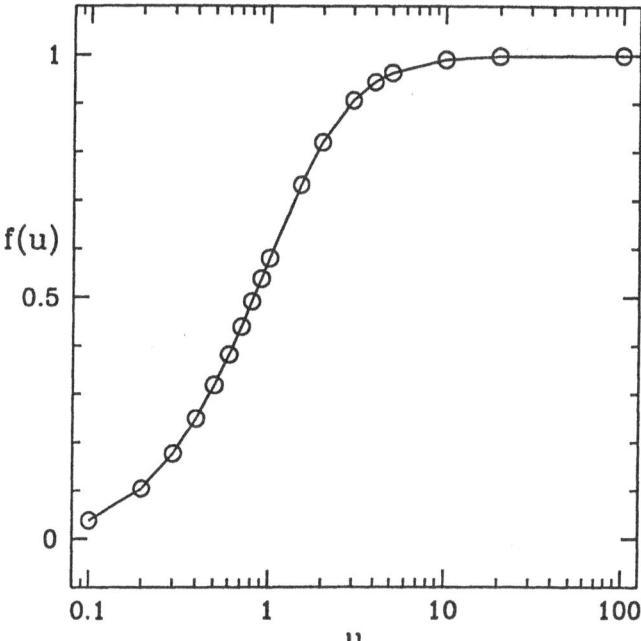

Fig. 2. Numerical factor f(u) relating ξ and ζ: $\zeta(s) = f(u)\xi(s)$, where $u = s/R$

the sample is diluted by more abundant and less correlated data for higher z. Using the formula in the equation (9) this effect can be quantified for each specific sample. In Bajtlik, Juszkiewicz, Bergeron (1995) detailed analysis of this effect as well as some additional effects contributing to diminishing of the correlation signal in 1-D catalogs will be presented. These effects are: flattened shape of clouds, allowing for a range of physical sizes of clouds, coherent peculiar motions on large scales. We will also present conclusions one can draw about the ratio of sizes of Ly-α and metal systems. Metal systems are correlated. If the large sizes of the Ly-α clouds (relative to their mean separations) are the cause of the apparent lack of the correlations between them we can hope to use this fact to learn about the ratio of sizes of both type of absorbers.

It is clear that the optimal observing strategy must be a compromise between two contradicting trends: it is favorable to look for correlations in a low redshift sample, as the correlation function is larger there. On the other hand there are fewer clouds at lower z. The optimal observational program should be to observe many lines of sight, each covering the same narrow redshift interval. The results will be presented in Bajtlik et al. (1995). Conclusions from this analysis will have applications to one dimensional (pencil beam) surveys of galaxies as well.

Acknowledgements. The author thanks J. Bergeron and R. Juszkiewicz for allowing the presentation of the results prior to publication. This research was partially supported by Polish grant KBN No 2 P304 022 06 and by PICS/CNRS No. 198 *Astronomie Pologne* Program. ESO supported the author's participation in the Workshop with funds from ESO C&EE Programme.

References

Bahcall, J.N. (1965). ApJ, 142, 1677

Bajtlik, S., (1993). The Evolution of the Intergalactic Medium and Ly-α Clouds, in Proceedings, Tetons Conference on Environment and Evolution of Galaxies, J.M. Shull, M.A. Thornson, Eds, Kluwer, p. 191

Bajtlik, S., Duncan, R.C. (1991). Proceedings of the ESO Mini-Workshop on Quasar Absorption Lines, ed. P.Shaver, E.J. Wampler, A.M. Wolfe (ESO Scientific Report No. 9 - February 1991), 35

Bajtlik, S., Juszkiewicz, R., Bergeron, J. (1995). To be published

Bond, J.R., Szalay, A.S., Silk, J. (1988). ApJ, 324, 627

Burbidge, E.M., Lynds, C.R., Burbidge, G.R.. (1966). ApJ, 144, 447

Carswell, R.F. (1995). This volume

Carswell, R.F., Morton, D.C., Smith, M.G., Stockton, A.N., Turnshek, D.A., Weymann, R.J. (1984). ApJ, 278, 486

Carswell, R.F., Rees, M.J. (1987). MNRAS, 224, 13P

Charlton, J.C., Salpeter, E.E., Hogan, C.J. (1993). ApJ, 402, 493

Christiani, S., D'Odorico, S., Fontana, A., Giallongo, E., Savaglio, S. (1995). to be published

Crotts, A.P.S. (1989). ApJ, 336, 550

Dinshaw, N., Foltz, C.B., Impey, C.D., Weymann, R.J., Morris, S.L. (1995). Nat, 373, 223

Dobrzycki, A., Bechtold, J. (1991). ApJ, 377, L69

Duncan, R.C., Ostriker, J.P., Bajtlik, S. (1988). ApJ, 345, 39

Foltz, C.B., Weymann, R.J., Roser, H.J., Chaffee, F.H. (1984). ApJ 281, L1

Ikeuchi, S., Ostriker, J.P. (1986). ApJ, 301, 522

Lanzetta, K.M., Bowen, D.V., Tytler, D., Webb, J.K. (1995). ApJ, submitted

Lu, L., Wolfe, A.M., Turnshek, D.A. (1991). ApJ 367, 19

Miralda-Escude, M. (1995). This volume

Ostriker, J.P., Bajtlik, S., Duncan, R.C. (1988). ApJ, 327, L35

Peebles, P.J.E. (1993). *Principles of Physical Cosmology*, Princeton University Press

Rees, M.J. (1986). MNRAS, 218, 25P

Sargent, W.L.W., Young, P.J., Boksenberg, A., Tytler, D., (1980). ApJS, 42, 41 (SYBT)

Sargent, W.L.W., Boksenberg, A., Staidel, C.C. (1988). ApJS, 68, 539

Smette, A., Surdej, J., Shaver, P.A., Foltz, C.B., Chaffee, F.H., Weymann, R.J., Williams, R.E., Magian, P. (1992). ApJ, 389, 39

Stockton, Lynds, R. (1966). ApJ, 144, 451

Wagoner, R.V. (1967). ApJ, 149, 465

Webb, J.K. (1987). PhD thesis, Cambridge University

Small-Scale Clustering in a Column-Density-Limited Sample of the Lyman-alpha Forest

Victor V. Chernomordik

Scientific Council on Cybernetics, Russian Academy of Sciences, 3A, 2-nd Baltiysky per., 125315, Moscow, Russia

Abstract. We have analyzed the clustering properties of Lyα forest lines in a HI column-density-limited sample over the redshift range $z = 2.7$–3.7. The sample has a relatively large lower cutoff column density ($N_c = 10^{14}\,\mathrm{cm}^{-2}$), and is based on published high-resolution spectra of three QSOs (0014+813, 0420−388, 2000−330). Two statistical tools (distribution of line intervals and two-point correlation function) were applied to discover the presence of small-scale clustering with velocity splittings $\sim$ 50–150 km/s at better than the 3σ level. Our analysis reveals also a gradual progression in clustering strength from virtually no clustering for weak lines to strong-line clustering. Marginal evidence for a change of cloud clustering properties with redshift is found: at $z \sim 2$ (high-resolution spectra of QSOs 2206 − 199 and 1101 − 264) similar clustering amplitude is observed for considerably smaller values of N_c than at $z \sim 3$. The importance of small-scale clustering for theoretical models of the Lyα forest is briefly discussed.

1 Introduction

Statistical characteristics of intergalactic Lyα clouds, in particular, their clustering properties can provide key constraints on theories of their physical nature and origin.

Previous studies have revealed a practically uniform velocity distribution for scales larger than 300 km/s (Sargent et al. 1980; Crotts 1989; Barcons and Webb 1991; Rauch et al. 1992).

Meanwhile, there were also some indications that for equivalent-width-limited samples with a relatively large cutoff equivalent width $W > W_c = 0.36\,\text{Å}$, Lyα lines cluster on smaller scales $200 < v < 400$ km/s (observed excess of the close line pairs $\xi \sim 0.3$–$0.5 \approx (2.5$–$3)\sigma$ (Webb 1987, Ostriker et al., 1988; Barcons and Webb 1991).

Recent results based on high-resolution spectra of two QSOs seem to be rather controversial (Rauch et al. 1992, 1993). On the one hand, no evidence for clustering on any scales was found for the total line sample of $0014 + 813$ spectrum ($z_{\text{em}} = 3.38$); on the other hand, there is marginal evidence for clustering at small scales 50– 150 km/s for the total line sample of $2206 − 199$ spectrum ($z_{\text{em}} = 2.56$).

The main purpose of our reanalysis of known high-resolution data is to extract information about a possible relationship between clustering properties of Lyα clouds and their HI column densities (Chernomordik 1995). Besides, some preliminary results concerning the redshift dependence of clustering amplitude are presented.

2   Correlation Characteristics of Lyα Forest lines

The basic sample of high-resolution data used was derived from the literature (Atwood et al. 1985, Carswell et al. 1987, Rauch et al. 1992). It has been summarized in Chernomordik 1995. There are a total 205 Lyα forest lines with $N_{\mathrm{HI}} \geq 10^{14}\,\mathrm{cm}^{-2}$ from 3 QSOs ($0014 + 813$, $0420 - 388$, $2000 - 330$) at redshifts $2.7 < z < 3.7$. To reveal the small-scale clustering, the sample and its subsamples have been analyzed by two statistical tools: the distribution of line intervals (DLI) and the two-point correlation function (TPCF) of the Lyα forest. Some results of our analysis are represented in Fig. 1. To simulate the influence of line blending on the DLI, we use the model of Ostriker et al. (1988).

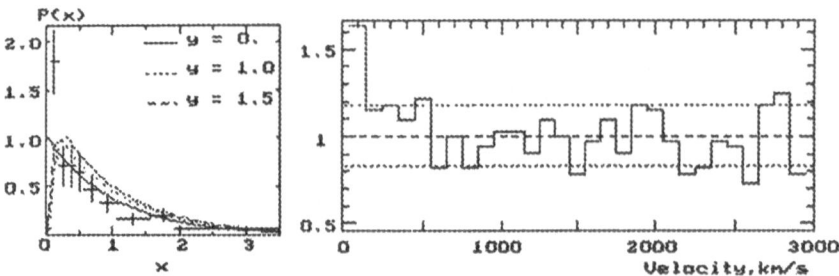

Fig. 1. Correlation characteristics of Ly alpha forest lines: (left panel) Distribution of line intervals ($0014 + 813$ & $0420 - 388$), where each interval is scaled to the mean value $x = \Delta z / \overline{\Delta z}$. Curves plotted show "blended Poisson" distributions for chosen values of the blending parameter y ($y = 0$ for simple Poisson distribution). The error bars are 1σ. (right panel) Two-point correlation function (total sample). The horizontal dashed line shows the mean level for a Poisson distribution. The horizontal dotted lines represent 1σ deviations from this mean level

Both the DLI and the TPCF show an evident peak near the origin (50–150 km/s). The pair number deviation from Poisson counting statistics is ~ 0.7, i.e. about 3.7σ for the total considered sample.

We note the sharpness of the peak in the Lyα forest splittings. A very small degradation in resolution will obscure the effect.

An absence of any clear signal on the scales ~ 100 km/s for the total line sample of 0014+813 found by Rauch et al. 1992 is likely to be a consequence of the smoothing influence of the more numerous and much less correlated weaker absorption lines. To confirm this, we analyze a set of column-density-limited subsamples of this total sample with varying cutoff N_c. The gradual decrease of clustering amplitude with reduction of N_c is really seen. In fact, the peak in the first bin vanishes for $N_c \sim 10^{13}\,\mathrm{cm}^{-2}$ (Fig. 2a).

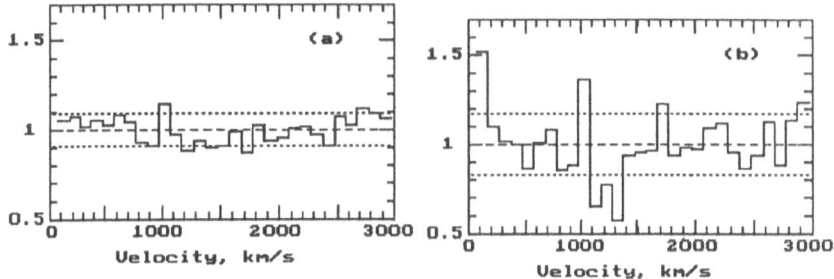

Fig. 2. Two-point correlation function, $N_c = 10^{13}\,\mathrm{cm}^{-2}$. The subsamples treated are based on the spectra of: (a) $0014 + 813$, and (b) $2206 - 199$ & $1100 - 264$

Interestingly, we note that Crotts 1989 has found significant clustering of Lyα clouds, observed in the relatively close lines of sight (transverse separation $\sim 0.5 - 0.7$ Mpc) at small velocity splittings (below 100 km/s) with a similar property: "a gradual progression in clustering strength from virtually no clustering for weak lines to strong-line clustering".

Marginal evidence for relationship between clustering amplitude of Lyα clouds and their redshifts can be found, if we consider, in addition, existing high-resolution spectra of two QSOs 2206-199 and 1101-264 ($z_{em} = 2.14$), with lower emission redshifts (Carswell et al. 1991; Rauch et al. 1993). At $z \sim$ 2 clustering in the Lyα forest is revealed for weaker lines ($N_c \gtrsim 10^{13}\,\mathrm{cm}^{-2}$) than at $z \sim 3$. This fact is illustrated by comparison of Figs. 2a and 2b.

3   Discussion and Conclusions

Both the minihalo model and the shocked-shell model of Lyα clouds can naturally explain the observed small-scale clustering (Ostriker et al., 1988).[1]

To discriminate between them, a detailed analysis of the relationship between cloud clustering properties and corresponding HI column densities and line redshifts is required. However, it is worth noting that growth of clustering strength with the increase of cloud HI column densities is expected in both models though for quite different reasons.

---

[1] In the latter case, to the first approximation, the TPCF equals to the inverse fraction of the space covered by explosions.

For the minihalo model it can be due to the growing role of gravitation. In the frames of the shocked-shell model, such a relationship can result from edge geometry of observed shell zones (line of sight is nearly tangential to the shell) attributed to the majority of Lyα forest lines with relatively large HI column densities $N_{HI} \gtrsim 10^{14}\,cm^{-2}$. Then, the observed excess of small velocity splittings in such sample is due to lines originating from the same explosion (Chernomordik 1988). In this case the growth of clustering strength for $N_{HI} \gtrsim 10^{14}\,cm^{-2}$ at $z \sim 3$ should be evidently accompanied by a break of the HI column density distribution in the same range. Recent observations of high redhift QSOs do reveal such a break (Petitjean et al. 1993).

The results of our analysis can be summarized as follows:

1. Small-scale clustering of the Lyα forest lines with relatively large HI column densities ($N_{HI} \gtrsim 10^{14}\,cm^{-2}$) at redshifts $z \sim 3$ has an amplitude: $\xi(50\text{–}150\,km/s) \sim 0.7$ (3.7σ excess).
2. Observed growth of small-scale clustering with the increase of the sample cutoff N_c at $z \sim 3$ can be naturally explained by the gradual progression in clustering amplitude with the strength of Lyα lines.
3. Marginal evidence for clustering variation with redshift is found: at $z \sim 2$ similar clustering amplitude is observed for considerably smaller values of N_c than at $z \sim 3$.
4. Correlation among Lyα clouds is much less pronounced than that for metal-line absorption systems, but characteristic scales for both types of clouds are similar.

Acknowledgements. The author is very grateful to Arlin Crotts for support.

References

Atwood, B., Baldwin, J.A., Carswell, R.F. 1985, ApJ, 292, 58.

Barcons, X., Webb, J.K. 1991, MNRAS, 253, 207.

Carswell, R. F., Webb, J. K., Baldwin, J.A., Atwood, B., 1987, ApJ, 319, 709.

Carswell, R.F., Lanzetta, K.M., Parnell, H.C., Webb, J.K., 1991, ApJ, 371, 36.

Chernomordik, V.V., 1988, SvA, 32, No. 1, 6.

Chernomordik, V.V., 1995, ApJ, 440, No. 2 (in press).

Crotts, A.P.S. 1989, ApJ, 336, 550.

Ostriker, J.P., Bajtlik, S., Duncan, R.C., 1988, ApJ, 327, L. 35.

Petitgean, P., Webb, J.K., Rauch, M., Carswell, R.F., Lanzetta, K., 1993,, MNRAS, 262, 499.

Rauch, M., Carswell, R.F., Chaffee, F.H., Foltz, C.B., Webb, J.K., Weymann, R.J. Bechtold, J., Green, R.F., 1992, ApJ, 390, 387.

Rauch, M., Carswell, R.F., Webb, J.K., Weymann, R.J., 1993,, MNRAS, 260, 589.

Sargent, W.L.W., Young, P.J., Boksenberg A., Tytler, D., 1980,, ApJS, 42, 41.

Webb, J.K., 1987, in it Observational Cosmology, IAU Symp. 124, ed. Hewitt A., Burbidge G., & L.-Z. Fang, (Dordrecht: Reidel), p. 803.

Discovery of a Lyα Cloud in a Cosmic Void

John Stocke[1], J. M. Shull[1], S. V. Penton[1], Megan Donahue[2], and C. Carilli[3]

[1] CASA, University of Colorado Boulder CO, 80309, USA
[2] Space Telescope Science Institute, Baltimore MD, 21218, USA
[3] Sterrewacht Leiden, Postbus 9513 2300 RA Leiden, The Netherlands

Abstract. In the course of a Hubble Space Telescope (HST) Goddard High Resolution Spectrograph (GHRS) program to discover local Lyα absorption lines in the UV spectra of bright ($V < 14.5$) AGN situated behind nearby galaxy voids, we have discovered one definite and two possible Lyα absorption lines located in voids. Despite the fact that our observations probe nearly equal pathlengths through local superclusters and voids, only one of the eight definite Lyα detections is within a void. Although the majority of local Lyα lines appear associated with supercluster structures, none of our detections is closer than 450 kpc to a bright galaxy. Thus there is no strong evidence for associating these clouds with individual galaxies. This is unlike the situation for higher equivalent width (EW) lines ($W_\lambda > 300$mÅ) where Lanzetta et al. (1995; L95) find most Lyα clouds closely associated with bright galaxies. Our results agree with those of Morris et al. (1993; M93) who found only a loose association of lower EW Lyα clouds with galaxies. These results suggest that at least some local Lyα clouds are truly intergalactic in origin.

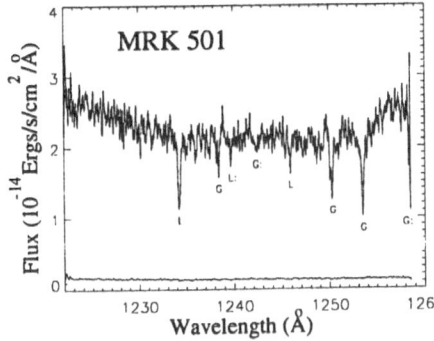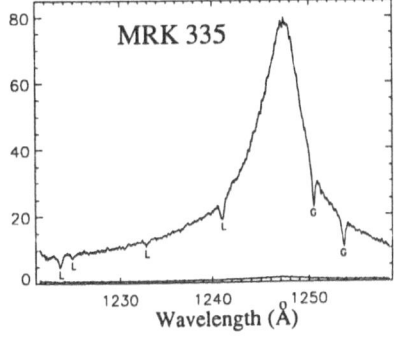

Fig. 1. Mrk501 and Mrk335 spectra and 1σ error vectors. G's and L's indicate Galactic metal and intervening Lyα lines. Colons indicate 3-4σ possible lines.

1 Observations

The GHRS+G160M grating was used to observe 3 bright AGN located behind well-defined voids based upon galaxy redshift surveys (CfA and Arecibo surveys). The pre-COSTAR setup yields an effective spectral resolution of 50 km/s and coverage of 1222-1259Å (1500-10,500 km/s for Lyα). Our high resolution spectra, two of which are shown in Fig. 1, yielded 4σ EW limits

of 20 mÅ for Mrk335, 40 mÅ for Mrk501 and 70 mÅ for IZw1. Eight defi-
nite ($> 4\sigma$) and three possible (3-4 σ) Lyα absorptions were detected with
EW=26-171 mÅ (log $N_{HI} \approx 12.7$–13.5).

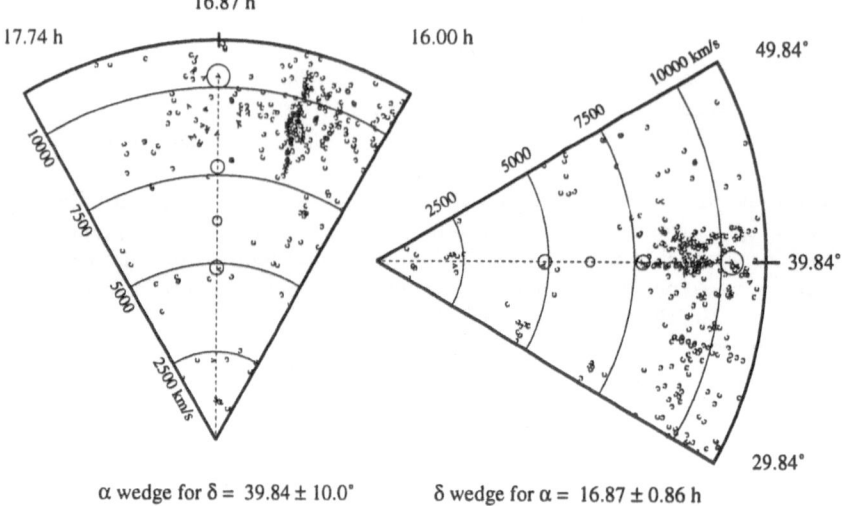

α wedge for $\delta = 39.84 \pm 10.0°$ δ wedge for $\alpha = 16.87 \pm 0.86$ h

Fig. 2. 10° RA and DEC "Pie Diagrams" for the Mrk501 sightline.

Because the sightlines probed by these observations are within the regions
of nearby galaxy redshift surveys, a comparison with the distribution of galaxies was immediately possible. One such comparison is shown in Fig. 2, "pie
diagrams" centered on the Mrk501 sightline with the target (largest circle),
definite Lyα lines (large circles) and possible Lyα line (small circle) posi-
tioned along the sightline (dashed line). To be along the sightline a "Merged
CfA Redshift Catalog" (Huchra 1993) galaxy (small c's and v's, if within 2°
of the sightline) must be along the dashed line in both the RA and DEC plots.
For example, even though the DEC plot seems to show the ∼7700 km/s Lyα
line close to a large cluster of galaxies (Abell 2199), the RA plot shows it to
be ∼10 Mpc ($H_o = 75$ km/s Mpc^{-1} used throughout) to the west and this
cloud is ∼16 Mpc foreground to the "Great Wall". The nearest known galaxy
to this Lyα cloud is 5.9 Mpc to the east, while the possible Lyα cloud at 6200
km/s has no galaxy within 10.5 Mpc. Pencil-beam galaxy surveys within 100
kpc (at 7700 km/s) of this sightline were conducted in the optical at Palomar
and using the HI 21cm line at Westerbork. No optical galaxy with B ≤ -16
and no object with HI mass ≥ 3×10^8 M$_\odot$ was found at the redshifts of
either the definite or possible void absorbers.

This was the only Lyα cloud out of eight definite detections found in
such a sparse galaxy region. Based upon the nearby galaxy survey data, we
find that the nearest neighbor galaxies to our eight Lyα clouds are 0.45-5.9
Mpc away. All definite detections but the one "void absorber" have galaxies

within ~2 Mpc and are plausibly associated with supercluster "walls and filaments"; e.g. the 4900 km/s Lyα line in Fig. 2 has a nearest galaxy 510 kpc away. Because the pathlengths probed through supercluster and void are nearly identical (10,800 km/s and 11,400 km/s respectively), there is statistical evidence that Lyα clouds avoid the voids (97% confidence level). But, despite their association with superclusters, the association of these clouds with individual galaxies is not obvious given the large nearest neighbor distances. Detailed models of the ionized gaseous halos around galaxies fail to produce column densities typical of our detections at impact parameters larger than 200 kpc (e.g., Dove & Shull 1994). Deep, pencil-beam optical and HI observations should be conducted in the vicinity of each Lyα cloud in order to make certain that no faint or low surface brightness galaxies exist close to these clouds.

2 Analysis

In order to investigate further the association of Lyα clouds with galaxies, we have formed a GHRS local Lyα sample consisting of 27 lines detected along 5 sightlines: the three observed by us plus those found toward 3C273 (M93) and PKS2155-304 (Bruhweiler et al. 1993). In all cases we have used only those Lyα lines whose vicinity has been well-scrutinized for galaxies, typically to absolute magnitudes of $B \leq -19$ and/or HI masses $\geq 10^9$ M$_\odot$. Fig. 3 shows the cumulative distribution of nearest galaxy neighbor distances for the GHRS Lyα sample (solid line) compared to the nearest galaxy neighbor distances for all CfA survey galaxies (dashed-dot line) compared to a Monte Carlo simulation of nearest galaxy neighbor distances for > 2600 clouds randomly placed along 1000 sightlines through the CfA survey region (dashed line).

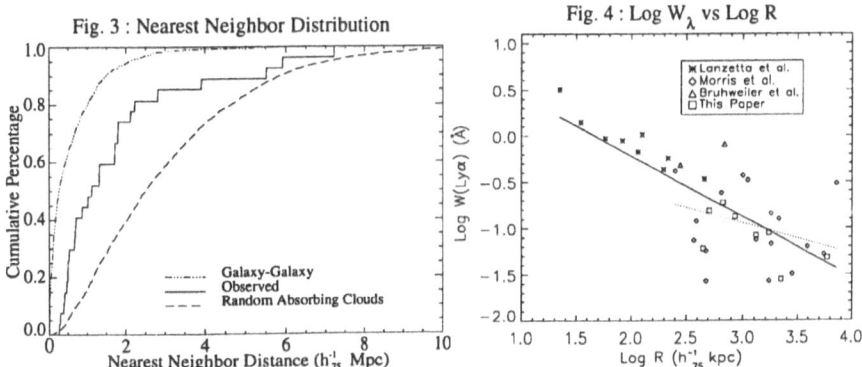

Fig. 3 : Nearest Neighbor Distribution

Fig. 4 : Log W$_\lambda$ vs Log R

The local Lyα lines exhibit neither a random distribution relative to galaxies (KS-test confidence level > 99.96%) nor as strong clustering with galaxies as galaxies with themselves (> 99.999% confidence level). These results are fully consistent with the M93 claim that the local Lyα clouds cluster only weakly with galaxies, but are inconsistent with the L95 results in which most high EW lines are within ~200 kpc of bright galaxies. Fig. 4 shows an augmented plot from L95, with an inverse correlation (solid line) between W_λ and nearest galaxy neighbor distance (R) for higher EW lines ($W_\lambda > 300$ mÅ) from HST/Faint Object Spectrograph (FOS) spectra. For the GHRS sample alone we find no such correlation (the best fit line (dots) has a slope half that of the L95 data and a correlation coefficient of only 0.34). The absence of L95 points in the lower left hand corner of Fig. 4 is not significant due to the EW limits of the FOS sample. The absence of GHRS points from that region is not yet statistically significant because the limited pathlength probed has not provided even a single galaxy impact parameter <300 kpc.

The analyses presented in Fig. 3 & 4 strongly suggest a division in absorption line systems similar to that found at high-z. The higher column density systems (typically but not exclusively metal-bearing) are associated with galaxies, but the lower column density systems (the Lyα forest) are only loosely associated with galaxies. A division of the GHRS sample into higher and lower EW lines at 85 mÅ finds that the lowest EW Lyα clouds have a distribution in R consistent with random placement with respect to galaxies. This suggests that the low EW lines detected by the GHRS may be a mixture of two or more types of clouds, only some of which are associated with galaxies (see Mo & Morris 1994). Our one definite Lyα detection in a galaxy void shows that some Lyα clouds are not associated with galaxies at all.

3   Conclusions

In the HST/GHRS spectra of three very bright AGN situated behind voids in the nearby distribution of bright galaxies, we have discovered one definite, and two possible, Lyα forest clouds in galaxy voids. But, statistically the local Lyα clouds seem to avoid the voids, with 7 of 8 Lyα clouds located in supercluster strings or filaments. And, since none of our Lyα detections are ≤ 450 kpc from bright galaxies, the relationship between these clouds and galaxies is still obscure. Specifically, the low EW Lyα clouds cluster much less strongly with galaxies than galaxies with themselves, but much stronger than expected at random. This is consistent with previous inferences about the local Lyα forest made by M93 on the basis of the 3C273 sightline, but inconsistent with results on higher EW lines recently presented in L95.

References

Bruhweiler, F.C. et al. 1993 ApJ 409, 199.
Dove J.B. & Shull, J.M. 1994 ApJ 423, 196.
Huchra, J.P. 1994 CfA Redshift Catalog.
Lanzetta, K., Bowen, D.V., Tytler, D. 1995 ApJ in press, and in this conf.(L95).
Mo, H.J. & Morris, S.L. 1994 MNRAS 269, 52.
Morris, S.L. et al. 1993 ApJ 419, 524 (M93).

The Observed Properties of Lyα Clouds as Signatures of Dark Matter Potential Wells

V.K. Khersonsky and D.A. Turnshek

Department of Physics and Astronomy, University of Pittsburgh

Abstract. The radial distributions of dark matter (DM) density and temperature in Lyα clouds seen in QSO spectra are discussed in the framework of a 'minihalo' model. Available observational data on the distribution of Lyα clouds in column density, in combination with assumptions about the hydrostatic equilibrium of DM and gas, and information provided by the equation of state of the gas, provide a reasonable basis for constraining the shapes of the DM density and temperature distributions. The analysis shows that DM density should be strongly peaked at the cloud center, while DM temperature should have a minimum in this region. In the framework of a minihalo model this work leads to the conjecture that an important process in minihalo formation may be some sort of gravitational-thermal instability in the DM gas.

One of the most reasonable models of Lyα clouds is the model developed by Rees (1986). In this model minihalos of DM are postulated to provide the gravitational potential wells required to stabilize the Lyα clouds which are hot and highly ionized. Ikeuchi, Murakami & Rees (1988) have developed this model. They made the assumption that the DM gas is in steady-state and introduced the concept of the virial temperature of DM. Here we point out that, in the context of certain assumptions, available observational data can be used to obtain some constraints on the radial distributions of DM density. The additional assumption that the DM gas is in hydrostatic equilibrium then places constraints on the radial distribution of the DM gas temperature. The observed distribution of Lyα clouds in column density, $d\mathcal{N}/dN_{HI} \propto N_{HI}^{-\beta}$ where $\beta \approx 1.5$ to 1.9, can be easily understood if the radial distribution of neutral gas in these clouds varies as $n_{HI} = n_{HI}(0)/[1+(R/R_*)^2]^\alpha$, where $\alpha = (\beta+1)/(\beta-1) \approx 1.83$. The values $n_{HI}(0)$ and R_* are model parameters. By applying results from the study of Black (1981) on the ionization equilibrium and equation of state ($P_g \propto \rho_g^{5/7}$) of a highly ionized rarefied gas one can (a) relate the distribution of HI and the radial distribution of total gas density, $\rho_g(R) = \rho_g(0)/[1 + (R/R_*)^2]^\delta$ with $\delta \approx 0.826$, and (b) take into account the contribution of DM to the hydrostatic equilibrium of the hot gas. This contribution is defined by the radial distribution of DM density, $\rho_{DM}(R)$, and can be derived from the hydrostatic balance equation. The result can be presented in the form

$$\frac{\rho_{DM}(R)}{\rho_{DM}(0)} = \frac{1}{\eta^2 - 1}\left[\frac{\eta^2}{3}\frac{3+(3-2\varepsilon)r^2}{(1+r^2)^{1+\varepsilon}} - \frac{1}{(1+r^2)^\delta}\right],$$

where $r = R/R_*$, $\eta = \sqrt{1 + \rho_{DM}(0)/\rho_g(0)}$, $\epsilon = 1 - 2\delta/7$, and $\rho_{DM}(0)$ is the DM density in the center of the potential well. Fig. 1 shows this dependence.

If we then assume that the DM in the potential well can be described in terms of the hydrostatic equilibrium of the DM gas, which satisfies the equation of state $P_{DM} = n_{DM}kT_{DM}$, the radial distribution of DM temperature, $T_{DM}(R)$, can be derived from the equation of DM hydrostatic balance, i.e.,

$$\frac{T_{DM}(r)}{T_{DM}(0)} = \frac{\rho_{DM}(0)}{\rho_{DM}(r)}$$

$$\times \left\{ 1 - \frac{\gamma}{2(\eta^2 - 1)} \left[\frac{\eta^2}{3} \left(2\epsilon\varphi_{2\epsilon}(r) + (3 - 2\epsilon)\varphi_{2\epsilon-1}(r) \right) - \varphi_{\delta+\epsilon-1}(r) \right] \right\}.$$

Here $\varphi_t(r) = [1 - (1 + r^2)^{-t}]/t$ and $\gamma \approx 1.21 c_g^2/c_{DM}^2$ is the ratio of the sound speeds squared in baryonic and DM gases. This dependence is shown in Fig.2.

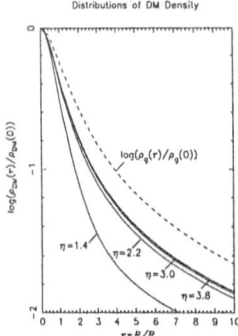

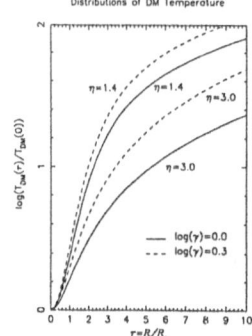

Fig. 1. Radial DM and gas (solid and dashed lines) density distributions.

Fig. 2. Radial DM temperature distributions.

The most interesting features of these dependences can be summarized as follows. First, the shape of the DM density is defined by only one free dimensionless parameter, η. The shape is peaked centrally larger than the gas distribution. Second, the shape of the DM temperature distribution depends on two dimensionless parameters, η and γ. The shape is a strong function of γ, i.e., the temperature of the DM gas. The DM temperature decreases toward the center of a cloud. This effect may be related to some sort of gravitational-thermal instability. In turn, this implies that our model of the observations is inconsistent with the idea that the DM consists of weakly interacting particles. See Khersonsky & Turnshek (1995) for additional details.

References

Black, J.H. 1981, MNRAS, 197, 553

Khersonsky, V.K, & Turnshek, D.A 1995, ApJ, March issue

Rees, M.J. 1986, MNRAS, 218, 25P

Ikeuchi, S., Murakami, I.,& Rees, M.J. 1988, MNRAS, 236, 21P

Redshift Dependence of the Number Density of Lyα Lines and Evolution of the Dark Matter Filamentary Structures

Jan P. Mücket[1], Patrick Petitjean[2], and Ron Kates[1]

[1] Astrophysikalisches Institut Potsdam, An der Sternwarte 16, D-14482 Potsdam, Germany

[2] Institut d'Astrophysique de Paris-CNRS, 98b Blvd Arago, F-75014 Paris, France

Abstract. We present results of N-body simulations, including effects of cooling and heating due to the background ionizing flux, aimed at modelling space distribution of the Lyα absorbers in the context of cosmological structure formation. The observed z-dependence of the number of Lyα lines per unit redshift is reproduced by the model within the uncertainties. The sharp decrease with time in the number density at high redshift is mainly due to the formation of structures whereas Lyα gas survives at low redshift because of the decreasing ionizing flux.

We have investigated the space distribution of the Lyα gas using N-body simulations of the formation and evolution of large-scale structures. We suggest that the Lyα absorption lines generally arise in gas associated with the potential wells of the rich filamentary structures that are ubiquitous in gravitational evolution dominated by the dark matter. We have improved and adapted a PM code (see Kates, Kotok & Klypin 1991; Klypin & Kates 1991) to include cooling of the baryonic material and the effects of photoionisation (Petitjean, Mücket & Kates 1995). The UV flux intensity is assumed to be proportional to the rate at which baryonic material cools below 5000 K in the simulation. This simulates an ionizing flux produced as a consequence of gas collapsing in the dense regions. The z-dependence of the ionizing flux intensity is found to be close to a power law $(1+z)^\beta$ with $\beta = (-1) - (-2)$ for z>5 and $\beta = 2$ below z=2.5. Between z=2.5 and z=5 the flux is approximately constant. The only free parameter is the normalisation which is found to be $F_0 = 10^{-21}$ erg s^{-1} cm^{-2} sr^{-1} Hz^{-1} at z=5.

To determine the time evolution of the number density of Lyα lines dN/dz, we have performed high resolution simulations (256^3 particles in 512^3 cells) saving at each time-step the distribution of the Lyα gas in the box. The size of the box was 25.6×25.6 Mpc^2. Then dN/dz is determined by averaging a large number of fictive lines of sight taken at random through the box. The result, after averaging over 100 sight lines, is shown in Fig. 1 for lines with N(HI)> 10^{14} cm^{-2}. The solid dots indicate the observed number of absorption lines per unit redshift (Bahcall, Bergeron, Boksenberg et al. 1993; Lu, Wolfe & Turnshek 1991). The number density of lines varies by about 25% from one individual line of sight to the other. It can be seen that the global behavior, a sharp decrease with time for $z > 2$ and a flattening at lower redshift, is well reproduced by the model. For the lower column density threshold, $N(HI) > 10^{13}$ cm^{-2}, the model yields a number density of 105 at $z = 0$ which is consistent with HST observations. In this case however the number density increases slowly with increasing z ($\gamma \approx 0.5$).

The simulations use the standard CDM power spectrum normalised at $\sigma(8h^{-1}$ Mpc). In this model, filamentary structures mainly form before $z = 2$. At smaller redshift, merging and clustering dominate. The z dependence of the number density reveals these two main stages of structure formation. The results are sensitive to the chosen power spectrum and we plan to investigate this in more detail. The results may be used as a discriminator between the various proposed power spectra.

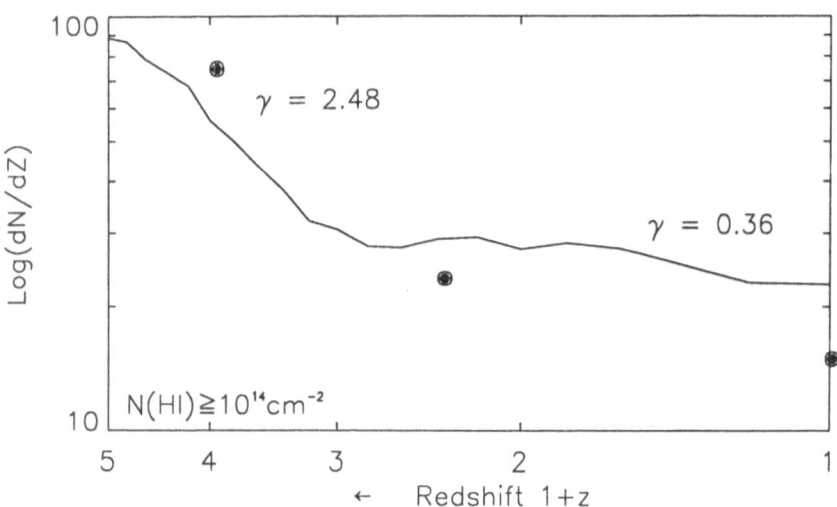

Fig. 1. z-dependence of the number density of Lyα lines.

References

Bahcall J.N., Bergeron J., Boksenberg A., et al., 1993, ApJS 87,1

Kates R.E., Kotok E.V., Klypin A.A., 1991, A&A 243,295

Klypin A.A., Kates R.E., 1991, MNRAS 251, 41p

Lu L., Wolfe A.M., Turnshek D.A., 1991, ApJ 367, 19

Petitjean P., Mücket J.P., Kates R., 1995, A&A (in press)

Hydrodynamical Simulations of Lyα Clouds

S. Murray[1] and P. Petitjean[2]

[1] Dept. of Astronomy, University of California, Berkeley, CA 94720, USA
[2] Institut d'Astrophysique de Paris - CNRS, 98bis Boulevard Arago, F-75014 Paris, France

Abstract. Very recent work by Petitjean et al. (1995) and Mücket et al. (1995) have shown that the z-dependence of the number density of Lyα absorption lines and the large number of lines detected by HST at low z both can be understood if the Lyman α gas follows the filamentary structures of the dark matter. In an attempt to understand the physical state of the gas, we have modified a 1D-hydrodynamical code to include photoionization and cooling of the baryonic material. We present here preliminary results showing under which conditions the gas can survive down to very low redshifts.

1 Numerical Code

We consider a spherical bound cloud of mixed dark matter and gas. The static potential is dominated by the dark matter, whose distribution is given by Bertschinger (1985) as $\rho_d = \rho_0 (r/r_0)^{-9/4}$ for $r_0 < r < r_{max}$. For numerical stability, we take $\rho_d = const.$ for $r < r_0 = 3$ kpc, while $\rho_d = 0$ for $r > r_{max} = 50$ kpc. The gas is taken to have primordial abundances, and its behavior is followed using a one dimensional Lagrangian hydrodynamics program (see MacLow & Shull 1986; Murray & Lin 1990). The time evolution of the ionization state is followed in detail. Ionization is due to collisions and photoionization resulting from illumination of the cloud by the ionizing background from QSOs and young galaxies. The latter is constant over the redshift range $2 < z < 5$, and then decreases as $J(\nu, z) = J(\nu_o, z = 2)(\nu/\nu_o)^{-\alpha}(1 + z/3)^{\beta}$ where $J(\nu_o, 2)$ is the flux at ν_o, the Lyman limit. We take $J(\nu_o, 2) \approx 10^{-21}$ erg cm^{-2} s^{-1} Hz^{-1} sr^{-1} as derived from study of the proximity effect (Bajtlik et al. 1988) although this value may be overestimated (Petitjean et al. 1992). The latter authors have also shown that the shape of the ionizing spectrum is an important parameter, and we consider $\alpha = 1$ and 3. For our standard model we use $\beta = 4$ (Bechtold et al. 1987) for $z < 2$.

Several processes are included in heating and cooling the clouds. Cooling rates are taken from Black (1981) and Hollenbach & McKee (1979). The full time dependence of the abundances of the ionization states of H and He, and of H$^-$ and H$_2$, which forms via H$^-$ in the gas phase, are computed, so that no assumptions are made regarding equilibrium. The total abundance of He is taken to be 0.1 by number relative to hydrogen. H$_2$ is important in that it allows the gas to cool to temperatures below about 6000 K (Murray & Lin 1990). Bremsstrahlung emission and cooling by Compton scattering of microwave background photons by electrons are also included.

The gas is initially given a large temperature $T \sim 10^5$ K, to simulate the effect of gas from the IGM falling into the potential well of the dark matter. The exact value of the temperature is not important because the gas rapidly cools to its equilibrium temperature under photoionization conditions. The initial gaseous density is taken to be constant at 10^{-4} or 10^{-3} cm^{-3}; the mass of baryons relative to dark matter within 50 kpc is therefore $0.1 \left(M_{dm}/3 \times 10^9 \text{ M}_\odot \right)$. The cloud is bounded by the pressure of an assumed IGM (Ikeuchi & Ostriker 1986), initially chosen to have the same pressure as the cloud, but which then drops due to cosmological expansion.

2   Preliminary Results

Simulations using a steep ionizing spectrum ($\alpha = 3$) and $M_{dm} = 5 \times 10^9$ M$_\odot$ show that (i) the radial velocity is important only in the very external layers of the cloud; (ii) it is possible to reproduce b values as low as 20 km s^{-1} for $N(\text{HI}) \sim 10^{13}$ cm^{-2}; however the mean value for such column density should be around 25 km s^{-1}; (iii) although the HI column density decreases by an order of magnitude from $z = 3.5$ to $z = 0$, the evolution is slow and a non-negligible fraction of the cloud remains observable at low z.

For $M_{dm} > 10^{10}$ M$_\odot$, a bound core with increasing column density remains until late times within about 10 kpc; in this core, low velocities and cooling lead to b values as low as 10 km s^{-1} in the high column density gas. When $M_{dm} = 5 \times 10^9$ M$_\odot$ and $\alpha = 1$, the greater rate of heating by photoionization of H, He, and He$^+$ results in higher cloud temperatures. The result is much more rapid expansion than in the first model above, with no observable material remaining after only a short time. As the heating is reduced for $z < 2$, the HI column density increases, but never exceeds 10^{12} cm^{-2}.

References

Bajtlik S., Duncan R.C., Ostriker J.P., 1988, ApJ, 327, 570
Bechtold J., Weymann E.J., Lin Z., Malkan M.A., 1987, ApJ, 315, 180
Bertschinger E., 1985, ApJS, 58, 39
Black J.H., 1981, MNRAS, 197, 553
Hollenbach, D., McKee, C. F., 1979, ApJS, 41, 555
Ikeuchi S., Ostriker J.P., 1986, ApJ, 301, 522
MacLow M.M., Shull J.M., 1986, ApJ, 302, 585
Mücket, J., Petitjean, P., Kates, R. (1995): This volume
Murray S.D., Lin D.N.C., 1990): ApJ, 363, 50
Petitjean, P., Bergeron, J., Puget, J.L. (1992): A&A, 265, 375
Petitjean, P., Mücket, J., Kates, R. (1995): A&A, in press

Clustering of the Ly$_\alpha$ Clouds in the Line of Sight to the z=3.66 QSO 0055–269

S. Cristiani[1], S. D'Odorico[2], A. Fontana[3], E. Giallongo[3] and S. Savaglio[4]

[1] Dipartimento di Astronomia, Università di Padova, vicolo dell' Osservatorio 5, I-35122 Padova, Italy
[2] European Southern Observatory, Karl-Schwarzschild-Str. 2, D-85748 Garching, Germany
[3] Osservatorio Astronomico di Roma, via dell'Osservatorio, I-00040 Monteporzio, Italy
[4] Dipartimento di Fisica, Università della Calabria, I-87036 Arcavata di Rende, Cosenza, Italy

Abstract. The spectrum of the Q0055-269 ($z = 3.66$) has been observed at the resolution of 14 km s^{-1} in the wavelength interval 4750–6300 Å. The statistical distribution of the Doppler parameter for the Ly$_\alpha$ lines is peaked at $b \simeq 23$ km s^{-1}. The column density distribution is described by a power-law with a break at $\log N_{HI} \simeq 14.5$. Significant clustering, with $\xi \simeq 1$ at $\Delta v = 100$ km s^{-1}, is detected for lines with $\log N_{HI} \geq 13.8$. Two voids of size ~ 2000 km s^{-1} are found in the spectrum with a probability of 2×10^{-4}.

1 The Clustering Properties of the Ly$_\alpha$ Clouds

No clustering in the velocity space has been detected so far for Ly$_\alpha$ lines on scales $300 < \Delta v < 30000$ km s^{-1} (Sargent et al. 1980, 1982; Webb & Barcons 1991). Their spatial distribution is different from that of the metal-line systems selected by means of the CIV doublet, that are known to cluster on scales at least up to 600 km s^{-1} (Sargent et al. 1988). Preliminary results at higher resolution seem to indicate weak clustering on smaller scales ($\Delta v = 50 - 300$ km s^{-1}, Webb 1987, Chernomordik, this conference).

The larger the density of the observed lines, the better the chances of detecting a clustering signal. For this reason we have looked for a QSO of redshift 3.66, Q0055-269, which has been observed at ESO La Silla with the NTT telescope and the EMMI instrument in the echelle mode (Cristiani et al., 1995). The weighted mean of the flux-calibrated spectra has a resolution $R \sim 22000$. The signal-to-noise ratio per resolution element at the continuum level ranges from $S/N = 12$ to 40 in the interval 4750–6300 Å.

Column densities and Doppler widths of the Ly$_\alpha$ lines and metal-line systems have been derived by a fitting procedure. The statistical distribution of the Doppler parameter for the Ly$_\alpha$ lines is peaked at $b \simeq 23$ km s^{-1}. The column density distribution is described by a power-law with a break or cutoff at $\log N_{HI} \simeq 14.5$. A featureless distribution is rejected with a probability of 99.94%.

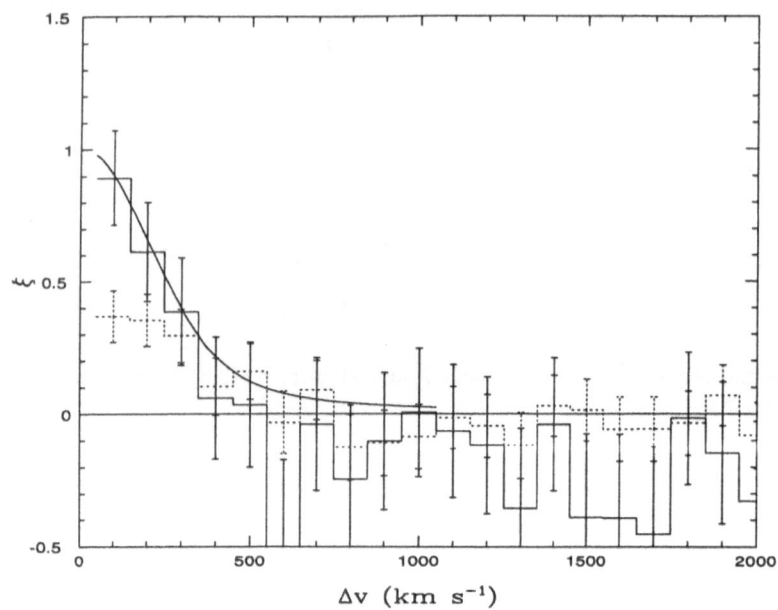

Fig. 1. Two-point correlation function for the Lyman−α lines in Q0055-269. The continuous histogram is for lines with $\log N_{HI} \geq 13.8$; the dotted histogram is for lines with $\log N_{HI} \geq 13.3$. The continuous curve is the model described in Sec. 2 and computed from Eq. (2) with $\gamma = 1.77$, $\sigma = 150$ km s^{-1}, $r_{cl} = 110$ kpc and $r_o = 280$ kpc at $z = 3.3$.

The two-point correlation function in the velocity space is defined as

$$\xi(v, \Delta v) = \frac{N_{obs}(v, \Delta v)}{N_{exp}(v, \Delta v)} - 1 \tag{1}$$

where N_{obs} is the observed number of line pairs with velocity separations between v and $v + \Delta v$ and N_{exp} is the number of pairs expected from a random distribution in redshift. In our line sample N_{exp} is obtained averaging 1000 numerical simulations generated according to the cosmological distribution $\propto (1 + z)^{\gamma}$, derived from the maximum likelihood analysis of the real line sample in the given interval of column density and redshift. In this way it is possible to correct for incomplete wavelength coverage due to gaps in the spectrum or occultation of weak lines due to strong complexes. No velocity splittings $\Delta v < 20$ km s^{-1} were included because of the intrinsic line blending due to the typical widths of the Lyα lines. The 1σ deviation from a random distribution is given by the $N_{obs}^{-1/2}$, a good approximation in the case of weak clustering $\xi \leq 1$. The resulting correlation function of Lyα lines in Q0055-269 with $N_{HI} > 13.3$ is shown in Figure 1. It is clear that a weak but significant signal is present with $\xi \simeq 0.34 \pm 0.06$ up to 350 km s^{-1}.

Exploring the variations of the correlation function as a function of the column density, an increase in the first velocity bin appears as the column density threshold is raised. The maximum significant signal is obtained for lines with $\log N_{HI} \geq 13.8$, for which the correlation function is $\xi = 0.89 \pm 0.18$ for $\Delta v = 100$ km s^{-1}. When computing ξ for lines in the interval $\log N_{HI} = 13.3 - 13.6$ (including in the simulated spectra gaps due to lines with $\log N_{HI} > 13.8$), no significant clustering is observed.

For comparison the correlation function of the Ly$_\alpha$ samples obtained from the QSO PKS 2126-158 (Giallongo et al., 1993) and 0014+81 (Rauch et al. 1992) has been computed. A significant clustering for $\Delta v = 100$ km s^{-1} appears when the samples are limited to strong lines (with a lower significance due to the smaller density of lines in the individual spectra): in PKS 2126-158 $\xi = 1.02 \pm 0.26$ is obtained for lines with $\log N_{HI} \geq 13.8$, in Q0014+81 $\xi = 0.85 \pm 0.30$ for lines with $\log N_{HI} \geq 14.1$.

The correlation found for the Ly$_\alpha$ cloud positions is less pronounced than for metal-line absorption systems or galaxies, but consistent with a scenario of gravitationally induced correlations, as expected in models where gravitation is an important confining agent (Mo, Miralda-Escudé, & Rees 1993). The correlation found would imply, in the standard CDM framework, that the Ly$_\alpha$ clouds responsible for the weakest absorption are abundant in underdense regions. The redshift autocorrelation function could be lower than the spatial autocorrelation.

2· The Dimensions of the Ly$_\alpha$ Clouds

At small velocity differences the three-dimensional spatial autocorrelation function ξ_r and the redshift autocorrelation function are related by the convolution (Heisler, Hogan, & White, 1989):

$$\xi_v = \int_0^\infty H\,dr\,\xi(r)\,P(v \mid r) \int_{r_{cl}}^\infty \frac{H\,dr}{\sigma} \left(\frac{r}{r_o} \right)^{-\gamma} \left[e^{-\frac{(Hr-v)^2}{2\sigma^2}} + e^{-\frac{(Hr+v)^2}{2\sigma^2}} \right] \quad (2)$$

where a Gaussian distribution of the peculiar motions with respect to the Hubble flow is assumed together with a power-law spatial correlation function of the galaxy-type.

At small velocity splittings, the redshift correlation scale r_o depends mainly on the cloud sizes r_{cl} and on the velocity dispersion. Although these quantities ·are poorly known, it is instructive to derive constraints on the cloud sizes and velocity dispersions from the observed correlation. Assuming an index $\gamma = 1.77$ for the power-law spatial correlation function, the fit shown in Fig. 1 provides upper limits on the cloud sizes as a function of the velocity dispersion and of the appropriate correlation scale. We obtain $r_{cl} \sim 100 - 190$ kpc for $\sigma_v = 50 - 160$ km s^{-1}, and $r_o \sim 245 - 420$ kpc, respectively. In the former case a good fit to the overall shape of the observed correlation function requires a very low value $\gamma = 1.1$. For $\sigma_v > 200$ km s^{-1} the redshift correlation function becomes too flat and very steep γ values are needed (e.g. $\sigma_v = 300$ km s^{-1} requires $\gamma = 4$).

3   Voids in the Ly$_\alpha$ Forest

Voids in the Ly$_\alpha$ forest provide a test for models of the large-scale structure, even if changes in the IGM due to fluctuations of the UV ionizing flux can be efficient in depleting neutral hydrogen along the line of sight.

Searches for megaparsec-sized voids have produced a couple of claims: a void with comoving size $\sim$ 30 Mpc in the spectrum of Q0420-388 (Crotts 1987,1989), whose statistical significance has been questioned (Ostriker, Bajtlik, & Duncan 1988); another one in Q0302-003, with a size $\sim$ 23 Mpc (Dobrzycki & Bechtold 1991). The statistical significance of a given void depends exponentially on the line density (Ostriker et al. 1988) and uncertainties in the line statistics strongly influence the probability estimate. High resolution data, less affected by blending effects, are ideal in this respect.

We have searched for gaps in a sample of Ly$_\alpha$ lines of Q0055-269 limited, to ensure statistical completeness and accuracy in the deblending, to $\log N_{HI} \geq$ 13.3 and $b \leq$ 30 km s^{-1}. Two regions devoid of such lines have been found, centered respectively at $\lambda \sim$ 5000, 5206 Å , with sizes $\Delta v \sim$ 2009, 2046 km s^{-1} (i.e. $\sim$ 20 Mpc). To establish the random probability of observing such gaps, a set of 50000 spectra have been simulated with the observed number ($N = 178$) of redshifts randomly generated in the same interval of the data (excluding the region within 8 Mpc from the quasar) according to the cosmological distribution. The probability of getting a gap larger than the maximum observed one is then $\simeq$ 0.02. The 2046 km/s gap is significant only to 2.2σ level, but the joint probability for the presence of the two observed gaps in the same spectrum is $\simeq 2 \times 10^{-4}$. Even if underdense regions are statistically significant in our spectrum, the filling factor is less than 10%.

References

Cristiani, S., D'Odorico, S., Fontana, A., Giallongo, E. & Savaglio, S. 1995, ESO
    prep. 1054
Crotts, A. P. S. 1987, MNRAS, 228, 41p
Crotts, A. P. S. 1989, ApJ, 336, 550
Dobrzycki, A., & Bechtold, J. 1991, ApJ, 377, L69
Giallongo, E., Cristiani, S., Fontana, A., & Trevese, D. 1993, ApJ, 416, 137
Heisler, J., Hogan, C. J., & White, S. D. M. 1989, ApJ, 347, 52
Mo, H. J., Miralda-Escudé, J., & Rees, M. J. 1993, MNRAS, 264, 705
Ostriker, J. P., Bajtlik, S., & Duncan, R. C. 1988, ApJ, 327, L35
Rauch, M., Carswell, R. F., Chaffee, F. H., Foltz, C. B., Webb, J. K., Weymann,
    R. J., Bechtold, J., & Green, R. F. 1992, ApJ, 390, 387
Sargent, W. L. W., Young, P. J., Boksenberg, A., & Tytler, D. 1980, ApJS 42, 41
Sargent, W. L. W., Young, P. J., & Schneider, D. P. 1982, ApJ, 256, 374
Sargent, W. L. W., Boksenberg, A., & Steidel, C. C. 1988, ApJS, 68, 539
Webb, J. K. 1987, in IAU Symposium 124, Observational Cosmology, ed. A. Hewett,
    G. Burbidge, & L. Z. Fang (Dordrecht: Reidel), 803
Webb, J. K., & Barcons, X. 1991, MNRAS, 250, 270

The Clustering of QSO Absorption-Line Systems

Paul Francis[1] and Shoba Veeraraghavan[2]

[1] University of Melbourne, School of Physics, Parkville Vic 3052, Australia
[2] NASA Goddard Space-Flight Center, Code 681, Greenbelt MD 20771, USA

Abstract. A growing body of mainly anecdotal evidence suggests that, at high redshifts, metal-line and damped Lyman-α absorption-line systems cluster very strongly, perhaps more strongly than bright galaxies today! We review this evidence, and present first results from a complete, quantitative survey of absorption-line clustering. We detect clustering of high column-density Lyman-α lines and C IV systems, and place an upper limit on the clustering of Mg II systems.

1 Background

Many models have been developed to account for the formation on large-scale structure. Most are normalised to fit the COBE fluctuation amplitude at the era of recombination, and the observed present-day clustering amplitude. However, they differ dramatically in their predictions for clustering at redshifts above one. Galaxy clustering at these high redshifts is currently impossible to measure directly. However, it is now well established, at least at low redshifts, that metal-line and high column-density Lyman-α lines are associated with galaxies. In principle we can therefore measure large-scale structure out to redshifts beyond four by studying the clustering of QSO absorption-lines.

2 Existing Observations

Many apparent clusters of QSO absorption-lines have been found (e.g. Jakobsen & Perryman 1992, Dinshaw 1995). Despite the spectacular nature of some of these clusters, they are of limited use in constraining cosmological models, as it is not known how typical such clusters are. Only a handful of papers have attempted to derive quantitative limits on the clustering of metal-line and high column-density Lyman-α absorption-lines. Heisler, Hogan & White (1989) studied C IV systems in an unbiased QSO sample and detected strong clustering out to scales of 50 Mpc, though their signal is dominated by absorption in a single, possibly anomalous QSO (Foltz et al. 1993). Williger et al. (1995) detect strong C IV clustering, albeit with low significance levels. Wolfe (1993) claims that Lyman-α emitting galaxies cluster strongly around damped Lyman-α systems.

3   The LBQS Clustering Survey

To quantitatively measure clustering of QSO absorption-lines, we compiled a complete sample of close pairs of QSOs drawn from a large QSO survey, the LBQS (Large Bright QSO Survey, Morris et al. 1991). All pairs of QSOs with sight lines less than 10 arcmin apart were re-observed to obtain high signal-to-noise ratio spectra.

Two clusters of high column-density Lyman-α absorption-lines were seen. The existence of two such clusters anywhere in the survey has a very low probability unless overdensities of > 30 on scales of 10 co-moving Mpc are common in the early universe. This result is published (Francis & Hewett 1993), and further observations of one of these clusters are reported by Francis et al. (1995).

One cluster of C IV absorption systems is seen. In the absence of any clustering, the probability of seeing one or more such clusters is $\sim 6\%$, so this result is consistent with, but does not require, strong clustering of C IV systems. At lower redshifts ($z \sim 0.8$) no Mg II system clusters are seen, constraining the overdensity of these systems on scales of 10 co-moving Mpc to be less than ~ 20.

4   Conclusions

Major conclusions are: (1) existing evidence suggests that galaxy clustering is very strong at redshifts above 1, (2) existing data is poor, and (3) that while QSO absorption-lines can be a powerful tool for measuring large-scale structure at high redshifts, large samples will be needed. Wide field fibre-fed spectrographs should make such large samples possible.

A fuller account of this work will be submitted to MNRAS.

References

Dinshaw, N. 1995, these proceedings
Foltz, C. B., Hewett, P. C., Chaffee, F. H., & Hogan, C. J. 1993, AJ, 105, 22
Francis, P. J. & Hewett, P. C. 1993, AJ, 105, 1633
Francis P. J. Woodgate, B. E., Warren, S. J. et al. 1995, these proceedings
Heisler, J., Hogan, C. J., & White, S. D. M. 1989, ApJ, 347, 52
Jakobsen, P., & Perryman, M. 1992, ApJ, 392, 432
Morris, S. L., Weymann, R. J., Anderson, S. F., Hewett, P. C., Foltz, C. B., Chaffee, F. H., Francis, P. J., & MacAlpine, G. M. 1991, AJ, 102, 1627
Williger, G. M., Hazard, C., Baldwin, J. A. & McMahon, R. G. 1995, in preparation
Wolfe, A. M. 1993, ApJ, 402, 411

Clustering in the Lyman-alpha Forest at $z > 4$

M.J. Irwin[1], L.J. Storrie-Lombardi[2], and R.G. McMahon[3]

[1] Royal Greenwich Observatory, Madingley Road, Cambridge CB3 0HE UK
[2] Institute of Astronomy, Madingley Road, Cambridge CB3 0HA UK
current address: UCSD-CASS, Mail Code 0111, La Jolla, CA 92093 USA
[3] Institute of Astronomy, Madingley Road, Cambridge CB3 0HA UK

Abstract. The APM Multicolour Survey for Bright $z > 4$ Objects resulted in the discovery of thirty-one quasars with $z \geq 4$ (Irwin, McMahon & Hazard 1991). High resolution echelle spectra have been obtained for 3 of these. These have been compared with simulated Lyα forest absorbers leading to constraints on both the shape and normalisation of the absorber distribution function and their spatial correlation. Excess power in the correlation function of the real quasars at velocity separations up to ≈ 500km/s suggests that the Lyman-alpha forest absorbers are spatially clustered on this scale. We have made detailed statistical studies of the Lyα forest region including: the evolution of D_A, the average continuum depression shortward of Lyα; the distribution function of the pixel-to-pixel variation of the Lyα absorption; and the number density and evolution of Lyman-limit absorbers, including damped Lyα systems (see Storrie-Lombardi et al., this volume and Storrie-Lombardi et al., 1995).

The traditional approach to investigating the correlation properties of the Lyman-alpha forest is based on the two-point correlation function $\xi(r)$ of resolved lines. An alternative to this which does not involve resolving and analysing individual lines is to examine the equivalent correlation function of the spectrum on a pixel-by-pixel basis such that $1 + \xi(r) = < x_i x_{i+r} > / < x_i >^2$ where $x_i = I_i / C_i$ is the fractional transmission of the spectrum I_i at pixel i, relative to the unabsorbed continuum, C_i and the expectation operator runs over some range of pixels indexed by i. $\xi(r)$ is closely related to the autocorrelation function, $\phi(r)$, of the transmission spectrum since $\phi(r) = < (x_i - < x_i >)(x_{i+r} - < x_i >) > / < x_i^2 >$ and hence $\xi(r) = \phi(r) < x_i^2 > / < x_i >^2$ Clearly, the main advantages of using direct correlation measures lie in their innate ability to take account of unresolved lines; their insensitivity to modest amounts of noise and the exact placing of the continuum; and the simplicity of the analysis. The only serious drawback is the necessity of using Monte Carlo simulations of the Lyman-alpha forest to interpret the shape of the correlation function. To eliminate problems due to Lyman limits, lower order Lyman lines, the proximity effect and the Lyβ, OVI quasar emission line the correlation function is only computed from a region covering 1050−1170Å in the quasar rest frame.

Figure 1(a) shows $\xi(r)$ calculated from a simulated Lyman-alpha forest region for a z=4.69 quasar. The forest was modeled using Voigt profiles with a random placement of absorbers satisfying a two power law N_{HI} distribution with a low column density index of $\beta = 1.7$ and cutoff at $N_{HI} = 2 \times 10^{12}$; a break at $N_{HI} = 10^{15}$ and a high column density index of $\beta = 1.3$ with a cut-off at $N_{HI} = 3 \times 10^{21}$. For absorbers placed randomly along the line-of-sight the expected shape of $\xi(r)$ is dominated by a peak at the origin extending to $\Delta v \lesssim 200$ km/s due to the average auto-correlation of the individual absorbers. If the column density distribution does not evolve with z then the shape of this peak is independent of redshift. The extended tail of lower level features, $\Delta v \gtrsim 200$ km/s, will vary and is caused by the chance superposition of independent absorbers. The height of the peak at the origin varies as $\approx (1 - D_A)^{-2}$ due to the absorbers and for real quasars will have a fractional noise contribution spreading over the first few bins. The effect of altering the velocity parameter, b, and using a different random set of absorbers is shown by the dashed line. The dot-dash line shows the effect of adding an extra damped absorber with $N_{HI} = 2 \times 10^{20}$. The solid line was generated using $< b > = 30$km/s with a Gaussian distribution of $\sigma_b = 10$km/s; whilst the dashed and dot-dash lines show the effect of an extended high velocity tail with $< b > = 40$km/s, $\sigma_b = 20$km/s and removing the low velocity end at < 10km/s. Figure 1(b) shows $\xi(r)$ for three quasars, 1108-0747 at z = 3.92 (continuous), 1033-0327 at z = 4.51 (dashed), 1202-0725 at z = 4.69 (dot-dash) from the APM sample, computed from CTIO and ESO 0.1 and 0.3Å resolution echelle spectra. There is an excess of power in the correlation function of the real quasars at velocity separations up to ≈ 500km/s suggesting that the forest absorbers are themselves spatially clustered on this scale.

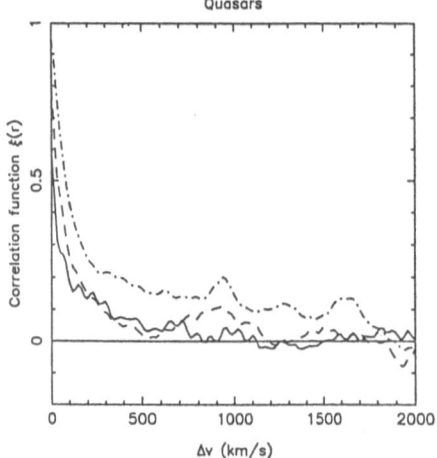

Fig. 1. $\xi(r)$ calculated from (a) simulations and (b) real quasars.

References

Irwin, M.J., McMahon, R.G. & Hazard, C., 1991, in *ASP Conf. Series, Vol. 21*, ed. Crampton D., (San Francisco: ASP), 117

Storrie-Lombardi, L.J., Irwin, M.J., McMahon, R.G., 1995, in preparation

Large-Scale Structure at High Redshift

Chris D. Impey and Nadine Dinshaw

Steward Observatory, University of Arizona, Tucson, AZ 85721, USA

Abstract. We present high resolution $(30\,\mathrm{km\ s^{-1}})$ echelle spectra of the wide QSO pair Tol $1037-2704$ and Tol $1038-2712$, as well as two neighboring QSOs. We confirm the existence of the five absorption complexes already reported and find additional complexes, many of which are split into multiple components with $\Delta v \simeq 50 - 1000\,\mathrm{km\ s^{-1}}$. We calculate the velocity and spatial correlation functions of the 19 CIV absorbers $(W_0 > 15\,\text{Å})$ distributed among the 4 lines of sight, and find significant correlation signal out to scales of $\sim 30\,h^{-1}$ Mpc $(h = H_0/100\,\mathrm{km}$ $\mathrm{s^{-1}}$ Mpc; $q_0 = 0.5)$ at $z \simeq 2$. The clustering amplitude on these scales is larger than is predicted by current theories of the formation of large scale structure.

1 Introduction

The large scale structure of the universe can be studied at high redshift using metal line QSO absorption systems, which are assumed to represent the halos of luminous galaxies. Recent evidence has pointed to very large-scale superclustering of the CIV absorbers. Heisler et al. (1989) detected significant signal in the two-point correlation function out to velocities of $\Delta v = 10,000\ \mathrm{km\ s^{-1}}$ from the large sample of CIV systems in 55 QSOs published by Sargent et al. (1988). The signal is dominated by a single large supercluster along the line of sight to the QSO $0237-233$. Recent pencil-beam surveys suggest that galaxies remain clustered on very large scales with an apparent regularity in their distribution on a characteristic scale of $128\ h^{-1}$ Mpc (Broadhurst et al. 1990), although the interpretation of this data is still disputed.

The wide QSO pair Tol $1037-2704$ $(z_{em} = 2.193)$ and Tol $1038-2712$ $(z_{em} = 2.331)$ exhibit a number of apparently correlated CIV absorption systems (Jakobsen et al. 1986; Sargent & Steidel 1987). Each spectrum contains at least five CIV complexes over the narrow redshift range $z = 1.88 - 2.15$. This represents an order of magnitude overdensity in the number of absorbers above the number expected from Poisson statistics. Each of the five complexes in Tol $1037-2704$ matches within $v \leq 2000\ \mathrm{km\ s^{-1}}$ one of the complexes in Tol $1038-2712$. The QSOs have an angular separation of $17.9'$ which corresponds to a comoving separation on the sky of ~ 12 Mpc. The favored hypothesis is that the two lines of sight intersect material associated with an intervening supercluster.

2   Observations and Line Identifications

High resolution spectroscopy of the QSOs Tol 1037−2704, Tol 1038−2712, Tol 1035−2737 and Tol 1029−2654 was obtained during 1992 February 29– March 3 UT using the echelle spectrograph and air Schmidt camera on the 4-m telescope of the Cerro Tololo Inter-American Observatory (CTIO). The 79 l mm^{-1} echelle grating and KPGL2 cross disperser were used with the Reticon CCD to give a wavelength coverage from $3810 - 5130$ Å in fifteen echelle orders. The spectral resolution was ~ 30 km s^{-1} FWHM, and exposures times were typically 5400 s.

A line list was constructed using only lines detected at or above the 5σ confidence level. The equivalent widths of the lines were measured in an interactive manner by marking the beginning and ending wavelengths over which to carry out the summation. The line centers were determined over the same wavelength range using a centroiding algorithm which weights each pixel by the depth of the line. For weak lines ($< 10\sigma$ in strength), the central wavelengths were derived by fitting Gaussian profiles unconstrained in width and central wavelength. Line identifications were done in an automated manner using algorithms similar to those described in Young et al. (1979). In addition, exhaustive searches by hand were also made for doublets of CIV and MgII, as well as NV, FeII and SiIV.

3   The Two-Point Correlation Function

The *velocity* correlation function is defined by the expression

$$\xi(v, \Delta v) = \frac{N_{obs}(v, \Delta v)}{N_{exp}(v, \Delta v)} - 1$$

where N_{obs} is the observed number of line pairs with velocity separations between v and $v + \Delta v$ and N_{exp} is the expected number of pairs for a random distribution of absorbers. $\xi(v)$ was calculated using the method of Heisler et al. (1989). Treating each redshift system in turn as a reference system, N_{obs} was estimated in each bin by counting the number of systems *in the same line of sight* which fell in the redshift interval(s) corresponding to the velocity separation represented by that bin. The expected number N_{exp} of pairs was derived from the catalog of CIV systems with $W_0 > 0.15$ Å of Sargent et al. (1988) by counting the number of systems which fell within larger ($\pm 25,000$ km s^{-1}) intervals – in which no clustering is evident – centered on the smaller redshift intervals in which N_{obs} was derived. Using the Sargent et al. catalog avoids having to assume a particular absorber evolution.

The velocity correlation function was calculated using the 19 CIV systems with $W_0 > 15$ Å contained in our sample. Possible associated systems within $\beta c \geq 5000$ km s^{-1} of the emission redshift have been excluded from the sample, and complexes with component separations $\Delta v \leq 500$ km s^{-1} have been collapsed into single systems. Since our sample contains an overdensity of systems compared to the Sargent et al. sample, we have normalized the observed correlation function by the average $\xi(v)$ of 1000 Monte Carlo trials of 19 randomly distributed absorbers with the number of systems along each line of sight as well as the redshift sensitivity intervals preserved. The usual expression for the uncertainty $(N_{exp})^{-1}$ in $\xi(v)$ is an underestimate of the true error because the pair count, N_{obs}, is not a Poisson variable (Olivier et al. 1993). Therefore the variance in $\xi(v)$ was calculated from the expression

$$\sigma^2(\xi) = \sigma^2(N_{obs})\left(\frac{1}{N_{exp}}\right)^2 = \frac{<N_{obs}^2> - <N_{obs}>^2}{N_{exp}^2}$$

where $\sigma^2(N_{obs})$ was estimated numerically from the same 1000 realizations.

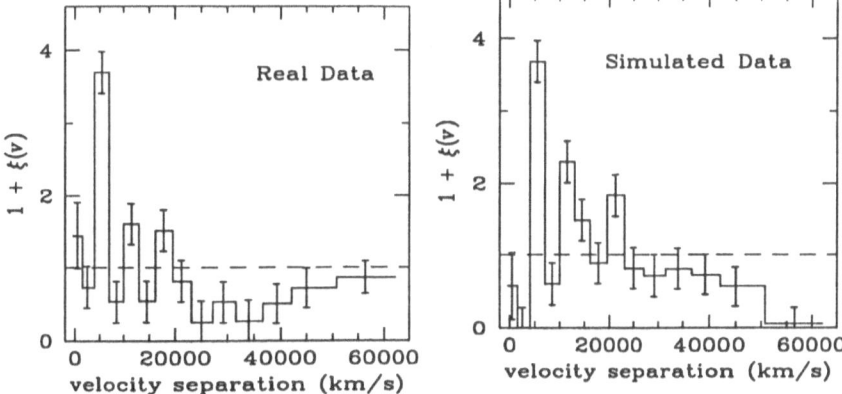

Fig. 1. The velocity correlation function for the 19 observed CIV systems (left panel) and for a simulated sample of systems (right panel). The error bars represent the 1σ uncertainties.

Figure 1 [left panel] shows significant signal in the correlation function for velocity separations corresponding to $4000 \leq \Delta v \leq 7000\,\mathrm{km\ s^{-1}}$. Upon close examination of the data, it is evident that there are 5 systems in Tol 1037–2704 and Tol 1038–2712 with roughly equal spacing. This feature has been pointed out by various authors (*e.g.* Jakobsen et al. 1986; Sargent & Steidel 1987). The spacing between each cluster of systems is roughly 5000 – 7000 km s^{-1}, consistent with the peak in $\xi(v)$. We placed systems with velocity separations of $\sim 5500\,\mathrm{km\ s^{-1}}$ and a velocity dispersion of 1000 km s^{-1} in the same redshift intervals of the data as a means of simulating this effect. The result of the simulations is shown in Figure 1 [right panel].

The peak in the velocity splitting histogram corresponds to a comoving spatial scale of ~ 30 Mpc. Since the power associated with virialized clusters on smaller scales has been removed, we can be confident that this highly significant peak represents power on supercluster scales. The *spatial* correlation function $\xi(r)$ was determined in a similar manner using only systems along different lines of sight in order to obtain a measure of the true spatial clustering. A signal of the same amplitude and scale is seen in Figure 2, although in this case the error bars are larger because of the sparse sampling of this pencil beam experiment. In standard CDM hierarchical clustering models, $\xi(r) < 1$ on comoving scales of 10 Mpc or larger at $z \sim 2$, in accord with N-body simulations (Evrard et al. 1994). In this case, we have focussed on a region known to have an overdensity of absorbers. However, if similar clustering signals are seen along other lines of sight, it will be considerable challenge to theories of large scale structure formation.

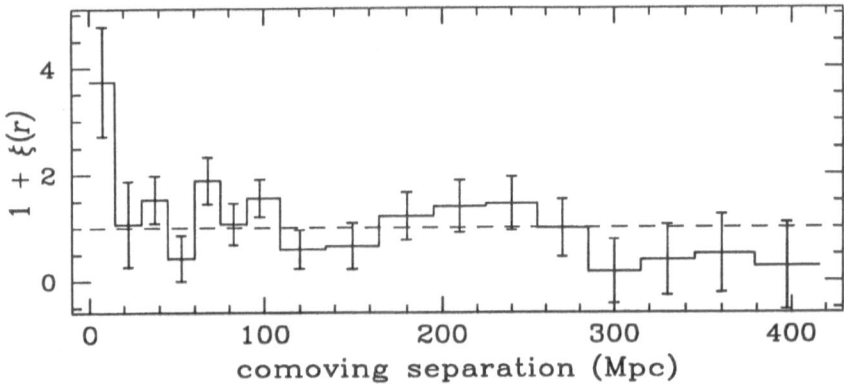

Fig. 2. The spatial correlation function for the 19 CIV absorbers. The error bars represent the 1σ uncertainties.

Acknowledgements. We acknowledge useful conversations with Craig Foltz, Craig Hogan, Julia Indik, and Gerry Williger, and excellent support from the technical staff at CTIO. This research was supported by NSF grant AST 90-01181.

References

Broadhurst, T.J., Ellis, R.S., Koo, D.C., & Szalay, A.S. 1990, Nature, 343, 726.
Evrard, A.E., Summers, F.J., & Davis, M. 1994, ApJ, 422, 11
Heisler, J., Hogan, C.J., & White, S.D.M. 1989, ApJ, 347, 52
Jakobsen, P., Perryman, M. A. C., Ulrich, M. H., Macchetto, F., & Di Serego Alighieri, S. 1986, ApJ, 303, L27
Olivier, S.S., Primack, J.R., Blumenthal, & Dekel, A. 1993, ApJ, 408, 17
Sargent, W. L. W., & Steidel, C. C. 1987, ApJ, 322, 142
Sargent, W. L. W., Boksenberg, A., & Steidel, C. C. 1988, ApJSupp, 68, 539

Investigating the Connection Between Lyα Absorbers and Large-Scale Galaxy Structures

Vicki Sarajedini[1], Richard F. Green[2], and Buell T. Jannuzi[2]

[1] Steward Observatory, University of Arizona, Tucson, AZ 85721
[2] NOAO, P.O. Box 26732, Tucson, AZ 85726

Abstract. We model the requirements on observational data that would allow an accurate determination of the degree of association between Lyman α absorbers and peaks in the redshift distribution of galaxies. We compare simulated distributions of low-redshift Lyman α absorption systems, constrained to be consistent with the distribution observed with HST, with the large-scale distribution of galaxies determined from pencil beam redshift surveys in order to investigate what observational data are required to 1.) map the large- scale distribution of galaxies and absorbers and 2.) allow a statistically significant test of the association of the two populations.

1 Sampling the Galaxy Distribution

We are interested in determining the minimum number of galaxies necessary to best estimate the location of large-scale structures in redshift space. Redshift surveys were used to determine the minimum number of galaxies having the largest possible fraction of members in the "peaked" regions of the galaxy distribution and having every peak in the true distribution represented by at least one member. The surveys used in our simulations are described in Sarajedini, Green and Jannuzi (1995 ApJ submitted) along with many of the details of these tests. The data from the literature consists of ∼200 galaxy redshifts from each of two sight lines. The distributions were used to define "peaks" and "voids" in redshift space producing a weighting function giving the highest weight of 1.0 to the peaked regions (Figure 1b). We simulated "observations" of galaxies in the redshift surveys through sub-cones having diameters up to the largest angular size of each survey containing redshifts out to z=0.4. Galaxies within each cone were weighted by the function in Figure 1b so that galaxies in or near a peak received a higher value than those in the voids. A mean function value was determined for the galaxies in each sub-cone. We limited the galaxies used in these simulations to $0.2 \leq z \leq 0.4$ and $M_b \leq$ -18 to increase the probability that a galaxy will lie within one of the distribution peaks.

Our simulations suggest that if at least ∼18 galaxies are observed (requiring an angular field of view of r≃10') a given galaxy has a ∼70% chance of lying within a peak in the true galaxy distribution and correctly identifies the location of a peak. All peaks within $0.2 \leq z \leq 0.4$ are located although ∼30% of the galaxies lie at redshifts where no true peak exists.

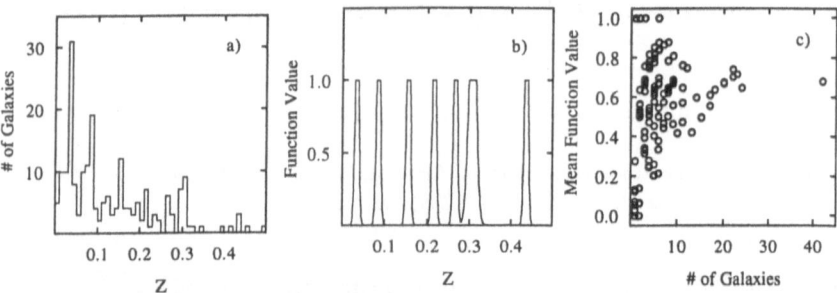

Fig. 1. a) The redshift distribution of the 1043+00 survey. b) Peak-defining function based on the distribution in Figure 1a. c) The number of galaxies observed per cone vs. the mean function value for the 1043+00 redshift survey.

2   Comparing the Galaxy and Lyα Cloud Distributions

To determine the number of sight lines necessary to show that the absorbers are uncorrelated with large-scale structures of galaxies, we first simulated a sample of Lyα absorbers following the simplified line distribution function as observed by Bahcall et al. (1993, ApJS 87, 1), so they have no velocity correlations on small scales but follow the observed trend of line-of-sight density evolution with redshift. Function values (from Figure 1b) are determined for these simulated Lyα lines to quantify their distribution with respect to the peaks in the galaxy distribution. These distributions of function values are compared through a KS test to the function values from subsets of galaxies drawn from the redshift surveys. We find that absorbers and galaxies must be observed from at least 12 sight lines within the redshift range z=0.2 to 0.4 to show that the 2 distributions are different at the 99% confidence level. Within this range there are an average of 3.9 absorbers per line list.

The aim of our last numerical experiment is to characterize to what accuracy we can determine the composition of an observed distribution of Lyα lines drawn from two parent distributions, one uncorrelated and one perfectly correlated with galaxy peaks, and also find the minimum number of Lyα lines which must be observed to do this. We generated large samples of Lyα absorption lines in these two populations. Function values for both sets of lines were determined to quantify how each population is distributed with respect to the true peaks in the galaxy distribution. Samples of Lyα lines containing mixtures of the two populations were then compared to the two parent populations using a KS test. As the number of Lyα lines in the sample increased, the range in apparent fractional composition values decreased as the true composition became more accurately known. We find that ~38 sight lines are required to sample enough Lyα absorbers within $0.2 \leq z \leq 0.4$ to determine the fraction of absorbers that are correlated with peaks in the galaxy distribution with 10% accuracy.

QSO Spectra and the Evolution of Structure

Andrei G. Doroshkevich[1],[2] and Victor I. Turchaninov[1]

[1] Keldysh Institute of Applied Mathematics Academy of Sciences Moscow, Russia
[2] Theoretical Astrophysics Center Blegdamsvej 17 Copenhagen O, Denmark

Abstract. The redshift distribution of Ly-α absorption lines in quasar spectra can be explained by the strong clustering of Ly-α clouds to the halo of structure elements of Large and Super Large Scale Structure (LSS and SLSS). It tests the evolutionary history of the LSS.

1 Introduction

As a model for the absorption spectra of quasars we examine intergalactic clouds concentrated around structure elements of LSS and SLSS. The paper is submitted to Ap.J.

The main quantitative characteristics of the LSS and SLSS were found by Buryak et al. (1994, hereafter BDF). Later the analysis of dynamical N-body simulations showed that the main fraction of galaxy filaments (perhaps, *all* filaments) are embedded in the dark matter (DM) sheets and there is the extended anisotropic DM and, probably, gaseous halo around galaxy filaments. Some DM sheets contain only the DM component and intergalactic gas. It could be considered as the dark elements (phantoms) of the LSS.

The possible connection between Ly-α lines and halos of galaxies was recently discussed by Salpeter (1993). Here we proceed further and assume that *all* Ly-α lines are related to the gaseous halos of the LSS and SLSS elements. The conception of such halos is the natural extension of the available information about the galactic halos and intergalactic medium in the clusters and superclusters of galaxies.

We assume that the population of Ly-α clouds can be associated with all types of structure elements. The main fraction of Ly-α clouds is, probably, concentrated into gaseous halos of galaxy filaments which are the most abundant elements of the LSS. Some part of the Ly-α lines may be related to clouds incorporated into the SLSS elements.

For saturated HI lines with the rest equivalent width $w \geq 0.5 \text{Å}$ two typical scales were found, namely, the characteristic radius of cloud halo, R_f, and the mean comoving size of SLSS cell, D_{SLSS}.

$$R_f \approx (0.3 - 0.4)h^{-1}Mpc \qquad D_{SLSS} \approx (80 \pm 12) \; h^{-1}Mpc$$

The analysis of the distribution of CIV doublets gives the size of the SLSS as

$$D_{SLSS} \approx (71 \pm 4)h^{-1}Mpc.$$

In this case the absorption in the halo of filaments is small. This fact can explain the different z - distribution of HI and CIV absorption lines because the evolution of LSS (filaments) and SLSS (superclusters) are very different.

2 Results

We can formulate our main results as follows:

1. A strong correlation between the absorption lines distribution and the spatial distribution of galaxies is demonstrated. The main fraction of lines could be formed in the gaseous halo of galaxy filaments and in the SLSS elements. Using the theory of the structure evolution (BDF) we obtain the quantitative description of the redshift distribution of absorption lines and, thus, relate the main parameters of this distribution to the observed and theoretical parameters of the structure.

2. The analysis of the saturated Ly-α line population reveals and tests the evolutionary history of the LSS. Our results show that the LSS has been formed at a period $z > 4$ and its later evolution agrees well with the theoretical description. The available observational data cover a wide interval of redshifts $0 \leq z \leq 4$ and, during this period, we do not see any evidence of strong perturbations in the structure evolution.

3. We argue that the redshift distribution of CIV doublets can be explained by their formation preferentially into SLSS elements (superclusters).

4. Three weak $Ly - \alpha$ absorption lines located in the central area of a huge void (Morris et al., 1993) may be considered as the first observational evidence of DM walls.

Acknowledgements. This paper was supported by Denmark's Grundforskningsfond through its support for an establishment of the Theoretical Astrophysics Center. We wish to acknowledge support from the Center of Cosmo-Particle Physics, Moscow.

References

Buryak, O.E., Doroshkevich, A.G., Fong, R. 1993, Ap.J., 434, 24 (BDF).
Morris, S.L., Weymann, R.J., Dressler, A. et al., 1993, Ap.J., 419, 524
Salpeter, E.E. 1993, AJ, 106, 1265

QSO Absorption Lines as a Cosmic Reference Frame

Michael Rauch

Observatories of the Carnegie Institution of Washington, 813 Santa Barbara St, Pasadena CA 91101 USA

We consider the possibility of measuring the motion of our solar system against a reference frame provided by high redshift objects like quasars and QSO absorption systems.

The Doppler effect caused by our motion relative to this cosmic reference frame (with velocity $\beta = v/c$) results in a difference between the observed redshift of an object, z', and its "true" redshift z as it would be measured by an observer at rest. This effect will lead to a distortion of the redshift distribution function dN/dz (i.e. the number of objects per unit redshift). A certain set of dN objects in redshift bin $[z, z+dz]$ and at an angle γ relative to the apex of the motion will be perceived by the moving observer as being located in $[z', z'+dz']$. Number conservation of objects gives

$$\frac{dN}{dz'}(z', \beta cos\gamma) = [1 - \beta cos\gamma]^{-1}\frac{dN}{dz}\left(z = \frac{z' + \beta cos\gamma}{1 - \beta cos\gamma}\right). \qquad (1)$$

Thus the observed distribution function dN/dz' appears shifted and compressed, and acquires a directional dependence when measured in our (moving) frame as compared to the dN/dz measured in the cosmic frame. It is this distortion that can be utilized to yield information about our motion (see Figure 1).

Assuming that the number of objects per unit redshift changes with z according to a power law with index κ, $dN/dz \propto (1+z)^\kappa$ we obtain

$$\frac{dN/dz}{dN/dz'} \propto (1 - \beta cos\gamma)^{\kappa+1}, \qquad (2)$$

so the measured dipole anisotropy will be the stronger the faster the number density of objects increases with redshift (i.e., the larger κ).

Lyman α forest absorption systems appear to be the class of high redshift objects most suitable for this sort of work, because they are numerous, only weakly clustered, and undergo a strong evolution with redshift. Using absorption systems rather than luminous objects we can also avoid selection effects and variations in detection probability across the sky which affect galaxy and QSO surveys.

The parameters of our motion (direction on the sky and amplitude of the velocity dipole) can be derived from a fit to an absorption redshift sample using a a maximum likelihood method (Rauch 1994).

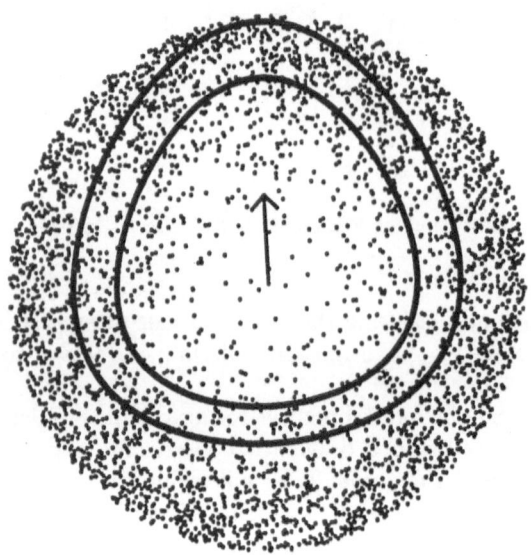

Fig. 1. Schematic plot showing how a circular annulus of constant observed redshift $[z', z' + dz']$ as defined by a moving observer would appear in the rest frame given by a population of high redshift objects. The radial coordinate is redshift z, and the number of objects per unit redshift is assumed to increase with redshift. The observer moving in the direction of the arrow and counting objects in this annulus finds larger numbers in the forward direction because (a) the region is stretched to higher redshifts where the objects are more abundant, and (b) the region is wider in the forward direction than in the opposite direction.

Using Lyman α absorption redshift samples available now (Lu et al. 1991, Bechtold 1994) we can put only an upper limit of v/c <0.05 on the velocity of our motion relative to the Lyα forest at $< z > = 2.9$. Simulations of absorption line spectra with various velocity dipoles indicate that at least of order 10^5 redshifts or 200 QSO high resolution echelle spectra (at z$\sim$ 3) will be required to reduce the one-sigma error to the value of the amplitude of the microwave background dipole (Kogut et al 1993) This is within reach of the coming generation of large optical telescopes. The redshift sample could be obtained as a by-product of QSO spectroscopy for other purposes.

References

Bechtold, J., 1994, ApJSuppl, 91, 1
Kogut, A. et al., 1993, ApJ, 419, 1
Lu, L., Wolfe, A.M., Turnshek, D.A., 1991, ApJ, 367, 19
Rauch, M., 1994, MNRAS, 271, 13

PART IX

THE UV BACKGROUND

PART IX

The Ultraviolet Extragalactic Background

Piero Madau

Space Telescope Science Institute, 3700 San Martin Drive, Baltimore, MD 21218

Abstract. We discuss the results of a detailed numerical study of the absorption and reprocessing of UV photons in a clumpy, photoionized universe, focusing on the effects of intervening discrete absorbers on the propagation of ionizing diffuse radiation. We compute the contribution to the metagalactic flux associated with H and He Lyα line emission, two-photon continuum and recombination radiation from the Lyα forest clouds and the Lyman-limit systems. In the light of new data and recent studies, we reexamine the contribution of quasars to the ionizing radiation field.

1 Introduction

The growing number of quasars with $z > 4$ observed in the past few years (Schneider et al. 1989), the use of the properties and evolution of QSO absorption systems as a diagnostic tool of the ambient physical conditions at early epochs (e.g., Charlton et al. 1993), and the recent detection of intergalactic ionized helium (Jakobsen et al. 1994), have all made the reionization of the universe a subject of much recent investigation and debate. It is believed that the integrated ultraviolet flux arising from QSOs and/or hot, massive stars in metal-producing young galaxies is responsible for maintaining the intergalactic diffuse gas and the Lyα forest clouds in a highly ionized state. Such a UV background may also be responsible for the ionization of the metal-rich QSO absorption systems (Vogel & Reimers 1993), and for producing the sharp edges of H I disks observed in nearby spiral galaxies (Maloney 1993). Photoionization by UV radiation may also inhibit the collapse of small-mass galaxies (Efstathiou 1992).

2 Cosmological Radiative Transfer

The radiative transfer equation in cosmology describes the evolution in time of the specific intensity J of a diffuse radiation field:

$$\left(\frac{\partial}{\partial t} - \nu \frac{\dot{a}}{a} \frac{\partial}{\partial \nu} \right) J = -3 \frac{\dot{a}}{a} J - c\alpha J + \frac{c}{4\pi} \epsilon, \tag{1}$$

where a is the cosmological scale parameter, c the speed of the light, α the continuum absorption coefficient per unit length along the line of sight, and ϵ is the proper space-averaged volume emissivity. Integrating equation (1) and averaging over all lines of sight yields the mean specific intensity of the radiation background at the observed frequency ν_o, as seen by an observer at redshift z_o:

$$J(\nu_o, z_o) = \frac{1}{4\pi} \int_{z_o}^{\infty} dz \frac{d\ell}{dz} \frac{(1+z_o)^3}{(1+z)^3} \epsilon(\nu, z) e^{-\tau_{eff}(\nu_o, z_o, z)}, \qquad (2)$$

where $\nu = \nu_o(1+z)/(1+z_o)$, $d\ell/dz = c/H_0(1+z)^{-2}(1+2q_0z)^{-1/2}$ is the line element in a Friedman cosmology, and $H_0 = 50h_{50}$ km s^{-1} Mpc^{-1} is the Hubble constant. In the following, we assume $h_{50} = 1$ and $q_0 = 0.1$.

3 Intergalactic Absorption

The *effective optical depth* τ_{eff} due to discrete absorption systems is defined as $\tau_{eff} = -\ln(\langle e^{-\tau}\rangle)$ where $\langle e^{-\tau}\rangle$ is the average transmission over all lines of sight (Paresce et al. 1980). For Poisson-distributed clouds we have

$$\tau_{eff}(\nu_o, z_o, z) = \int_{z_o}^{z} dz' \int_0^{\infty} dN_{\mathrm{HI}} f(N_{\mathrm{HI}}, z')(1 - e^{-\Delta}), \qquad (3)$$

where $f(N_{\mathrm{HI}}, z')$ is the redshift and column density distribution of absorbers along the line of sight. The continuum optical depth through an individual cloud is $\Delta = N_{\mathrm{HI}}\sigma_{\mathrm{HI}}(\nu') + N_{\mathrm{HeI}}\sigma_{\mathrm{HeI}}(\nu') + N_{\mathrm{HeII}}\sigma_{\mathrm{HeII}}(\nu')$, where $\nu' = \nu_o(1+z')/(1+z_o)$, and $\sigma_{\mathrm{HI}}, \sigma_{\mathrm{HeI}}$, and σ_{HeII} are the hydrogen and helium photoionization cross-sections. For the redshift and column density distribution of intervening absorbers we take

$$f(N_{\mathrm{HI}}, z) = \begin{cases} N_1 N_{\mathrm{HI}}^{-1.5}(1+z)^{2.46} & (10^{13} < N_{\mathrm{HI}} < 1.59 \times 10^{17}\,\mathrm{cm}^{-2}); \\ N_2 N_{\mathrm{HI}}^{-1.5}(1+z)^{1.55} & (1.59 \times 10^{17} < N_{\mathrm{HI}} < 10^{20}\,\mathrm{cm}^{-2}), \end{cases} \qquad (4)$$

where $N_1 = 2.9 \times 10^7$ and $N_2 = 5.4 \times 10^7$ (Press & Rybicki 1993; Storrie-Lombardi et al. 1994; Tytler 1995).

4 A QSO-Dominated Background

We shall estimate the integrated flux from QSOs following Pei (1995), who has recently derived an analytical model which fits well the empirical LF estimated by Hartwick & Shade (1990) and Warren et al. (1994). The proper volume emissivity can be written as $\epsilon_Q(\nu, z) = \epsilon_Q(\nu_B, 0)(1 + z)^{\alpha+2}(\nu/\nu_B)^{-\alpha} \exp[-z(z - 2z_*)/2\sigma_*^2]$, where $\epsilon_Q(\nu_B, 0)$ is the extrapolated $z = 0$ emissivity at the reference frequency $\nu_B = c/4400\text{Å}$, $\epsilon_Q(\nu_B, 0) \simeq 7.7 \times 10^{32}$ ergs Gpc^{-3} s^{-1} Hz^{-1}, and α is the slope of the quasar power-law spectrum. This Gaussian form can fit reasonably well the observational data in the entire range $0 < z < 4.5$; the evolution of QSOs reaches a maximum at

$z \sim 2.8$ and declines at higher redshifts. We shall adopt the following model
for the "typical" quasar spectral energy distribution:

$$f(\nu) \propto \begin{cases} \nu^{-0.3} & (2000 < \lambda < 4400\text{Å}); \\ \nu^{-0.7} & (1216 < \lambda < 2000\text{Å}); \\ \nu^{-1.4} & (60 < \lambda < 1216\text{Å}); \\ \nu^{-1.6} & (\lambda < 60\text{Å}). \end{cases} \tag{5}$$

This is based on observations by Sargent et al. (1989), Francis et al. (1991),
and Tytler (1995).

5 Recombination Radiation from Discrete Absorbers

The attenuation due to the accumulated H I and He II photoelectric absorption by optically thin Lyα clouds and optically thick Lyman-limit systems
will cause a large reduction of the background ionizing flux. At the same
time, intervening absorbers will produce a significant contribution to the local ionizing emissivity. The specific emissivity due to radiative recombinations
within the absorbing clouds can be expressed as

$$\epsilon_r(\nu, z) = 4\pi\nu \, p(\nu) \frac{\alpha^{eff}}{\alpha} \int_{\nu_t}^{\infty} \frac{d\nu'}{\nu'} J(\nu', z) \frac{d\tau_{eff}}{d\ell}. \tag{6}$$

The above formula simply states that the number of recombination photon-
s emitted is proportional to the number of photons absorbed at any given
redshift. Here, α^{eff} is the effective recombination coefficient for the relevan-
t atomic transition, while α is the total recombination coefficient (α_A for
optically thin clouds, α_B for optically thick clouds). The function $p(\nu)d\nu$
represents the normalized probability per recombination that one photon is
emitted in the range ν to $\nu + d\nu$. In our calculations we include: a) H I and
He II Lyα recombination radiation and two-photon continuum from optically
thick absorbers; and 2) H I and He II continuum recombination radiation to
the ground state and to the 2 ^2L level, plus Lyα line emission and two-photon
continuum from optically thin clouds (Haardt & Madau 1995).

6 Results

In Figure 1 we show the spectrum of the ionizing background radiation field
at $z = 0$ (lower set of curves) and $z = 3$ which results from the numerical integration of equation (2), as modified by the absorption of intervening systems
(*dashed lines*). The solid curves include the contribution due to radiative
recombinations within the absorbing clouds. Note, in the $z = 3$ spectrum,
the large peaked contribution at $\lambda \geq 304$Å and ≥ 1216Å due to He II and
H I Lyα emission. The background intensity in the limit of a perfectly transparent ($\tau_{eff} = 0$) universe is also shown for comparison (*dotted lines*). In

the absorption plus reemission case, the integrated intensity at the hydrogen Lyman edge, J_{912}, increases by a factor ~ 27 between $z = 0$ and $z = 3$, where it peaks at a value of $J_{912} = 6 \times 10^{-22}$ erg cm^{-2} s^{-1} Hz^{-1} sr^{-1}. This is about a factor of 2.5 smaller than the value in the optically thin limit, and about a factor of 1.8 higher than the value in the absorption no-reemission case. The impact of these calculations on the ionization and thermal state of intergalactic material will be discussed in Haardt & Madau (1995).

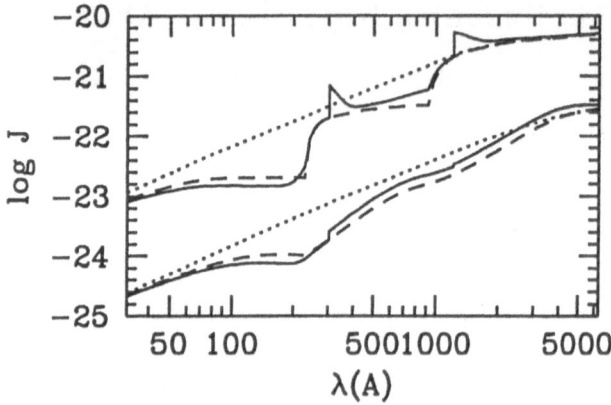

Fig. 1. Spectrum of a QSO-dominated background at $z = 0$ and 3.

References

Charlton, J. C., Salpeter, E. E., Hogan, C. J., 1993, ApJ, 402, 493.
Efstathiou, G., 1992, MNRAS, 256, 43P.
Francis, P. J., et al., 1991, ApJ, 373, 465.
Haardt, F., Madau, P., 1995, ApJ, submitted.
Hartwick, F. D. A., Schade, D., 1990, ARA&A, 28, 437.
Jakobsen, P., et al., 1994, Nature, 370, 35.
Maloney, P., 1993, ApJ, 414, 41.
Paresce, F., McKee, C., Bowyer, S., 1980, ApJ, 240, 387.
Pei, Y. C., 1995, ApJ, 438, 623.
Sargent, W. L. W., Steidel, C. C., Boksenberg, A., 1989, ApJ Suppl, 69, 703.
Schneider, D. P., Schmidt, M., Gunn, J. E., 1989, AJ, 98, 1507.
Storrie-Lombardi, L. J., et al., 1994, ApJ, 427, L13.
Tytler, D., 1995, this proceedings.
Vogel, S., Reimers, D., 1993, A&A, 274, L5.
Warren, S. J., Hewett, P. C., Osmer, P. S., 1994, ApJ, 421, 412.

Effects of the Photoionizing Background on Galaxy Formation

Masashi Chiba[1] and Biman B. Nath[2]

[1] Astronomical Institute, Tohoku University, Sendai 980-77, Japan
[2] Inter University Center for Astronomy and Astrophysics, Post Bag 4, Ganeshkhind, Pune 411007, India

Abstract. We investigate the effects of the observed UV background radiation on galaxy formation. In the CDM model, we use the criterion that self-shielding of the gas in halos against the photoionizing background radiation is essential for star formation to calculate the mass function of galaxies and its evolution with time based on both Press-Schechter and the peaks formalisms. We further discuss the implication of this biasing mechanism for the number counts of galaxies.

1 Introduction

The observations of QSO absorption-line systems suggest the existence of the UV background radiation with intensity (at the Lyman limit) $J \sim 10^{-22}$ to 10^{-21} ergs cm^{-2} s^{-1} sr^{-1} Hz^{-1} permeated the universe at redshifts $z \lesssim 5$. Since protogalactic gas in halos with $T \lesssim 10^{5.5}$ K cools mainly by line cooling of H and He$^+$ atoms, the cooling efficiency of the gas is diminished in the presence of the ionizing photons (Efstathiou 1992). As a result, objects with only large density contrasts can self-shield against the UV radiation, thereby allowing the shielded neutral cores to cool and form stars – i.e., biasing galaxy formation. In the present contribution, we attempt to answer the questions: to what extent does the UV background inhibit galaxy formation, and how much it affects the mass function of galaxies and the evolution of the mass function with time ? (see Chiba & Nath 1994 for details).

2 Methods and Results

For objects with large density contrast, there exists a self-shielded neutral core with mass M_c. The gas inside the core can cool fast and thus form stars, since no UV photons can reach inside. The condition for objects to have a core $M_c > 0$ depends on the UV intensity J_{-21} ($\equiv J/10^{-21}$) and the density contrast. In the context of CDM model with $\Omega = 1$ and $h_{50} = 1$, we derive the threshold for the amplitude of density perturbation (in units of rms density fluctuation), ν^{UV}, to have $M_c > 0$, as functions of J_{-21} and the baryon mass M_b. Furthermore, objects have to collapse and cool by the given epoch z, thereby defining the thresholds for density perturbation $\nu^{collapse}$ and ν^{cool}, respectively. Thus the criterion for galaxy formation may be expressed as $\nu(M_b) \geq \nu_{th}(M_b)$, where $\nu_{th} = \max(\nu^{collapse}, \nu^{cool}, \nu^{UV})$.

Having obtained $\nu_{th}(M_b)$, it is possible to calculate the mass function of galaxies and its evolution, given the UV flux as $J_{-21}(z) = const.$ for $z \geq 2$ and $const. \times [(1+z)/3]^\alpha$ for $z < 2$. If the present J_{-21} is as small as ~ 0.1, the generalized Press-Schechter (PS) mass function (Peacock & Heavens 1990) at $z = 0$ is not greatly influenced by the UV flux. However the time evolution of PS mass function is subject to the interesting change: although the number of collapsed small galaxies with $10^8 \lesssim M_b \lesssim 10^{10} M_\odot$ is usually decreasing with time due to hierarchical merging (Fig.1a), the trend is *reversed* (the number of low-mass galaxies is greater at lower z) due to the decreasing UV flux with time after $z \sim 2$ (Fig.1b) for $J_{-21}(z = 2) = 3.0$ and $\alpha = 3.1$]. It means that the intense ionizing photons at higher z suppress the formation of (low-mass) premergers.

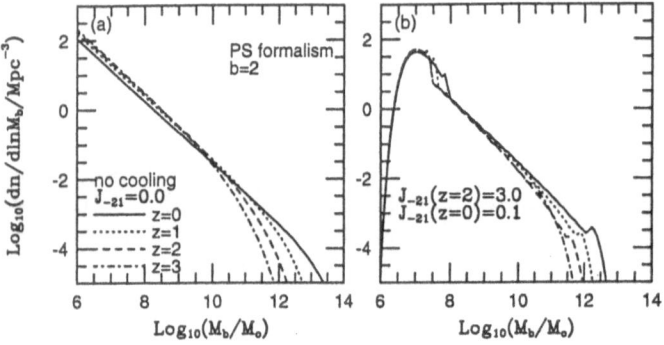

Fig. 1. (a) Time evolution of the PS mass function for all collapsed objects from $z = 3$ to 0. (b) for cooled and self-shielded objects

If the UV background also decreases the amplitude of luminosity function by reducing the number of pregalactic objects at higher z, it is suggested that the result of galaxy number counts cannot be explained by the model of strong number evolution of galaxies via merging with the high-density $\Omega = 1$ universe.

References

Chiba M., Nath B.B., 1994, ApJ, 436, 618.
Efstathiou G., 1992, MNRAS, 256, 43P.
Peacock, J.A., Heavens, A.F., 1990, MNRAS, 243, 133.

The Helium Gunn-Peterson Test

Peter Jakobsen

Astrophysics Division, Space Science Department of ESA, ESTEC, 2200 AG Noordwijk, The Netherlands

Abstract. Thanks to the gain in sensitivity of the refurbished HST, considerable progress has recently been made toward the goal of extending the study of quasar absorption systems to the lines of neutral and singly ionized helium. Strong intervening intergalactic HeII $\lambda304$ Gunn-Peterson absorption has now been detected in the far-UV spectra of two quasars: Q0302−003 ($z = 3.29$) and PKS 1935−692 ($z = 3.18$). Weak HeI $\lambda584$ lines have also been detected in several metal line systems in HS1700+6416 ($z = 2.72$). The background to these results and their possible implications are outlined. The helium observations not only mark the likely detection of the diffuse intergalactic medium, but also provide strong qualitative confirmation of both Big Bang nucleosynthesis theory and current ideas concerning the ionization state of intergalactic gas at high redshift.

1. Introduction

One "Holy Grail" of observational cosmology and ultraviolet space astronomy that has been with the HST project since its inception (e.g. Bahcall & Salpeter 1965), is the goal of validating Big Bang nucleosynthesis theory and detecting the elusive intergalactic medium (IGM) by extending the classical Gunn & Peterson (1965) test to the HeI $\lambda584$ and HeII $\lambda304$ main resonance lines of neutral and singly ionized helium.

Since the discovery of quasar absorption lines in the mid 1970s, the agenda of this program has become increasingly more sophisticated – as has our appreciation of the many practical problems of observing quasars the very short extreme ultraviolet rest wavelengths required to detect helium foreground absorption. Because intergalactic gas is believed to be highly photoionized, the focus today is primarily on the HeII ion. Although the ultimate goal of resolving the individual HeII lines of the Lyman forest may well turn out to lie beyond even the capabilities of even the refurbished HST, dramatic progress has recently been made in the technically less demanding first step of the program – namely detecting the combined redshift-smeared HeII line absorption of both the diffuse IGM and the Lyman forest clouds.

2. The Ionization State of Intergalactic Helium

A seminal discussion of the physical state of intergalactic helium was given by Sargent, Young, Boksenberg & Tytler (1980), who pointed out that if the Lyman forest clouds are kept photoionized by the integrated background of quasars, then one clear prediction is that the helium present in these clouds and in the surrounding ambient IGM should also be ionized – and therefore primarily detectable through absorption in singly ionized HeII rather than neutral HeI.

That notion that the forest clouds are photoionized has since received indirect observational support in the form of the large $b \simeq 30$ km s^{-1} average velocity width of the Lyman forest lines, (Carswell et al. 1987, 1991), and the discovery of the so-called "inverse" or "proximity" effect (Carswell et al. 1982; Murdoch et al. 1986). Through detailed quantitative analysis of the latter (Bajtlik, Duncan & Ostriker 1988; Lu, Wolfe & Turnshek 1991; Bechtold 1994) one infers that the intensity of the intergalactic ionizing background in the redshift range $1.7 \lesssim z \lesssim 3.8$ is of order $I_{\nu_H} \approx 10^{-21}$ ergs s^{-1} cm^{-2} sr^{-1} Hz^{-1}.

Although the origin of this ionizing flux is a topic of some debate (Bechtold et al. 1987; Miralda-Escudé & Ostriker 1990,1992; Madau 1992), it is customary to assume that its spectrum is a simple power-law: $I_\nu \propto \nu^{-\alpha}$. The equations governing the relative ionization of hydrogen and helium can then be written (Osterbrock 1989):

$$\left[\frac{\text{HI}}{\text{H}}\right] \simeq \frac{1}{U}\left(\frac{\alpha_{\text{HI}}}{c\sigma^o_{\text{HI}}}\right)\left[\frac{\alpha+3}{\alpha}\right] \tag{1}$$

$$\left[\frac{\text{HeII}}{\text{He}}\right] \simeq \frac{1}{U}\left(\frac{\alpha_{\text{HeII}}}{c\sigma^o_{\text{HeII}}}\right)\left[\frac{\alpha+3}{\alpha}\right]\left(\frac{\nu_{\text{HeII}}}{\nu_{\text{HI}}}\right)^\alpha \tag{2}$$

$$\left[\frac{\text{HeI}}{\text{He}}\right] \simeq \left[\frac{\text{HeII}}{\text{He}}\right]\frac{1}{U}\left(\frac{\alpha_{\text{HeI}}}{c\sigma^o_{\text{HeI}}}\right)\left[\frac{\alpha+3}{\alpha}\right]\left(\frac{\nu_{\text{HeI}}}{\nu_{\text{HI}}}\right)^\alpha \tag{3}$$

where α_{HI}, α_{HeI} and α_{HeII} are the HI, HeI and HeII recombination coefficients; σ^o_{HI}, σ^o_{HeI} and σ^o_{HeII} the matching photoionization cross sections at threshold at $\lambda = 912$ Å, $\lambda = 504$ Å and $\lambda = 228$ Å, respectively; and $U \simeq \frac{4\pi}{hc\alpha}(I^o_{\nu_{\text{HI}}}/n_{\text{H}})$ is the ionization parameter of the forest clouds.

For $\alpha \sim 1$ and the value of $U \sim 10^{-1}$ anticipated for the Lyman forest clouds (Sargent et al. 1980; Carswell 1988), the equations above predict [HI/H] $\sim 10^{-4}$, [HeI/He] $\sim 10^{-6}$ and [HeII/He] $\sim 10^{-2}$. Hence bulk of the gas in the forest clouds is predicted to be in the form of fully ionized HII and HeIII – with the residual species giving rise to detectable absorption representing only a negligible fraction of the total mass present in the clouds.

One unfortunate consequence of this high ionization is that it makes it unlikely that the absolute helium abundance of the forest clouds can ever be determined to an accuracy of 10-20% of interest to Big Bang nucleosynthesis calculations (Walker et at. 1991). In this sense the helium Gunn-Peterson test is only a qualitative – albeit still critical – test of Big Bang theory. The emphasis today is therefore just as much on validating the above expectations concerning the high ionization state of the intergalactic gas by determining the relative strengths of the absorption in HI, HeI and HeII.

The anticipated strength of the HeI absorption in the forest clouds can be estimated from the predicted density ratio of residual neutral hydrogen and helium:

$$\left[\frac{\text{HeI}}{\text{HI}}\right] \simeq \left[\frac{\text{He}}{\text{H}}\right]\left[\frac{\text{HeII}}{\text{He}}\right](\frac{\alpha_{\text{HeII}}}{\alpha_{\text{HII}}})(\frac{\sigma^o_{\text{HI}}}{\sigma^o_{\text{HeI}}})\left(\frac{\nu_{\text{HeI}}}{\nu_{\text{HI}}}\right)^\alpha \approx 0.1 \times 1.8^\alpha \left[\frac{\text{HeII}}{\text{He}}\right] \quad (4)$$

For a standard Big Bang helium abundance relative to hydrogen of [He/H] $\simeq$ 8% by number, equation (4) predicts that the abundance of neutral helium in the Lyα forest clouds should be far lower than that of neutral hydrogen; i.e. [HeI/HI] $\sim 10^{-3}$. The HeI $\lambda584$ forest lines are therefore expected to be extremely weak compared to Lyα.

In contrast, absorption due to singly ionized helium is predicted to be universally strong:

$$\left[\frac{\text{HeII}}{\text{HI}}\right] \simeq \left[\frac{\text{He}}{\text{H}}\right](\frac{\alpha_{\text{HeIII}}}{\alpha_{\text{HII}}})(\frac{\sigma^o_{\text{HI}}}{\sigma^o_{\text{HeII}}})\left(\frac{\nu_{\text{HeII}}}{\nu_{\text{HI}}}\right)^\alpha \approx 2.0 \times 4^\alpha \quad (5)$$

It follows that for an ionizing background spectrum with slope in the range, $\alpha \simeq 0.5 - 2$, the column density of once ionized helium through a given Lyman forest cloud is predicted to exceed that of neutral hydrogen by a factor [HeII/HI] $\approx 5 - 40$. Note that softer photoionizing spectra lead to larger predicted [HeII/HI] ratios. For instance, if the ionizing background flux does not extend down to the HeII photoionization threshold at $\lambda228$, and is therefore only capable of ionizing helium once, then the relative abundance of HeII can reach values as high as [HeII/HI] $\simeq$ [He/H]/[HI/H] $\approx 10^3$.

Since the predicted values of [HeI/HI] and [HeII/HI] given by equations (4) and (5) depend only on the shape of the ionizing spectrum, the above considerations apply equally well to the case of the surrounding diffuse intergalactic medium. In particular, it has long been appreciated (Sargent et al. 1980) that a diffuse IGM could reveal its presence through strong redshift-smeared Gunn-Peterson absorption in the HeII $\lambda304$ line, in spite of no such absorption having been detected in HI (Steidel & Sargent 1987, Schneider, Schmidt & Gunn 1991, Webb et. al 1992, Giallongo, Cristiani & Trevese 1992) or HeI (Green et al. 1980; Tripp, Green & Bechtold 1990; Beaver et al. 1991; Reimers et al. 1989; Tripp, Bechtold & Green 1994).

3.   The Intervening Absorption Problem

From the considerations of the previous section, it is clear that the prime
transition of interest for the helium Gunn-Peterson test is the HeII $\lambda 304$ resonance line of singly ionized helium. Unfortunately, in practice, observations
of this line are bedeviled by several problems.

For one, because of the short wavelength MgF_2 cutoff of the HST optics,
the HeII $\lambda 304$ line is only accessible with HST in the case of the most remote
– and therefore faintest – quasars at $z > 3$. Equally important, the resonance
lines of HeI and HeII all fall at wavelengths well below the Lyman limit at
$\lambda = 912$ Å. The background light from $z > 3$ quasars at these wavelengths is
therefore subject to photoelectric absorption in neutral hydrogen encountered
along the line of sight.

The realization that quasar absorption systems contain enough neutral
hydrogen to make the high redshift universe optically thick at extreme ultraviolet energies came about only gradually, and first emerged in the context of
attenuation of the UV background (Paresce, McKee & Bowyer 1980; Bechtold et al. 1987; Miralda-Escudé & Ostriker 1990; Meiksin & Madau 1993).
A discussion with emphasis on the helium Gunn-Peterson problem was given
by Møller and Jakobsen (1990), who analysed the character and statistics of
the cumulative photoelectric absorption due to both the Lyman forest and
Lyman limit classes of absorption system (see also Zuo & Phinney 1993).

Although Lyman forest systems typically have HI column densities of
order $N_{HI} \simeq 10^{15}$ cm^{-2}, and are therefore individually optically thin in the
Lyman continuum, the accumulated redshift-smeared absorption from the
large number of forest clouds intercepted out to high redshift (~ 400 out to
$z \sim 3$), leads to a characteristic "Lyman valley" shaped average absorption
spectrum that reaches optical depth of order unity near ~ 650 Å in the quasar
rest frame. Just as all quasars show Lyman forest lines, all quasars will be
subject to this matching continuum absorption. The Lyman valley was first
observed in the archetypical "clear" $z \simeq 2.72$ quasar HS1700+6416 (Reimers
et al. 1989, 1992).

The much rarer, but higher column density, "Lyman limit" absorption
systems (Tytler 1982; Bechtold et al. 1984; Lanzetta 1988; Sargent, Steidel
& Boksenberg 1989; Bahcall et al. 1993) present a far more serious obsta-
cle for observing high redshift quasars at short emitted extreme-UV wavelengths. In particular, the densest of such systems with column densities
$N_{HI} > 10^{19}$ cm^{-2} completely absorb the spectrum of the background quasar
at wavelengths shortward of the redshifted Lyman edge of the intervening
cloud. As shown by Møller and Jakobsen (1990), the observed statistics of
Lyman limit systems imply that photoelectric absorption in these systems is
expected to quench completely the received UV flux from the vast majority
of quasars at $z > 3$. However, in contrast to the Lyman forest, this absorption component arises in a modest number of clouds (on average ~ 6 systems
with $N_{HI} > 10^{17}$ cm^{-2} are intercepted out to $z \sim 3$) and therefore exhibits

large fluctuations between different sightlines. The probability of a random line of sight out to $z > 3$ experiencing less than 90% absorption at redshifted HeII $\lambda 304$ is predicted to be of order $\sim 5\%$. This probability, combined with the fact that $z > 3$ quasars tend to be faint ($V \simeq 18$), implies that the whole sky probably contains at most a few hundred $z > 3$ quasars with residual far-UV fluxes bright enough to serve as HeII background sources for HST (Picard & Jakobsen 1993). The first and foremost challenge of the helium Gunn-Peterson program is therefore to locate such an object.

4. The Search for "Clear" Quasars

The general prognosis concerning the scarcity of UV-bright quasars at high redshift is borne out by the fact that to date a total of only 17 quasars have ever been detected at a wavelength shortward of the emitted HeI $\lambda 584$ line by either IUE or HST. Of these, 13 (including HS1700+6416 in which HeI $\lambda 584$ lines have been detected; see below) have redshifts $z < 3$ and are therefore not suitable for detecting HeII $\lambda 304$ absorption with HST (Reimers et al. 1992; Beaver et al. 1991; Tripp et al. 1990, 1994; Reimers et al. 1994).

  Since the launch of HST, there have been two systematic searches for rare $z > 3$ quasars whose brightness and exceptionally clear line of sight conspire to make detection of redshifted HeII absorption feasible: the FOC objective prism survey (Jakobsen et al. 1993) and the FOC "snapshot" imaging survey of David Tytler and coworkers. Between them, these two surveys have sampled the far-UV fluxes more than 110 $z > 3$ quasars selected from the literature. So far only four $z > 3$ quasars have been detected spectroscopically down to emitted HeI $\lambda 584$, of which two unfortunately display Lyman limit systems just longward of the HeII $\lambda 304$ line (Beaver et al. 1992; Jakobsen et al. 1993). Nonetheless, not least thanks to the gain the in sensitivity provided by COSTAR, both surveys have now met with success in that they each have uncovered an object against which the first long awaited detections of intergalactic HeII Gunn-Peterson absorption have been made.

5. Detections of Helium Absorption

Although different in nature, the three cases of HeI and HeII absorption in quasars obtained so far confirm the theoretical expectations concerning the ionization state of intergalactic helium outlined above. Weak absorption lines due to HeI $\lambda 584$ have been detected in several metal line systems in HS1700+6414, while strong redshift-smeared HeII $\lambda 304$ Gunn-Peterson absorption has been detected in the two $z > 3$ quasars Q0302−003 and PKS1935−692.

5.1 HeI Line Absorption in HS1700+6414

No discussion of helium absorption in quasars would be complete without mention of HS1700+6416 – the only object in which HeI lines have so far been detected (Reimers et al. 1992). This quasar – which has become something of Rosetta Stone for UV absorption line studies – is highly remarkable in several regards. For one, at $V \simeq 16$ it is extremely bright for its redshift of $z = 2.72$. Moreover, the line of sight to this quasar happens to be unusually clear. Because of this, HS1700+6416 is exceptionally bright in the ultraviolet.[1]

The UV continuum of HS1700+6416 was first detected by Reimers et al. (1989) using IUE, and later re-observed at higher resolution and signal-to-noise ratio using the FOS on board HST (Reimers et al. 1992) Unfortunately, at $z = 2.72$ the redshift of HS1700+6416 is too small to bring the HeII $\lambda 304$ line of ionized helium within range of HST (but within range of future missions such as Astro-2 (Davidsen 1994) and FUSE). However, the HeI $\lambda 584$ line is accessible at $1 < z < 2.7$.

A detailed analysis of the HeI absorption in HS1700+6416 has been carried out by Reimers & Vogel (1993). In agreement with the theoretical expectations, HeI $\lambda 584$ lines corresponding to the Lyα forest are too weak to be detected in the FOS spectrum of HS1700+6416. Also – as is the case for all UV bright quasars observed so far – there is no evidence of a HeI $\lambda 584$ Gunn-Peterson trough. Weak HeI lines are, however, detected in the case of four denser ($N_{HI} \sim 10^{17} \mathrm{cm}^{-2}$) metal line systems seen in the spectrum at $z \simeq 1.85 - 2.43$.

Reimers & Vogel (1993) attempted to model the ionization of these systems by matching photoionization calculations to the observed ionization ratios of the detected heavy elements. According to their analysis, the values of [HI/HeI]$\simeq 1 - 5 \times 10^{-2}$ inferred for the absorbers – although indeed low as expected for a photoionized gas – are still a factor of ~ 5 higher than predicted for the model that best fits the ionization of the heavy elements, assuming a cosmological helium abundance. As discussed by Reimers & Vogel (1993), this discrepancy is not likely to imply a high helium abundance, but almost certainly merely reflects uncertainties in our understanding of the detailed ionization processes at work in quasar absorption systems.

Although the four systems in HS1700+6416 in which HeI absorption is seen are metal line systems – i.e. not possibly primordial Lyman forest systems – it is still reassuring that these clouds of mildly stellar-processed material do contain helium, and that the neutral helium lines detected are indeed weak as anticipated.

---

[1] Picard & Jakobsen (1993) have estimated that this fortuitous combination of a very luminous quasar lying along an unusually clear line of sight is sufficiently rare that the whole sky should contain less than ~ 20 quasars at redshifts $2 < z < 3$ comparable in brightness to HS1700+6416 in the UV. Nevertheless, Reimers et al. (1994) have recently identified several further examples of such objects.

5.2   HeII Gunn-Peterson Absorption in Q0302−003

The quasar Q0302−003 ($z = 3.29$, $V \simeq 18.4$) was identified during the course of the FOC prism survey, and was later re-observed with the COSTAR-corrected FOC. The low resolution FOC far-UV prism spectrum of this quasar shown in Figure 1 reveals what is almost certainly intense redshift-smeared line absorption shortward of the emitted HeII $\lambda304$ line (Jakobsen et al. 1994).

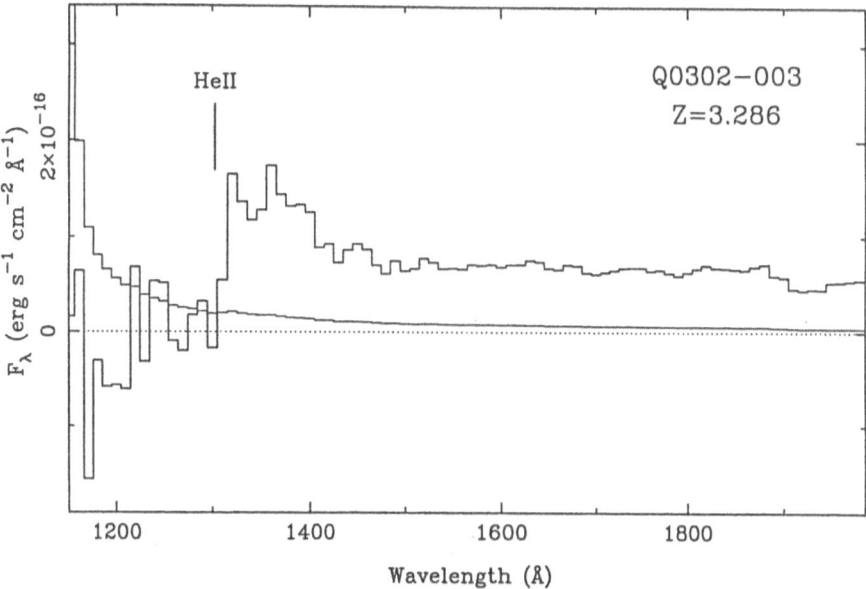

Fig. 1. FOC far-ultraviolet prism spectrum of the $z = 2.786$ quasar Q0302−003. The thin solid line gives the 1σ uncertainty per $\Delta\lambda = 10$ Å wavelength bin. The position of the HeII $\lambda304$ line in the quasar restframe is marked.

While the mere detection of the anticipated singly ionized intergalactic helium absorption provides dramatic confirmation of both Big Bang nucleosynthesis theory and the notion that intergalactic gas is highly ionized, several important questions remain to be answered. For one, it is not clear at present what fractions of the detected HeII absorption are due to HeII forest lines and the IGM proper. The FOC prism spectrum of Q0302−003 is of too low spectral resolution to distinguish directly between the absorption from unresolved HeII $\lambda304$ line-blanketing from the Lyman forest clouds and true HeII Gunn-Peterson absorption from a smooth intergalactic medium.

The spectrum of of Q0302−003 is seemingly completely absorbed on the blue side of the redshifted HeII line, with no flux detected below the edge at the sensitivity limit of the 3 hour FOC integration. This places a lower limit on the total effective HeII optical depth of $\tau_{\mathrm{HeII}} > 1.7$ (90% confidence). As discussed in detail by Jakobsen et al. (1994) (see also Madau & Meiksin 1994), this quantitative limit on the total HeII absorption is far larger than the corresponding wavelength-averaged HI optical depth of $\tau_{HI} \simeq 0.35$ measured shortward of Lyα at $z \simeq 3.2$ (Steidel & Sargent 1987; Schneider, Schmidt & Gunn 1991; Webb et al. 1992), which suggests that there may be a significant contribution from true Gunn-Peterson absorption in the IGM. However, the contribution to τ_{HeII} from unresolved HeII λ304 forest absorption depends critically on both the ionization of the forest gas and its detailed distribution in column density. For example, values of τ_{HeII} exceeding unity could in principle be reached in an abundant population of low column density forest clouds of extremely high HeII to HI ratio: [HeII/HI]$> 10^3$. In this situation – where the forest clouds are highly ionized in hydrogen ([HI/H]$< 10^{-4}$), but only singly ionized in helium ([HeII/He]~ 1) – forest systems with HI column densities as low as $N_{HI} \simeq 10^{12} \text{cm}^{-2}$ would be exceeding weak in Lyα, but strong in HeII λ304. The recent discovery of the existence of such low column density Lyman forest systems (Tytler 1994) raises the possibility that the net forest absorption in HeII λ304 could be considerably stronger than in Lyα.

The lack of knowledge of the ionization conditions in the plasma giving rise to the HeII absorption in Q0302−003 obviously also affects the quantitative estimates of the total amount of baryonic matter detected. As discussed by Jakobsen et al. (1994), even if due entirely to the IGM, the detected intergalactic plasma is in all likelihood very tenuous ($\Omega \sim 10^{-5} - 10^{-2}$) and not likely to conflict with constraints on the baryonic content of the universe stemming from considerations of nucleosynthesis theory.

In any case, the intensity of the detected HeII absorption with respect to Lyα implies that the effective spectrum of the ionizing background at $z \sim 3.2$ must be very steep (Miralda-Escudé 1993; Jakobsen et al. 1994; Madau & Meiksin 1994) – which, in turn, would imply that the high ionization species such as CIV and SiIV seen in metal line systems must arise from local ionization processes. Another intriguing question is whether the HeII absorption displays a "proximity effect" analogous to that seen in Lyα (Zheng & Davidsen 1994; Davidsen 1994)

5.3   HeII Gunn-Peterson Absorption in PKS 1935−692

The quasar PKS 1935−692 ($z = 3.17$, $V \simeq 18.8$) was identified during the FOC/HST snapshot survey, and was recently re-observed using the FOS. The spectrum of PKS 1935-692 also shows clear signs of apparent Gunn-Peterson absorption trough shortward of emitted HeII λ304 (Tytler 1994).

The most intriguing aspect of this second detection of intergalactic HeII absorption, however, is that the absorption detected in PKS 1935−692 appears to be weaker than that seen at slightly higher redshift toward Q0302−003. The inferred total HeII optical depth toward PKS 1935−692 is only of order $\tau_{HeII} \simeq 1$. If confirmed, this difference in HeII opacity along the two lines of sight probed by Q0302−003 and PKS 1935-692 may imply that the universe was only in the process of becoming re-ionized in helium at a redshift as late as $z \sim 3.2$. Either the ionization of the intergalactic medium changed very rapidly between $z \simeq 3.3$ and $z \simeq 3.1$, or, more likely, the HeII opacity is highly patchy at these redshifts – as might be expected if quasars themselves are responsible for re-ionizing the universe (Shapiro, Giroux & Babul 1994; Madau & Meiksin 1994). Needless to say, the prospect of pinpointing the epoch of helium re-ionization has far-reaching ramifications for our understanding of the thermal history of the early universe and the nature of the absorbing gas.

6. Things to Come

The first detections of intergalactic helium absorption obtained so far, while having provided a first glimpse of the intergalactic medium and qualitatively confirmed the theoretical expectations concerning its ionization, also raise several important questions. Hopefully, these will be addressed by further observations with HST and other space observatories in the coming years. However, given the severe paucity of suitably UV-bright background quasars, and the fact that the observations in question lie at the very limit of the capabilities of even the refurbished HST, progress is not likely to come easily on this topic.

References

Bahcall, J. N. et al. 1993 ApJS 87,1

Bahcall, J. N., & Salpeter, E. E. 1965 ApJ 142, 1677

Bajtlik, S., Duncan, R. C., & Ostriker, J. P. 1988 ApJ 327,570

Beaver, E. A, Burbidge, E. M., Cohen, R. D., Junkkarinen, V. T., Lyons, R. W., Rosenblatt, E. I., Hartig, G. F., Margon, B., & Davidsen, A. F. 1991 ApJ 377, L9

Beaver, E. A, Burbidge, E. M., Cohen, R., Junkkarinen, V., Lyons, R., & Rosenblatt, E. 1992, in Science with the Hubble Space Telescope, ed. P. Benvenuti & E. Schreier (Garching: European Southern Observatory), p.53

Bechtold, J. 1994 ApJS 91, 1

Bechtold, J., Green, R. F., Weymann, R. J., Schmidt, M., Estabrook, F. B., Sherman, R. D., Wahlquist, H. D., & Heckman, T. M. 1984 ApJ 281, 76

Bechtold, J., Weymann, R. J., Lin, Z., & Malkan, M. A. 1987 ApJ 315, 180

Carswell, R. F., 1988, in QSO Absorption Lines, Probing the Universe, ed. J. C. Blades, D. Turnshek & C. A. Norman (Cambridge: Cambridge University Press), p91

Carswell, R. F., Lanzetta, K. M., Parnell, H. C & Webb, J. K. 1991 ApJ 371, 36

Carswell, R. F., Webb, J. K., Baldwin, J. A. & Atwood, B. 1987 ApJ 319, 709

Carswell, R. F., Whelan, J. A. J., Smith, M. G., Boksenberg, A., & Tytler, D. 1982 MNRAS 198,91

Davidsen, A. 1994 (these proceedings)

Giallongo, E., Cristiani, S., & Trevese, D. 1992 ApJ 398, L9

Green, R. F., Pier, J. R., Schmidt, M., Estabrook, F. B., Lane, A., & Wahlquist, H. D. 1980 ApJ 239, 483

Gunn, J. E., & Peterson, B. A. 1965 ApJ 142, 1633

Jakobsen, P. et al. 1993 ApJ 417, 528

Jakobsen, P., Boksenberg, A., Deharveng, J. M., Greenfield, P., Jedrzejewski, R., & Paresce, F. 1994 Nature 370,35

Lanzetta, K. M. 1988 ApJ 332, 96

Lu, L., Wolfe, A. M., & Turnshek, D. A. 1991 ApJ 367, 19

Madua, P. 1992 ApJ 389, L1

Madau, P. & Meiksin, A. 1994 ApJ 433, L53

Meiksin, A. & Madau, P. 1993 ApJ 412, 34

Miralda-Escudé, J. 1993 MNRAS 262, 273

Miralda-Escudé, J., & Ostriker, J. P. 1990 ApJ 350, 1

Miralda-Escudé, J., & Ostriker, J. P. 1992 ApJ 392, 15

Murdoch, H. S., Hunstead, R. W., Pettini, M., & Blades, J. C. 1986 ApJ, 309

Møller, P., & Jakobsen, P. 1990 A&A 228, 299

Osterbrock, D. E. 1989, Astrophysics of Gaseous Nebulae and Active Galactic Nuclei (Mill Valley: University Science Books), p23

Paresce, F. McKee, C., & Bowyer, S. 1980 ApJ 240, 387

Picard, A. & Jakobsen, P. 1993 A&A 276, 331

Reimers, D., Clavel, J., Groote, D., Engels, D., Hagen, H. J., Naylor, T., Wamsteker, W., & Hopp, U. 1989 A&A 218, 71

Reimers, D., Rodriguez-Pascual, P., Hagen, H. J., Wisotski, L. 1994 A&A 293, L21

Reimers, D. & Vogel, S. 1993 A&A 276, L13

Reimers, D., Vogel, S., Hagen, H. J., Engels, D., Groote, D., Wamsteker, W., Clavel, J., & Rosa, M. R. 1992 Nature 360, 561

Sargent, W. L. W., Steidel, C. C., & Boksenberg, A. 1989 ApJS 69, 703

Sargent, W. L. W., Young, P. J., Boksenberg, A., & Tytler, D. 1980 ApJS 42, 41

Schneider, D. P., Schmidt, M. & Gunn, J. E. 1991 AJ 101, 2004

Shapiro, P. R., Giroux, M. L., & Babul, A. 1994 ApJ 427, 25

Steidel, C. C., & Sargent, W. L. W. 1987 ApJ 318, L11

Tripp, T. M., Bechtold, J., & Green, R. F. 1994 ApJ 433, 533

Tripp, T. M., Green, R. F., & Bechtold, J 1990 ApJ 364, L29

Tytler, D. 1982 Nature 298, 427

Tytler, D. 1994 (these proceedings)

Walker, T. P., Steigman, G. Schramm, D. N., Olive, K. A. & Kang, H.-S. 1991 ApJ 376, 51

Webb, J. K., Barcons, X., Carswell, R. F. & Parnell, H. C. 1992 MNRAS 255, 319

Zheng, W., & Davidsen, A. 1994 ApJ (in press)

Zuo, L., & Phinney, E. S. 1993 ApJ 418, 28

New Results on the Gunn-Peterson Effect at High Redshift

E. Giallongo[1], S. D'Odorico[2], A. Fontana[1], S. Savaglio[3], S. Cristiani[4], P. Molaro[5]

[1] Osservatorio Astronomico di Roma, I-00040 Monteporzio, Italy
[2] ESO, Karl-Schwarzschild-Str. 2, D-85748 Garching, Germany
[3] Dipartimento di Fisica, Università della Calabria, Italy
[4] Dipartimento di Astronomia, Università di Padova, I-35122 Padova, Italy
[5] Osservatorio Astronomico di Trieste, via G.B. Tiepolo 11, I-34131 Trieste, Italy

Abstract. Attempts to measure the optical depth due to a diffuse intergalactic medium (IGM) in the spectra of high redshift quasars (GP test) within the ESO key program on the intergalactic medium are reviewed. It is shown that there is no evidence for any Gunn-Peterson effect up to the highest redshifts observable in quasar spectra. The Lyα line statistics consistent with this limit is not able to explain the strong absorption observed at the HeII forest in one QSO, leaving room for a true HeII GP effect. The HI/HeII GP ratio implies a steep ionizing UV background at $z \sim 5$ and ionizing sources of stellar origin.

1 Introduction

The optical depth observed in spectra of high redshift quasars shortward of their Lyα emission is due to the presence of cosmologically distributed neutral hydrogen along the line-of-sight as suggested by Gunn & Peterson in 1965. The value they found was so low that it is generally assumed that the intergalactic medium is highly ionized rather than almost completely absent.

From an observational point of view, the Gunn-Peterson (GP) optical depth $\tau_{GP} = -ln(I_c/I_{extr})$ is the ratio between the local continuum level measured shortward of the quasar Lyα emission and the extrapolated continuum defined longward of the Lyα emission, where the quasar continuum emission is unaffected by intergalactic hydrogen absorption. In fact, most of the absorption is due to numerous and narrow absorption lines (the so called "Lyα forest") interpreted as HI Lyα absorption from intervening intergalactic clouds along the line-of-sight. Thus, the estimate of the true continuum level needed for the measure of τ_{GP} in the Lyα forest is made difficult by the large number of absorption lines present at high redshifts. In the next section, we briefly review previous measures on the GP effect that rely on the knowledge of the statistical properties of the Lyα lines as a function of redshift,and discuss the new results obtained by the authors within the ESO Key Project on the intergalactic medium at high z.

2   Estimates of the GP Optical Depth

Steidel & Sargent (1987) have shown that an upper limit to the diffuse HI absorption, relying on the knowledge of the average line absorption, can be obtained by subtracting the contribution of the Lyα lines to the total absorption observed between Lyα and Lyβ emissions in low-resolution QSO spectra. They estimated a value $\tau_{GP} \simeq 0.02 \pm 0.03$ at $z = 2.6$. Giallongo & Cristiani (1990) and Cristiani et al. (1993) compared all the available data with simulations of the average absorptions in synthetic QSO spectra computed on the basis of the known statistics of the Lyα lines. They showed that there was no evidence for a GP effect (i.e. $\tau_{GP} < 0.1$) up to $z = 5$. A more refined analysis, suggested by Jenkins & Ostriker (1991), consists in the relative frequency distribution of the transmitted fluxes. This way the simulated spectra, which depend on the line statistics, can be compared with a distribution of intensities rather than a single average value. In contrast, the observed distribution is influenced by the instrumental resolution. Thus, high resolution spectra are needed to derive a more sensitive test. Webb et al. (1992) applied a similar method to the high resolution spectrum of 0000-26 at $z_{em} = 4.1$. The inferred value was based on the slope β of the column density distribution $\propto N_{HI}^{\beta}$ of the weak lines. They found $\tau_{GP} = 0.04$ for a steep N_{HI} distribution with $\beta = -1.7$. A null optical depth required a flat N_{HI} distribution with $\beta = -1.3$ extrapolated down to $\log N_{HI} = 12$. It is important to note in this respect that high resolution data at 14 km/s in the redshift interval $z = 2.9 - 3.6$ (Giallongo et al. 1993, Cristiani et al. 1995) indicate a flat N_{HI} power-law with $\beta = -1.4, -1.5$ in the range $13.3 \leq \log N_{HI} \leq 14.5$ with a cutoff or a steepening at larger column densities. This favours the Webb et al. solution with $\tau_{GP} = 0$ at $z_{abs} = 3.8$.

A more stringent and direct upper limit to the GP effect has been given by Giallongo et al. 1992,1994 in the framework of an ESO key program devoted to the study of the intergalactic medium at high z. They used flux calibrated spectra obtained at relatively high resolution (14-40 km/s) and extending up to about 10000 Å   (PKS 2126-158 $z_{em} = 3.3$; BR 1202-07 $z_{em} = 4.7$, respectively). The high resolution allows a better evaluation of the continuum shape and a direct selection of regions in the Lyα forest which are free of strong absorption lines. The ratio I_c/I_{extr} in the Lyα forest was computed, giving an average optical depth at $z = 4.3$ $\tau_{GP} = 0.02 \pm 0.03$ where the error is due to the noise in the spectrum and to the slope uncertainty in the extrapolated continuum. The same method applied to the QSO PKS 2126-158 gives $\tau_{GP} \simeq 0.01 \pm 0.03$ at $z = 3$.

Using the Jenkins & Ostriker method, the histogram of the relative intensity in the Lyα forest of PKS 2126-158 in the range 4850–5000 Å is shown in Fig. 1a (thick line). The thin line is derived from a synthetic spectrum whose line distribution parameters are obtained from a maximum likelihood analysis on the high resolution line sample used by Cristiani et al. (1995). A best fit was obtained extrapolating the flat N_{HI} power-law distribution with slope $\beta = -1.45$ down to $\log N_{HI} = 12.2$ with the addition of a GP optical depth $\tau_{GP} = 0.015$. The intensity histogram of the Lyα forest in BR 1202-07 is shown in Fig. 1b (thick line). The range is 6200-6450 Å, i.e. $z \sim 4.1 - 4.3$. The thin line is the extrapolation to $z = 4.3$ of the same N_{HI} distribution obtained at $z = 3$. This extrapolation requires a steep power-law evolution in z with slope $\gamma = 4.2$. Independent estimates by Zuo & Lu (1993) are consistent with increasing redshift evolution of the Lyα lines for $z > 4$. A best fit solution is found for low N_{HI} thresholds in the range $\log N_{HI} = 12.6 - 12.8$ and with $\tau_{GP} = 0 - 0.04$. The GP optical depth found in this way is consistent within 1σ with the value measured in the Lyα forest by Giallongo et al. (1992,1994) selecting regions of width 3-4 Å which are free of strong absorption lines. Fig. 2a shows that intervals between Lyα lines of this width are not rare voids, but are just expected from a poissonian distribution of the same synthetic lines ($\log N_{HI} > 12.6$) which fit the observed histogram of the pixel intensity.

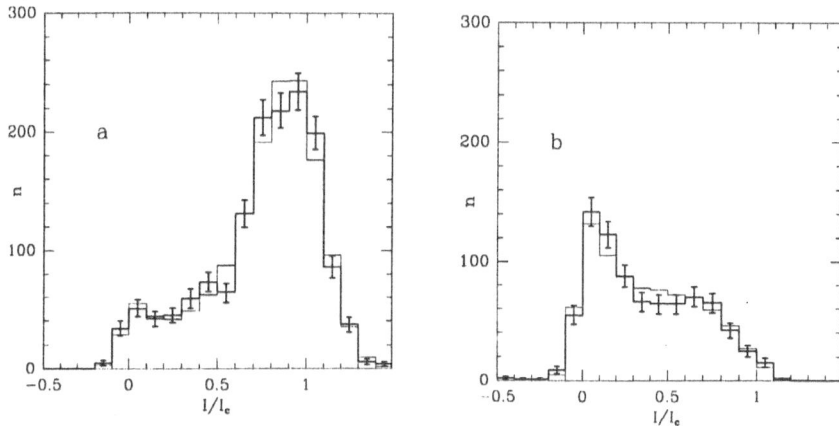

Fig. 1. Intensity histograms of the Lyα forest in PKS 2126-158 a), and BR 1202-07 b), (thick lines). Best fit intensity distributions from corresponding synthetic spectra are also shown (thin lines). In a) $\tau_{GP} = 0.015$; in b) $\tau_{GP} = 0$.

It is interesting to show the contribution of the same synthetic line distributions to the average optical depth at the HeII Lyman-α forest, assuming an intensity ratio S_L of the UV background between the Lyman limit and the HeII edge $\sim 10^2$. The predicted intensity distribution at $z = 3.3$ is shown in Fig. 2b. The corresponding average $\tau_{HeII} = 1.2$ so obtained, can be compared with the strong absorption ($\tau_{HeII} = 3.2$) observed for one line-of-sight at the same z by Jakobsen et al. (1994). Increasing S_L to values $\sim 10^3$ does not change appreciably the average HeII optical depth ($\tau_{HeII} = 1.6$, (see also Madau & Meiksin 1995)). The two HI line distributions used to fit the data at $z = 4.1 - 4.3$ give a contribution in the HeII forest $\tau_{HeII} = 2.5 - 2.1$, (with $S_L = 10^2$), respectively. Thus, the Lyα line population can not reproduce the observed HeII optical depth and a true $\tau(HeII)_{GP} \sim 1.8$ optical depth is

suggested (using the best fit value given by Jakobsen et al. 1994). This high
value remains consistent with the small upper limit for the GP optical depth
found in the HI Lyα forest if $S_L \sim 200$.

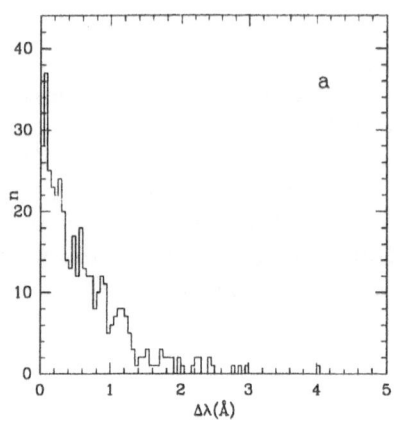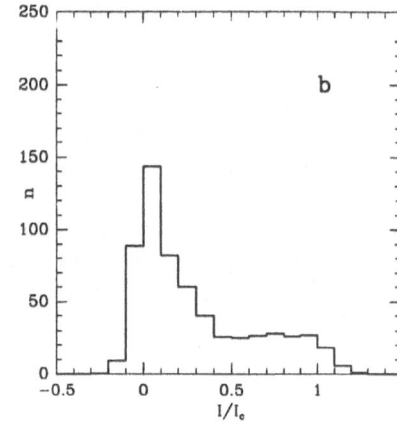

Fig. 2. a) Interval distribution in a synthetic spectrum ($\log N_{HI} \geq 12.6$) which fits
the intensity histogram in Fig. 1a. b) Intensity histogram in the HeII Lyα forest of
the same synthetic spectra as in Fig. 1a.

A dominant starlight contribution to the UVB at $z \sim 5$ seems to be
favoured both by the high value of the UVB at the Lyman limit (needed to
explain the small value $\tau_{GP} < 0.04$ found at $z \sim 4.5$) and by the amount of
steepening of the UVB between HI Lyman limit and the HeII edge (needed
to explain the high value $\tau(HeII)_{GP} \sim 1.8$ at $z = 3.3$).

References

Cristiani, S., D'Odorico, S., Fontana, A., Giallongo, E., & Savaglio, S. 1995, MN-
    RAS, in press
Cristiani, S., Giallongo, E., Buson, L. M., Gouiffes, C., La Franca, F. 1993, A&A,
    268, 86
Giallongo, G. & Cristiani, S. 1990, MNRAS, 247, 696
Giallongo, E., Cristiani, S., & Trèvese D. 1992, ApJ, 398, L9
Giallongo, E., Cristiani, S., Fontana, A., & Trevese, D. 1993, ApJ, 416, 137
Giallongo, E., D'Odorico, S., Fontana, A., McMahon, R. G., Savaglio, S., Cristiani,
    S., Molaro, P., & Trevese, D. 1994, ApJ, 425, L1
Gunn, J. M., & Peterson, B. A. 1965, ApJ, 142, 1663
Jakobsen, P., Boksenberg, A., Deharveng, J. M., Greenfield, P., Jedrzejewski, &
    Paresce, F. 1994, Nature, 370, 35
Jenkins, E. B., & Ostriker, J. P. 1991, ApJ, 376, 33
Madau, P. & Meiksin, A. 1995, ApJ Letters, in press
Steidel, C. C., & Sargent, W. L. W. 1987, ApJ, 318, L11
Webb, J.K., Barcons, X., Carswell, R.F., & Parnell, H.C. 1992,MNRAS,255,319
Zuo, L., & Lu, L. 1993, ApJ, 418, 601

The HeII Proximity Effect

Avery Meiksin

University of Chicago, 5640 S. Ellis Ave., Chicago, IL 60637, USA

Abstract. A major uncertainty in the interpretation of the recent *HST* detection of He II Lyα resonant absorption in Q0302–003 (Jakobsen *et al.* 1994), is the contribution to the opacity arising from line-blanketing by the Lyα forest. It is shown how moderate resolution UV spectra may be used to distinguish between the case of total opacity dominated by line-blanketing from that including a substantial component from a diffuse medium. The method permits a direct measurement of the total baryonic density of the IGM.

1 Introduction

Recently, the *HST* FOC detected absorption by the He II 304A Lyα resonant line in the spectrum of Q0302–003 at $z = 3.286$ (Jakobsen *et al.* 1994). They determine a 90% lower bound on the optical depth of $\tau_{304} > 1.7$. The amount of the absorption due to a diffuse component of the IGM is made uncertain by the amount of He II line-blanketing by the Lyα forest. While a contribution from the diffuse component of $\tau_{D,304} \sim 1$ may reasonably be expected for the estimated baryon density produced in Big Bang nucleosynthesis and the integrated contribution of QSOs to the ionization of He II , the amount of blanketing depends critically on the number and dynamics of Lyα forest clouds below the threshold for detection in H I . It is possible that the measured absorption may be dominated completely by line blanketing (Madau & Meiksin 1994). A method for separating these two sources of opacity is presented here.

2 Resonant Opacity Near QSO

The effective opacity due to line-blanketing by the Lyα forest is given by $\tau_{\mathrm{eff}} = [(1 + z)/\lambda_\alpha] \int (\partial^2 N/\partial W \partial z) W dW$, where λ_α is the rest wavelength of Lyα , and the adopted rest equivalent width distribution for H I is $\partial^2 N/\partial W \partial z = 91 \exp(-W/W_*)(1 + z)^{2.75}$ for $0.06 < W < 2$Å, and $18(W/W_*)^{-1.5}(1 + z)^{2.75}$ for $W_{\mathrm{min}} < W < 0.06$Å, where $W_* = 0.13$ A, based on Kulkarni *et al.* (1994). The equivalent width in He II for an absorber is related to that in H I through the curve-of-growth. The amount of line-blanketing in He II depends critically on the adopted lower cut-off W_{min}, and on the ratio of the He II line width to that of H I (thermally vs. velocity broadened lines). A value of $W_{\mathrm{min}} = 0.0025$ A is assumed here, with thermally broadened lines.

Because line-blanketing is dominated by saturated lines, with equivalent widths depending only logarithmically on the density of He II in the clouds, while absorption by the diffuse medium depends linearly on the diffuse IGM He II density, the amount of diffuse absorption is much more sensitive to the presence of the QSO than is the line-blanketing. This difference in behavior may be exploited to detect the presence of diffuse absorption.

The opacities are illustrated in Figure 1. A source of intensity $L_\nu = 10^{31}(\nu/\nu_L)^{-1.5}\,\mathrm{erg\,s^{-1}\,Hz^{-1}}$, comparable to that of Q0302–003, turns on at $z = 5$ into a diffuse medium with $\Omega_D = 0.01$. The medium was preionized at $z = 5.5$ by a metagalactic UV field of intensity $10^{-21}\,\mathrm{ergs\,cm^{-2}\,s^{-1}\,Hz^{-1}\,sr^{-1}}$ at the H I Lyman edge, and $3 \times 10^{-24}\,\mathrm{ergs\,cm^{-2}\,s^{-1}\,Hz^{-1}\,sr^{-1}}$, that may be provided by QSOs, at the He II Lyman edge. Without a contribution from d-iffuse absorption, the opacity would rise rapidly near the QSO to the ambient value of 1.7 (*dashed line*). Including a small component of diffuse absorption results in a gentler increase in opacity toward shorter wavelengths. This signature would be detectable by a UV spectrograph of moderate resolution, $3 - 5$ A. The relative contributions may then be obtained by fitting to the total opacity. Since the QSO flux is measurable, the value for the diffuse component may be combined with the equation of ionization equilibrium to solve for the density of the diffuse IGM.

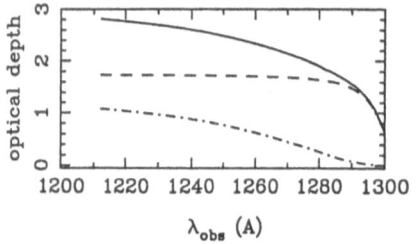

Fig. 1. He II 304A opacity near a QSO. Shown are the total optical depth from He II 304A resonant absorption (*solid line*), the contribution from line-blanketing (*dashed line*), and the contribution from diffuse absorption (*dot-dash line*), as a function of observed wavelength for a QSO at $z = 3.286$.

References

Gunn, J. E., & Peterson, B. A. 1965, ApJ, 142, 1633

Jakobsen, P., Boksenberg, A., Deharveng, J. M., Greenfield, P., Jedrzejewski, R., & Paresce, F. 1994, Nature, 370, 35

Kulkarni, V. P., Huang, K., Green, R. F., Bechtold, J., Welty, D. E., & York, D. G. 1994, in preparation

Madau, P., & Meiksin, A. 1994, ApJ, 433, L53

On the He II Absorption Towards Q0302-003

Mark A. Fardal, Mark L. Giroux, and J. Michael Shull

JILA/CASA, University of Colorado, Boulder, CO 80309, USA

Abstract. The recent HST observation of the spectrum of the $z = 3.286$ quasar Q0302-003 (Jakobsen et al. 1994) shows an absorption edge at the redshifted wavelength of He II 304 Å. One cannot yet distinguish between contributions from discrete Lyα forest clouds and a smoothly distributed intergalactic medium (IGM). We model the contributions from these sources of He II absorption, including the distribution of line widths and column densities and the "He II proximity effect" from the quasar. Our models include a self-consistent treatment of the He II opacity of the universe as a function of the intrinsic ionizing source spectrum. The He II edge can be fully accounted for by the Lyα forest, for reasonable distributions of line widths and column densities, provided that the ionizing background sources have spectral index $\alpha > 1.9$. Even with some contribution from a diffuse IGM, it is difficult to account for the observed edge with a "hard" source spectrum ($\alpha < 1.3$). The proximity effect will increase the relative contribution of the clouds to $\tau_{\text{He II}}$ near the quasar ($z \approx z_q$). Higher resolution observations that characterize the change in transmission as $z \to z_q$ would be useful. These same observations may resolve line-free gaps in the continuum and set limits on the diffuse IGM.

1 He II Line Blanketing

To model the absorption from the Lyα forest, we use the distribution in H I column density ($N_{\text{H I}}$) and Doppler width (b) of Press & Rybicki (1993). This is a power law between lower and upper limits N_l and N_u. We modify the background spectrum to take into account the different opacities at the H and He ionization thresholds. Varying N_l or N_u has a large effect on the line opacity, N_l directly and N_u indirectly by modifying the background spectrum. The b distribution is not very important; a good approximation is to replace b by its mean value.

2 The He II Proximity Effect

The flux from the quasar itself will ionize the gas if it is close enough. For Q0302-003, this should happen around 1280 Å. The opacity from the IGM is inversely proportional to the flux, but the Lyα forest opacity is less sensitive because many of the lines are saturated, the same reason it does not automatically dominate the total opacity in the first place. Possible models for the flux near the He II edge are shown in Fig. 1. Much more sensitive observations could determine the ionizing background using this effect.

3   Results

We require the average transmission from $1250-1300$ Å to satisfy $\langle T \rangle < e^{-1.7}$, in accordance with Jakobsen et al.'s 90% confidence limit. When H I GP limits are included, we get limits on the allowed parameter space. If the lines are velocity broadened and/or extend down to 10^{12} cm^{-2} they can provide all of the He II opacity, consistent with Madau & Meiksin (1994). The proximity effect tightens the constraint on α by about 0.2; soft source spectra ($\alpha > 1.5$) are favored. We also suggest the possibility of small gaps between the lines, observable in higher-resolution observations.

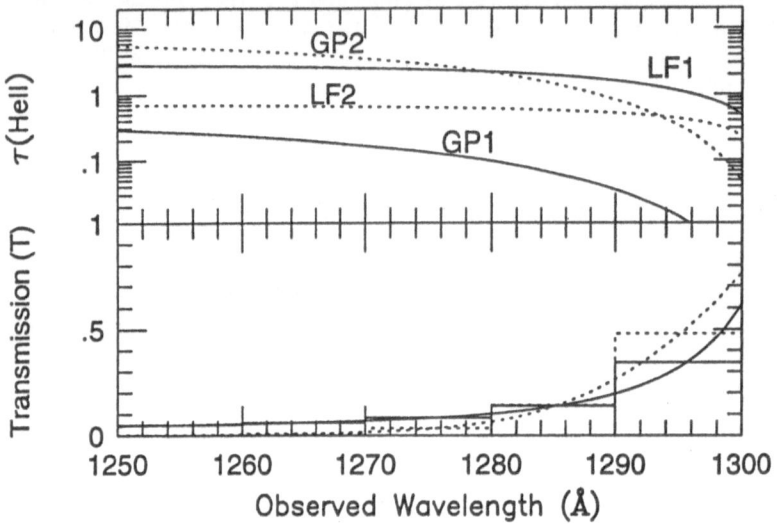

Fig. 1. Shows the increase in the flux as $z \to z_q$. Two different models with $\langle T \rangle = e^{-2.0}$ are shown. Model 1 (solid) has parameters $\tilde{\Omega}_{IGM} = 0.003$, $N_l = 10^{12}$, $N_u = 10^{16}$, and velocity-broadened b's. Model 2 (dotted) has $\tilde{\Omega}_{IGM} = 0.02$, $N_l = 10^{13}$, $N_u = 10^{10}$, and thermal b's. Here $\tilde{\Omega}_{IGM} \equiv h^{3/2} T_4^{0/8} J_{-21}^{-1/2} \Omega_{IGM}$.

Acknowledgements. This work was supported by NASA grants NAGW-766 and NAGW-1479.

References

Jakobsen, P., Boksenberg, A., Deharveng, J.M., 1994, Nature, 370, 35
Madau, P., Meiksin, A., 1994, ApJ, 433, L53
Press, W.H., Rybicki, G.B., 1993, ApJ, 418, 585

Active Galactic Nuclei as Cosmological Probes

Ewa Szuszkiewicz

International School for Advanced Studies, via Beirut 2-4, 34013, Trieste, Italy

Abstract. Our aim is to study properties of the intergalactic medium (IGM) using the spectra of active galactic nuclei (AGN). We plan to construct a self-consistent stationary model of the universe at high red shift and to calculate constraints on the shape and amplitude of the ionizing background radiation, the physical properties of absorbing clouds, and the abundance and ionization states of hydrogen, helium and heavy elements in the intergalactic matter. As a starting point we discuss here the properties of quasars and active galaxies which are most relevant for our purposes.

1 Observations of Distant Parts of the Universe

High red shift quasars reveal information about important aspects of the universe. They are expected to be useful probes of geometrical differences between various cosmological models and their existence shows that galaxy-sized objects had formed at early epochs ($z \approx 5$) which is an important constraint on models for galaxy formation (Rees 1992). They can also be used to probe the physical state of the nonluminous matter (IGM) throughout the history of the universe (as discussed elsewhere in these proceedings). The abundance of smoothly distributed intergalactic hydrogen and helium can be deduced from the relevant Gunn-Peterson tests. The absorption lines observed in quasar spectra provide information about the distribution, structure and abundance of H, He and heavy elements in clouds irradiated by the intergalactic flux of far ultraviolet photons. Here we discuss what we should know about the probes themselves - the active galactic nuclei - to be able to reach a meaningful conclusion about the content of the universe at high red shift.

2 Quasar Spectra and Opacities in the Universe

The absence of Gunn-Peterson H Ly α absorption troughs in the spectra of quasars implies that the intergalactic gas distributed smoothly on cosmological scales is either neutral (but contained an undetectable small fraction of the baryon density) or is highly ionized. Observational estimates of the intergalactic flux of hydrogen ionizing photons generally exceed the most recent determinations of the integrated contribution from quasars. Various other photon sources at high red shift have been proposed (for an authorative summary, see Sciama 1993). The measurements of helium abundance in the high red shift universe are based on the HeI resonance lines and the HeII Gunn-Peterson effect. At present there is only one suitable object for such studies, namely HS 1700+6416 (Reimers & Vogel 1993). Therefore not much is known about helium content in the high-z universe.

The intergalactic ionizing flux plays a crucial role in studies of IGM properties. Its determination was performed by integrating the radiation of known UV emitters (see, Bechtold et al. 1987) and by using the distribution of clouds near the quasar (proximity effect; Bajtlik et al. 1988). Both methods assume knowledge of the luminosity function, the intrinsic flux of the considered photon source and the absorption of the intervening medium. Recent work on the quasar luminosity function is reviewed by Boyle (1994). For lack of a better understanding of quasar spectra, a broken power law has commonly been used. The intervening absorption was modelled by Meiksin & Madau (1993) among others. These three components, necessary for studying the IGM, are not independent from each other, they are coupled and this is a source of quite pronounced uncertainties in the calculations (see discussion in Bechtold et al. 1987). The steepness of the quasar luminosity function, for example, is enhanced by the effects of intervening continuum absorption and leads to the number of observable quasars being strongly dependent on the limiting sensitivity and assumed quasar spectrum. A number of effects can contribute to the steepening of the quasar continua shortward of Ly α 1200Å.

We propose here to improve on one of the key elements mentioned above, namely on the intrinsic far and extreme-UV portions of quasar spectra. They are believed to be dominated by a "big blue bump" component that arises as thermal emission from an accretion disk surrounding a central massive black hole. The present disk models agree well with the recent observations in the far-UV and soft X-ray energies (Szuszkiewicz et al. 1995).

Progress in modelling the central engine of AGN in the framework of accretion disk theory gives a better understanding of the detailed shape of the continuous spectral energy distributions of quasars. It provides the appropriate intrinsic spectra for the absorption line studies.

References

Bajtlik S., Duncan R.C., Ostriker J.P., 1988, ApJ, 327, 570

Bechtold J., Weymann R.J., Lin Z., Malkan M.A., 1987, ApJ, 315, 180

Boyle B.J., 1994, in *The Nature of Compact Objects in AGN*, eds. Robinson A. & Terlevich R.J., (Cambridge: Cambridge University Press), p. 110

Meiksin A., Madau P., 1993, ApJ, 412, 34

Rees M.J., 1992, in *Physics of Active Galactic Nuclei*, eds. Duschl W.J. & Wagoner S.J., (Heidelberg: Springer-Verlag), p.662

Reimers D., Vogel S., 1993, A&A, 276, L13

Sciama D.W., 1993, *Modern Cosmology and the Dark Matter Problem*, (Cambridge: Cambridge University Press), Chapter 7

Szuszkiewicz E., Malkan M.A., Abramowicz M.A., 1995, ApJ, submitted

PART X

FORMATION

OF ABSORBING SYSTEMS

Models of Lyα Forest Clouds

Jane C. Charlton

Astronomy and Astrophysics Department, Davey Laboratory, Pennsylvania State University, University Park, PA, 16802, USA

Abstract. The classic question regarding Lyα forest clouds is whether they are confined by the pressure of a hot, diffuse surrounding medium, or by the gravity of dark matter mini-halos. This paper reviews these basic models for forest clouds, considering spherical and slab geometries. At high redshift, the clouds are still likely to be in the process of formation, and it seems essential to consider them as dynamic structures in a cosmological context. At low redshift, observations of large cloud sizes (indicated by their covering of both lines of sight toward quasar pairs) have recently reshaped our view of the forest. Galaxy-like disks/slabs, which would be gravity confined near the center, but pressure confined in their outer regions (extending out to hundreds of kpcs), may be responsible for the low z forest. The ultimate view of the identity of the forest clouds and their relationship to galaxies is likely to involve a synthesis of many of the models discussed here.

1   Introduction

A review of models of any type of object should naturally begin with a definition of the object. Yet, it is not simple to draw a picture of what we mean by a Lyα forest cloud. The picture is still being adjusted by knowledge of new data, and further, the clouds may not represent the same population over all time. In fact, it is likely that different experts in the field would have qualitatively distinct views of the types and distributions of the structures in which the Lyα clouds arise.

Why is it that the fundamental aspects of Lyα forest clouds have not yet been clarified? It is because knowing the number of clouds that intersect a given line of sight per unit redshift, z, and in some interval of neutral hydrogen column density, N_H, still allows many possibilities. A large density of small clouds or a few clouds that cover a large area could yield this same number per unit z. Also, different N_H can be due to various lines of sight through the same object, or to different types (masses) of objects.

The picture that we would draw of a Lyα forest cloud is determined by our view of: 1) whether and how the cloud is confined (eg. by gravity or pressure), 2) how the cloud is formed, 3) the cloud geometry (eg. sphere, slab, or filament), 4) the relationship to galaxies. (Is it a separate population or does it in some way form a continuum with galaxies?) Observational constraints from lensed and double quasar lines of sight play a major role in shaping our view (Smette 1994; Bechtold et al. 1994, Dinshaw et al. 1994; Elowitz et al. 1994). By observing 4 forest clouds in common between lines of sight

to the pair 0107-025A,B, Dinshaw et al. (1994; these proceedings) were able
to place a model-independent lower limit of $160h^{-1}$kpc on the cloud radii
$(0.5 < z < 0.9)$ assuming spherical geometries. Measurements of metals in
the forest clouds (Tytler 1994, these proceedings) and detailed information
on Doppler b parameters and line profiles will also shape our view.

2   Pressure vs. Gravity; Sphere vs. Slab

This review begins by considering the four traditional models for Lyα forest
clouds. These models consider pressure and gravity confinement, each in both
a spherical and a slab geometry. For another perspective on models of forest
clouds see the recent review by Bajtlik (1993).

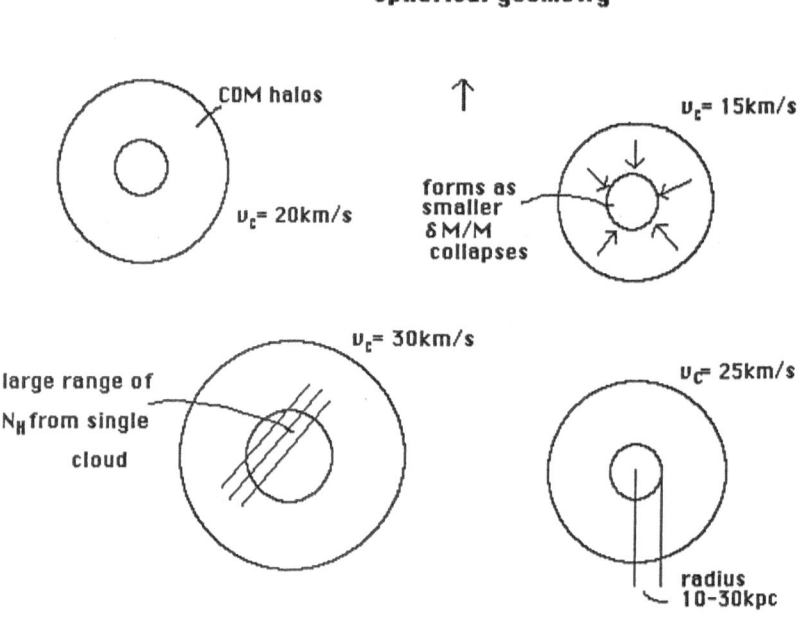

Fig. 1. Original minihalo

2.1   Original Minihalo Model

In this model the clouds were taken to be spheres in hydrostatic equilibrium, confined by the gravity of a cold dark matter "mini-halo" (Rees 1986; Ikeuchi 1986; Ikeuchi et al. 1988). The confining halos were envisioned to have smaller masses than ordinary galaxy halos, with velocity dispersions of the order of tens of km/s. A "cartoon" view of this model is given in Fig. 1. In these spherical clouds, the density falls off rapidly with radius, so that it is possible to produce a large range of N_H from a single cloud. This model has considerable predictive power as demonstrated in some recent papers (Miralda-Escude & Rees 1993; Mo et al. 1993). For current views on the mini-halo model see the article by Rees (1994) in these proceedings.

2.2   Original Pressure Confined Model

This model was discussed in the classic paper by Sargent et al. (1980) and is illustrated in Fig. 2. Spherical clouds are confined in hydrostatic equilibrium by the pressure of a hotter, but more diffuse, exterior medium. The product of the density and the temperature of the cloud must balance with the external pressure. The pressure could be local to the clouds (in surrounding hot halos), or at the opposite extreme it could be a uniform intergalactic medium (IGM). This model has been developed by Ikeuchi and Ostriker (1986) and by Baron et al. (1989).

The pressure confined spherical cloud model has been found to have a serious problem. For a constant value of the external pressure, P_{ext}, the total number density n_c of H (including both ionized and neutral) within the cloud is constant for all clouds. Thus the large range of N_H for the forest cloud population (10^{13} - 10^{17}cm^{-2}) can only be a result of an unrealistically large range of cloud masses (Williger & Babul 1992). This can be demonstrated as follows.

For clouds in ionization/ recombination equilibrium we can write:

$$\alpha_{rec} n_c N_{tot} = \zeta N_H \qquad (1)$$

where N_{tot} and N_H are the total (ionized plus neutral) and neutral column densities, α_{rec} is the recombination coefficient, and ζ is the ionization rate due to the incident extragalactic background radiation. For a constant n_c we have $N_{tot} \propto N_H$. Using $M_{tot} \propto N_{tot} R^2$ we find that $N_H \propto M_{tot}^{1/3}$. Thus a range of three orders of magnitude in N_H requires a range of nine orders of magnitude in M_{tot}.

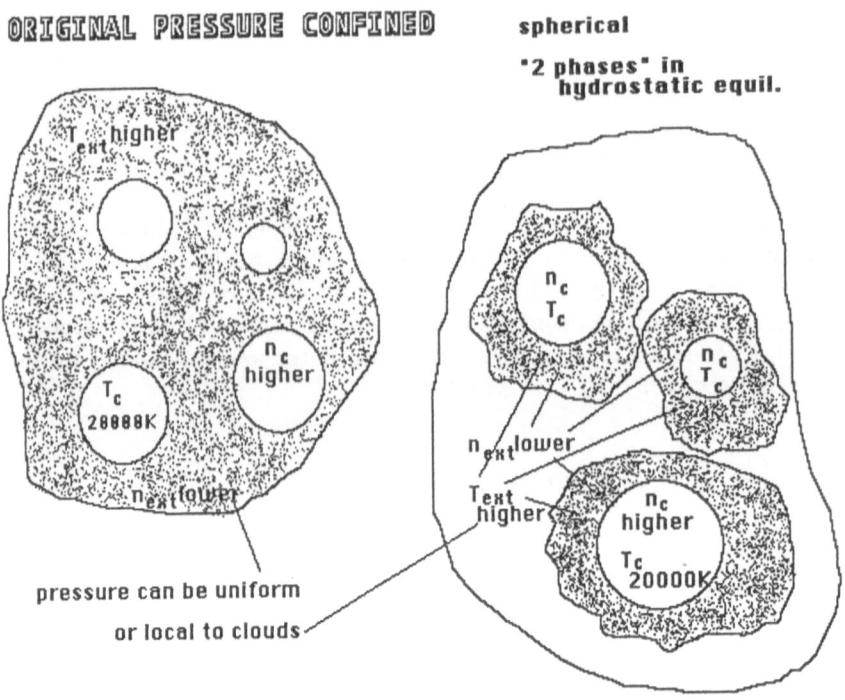

Fig. 2. Original pressure confined

2.3 Pressure Confined Slabs

In a spiral galaxy we have a disk that has an outer region that contains gas, but no stars. Is it essential to the formation of a gaseous disk that stars will form at its center, and that it will become a galaxy? It is certainly not obvious that we could not form disks which have column densities below the threshold for star formation. The resulting scheme for Lyα forest clouds is illustrated in Fig. 3. This raises the question of what fraction of the forest clouds are associated with ordinary galaxies. At low redshifts $\sim 50\%$ of the forest clouds have a luminous galaxy within $160h^{-1}$kpc (Lanzetta et al. 1994). However, at high redshift the larger numbers of forest clouds suggests that ordinary galaxies do not provide a sufficient cross section.

The model of pressure confined slabs was considered by Barcons and Fabian (1987). The slab geometry is advantageous in that it lends itself to large cloud dimensions, thus satisfying the observational size constraints mentioned in the introduction. The external pressure can be large enough to confine the clouds in the thin dimension, yet they can still be quite extended. However, the problem of requiring a huge range of cloud masses is merely rephrased as the problem of needing a variation of nine orders of magnitude in N_{tot} over the radius of the slab. Furthermore, confining forest clouds with $N_H > 10^{16}$cm^{-2}

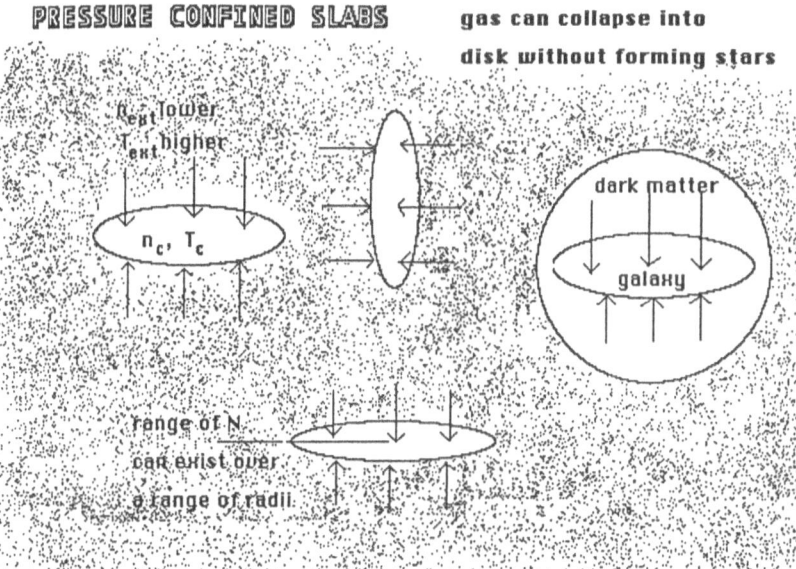

Fig. 3. Pressure confined slabs

would require pressures that are likely to exceed constraints placed by the COBE y-parameter limits (inverse Compton scattering of microwave background photons off of the hot electrons) (Mather et al. 1994).

2.4   Gravity Confined Slabs

For completeness, we mention a fourth simple model, illustrated in Fig. 4, in which the mini-halos confine gas in a slab geometry. The velocity dispersions of these halos would still be less than those of ordinary galaxies, and the central gas column densities would be smaller. Large cloud dimensions are inconsistent with this simple class of models.

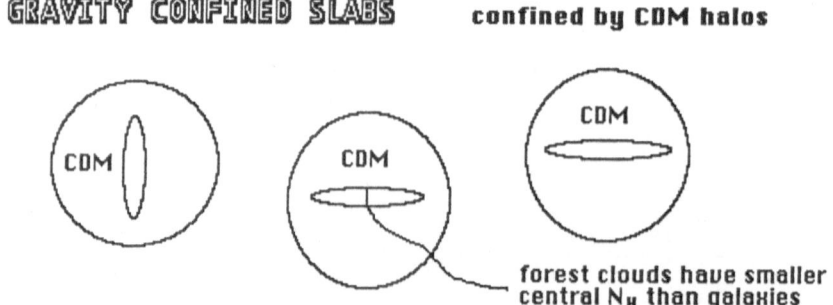

Fig. 4. Gravity confined slabs

3 Mixed Pressure and Gravity Confinement (Slab Geometry)

It is possible, and in fact natural, that more than one mechanism is responsible for cloud confinement. Here we shall discuss a particular class of models in which forest clouds are produced by a single type of structure, a slab or disk which has its inner regions (with higher N_H) gravity confined and its outer regions (with smaller N_H) pressure confined. In this way it is possible to have very large disk sizes without an unreasonable range of N_{tot} over the radius of the disk.

A particular example of this class of models has been discussed by Salpeter (1993) and further developed in Hoffman et al. (1993), Salpeter (1994), and Salpeter and Hoffman (1995). In this example, a hypothetical class of galaxies is responsible for producing the majority of the Lyα forest clouds. The idea is that the disks of these "Vanishing Cheshire Cat" galaxies have central column densities somewhat smaller than those of ordinary galaxies (perhaps a factor of ten smaller). In order to produce the observed covering factor of forest clouds at high redshifts these systems must be more abundant than ordinary galaxies (at least by a factor of 10). The column density would be exponential near the disk center, but would switch to a power law form further out, which could continue out to radii of a couple of hundred kpc. (In the outer disk regions it may be best to think of a collection of high velocity clouds as opposed to a smooth disk (Salpeter 1994).) This idea is partly motivated by 21cm observations of extended disks in dwarf galaxies (Hoffman et al. 1993). The large cloud radii that were envisioned in this class of models are quite consistent with the observed numbers of coincidences and anti-coincidences of Lyα forest lines along adjacent lines of sight toward double quasars These observations give a lower limit of $70h^{-1}$kpc at $z \sim 2$ down to a column density of approximately $10^{14.5}\text{cm}^{-2}$ (Bechtold et al. 1994; Dinshaw et al. 1994). Even larger sizes are likely since the clouds will probably extend to smaller N_H at larger distances from cloud center. Most classes of models have difficulties producing such large cloud sizes (Bechtold et al. 1994; Dinshaw et al. 1994).

The evolution of the typical "Vanishing Cheshire Cat" is illustrated in Fig. 5. At high redshifts the inner portions of the disks could produce many of the lower column density damped systems (the highest columns would only come from ordinary galaxies), slightly larger radii would produce Lyman limit systems, and the outer disk would be responsible for the forest clouds. Star formation would be delayed until intermediate redshifts in these lower column density disks, and the first generation of stars would be responsible for ejecting the majority of the gas (as in Babul and Rees (1992)). The damped systems, and some of the Lyman limit systems contributed by this population, would disappear and the outer disk (the smile of the cat) would be left behind today as forest clouds. Observations of low redshift Lyman limit systems do suggest that there are relatively fewer with $N_H > 10^{18}$cm^{-2} (as compared to the number in the range $10^{17} < N_H < 10^{18}$cm^{-2}) at recent times (Storrie-Lombardi et al. 1994). The inner region would be best described as a red population of very low surface brightness galaxies at the present time. With time, the transition radius between the gravity and pressure confined regions of the disk changes, subject to changes in the external confining pressure and the extragalactic ionizing radiation.

4   Pressure vs. Gravity

The value of the external pressure (whether it be local to the clouds or universal) determines the boundary between the inner, gravity confined and the outer, pressure confined regions of a slab. Balance of the forces in hydrostatic equilibrium can be written in the "half-slab approximation" (see Charlton et al. 1993; 1994) as:

$$\frac{\pi}{2}Gm_H^2\eta N_{tot}^2 + P_{ext} = 2\,n_{tot}\,k\,T. \tag{2}$$

Here, η is the contribution to gravity due to dark matter relative to that of ordinary gas. The gravitational and pressure forces are equally important at the specific value of

$$N_{tot} = N_1 = \left(\frac{2}{\pi}\frac{P_{ext}}{Gm_H^2\eta}\right)^{1/2}. \tag{3}$$

For $N_{tot} < N_1$ the slab is pressure confined, and using (1) we find that $N_H \propto N_{tot}$. In the gravity confined regime (for inner radii where $N_{tot} > N_1$) the specific relationship depends on the distribution of dark matter around clouds of different N_{tot}, but roughly $N_H \propto N_{tot}^3$.

The different relationships between N_H and N_{tot} translate to a prediction for the distribution function for the clouds (the number of clouds with a given

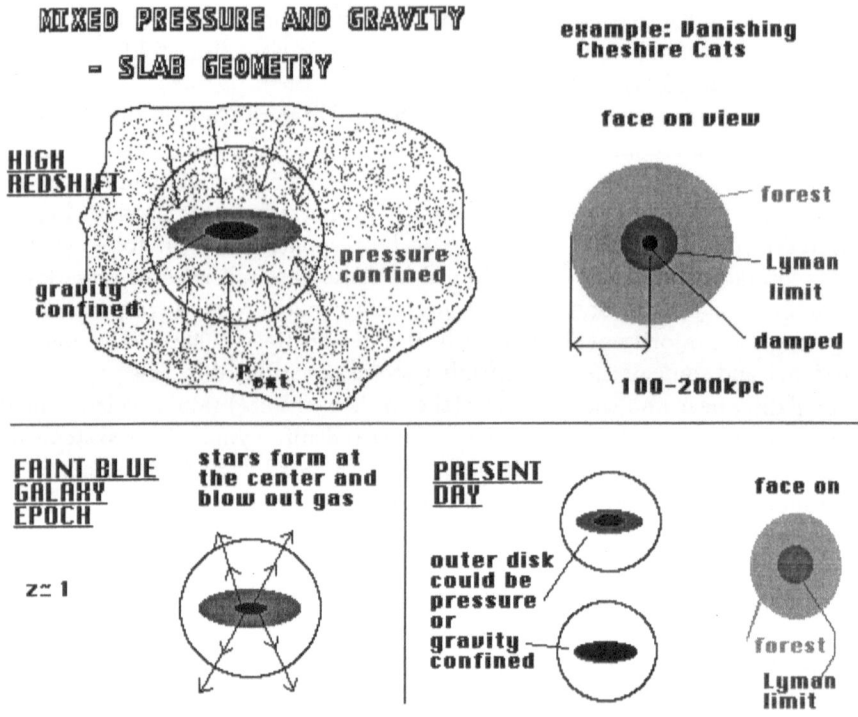

Fig. 5. Mixed pressure and gravity - slab geometry

column density). Specifically we know that the total column density distribution $g(N_{tot})$ is related to the neutral column density distribution $f(N_H)$ by:

$$g(N_{tot})dN_{tot} = f(N_H)dN_H. \qquad (4)$$

Each distribution is taken to have a power law form:

$$g(N_{tot}) \propto N_{tot}^{-\beta} \qquad (5)$$

$$f(N_H) \propto N_H^{-\epsilon}. \qquad (6)$$

Then we find, from the relationships between N_H and N_{tot} in the two regimes, that:

$$\beta = \epsilon \qquad \text{pressure confined}$$

$$\beta = \frac{\epsilon + 2}{3} \qquad \text{gravity confined.} \qquad (7)$$

For $\epsilon = 1.8$, we find that $\beta = 1.8$ in the pressure dominated regime and $\beta = 1.3$ in the gravity dominated regime. This type of shape for $f(N_H)$ is apparent in the data of Petitjean et al. (1993a) (although some other studies do not agree (Kulkarni et al. 1994; Meiksin & Madau 1993)). If we assume the change in slope of $f(N_H)$ to be due to this transition, the value of N_H at which the turnover occurs can be used to derive a value of the external pressure. In terms of the observed N_H value, N_{H1obs}, combining (1) and (3), the pressure is given by:

$$\frac{P_{ext}}{k} = 5.1\eta_1^{1/3} \left(\frac{\zeta}{2.7 \times 10^{-12}\text{s}^{-1}} \times \frac{N_{H1obs}}{2 \times 10^{15}\text{cm}^{-2}} \right)^{2/3} \text{cm}^{-3}\text{K} \qquad (8)$$

The parameter η_1 relates to the ratio of dark matter to gas, and is expected to be of order 10. Thus at a redshift of 2.5 (where most of the Lyα forest data exists) the value is approximately $P_{ext}/k = 10\text{cm}^{-3}\text{K}$.

The specific non-power law shape of the observed $f(N_H)$ distribution was predicted by the transition from gas pressure confinement to dark matter gravity confinement, however a more general fact has also been demonstrated. If the mechanism of confinement changes at some column density, the distribution function of the clouds should also have a break at that column density. As more lines of sight are observed with high resolution (HIRES) by the Keck telescope it will be possible to accurately determine $f(N_H)$ and to chart its evolution with time.

It should be noted that this explanation of deviations from a power law distribution of the numbers of forest clouds is not unique. Petitjean et al. (1993a) have proposed that the change in shape at N_H of 10^{15} or 10^{16}cm^{-2} is due to a transition between metal-poor and metal rich systems. This is based on a detailed study of photoionization models of pressure confined, spherical clouds, with density profiles determined by gravity. The Petitjean et al. (1993b) models are an example of another class: a spherical geometry, including external pressure and gravity.

5   Cloud Structure and Formation Models

It seems likely that the simple models discussed above are in several ways unrealistic. Meiksin (1994) has performed one-dimensional hydrodynamic calculations in which he solves for the cloud structure in the context of the mini-halo model (in both spherical and slab geometry). These calculations do not assume the clouds to be in hydrostatic or in thermal equilibrium. Also, Meiksin consider the fact that the clouds are in a cosmological setting, i.e. matter is constantly accreting from outside the cloud. Murikama and Ikeuchi (1993) performed similar calculations, but did not find the same layered structure because their simulations were not performed in a cosmological context. As illustrated in Fig. 6, the larger clouds have an extent of 30kpc in radius, and they have a three phase structure. There is an inner core in hydrostatic and thermal equilibrium, a transition layer that is not in thermal equilibrium, and an outer accretion layer that joins onto the Hubble expansion. The core region is responsible for the higher column density forest

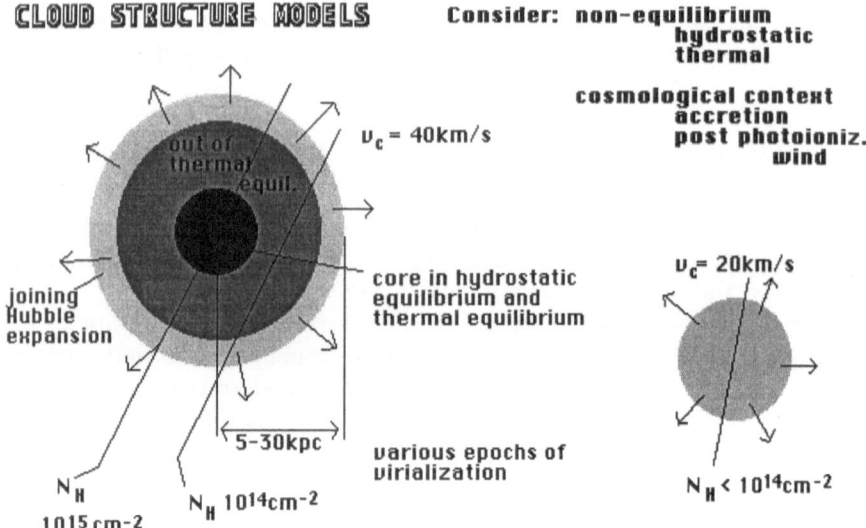

Fig. 6. Cloud structure models

clouds (N_H of 10^{14} or 10^{15}cm^{-2}), and the outer regions for those with lower column densities.

It has recently become possible to form Lyα forest clouds in the context of specific structure formation models using a "shock capturing" cosmological hydrocode (Cen et al. 1994). A specific application with initial conditions of cold dark matter and a cosmological constant is described by Miralda-Escude (these proceedings) so it will not be covered in detail here. However, the basic features are worth noting, because the picture differs substantially from the previous sketches. A schematic diagram is given in Fig. 7. The structures that would give rise to the lower N_H forest clouds (10^{14}cm^{-2}) are often part of a filamentary, sheetlike network. Within the sheets there can be larger density enhancements that produce somewhat larger column densities. The sheet-like clouds are confined roughly equally by gravity and by the ram pressure of infalling gas.

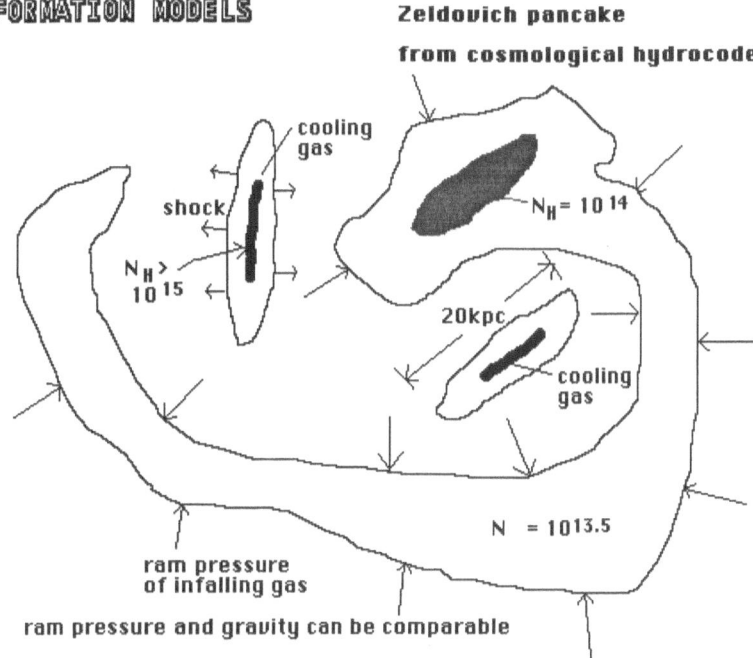

Fig. 7. Formation models

6   The Death of Pressure Confinement is Premature

The predictive power and the successes in matching observations (Miralda-Escude & Rees 1993; Mo et al. 1993) seems to have led to a focus on gravity confined as opposed to pressure confined models. As pointed out by Rees (these proceedings) this eliminates the need for a hot intergalactic medium. On the other hand, the large sizes of clouds imply that another mechanism is needed in addition to gravity. It is not all that implausible that the inter-galactic medium could be dense and hot enough to affect confinement of the low N_H end of the forest.

In particular, it is worth noting that feedback from galaxy formation can have a substantial effect on heating the IGM (Cen & Ostriker 1993; Shapiro, Giroux, & Babul 1994). In fact the pressures produced by these cold dark matter based studies are large enough to dominate gravity for the confinement of small N_H clouds. The IGM produced by this mechanism will be complex and will be hotter near to galaxies than in the voids. It is plausible that $\Omega_{IGM} = .02$ and $T = 10^6$K in the vicinity of clouds, and that would confine the forest by pressure up to column densities of $N_H = 10^{15}$cm^{-2}.

Alternatively, the confining pressure for large slab-like forest clouds could be due to hot gaseous halos that surround these clouds. In this case the confining pressure may increase with time. Alternatively, if the IGM was heated at a high redshift, then adiabatic cooling will lead to a decrease of the pressure with time.

7   Evolution of Lyα Clouds

The number of clouds as a function of N_H and z can in principle be used to constrain the models. The latest measurements of the rate of change of the ionizing background radiation J and of the evolution of the number of clouds were described in Bechtold's contribution to these proceedings. In that contribution, having a better knowledge of J allowed a better constraint on the simple equilibrium models. The goal of this section is not to provide a detailed interpretation of that data, but rather to explain the effect that numerous factors will have on cloud evolution. Let us quantify the evolution of clouds, with an equivalent width greater than some limiting value, as $N(z) \propto (1 + z)^\Gamma$.

The ionizing UV background (likely due to quasars and young galaxies) decreases over the time that we observe the forest clouds. This pushes each cloud to a higher value of N_H and thus (since there are more lower N_H clouds than higher) will lead to more clouds as time goes by (Γ decreases).

The external pressure could decrease adiabatically with time, or it could increase due to feedback from galaxies. A decreasing pressure results in fewer clouds confined by pressure, but the ones that are will have smaller densities n_{tot} due to expansion. This results in a larger decrease in the number of recombinations relative to the decrease in number of ionizations (see (1)). Thus N_H decreases for each pressure confined cloud and there are fewer clouds (a larger Γ) as time goes by.

Several other factors also affect evolution, and each is very poorly known. If clouds are expanding freely (not in equilibrium) it is expected that this effect would destroy the small clouds. Thus for those clouds Γ would increase. Mergers will provide a shift in the distribution from more smaller clouds to more larger clouds. If small clouds are not constantly formed to replenish the supply this would lead to an increase in Γ as well. Finally, if the number of new clouds that form decreases with time, then obviously Γ can increase for that reason. It will undoubtedly be quite difficult to sort out all these effects.

The decrease in the ionization rate due to the extragalactic background radiation is expected to affect the observed sizes of individual clouds. Assume that the total column density distribution as a function of radius in a single cloud, $N_{tot}(R)$, does not change with time. As time goes on, a given N_{tot} will correspond to a larger value of N_H, i.e. the cloud becomes more neutral. This also means that a given N_H is found at a larger radius in the cloud at later times when the ionization rate has decreased. This is general conclusion for any model if ζ is decreasing. For a given detection threshold of column density (equivalent width), the apparent radius of a forest cloud, detected by its neutral Hydrogen content, is expected to increase with time. This is consistent with constraints on forest cloud sizes from quasar double lines of sight (Dinshaw et al. 1994; Bechtold et al. 1994; Dinshaw (these proceedings)).

8   Which Came First?

In closing, I'd like to pose a question that provides a reminder of how little we know for certain about the formation and evolution of the forest. The question is: which comes first, the galaxy, the Lyα clouds, or the IGM? Of course, one might say that, by definition, the IGM must have come first. However, in this case, let us consider the majority of the gas in the IGM, whether it was primordial or reproduced by galaxies, and when and if it was heated.

In fact, it is possible to envision three perfectly reasonable scenarios that produce the three possible orderings of these events. For example, in a simple, hierarchical structure formation model there was first an IGM, it clumped together small scales to form Lyα forest clouds, and these gradually merged together to make larger structures. Alternatively, the IGM could still be primarily leftover from galaxy formation, but Jean's mass constraints could lead to galaxy formation, followed by later fragmentation to form forest clouds, or ejection of them as proposed by Wang (1994). Finally, perhaps galaxies formed first and efficiently, and produced a considerable, moderately hot IGM by hydrodynamic feedback. This IGM may have been necessary to confine the Lyα forest clouds so that they were the last structures to form.

Surely, many other general scenarios can be proposed. It is clear that no single, simple model provides an adequate description of the Lyα forest. However, a combination of these ideas does seem to have a ring of truth. At high redshifts it is almost certainly the case that the forest clouds are dynamic structures, best explored by hydrodynamic models such as those of Cen et al. (1994). However, at low redshifts they are astoundingly large, coherent structures. Such structures seem to require some combination of gravity and pressure confinement. In addition, if the forest clouds were produced by objects with a radius of 100kpc, at a redshift of three or four they would have to be nearly overlapping to produce the observed covering factor. Some aspects of the simple models may be relevant for describing the low redshift structure and evolution of the forest clouds. Undoubtedly, the theoretical models that are shaped by these simple ideas will soon have overwhelming amounts of new data to confront.

Acknowledgements. This work was supported by NASA grant NAGW-3571. The ideas presented have been shaped by collaborations with Edwin Salpeter, Suzanne Linder, and Craig Hogan. I'd also like to thank Chris Churchill for his insightful criticisms of this manuscript.

References

Babul, A., Rees, M.J., 1992, MNRAS, 255, 346.

Bajtlik, S., 1993, in The Environment and Evolution of Galaxies, eds. Shull, J. M. & Thronson, H. A., (The Netherlands: Kluwer), p. 191.

Barcons, X., Fabian, A.C., 1987, MNRAS, 224, 674.

Baron, E., Carswell, R.F., Hogan, C.J., Weymann, R.J., 1989, ApJ, 337, 609.

Bechtold, J., Crotts, A.P.S., Duncan, R.C., Fang, Y., 1994, ApJ, 437, L83.

Cen, R., Miralda-Escudé, J., Ostriker, J.P., Rauch, M. 1995, ApJ, in press.

Cen, R., Ostriker, J.P., 1993, ApJ, 417, 404.

Charlton, J.C., Salpeter, E.E., Hogan, C.J., 1993, ApJ, 402, 493.

Charlton, J.C., Salpeter, E.E., Linder, S.M. 1994, ApJ, 430, L29.

Dinshaw, N., Impey, C.D., Foltz, C. B., Weymann, R. J., Chaffee, F. H., 1994, ApJ, 437, L87.

Elowitz, R.M., Green, R.F., Impey, C.D., 1995, ApJ, in press.

Hoffman, G.L., Lu, N.Y., Salpeter, E.E., et al., 1993, AJ, 106, 39.

Ikeuchi, S., 1986, As&SS, 118, 509.

Ikeuchi, S., Murikama, I., Rees, M.J., 1988, MNRAS, 236, 21p.

Ikeuchi, S., Ostriker, J.P., 1986, ApJ, 337, 609.

Kulkarni, V.P., Huang, K.-L., Green, R. F., et al., 1994, ApJ, submitted.

Lanzetta, K.M., Bower, D.V., Tytler, D., Webb, J.K., 1995, ApJ, in press.

Mather, J. C. et al., 1994, ApJ, 420, 439.

Meiksin, A., 1994, ApJ, 431,109.

Meiksin, A., Madau, P., 1993, ApJ, 412, 34.

Miralda-Escudé, J., Rees, M.J., 1993, MNRAS, 260, 617.

Mo, H.J., Miralda-Escudé, J., Rees, M.J. 1993, MNRAS, 264, 705.

Murikama, I., Ikeuchi, S., 1993, ApJ, 409, 42.

Petitjean, P., Webb, J.K., Rauch, M., et al., 1993a, MNRAS, 262, 499.

Petitjean, P., Bergeron, J., Carswell, R.F., Puget, J.L., 1993b, MNRAS, 260, 67.

Rees, M.J., 1986, MNRAS, 218, 25p.

Salpeter, E.E., 1993, AJ, 106, 1265.

Salpeter, E.E., 1994, preprint.

Salpeter, E.E., Hoffman, G.J., 1995, ApJ, in press.

Sargeant, W.L.W., Young, P.J., Boksenberg, A., Tytler, D., 1980, ApJSupp. 42, 41.

Shapiro, P.R., Giroux, M.L., Babul, A., 1995, ApJ, in press.

Smette, A., Surdej, J., Shaver, P.A. et al., 1992, ApJ, 389, 39.

Storrie-Lombardi, J. L., McMahon, R. G., Irwin, M. J., Hazard, C., 1994, ApJ, 427, L13.

Wang, B., 1995, ApJ Letters, in press.

Williger, G. M., & Babul, A., 1992, ApJ, 399, 385.

Sizes and Dynamics of the Lyman Forest Clouds

Martin J. Rees

Institute of Astronomy, Madingley Road, Cambridge CB3 0HA, England

Abstract. This paper discusses further implications of the hypothesis that the Lyman forest is due to gas confined within (or falling into) the potential wells of dark 'minihalos'. The number of weak lines now being found confirms that the clouds must be associated with objects more numerous and less massive than galaxies . Most of the lines (those of low column density) arise from gas that is too diffuse to have achieved either dynamical or thermal equilibrium; these lines would not display Voigt profiles. The factors determining the temperature are mentioned, along with the interpretation of correlations between neighbouring lines of sight (which it may be too naive to interpret as 'cloud sizes'). A subset of the high-z absorption lines, and the majority of those at low z, would arise from pressure-confined gas in larger halos. The potential and problems of future computer simulations is briefly discussed.

1 Introduction: Hierarchical Clusters

Two decades of effort by many theorists to understand the Lyman forest have bequeathed us a great deal of historical baggage, which confuses the subject even more than necessary. These absorption lines now seem a natural consequence of current ideas on hierarchical formation of galaxies and other structures – if they had only just been discovered, this is surely the interpretation that would suggest itself to most of us. But when they were discovered, far less was known about the high-redshift universe, and there were fewer constraints on, for instance, a hot intergalactic medium.

In hierarchical ('bottom up') cosmogonies, the first structures to form are on subgalactic scales. If the dominant matter is non-baryonic (as, for instance, in CDM models), there is no definite lower limit to the scales of inhomogeneities in the overall mass distribution. However, the gas is only influenced by perturbations on scales comparable with (or larger than) the Jeans length. Until the very first structures form, the gas would be largely neutral, and even colder than the 2.7 (1+z) K of the primordial background radiation. However, before the epoch corresponding to z=5, enough UV seems to have been produced to photoionize whatever fraction of the baryons then remain in intergalactic space.

As soon as the dark matter starts to cluster on scales above the Jeans length of the gas, the intergalactic medium no longer produces a smooth Gunn-Peterson trough, but the nonuniformities in the density field (and in the velocity field as well) would imprint weak Lyman forest lines on any UV continuum passing through it. The characteristic scales and column densities are then determined by the clustering of the dark matter, modified by the pressure gradients in the gas.

Gravitational confinement models (Rees 1986, 1988) predict that the first clouds form when that scale of structure goes non-linear. In, for instance, the CDM model, this happens at redshifts of 5 or more, somewhat before galaxies form. (The alternative pressure-confinement model, popular in the early 1980s, does not naturally account for this scale.) I shall offer a few footnotes to Jane Charlton's excellent review in these proceedings. Note that a photoionized mixture of H and He never gets hotter than a few times 10^4 K, however intense the UV background is. The relevant minimum scale for incipient structures relevant to the Lyman forest is therefore the Jeans length at 10^4 K .

2   Lines of Low Column Density

If a homogeneous IGM has a Gunn-Peterson optical depth of (say) 0.1, then a line of optical depth unity is formed by any material extending over more than a Jeans length that is compressed one-dimensionally by more than a factor 10. Most of the lines revealed in the latest high-resolution spectra, such as have been described by Tytler at this meeting, correspond to HI column densities of only a few times 10^{12} cm^{-2} and optical depths no more than unity even in the line centre. The low column density lines can come from gas that is 10 times, or at most 100 times, denser than the mean IGM.

This straightforward deduction immediately implies that the regions responsible for the weak lines are in a dynamical state – any cloud that has collapsed and achieved virial equilibrium must have at least several hundred times the mean density, and a line of sight through it would intercept an HI column density of more than 10^{15} cm^{-2}. The gas that creates the weak lines is being pulled together by gravity, but has not yet evolved into a quasi-static virialised cloud (and maybe never will). In practice, of course, the collapse is more complicated – it is certainly not in general spherical. When the initial fluctuations have the form expected in CDM models, where the amplitude increases only very slowly towards smaller scales, we generically expect collapse to a sheet or pancake, several times larger than the eventual virialised cloud (See the following paper by Jordi Miralda-Escude.)

In principle, CDM calculations can take account of the dark-matter fluctuations on a wide range of scales, and compute the gravitational potential. But the infall of gas into the shallow potential wells of pregalactic 'minihalos' is influenced by pressure; this pressure in turn depends on what adiabat the gas is on. The temperature is determined primarily by the heat input from photoionization, but the recombination times at intergalactic densities are so long that adiabatic cooling in the expanding IGM (and subsequent adiabatic heating when a proto-cloud has started to collapse) affects the temperatures;

so also, at high redshifts, does Compton cooling on the microwave background (see Miralda-Escude and Rees, 1994 for some illustrations of such effects). The temperature therefore depends on when the gas was first ionized, as well as on the UV spectrum slope. The gas temperature in the IGM affects the dynamics of infall into shallow potential wells; also, the thermal conditions in the gas responsible for low-N lines may depend on the history of the gas, rather than being determined by instantaneous photo-ionization equilibrium.

3   Infall into Minihalos and Higher-Mass Systems

In 'minihalos', which we may define as systems with virial velocities of no more than 50 km/sec, any photoionized gas is only marginally supersonic – pressure gradients cannot be neglected, and the details of what happens may be sensitive to what adiabat the gas is on. The intensity of the UV background obviously controls the HI fraction in a given 'cloud'; but – more than that – the 'clouds' are the outcome of dynamical processes that themselves depend on the UV spectrum and intensity.

When the gas is falling into larger and deeper potential wells characterised by a virial temperature $T_{\text{virial}} \gg 10^4$ K, pressure gradients are dynamically negligible until it gets shocked. The shocks, with generically sheet-like structure, boost the gas temperature to a substantial fraction of T_{virial}; the shocked gas may then cool radiatively to the photoionization equilibrium temperature. If there is time for this cooling to occur, there will be a central sheet at the photoionization temperature sandwiched between two layers of hot gas shocked up to T_{virial}; this hot gas is itself confined by the ram pressure of material falling in. (This model first became familiar to astrophysicists through the scenarios for large scale structure formation that were investigated by the Moscow group in the 1970s.)

Some interesting scaling laws emerge if we compare the collapse (at the same epoch, and therefore with the same density) of regions with different length-scale R. The virial temperature then scales as R^2; so also, therefore, does the density of the cooled material in the pancake. The fraction of neutrals, therefore scales as R^2. Provided that most of the shocked material has time to cool down (as it would in this case, unlike in the bigger 'pancakes' relevant to large-scale structure), the neutral column density N goes as R^3, the extra power coming from the path length. This assumes that the system flattens, but does not collapse at all in the plane of the "pancake"; if the pancake shrinks by a factor x, then the total density rises by a further factor x^2 and the density of neutrals as x^4. A plausible range in R of around 10, corresponding to mass-scales between (in round numbers) 10^9 and 10^{12} solar masses, coupled with collapse factors x in the range from ~ 1 to ~ 3, is therefore sufficient in itself to cover the entire range of column densities up to those that induce self-shielding. (It is, in particular, possible that some Lyman limit systems could be much more extended than galaxies.)

4 Lines of Higher Column Density, and Cloud Sizes

The foregoing very simple considerations suggest two different origins for lines of column densities in the range $10^{15} - 10^{17}$ cm^{-2}: (i) Some might correspond to the virialised cores of minihalos. (Recall that completely standardised minihalos with the density profile of an isothermal sphere can account naturally for a slope of order -5/3 in the number versus column density relation.) or (ii) Others could be due to pressure-confined gas in larger-scale systems of high virial temperature, with bigger N corresponding to systems with higher T_{virial}. (There would, of course, be some dependence on impact parameter, just as for the minihalos, though this is hard to calculate.)

Both types of lines should exist (and there is in any case not a sharp demarcation between the two). However, pressure confinement may predominate at later epochs or lower redshifts, when the characteristic scale of structure is larger. Indeed, at the low redshifts probed by the HST, all the lines may be associated with galactic scales. (Note that in a galaxy like ours there is a two-phase medium in the halo. Cool phase material shows up in the 21 cm line. However, a similar cloud further out in the halo, where ambient pressure is lower, would be photoionized: its HI column density would be far too low to be detected by 21cm emission, but clouds of such gas, either pressure-confined or associated with a small satellite system, could extend out to 100 kpc or even beyond. An association of low-z absorption features with a galaxy near the line of sight would not be surprising.)

The evidence on 'cloud sizes' is another constraint on models, and indeed a discriminant between (i) and (ii) above.

If the clouds were like discs with sharp edges, then the chance that a cloud would intercept one line of sight and not another displaced by r would be of order r/R. If n lines are seen, and all are common, then the sharp edged model implies that R must be at least n times larger than r. For sensitive observations n may be very large, and this argument then leads to the inference that the clouds must be exceedingly large.

But such inferences can be rather misleading. Although they may have a characteristic scale R, gravitationally-confined or collapsing clouds would not have sharp boundaries. Whenever a cloud produces a spectral feature along one line of sight, there would also be a feature along any other line of sight separated by $< R$. If the lines revealed column densities that agreed with a precision of (say) 10 per cent, then the lower limit to the size would be 10 times larger. This limit could be improved by higher-sensitivity observations to search for even smaller differences than 10 per cent; but more lines with the same sensitivity (i.e. simply increasing n) would not help. We could still only say that the lines of sight were separated by less than 0.1 of the cloud scale length – in other words, that the clouds were at least ten times larger than the separation of the lines of sight.

Model (i) above predicts generally smaller sizes than model (ii). It also predicts smaller sizes for the high-column-density lines, whereas (ii) predicts the opposite

When the two lines of sight have separations of order an arc minute (e.g. Dinshaw et al. 1994) , they will display correlations due to structures whose scale is of order a Mpc. Such structures, proto-galaxies or proto-groups, would have typical velocity dispersions of 100-300 km/sec (less if they are just turning around). It would be very surprising if there were a close detailed correlation between the column densities in this case.

5   Line Profiles and the Gunn-Peterson Effect

The line densities in the spectra of the high-z quasars are so high that the weak systems almost overlap. This is simply a consequence of the gravitationally- induced clumping of the medium. Some lines are weaker than others not primarily because the responsible clouds are smaller, but because they are less overdense. (This is a contrast with pressure-confined models.)

The lines are so close that we shouldn't think in terms of widely-separated discrete clouds with a uniform medium between them. That is unrealistic – just as it would be to model an ocean surface as completely flat except for a few isolated high waves. The density is irregular everywhere along any line of sight: overdense regions are separated by only a few Jeans lengths; between them, the density falls below that of a uniform IGM. It is therefore hard to define the continuum with the requisite precision to measure the Gunn-Peterson decrement. Indeed, if we try to do so we would be measuring the absorption due to the regions of lowest density. These regions would be several times less dense than the mean, because matter tends to drain into the overdense regions. Allowance for this is essential in inferring, for instance, the UV background, especially as the Gunn-Peterson absorption, for a given photoionization rate, goes as the square of the gas density. There is a further correction, which also tends to reduce the depth of the absorption, because the velocity gradient in the underdense regions will be slightly larger than for the mean Hubble flow. (For these reasons, the postulate that the Gunn Peterson depth due to the mean IGM is 0.1 (cf the end of section 1) is compatible with the lower estimates that are actually quoted, since these refer to inter-cloud regions several times below the average density.)

We expect deviations from Voigt profiles, because the lines will be Doppler-broadened partly by bulk motions. As explained in section 2, these motions are definitely expected for the low-N lines, which come from infalling gas too diffuse to have attained virial equilibrium. It is therefore important to set the lowest possible upper limit to the thermal broadening, and therefore to T, as a function of N, z, etc. It should be realistic to do this with the signal-to-noise now available with the Keck Telescope. (Note that there may be a population of weak thermally-broad lines due to gas that has been adiabatically compressed but has not radiatively cooled.) Line profiles should give at least an upper limit to the temperature, as a function of redshift. This is likely to prove the best evidence on the spectrum of the UV background, and on the epoch ($z \simeq 5$, or $z \gtrsim 10$?) when the heat input switched on and the IGM was reionized.

6   Evolution with Redshift

Finally, let me summarise the prospects and the limitations of attempts to model the Lyman forest and its evolution. Gravitational clustering of dark matter will be computed with higher and higher resolution. It should, up to a point, be possible also to model the gas dynamics, including shocks and cooling. If the gas were initially cold, and heated only by shocks when it fell into a bound system or formed a pancake, then the problem would be well-posed, and calculations would be feasible throughout all the non-linear stages. However the pressure of the gas, once it is photoionized, is not negligible. And we know that it was indeed photoionized before the cosmic eras that observations can so far probe. The dynamics of the gas then cannot be computed unless one knows the heat input via photoionization.

What determines this heat input? It cannot precede the formation of first-generation bound systems (unless it is due to decaying particles). In principle, the formation of the first stars and quasars is determined by the initial conditions. But there is no realistic chance of modelling the internal processes of gas dynamics and formation of stars (or central 'active nuclei') within bound systems with the confidence that would be needed in order to predict the UV background as a function of z. This must therefore be an extra input into the calculation. Models of how cloud properties evolve will never be strictly deductive, any more than calculations of galaxy properties will be, because there is no hope of including all the physics and feedback effects from star formation, etc. We not only require initial conditions, but also the gas temperature (or UV background spectrum) as a function of z; the latter will have to be inferred observationally. With this input, it should be possible to calculate other features of the Lyman forest that can be tested against observations: for instance line profiles and column density distributions: also the angular correlation properties, and how these depend on z.

If the UV came from ordinary massive stars in early-forming systems of subgalactic scale, there would be concurrent widely-dispersed production of heavy elements. It would therefore not be surprising if the IGM, and the high-z Lyman forest, were already p pervaded by a detectable abundance of some heavy ions such as CIV, NV and OVI. Evidence on these heavy ions offers another probe of pregalactic history. For a given UV background, the properties of the observed lines all depend on the form of the initial fluctuations, and on the nature of the non-baryonic dark matter. The Lyman forest thereby offers several possible tests of structure formation models, being specially sensitive to effects on galactic and subgalactic scales. The observed properties of the lines may then be used to discriminate among the different thermal histories, as well as serving as a probe for structure evolution at all epochs back to the highest-z quasars.

Acknowledgements. I am grateful to Jordi Miralda-Escude for many interesting discussions on these subjects.

References

Cen R., Miralda-Escude J., Ostriker J.P., Rauch M., 1995, ApJL (in press)
Dinshaw, N., et al., 1994, ApJL 437, L87
Miralda-Escude, J., Rees, M.J., 1994, MNRAS 266, 343
Rees, M.J., 1986, MNRAS 218, 25
Rees, M.J., 1988, in "Quasar Absorption Lines", ed C. Blades et al. (CUP)

Hydrodynamic Simulations of the Lyman Alpha Forest in a Theory of Gravitational Collapse

Jordi Miralda-Escudé[1], Renyue Cen[2], Jeremiah P. Ostriker[2], and Michael Rauch[3]

[1] Institute for Advanced Study, Princeton, NJ 08540, USA
[2] Princeton University Observatory, Peyton Hall, Princeton University, Princeton, NJ 08544, USA
[3] The Observatories of the Carnegie Institution of Washington, 813 Santa Barbara Street, Pasadena, CA 91101, USA

Abstract. Using hydrodynamic simulations, it was shown in Cen et al. that the existence of the Lyα forest is predicted in theories of hierarchical gravitational collapse, as a result of the collapse of photoionized gas in structures with the form of sheets, filaments and halos over a wide range of scales. Here, we present additional preliminary results that we are obtaining on this work.

1 Introduction

Theories of hierarchical gravitational collapse with dark matter, were initial density fluctuations cause structures to collapse and merge at successively larger scales, can successfully explain the observations of galaxy clustering and the fluctuations in the microwave background. The origin and nature of the fluctuations are still not clear, and various theories are possible (e.g., see reviews by Ostriker 1993, and White, Scott, & Silk 1994).

Photoionized gas in the intergalactic medium must have collapsed in the past in objects on smaller scales than present-day galaxies and clusters, and it should yield absorption lines. This idea was first suggested as the minihalo model (Rees 1986; Ikeuchi 1986), were the gas was assumed to be in hydrostatic equilibrium in dark matter potential wells, and in thermal equilibrium with the ionizing background. Alternative models of expanding and contracting clouds were proposed in the same context by Bond, Szalay, & Silk (1988). Actually, in such theories of gravitational collapse, most of the lines in the Lyα forest must arise from gas that is not in dynamical equilibrium, but is in the process of collapsing (Miralda-Escudé & Rees 1993).

The hydrodynamic simulations of Cen et al. (1994) revealed a picture were the gravitationally collapsing gas is shocked in structures that are similar to Zel'dovich pancakes (Sunyaev & Zel'dovich 1972), and multiple shocks continue to take place as the gas collapses to higher densities. This process should occur over a wide range of scales (not just a restricted range as in the original minihalo model). The main assumption in these simulations is that the intergalactic medium is photoionized by a uniform background, and that it is not affected by any energetic ejection of matter from the collapsed objects. The observed spectra of Lyα absorption can be directly simulated, and the first results indicate that the characteristics of the predicted absorption lines generally agree with the observed ones. It should be emphasized that this is a new prediction of the theories of formation of structure by hierarchical gravitational collapse, which were invented for reasons that had nothing to do with the observations of the Lyα forest.

2   Some Results from the Simulations

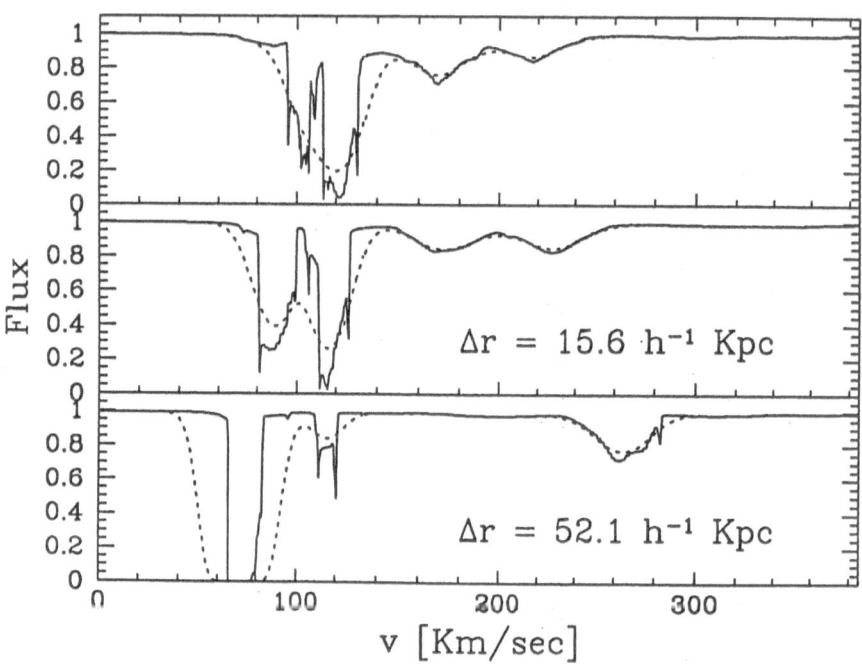

Fig. 1. Example of three Lyα spectra along parallel rows. The last two are separated from the first by the indicated transverse distances Δr. The dotted lines are the spectra that would be observed, and the solid ones include only the hydrodynamic motions, but not the thermal broadening.

Figure 1 shows three simulated spectra obtained from three parallel rows along a simulation of a CDM model with a cosmological constant $\Lambda = 0.6$, at $z = 3$, using a box with periodic boundary conditions of length $3h^{-1}\,\mathrm{Mpc}$ (see Cen et al. 1994 for further details). This corresponds to a period of

384 Km s^{-1} for these spectra. The second and third rows are separated from
the first by a transverse distance Δr as indicated in the figure, in proper
units. The solid lines are the spectra that would be seen without including
thermal broadening, and the dotted lines include thermal broadening. The
solid lines show several peaks in the optical depth which are due to velocity
caustics. However, these are typically smoothed by thermal broadening, and
the high contrast between the lines and any Gunn-Peterson effect arise from
the difference in density between collapsed regions and voids, and the fact
that the neutral fraction in photoionized gas is proportional to the density.

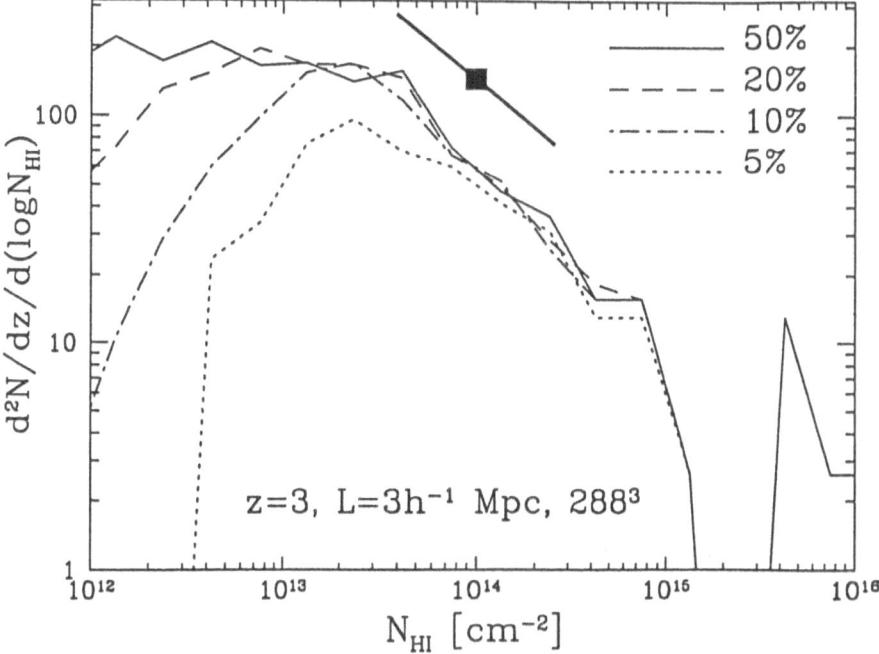

Fig. 2. Column density distribution within intervals where the flux is below a
threshold F_0, such that the covering fraction of the intervals in the spectra has the
values indicated. The fluctuations (specially at high N_{HI}) arise from noise due to
having selected only 300 random spectra.

Figure 2 shows the column distribution of lines, from a total of 300 randomly chosen rows. As an algorithm to detect the lines, we have selected all
intervals in the spectra where the flux is continuously below a given threshold, F_0, and we have calculated the column density within each such interval, corresponding to the integrated optical depth. We then select those flux
thresholds which result in a given covering fraction of all the spectra by the
selected intervals. The column density distribution is shown for four values of

the covering fraction (e.g., for the solid line, the flux threshold used was such that the intervals defining the lines cover half of the spectra). In the theory we are considering line profiles are generally not Voigt profiles since the gas can have an arbitrary distribution of velocity and temperature. Thus, fitting the line shapes as superposed Voigt profiles is not necessarily the best way to analyze the absorption lines. The point at $N_{HI} = 10^{14}\,cm^{-2}$ is taken from Petitjean et al. (1993), and the solid line indicates a slope of $\beta = 1.7$. The number of predicted lines can agree better with observations by declining the intensity of the ionizing background assumed in the simulation, which shifts the curves to the right of the figure.

The column density shows a turnover at $N_{HI} \simeq 10^{13.5}\,cm^{-2}$. This is related to the Jeans scale; column densities below the turnover are produced only by sheets on scales smaller than the typical scales collapsing at the epoch of the simulation. Such a turnover, and a steepening towards higher column densities, has been seen in recent observations (e.g., Petitjean et al. 1993).

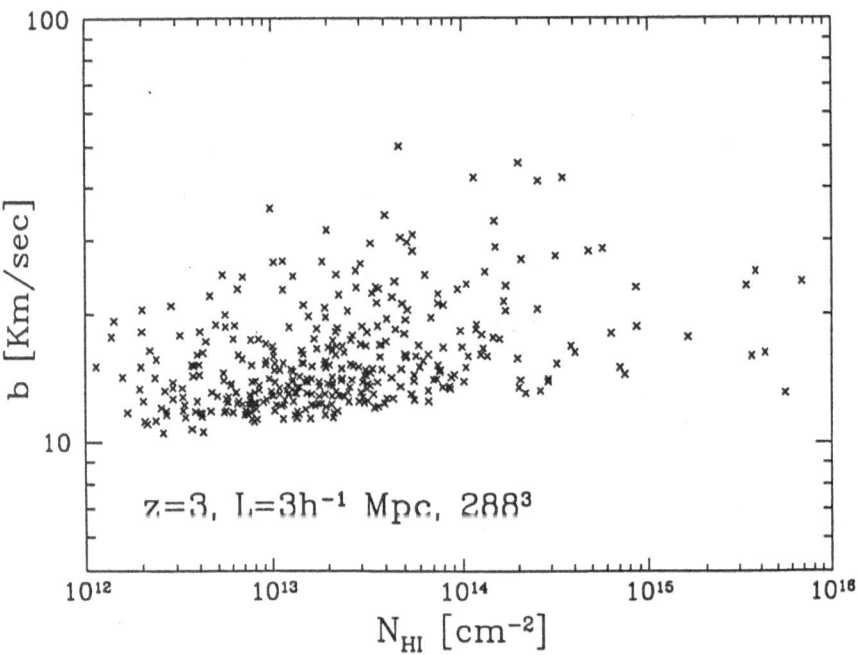

Fig. 3. Column densities and b-parameters of 384 lines detected in 300 spectra from the simulation like those in Fig. 1

In Figure 3, we show the column densities and b-parameters of all the lines detected in the same 300 rows. The b-parameters were found given the widths of the intervals defined above, and the maximum of the optical depth within the intervals, assuming a Voigt profile. This gives similar values to the ones obtained by detailed fitting to Voigt profiles, as described in Cen et al. A flux threshold of 0.75, relative to an assumed unabsorbed continuum, has been used. They are mostly in the range $12 - 20 \, \mathrm{Km \, s^{-1}}$, which is somewhat lower than observed (e.g., Cristiani et al., Tytler, these proceedings). However, this can be modified depending on several physical processes that can change the temperature of the intergalactic medium, as described in Cen et al. We have also started to analyze a larger simulation on a box of length $10h^{-1}$ Mpc, and we find that the addition of large scale power does not alter very much the distribution of b-parameters in this model.

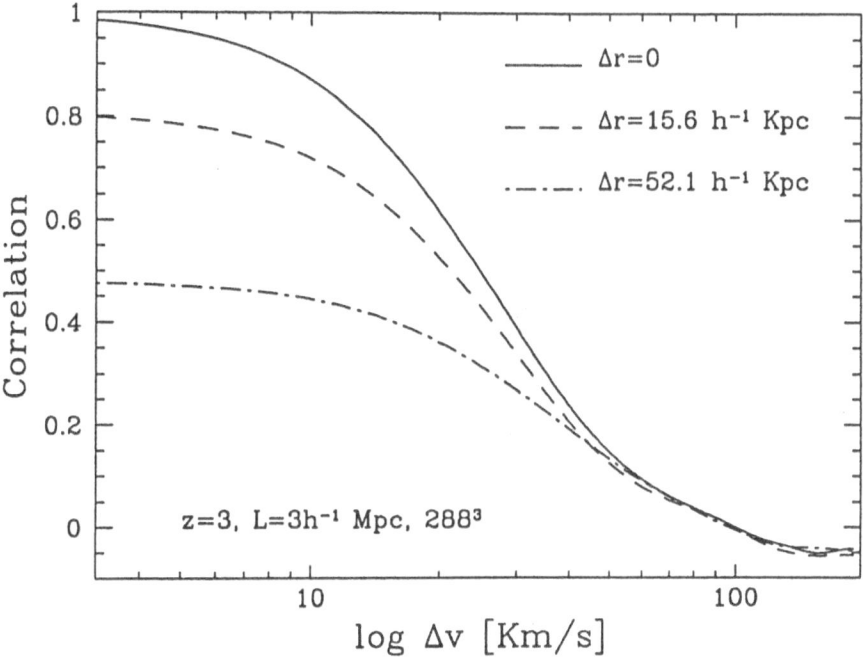

Fig. 4. Autocorrelation (solid line) and cross-correlation of the intensities on rows separated by the proper distances indicated

Finally, Fig. 4 shows the autocorrelation of the intensity for the same 300 spectra as the solid line, and the cross-correlation of intensities in spectra from parallel rows separated by the transverse distances that are indicated. At a transverse distance of $\sim 50h^{-1}$ Kpc (in proper units), there is still a significant correlation, which is encouraging given the recent observations on the transverse size of the Lyα clouds (e.g., Smette et al. 1992; Bechtold et al. 1994; Dinshaw et al. 1994).

The intensity correlation is mostly sensitive to the correlation of weak lines, which can cover a substantial part of all the spectrum. These lines are caused in sheets and filamentary structures, which arise from caustics were the gas is shocked. These caustics are not a result of the resolution of the numerical simulation, but are physically present in the distribution of the photoionized gas as a result of the Jeans scale, which introduces a natural cutoff in the power spectrum of density fluctuations for the gas. Thus, a natural prediction in this theory is that the column densities of the weak lines should be correlated over relatively large transverse distances, even on scales where the gas is not in pressure equilibrium. Stronger lines must arise from halos, which are of smaller size; the line coincidences on the scales reported by Bechtold et al. and Dinshaw et al. can only be understood in terms of the correlation function of different objects producing the lines, rather than as a measure of the size of these halos.

Acknowledgements. JM is grateful for financial support from the W. M. Keck Foundation and NASA grant NAG-51618.

References

Bechtold, J., Crotts, A. P. S., Duncan, R. C., & Fang, Y. 1994, ApJ, 436, X

Bond, J. R., Szalay, A. S., & Silk, J., 1988, ApJ, 324, 627

Cen, R. Y., Miralda-Escudé, J., Ostriker, J. P., & Rauch, M. R. 1994, ApJ, 437, L9

Dinshaw, N., Impey, C. D., Foltz, C. B., Weymann, R. J., & Chaffee, F. H. 1994, ApJ, 436, X

Ikeuchi, S. 1986, Astrop. & Spa. Sci., 118, 509

Miralda-Escudé, J., & Rees, M. 1993, MNRAS, 260, 624

Ostriker, J. P. 1993, ARAA, 31, 689

Petitjean, P., Webb, J. K., Rauch. M., Carswell, R. F., & Lanzetta, K. 1993, MNRAS, 262, 499

Rees, M.J. 1986, MNRAS, 218, 25p

Smette, A., Surdej, J., Shaver, P. A., Foltz, C. B., Chaffee, F. H.,

Sunyaev, R. A., & Zel'dovich, Ya. B. 1972, A&A, 20, 189

White, M., Scott, D., & Silk, J. 1994, ARAA, 32, 319

Uncertainties in the Interpretation of the Lyman–Alpha Forest Lines

S. A. Levshakov[1] and W. H. Kegel[2]

[1] Department of Theoretical Astrophysics, A. F. Ioffe Physico-Technical Institute, St.Petersburg 194021, Russia
[2] Institut für Theoretische Physik der Universität Frankfurt am Main, Robert-Mayer-Str. 8-10, 60054 Frankfurt am Main, Germany

Abstract. We study the formation of hydrogen HI Lyα forest lines in turbulent media accounting for the effects of a finite correlation length, l, in the velocity field. This is in contrast to the usual procedure, in which line formation is treated in the microturbulent limit ($l \equiv 0$). Our results show clearly that a finite velocity correlation length ($l > 0$) affects strongly (i) the profile of the Lyα line, (ii) its equivalent width, and (iii) the intensity ratios of Lyα and higher order Lyman lines. We argue that the column density distribution function $f(N)$ has to be *reexamined*. A deficit of absorption systems revealed in the Lyα forest in the range 10^{14} cm$^{-2} \leq N \leq 10^{16}$ cm^{-2} may simply be an *artifact*.

1 Results

The radiative transfer in a turbulent medium with $l > 0$ can be described in terms of the statistical properties of the velocity field (see e.g. Levshakov & Kegel, 1994). In this case one has to specify 4 parameters : total column density N, thermal, v_{therm}, and rms turbulent, σ, velocities, and the ratio L/l, where L is the geometrical size of a cloud.

We computed HI Lyα mesoturbulent profiles for $L/l = 5$, $T_{kin} = 10^4$ K, and different hydrogen column densities N and then fitted them by Voigt profiles to estimate possible uncertainties in the determination of N and v_{dop}, i.e. we considered the calculated profiles as "measured" ones and determined "apparent" column densities.

In Fig. 1, the open circles represent our mesoturbulent intensities while the best fitting Voigt profiles are shown by solid lines. Each panel lists initial N, v_{dop} and output $N^\star$, $v_{\text{dop}}^\star$ parameters for HI column densities in the range $10^{14} - 10^{19}$cm^{-2}.

As seen, mesoturbulent Lyα profiles for $N \leq 10^{17}$ cm^{-2} can be fitted satisfactorily by Voigt profiles. However, the values of $N^\star$ and $v_{\text{dop}}^\star$ derived by profile fitting show increasing deviations from the initial parameters with increasing N. The ratio $N/N^\star$ changes from 1.25 at $N = 10^{14}$ cm^{-2} to about 200 at $N = 10^{17}$ cm^{-2}.

From the results in Figure 1 it is evident that the determination of v_{dop} and N might be *ambiguous* for the Lyα forest lines (see Levshakov & Kegel, 1995, for details).

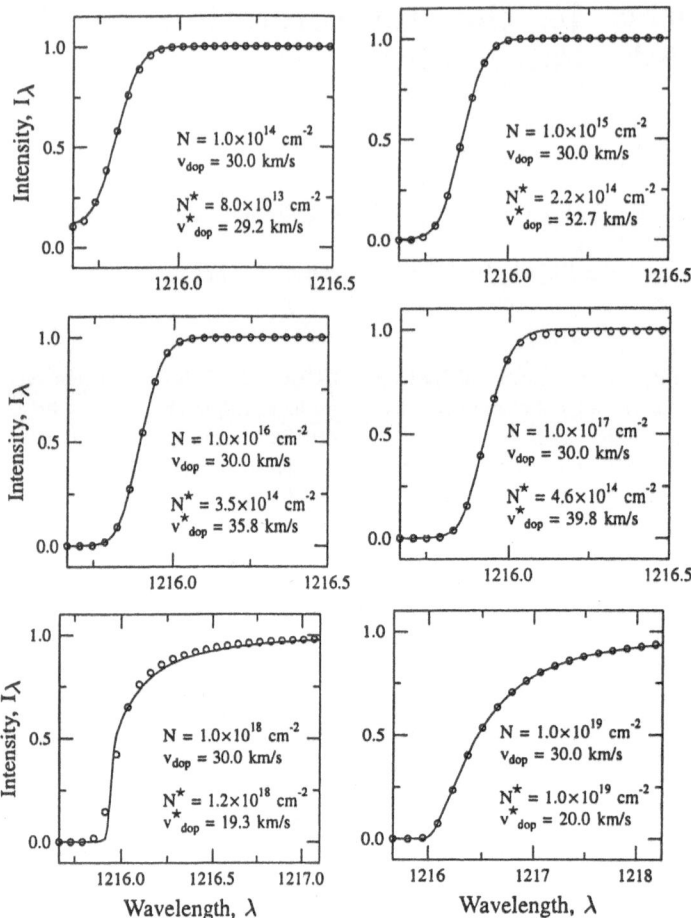

Fig. 1. Synthetic mesoturbulent spectra of the HI Lyα line for $T_{kin} = 10^4$ K, $L/l = 5$, $v_{dop} = 30$ km s^{-1}, and $N = 10^{14} - 10^{19}$ cm^{-2} (*open circles*). The best fitting Voigt profiles are shown by *solid lines*. Corresponding best fitting parameters $N^\star$ and $v^\star_{dop}$ are listed in each panel.

Acknowledgements. SAL was supported by the ESO C&EE Programme (grant D-06-005) .

References

Levshakov S. A., Kegel W. H., 1994, MNRAS, 271, 161.
Levshakov S. A., Kegel W. H., 1995, MNRAS, *submit..*

Environmental Effects on a Lyman–Alpha Cloud Confined by a Dark Halo

Izumi Murakami[1] and Satoru Ikeuchi[2]

[1] CITA, University of Toronto, Toronto, Ontario M5S 1A7, Canada
[2] Department of Earth and Space Science, Faculty of Science, Osaka University, Toyonaka, Osaka 560, Japan

Abstract. We study the interaction of an intergalactic cloud confined by a dark halo, so-called minihalo, with supersonic gas flow to consider the evolution of the gas cloud affected by ram pressure using two-dimensional hydrodynamical calculations. Using the results, we examine the effect of a blast wave caused by the activity of quasars and the effect of clustering on minihalos. We find that the blast wave blows out the gas from minihalos and that minihalos are stripped of gas due to their motion in the cluster. Such minihalos will not be observed as Ly α absorption lines.

1 Introduction

Minihalo model is one of theoretical models to explain Ly α forest. In this model, an intergalactic gas cloud confined by the gravity of a CDM dark halo is observed as a quasar absorption line (Rees 1986; Ikeuchi 1986; Murakami & Ikeuchi 1993, 1994b, and references therein). Even though all CDM halos are expected to cluster, observed Ly α forests do not show strong spatial correlation (Webb 1987).

Here we try to explain the absence of the spatial correlation with a picture that minihalos will lose their gas when they cluster, so that they will not be observed. We perform two-dimensional hydrodynamical calculations in which minihalos are relatively exposed to supersonic gas flow and estimate the condition and time scale of gas stripping. Models and results are described in detail by Murakami & Ikeuchi (1994a).

2 Results

Figure 1 shows the time evolution of density distribution for a model minihalo (density ratio of gas to dark matter at the center, $C_i = 0.003$; uniform dark matter density; gas mass, $5.7 \times 10^6 M_\odot$; and a supersonic flow of Mach number 8 and density $1.2 \times 10^{-29} \mathrm{g\ cm^{-3}}$). The central region remains for a while but is gradually stripped of its gas. Finally all gas is stripped from the minihalo. The condition of stripping depends on the ratio of ram pressure to thermal pressure of the minihalo and on the ratio of the flow velocity to the escape velocity of the minihalo. All gas is eventually stripped when both ratios are larger than unity.

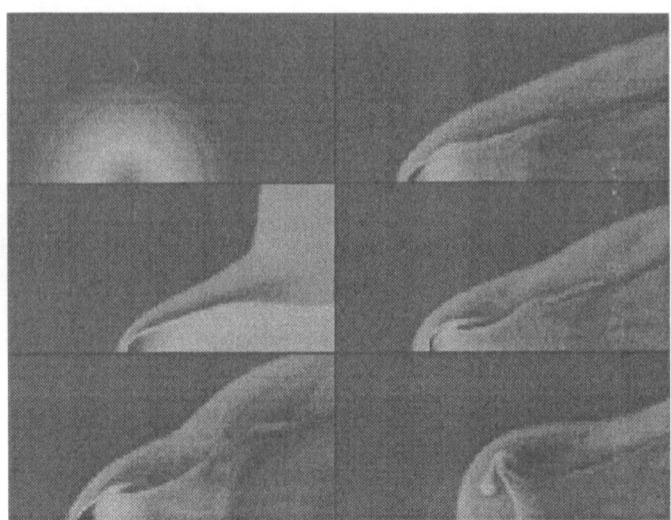

Fig. 1. Gas density distribution. The cloud center has the highest density and the ambient gas has the lowest density. Each panel is for different time: $t = 0$ (left top); 5.22×10^8 yr (left middle); 9.49×10^8 yr (left bottom); 1.42×10^9 yr (right top); 1.85×10^9 yr (right middle); and 2.26×10^9 yr (right bottom).

Using a simple model for a cluster of minihalos, we expect that the time scale of stripping is shorter than the Hubble time and that clustered minihalos are stripped of their gas easily. Hence, minihalos with high spatial correlation may not be observed as absorption lines.

Considering the effect of blast wave from a quasar, gas clouds within $2 \sim 3$ Mpc from a quasar are destroyed due to supersonic gas flow caused by an explosion with energy 10^{63} erg. This effect may account for the proximity effect of Ly α forests.

References

Ikeuchi S., 1986, A& SS, 118, 509

Murakami I., Ikeuchi S., 1993, ApJ, 409, 42

Murakami I., Ikeuchi S., 1994a, ApJ, 420, 68

Murakami I., Ikeuchi S., 1994b, ApJ, 421, L79

Rees M.J., 1986, MNRAS, 218, 25p

Webb J.K., 1987, in *Observational Cosmology*, eds. Hewitt A., Burbidge G., & Fang L.-Z., (Dordrecht: Reidel), p. 803

Photoionization Edges of Galaxies and the HI Column Density Distribution

Mark A. Fardal and Philip R. Maloney

JILA/CASA, University of Colorado, Boulder, CO 80309, USA

Abstract. We look for a photoionization feature in the H I column density distribution. Our model assumes the H I lines come from rotationally-supported disks in dark matter potential wells (i.e., galaxies), and the outer edges of these disks are ionized by the metagalactic radiation field. The predicted result is a flattening of the distribution from $N_{\rm H\,I} \sim 10^{17.5}$ cm^{-2} to $N_{\rm H\,I} \sim 10^{20.5}$ cm^{-2}. Our preliminary results do not show strong evidence for this feature.

1 The Model

The H I distribution in present-day galaxies is observed to cut off very sharply beyond total columns of $\sim 3 \times 10^{19}$ cm^{-2}. This can be explained by photoionization by a metagalactic ionizing background (Maloney 1993). The column density at which this cutoff should occur is roughly $N_{cr} \sim (\Phi H/\alpha_B)^{1/2} \sim 2 \times 10^{19}(\Phi/10^4)^{1/2}$ cm^{-2} for typical galaxies today, where Φ is the flux of ionizing photons in cm^{-2} s^{-1} and H is the scale height of the galaxy. The rapid decline in $N_{\rm H\,I}$ which occurs around N_{cr} will slow when the gas becomes optically thin, *i.e.*, at a neutral column of $\sim 10^{17}$ cm^{-2}, although this has not yet been observed.

The present-day value for N_{cr} implies an ionizing flux of $\Phi \sim 10^4$ cm^{-2} s^{-1}. However, the ionizing background is believed to be about $30-100$ times stronger at $z \sim 2$. Hence we expect a feature in the $N_{\rm H\,I}$ distribution starting around 10^{17} cm^{-2} and extending up to 10^{21} cm^{-2} or so. Although we have modeled the distribution in the specific context of disk galaxies, this feature is expected in almost any model of the high-$N_{\rm H\,I}$ systems.

We assume the total gas column within a galaxy follows a power law, $N_H = 10^{20.2} (R/R_0)^{-\alpha}$ cm^{-2}. The variable parameters in the model are the flux of ionizing photons Φ, R_0, and α. We populate the universe at $z \sim 2$ with disk galaxies, drawn from the present-day luminosity function, and calculate their ionization structure. The resulting $N_{\rm H\,I}$ distribution is then averaged over random orientations, assuming the disks are thin so that $N_{\rm H\,I} \propto \sec\theta$ and the area $\propto \cos\theta$.

We fit this model to the $N_{\rm H\,I}$ distribution of Petitjean et al. (1993). The Lyman limit systems from $10^{17.7}$ to $10^{20.5}$ fall into a single bin, which covers the most interesting range in $N_{\rm H\,I}$ for the photoionization feature. However, there should still be an offset in the distribution between the systems with $N_{\rm H\,I} \lesssim 10^{17}$ cm^{-2} and those with $N_{\rm H\,I} \gtrsim 10^{20}$ cm^{-2}.

2 Results

One such fit is shown in Fig. 1. The model is a poor fit ($\chi_r^2 = 4.2$), although it is a much better fit than a power law over the same range. Most of the discrepancy between the model and the data comes from the region $10^{14.5}$ cm^{-2} to 10^{16} cm^{-2}, where there is a depression in the data. This feature could result from the overlap of two different populations at $N_{HI} \sim 10^{16}$ cm^{-2}. In this case our fit should be restricted to columns above 10^{16} cm^{-2}, and over this region a simple power law is *better* than our complicated model.

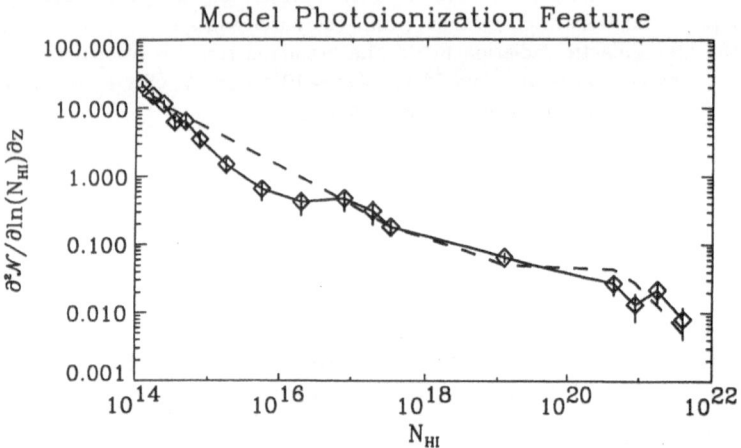

Fig. 1. Fit (*dashed*) to column density distribution of Petitjean et al. (1993) (*diamonds*), with parameters $\Phi = 1.5 \times 10^5$ cm^{-2} s^{-1} , $R_0 = 35$ kpc, and $\alpha = 1.3$.

The present lack of evidence for the photoionization feature could mean that Φ is $\lesssim 3 \times 10^5$. It could also mean that the data poorly constrain the shape of the N_{HI} distribution. In particular, better statistics on the damped Lyα lines would help to constrain or rule out our model.

Acknowledgements. MF thanks the organizers and the CU-Boulder Dean's Small Grant Program for enabling him to attend the conference. This work was supported by NASA grants NAGW-766 and NAGW-1479.

References

Maloney, P.R., 1993, ApJ, 414, 41
Petitjean, P., Webb, J.K., Rauch, M., Carswell, R.F., Lanzetta, K., 1993, MNRAS, 262, 499

Hot Galactic Halos and Lyman Limit Systems of Quasars

Amancio C. S. Friaca[1] and Sueli M. Viegas[2]

[1] Royal Greenwich Observatory, Madingley Road, Cambridge CB3 0EZ, UK
[2] Instituto Astronômico e Geofisico, USP, Caixa Postal 9638,
01065-970 São Paulo, SP, Brazil

Abstract. As an alternative to the standard model in which the absorption-line systems of quasars are photoionized by a metagalactic radiation field, we develop a model in which the ionizing radiation is assumed to be local, arising from to the EUV and soft X-ray emission produced by the hot gas in the galactic halo, as predicted by a chemo-dynamical model for the evolution of elliptical galaxies. The radiation emergent from the hot galactic halo has the optimal shape to explain simultaneously the presence of He I as well as the ratios of several ionization stages of O in the LLSs of the QSO HS 1700+6416.

1 Hot Halos as the Source of the Ionizing Radiation

Observations with the Hubble Space Telescope (Vogel & Reimers 1993; Reimers & Vogel 1993) of LLSs of the QSO HS1700+6416, revealing the simultaneous presence of He I and of highly ionized ions of CNO, have posed serious difficulties for the standard model for absorption-line systems of quasars in which they are ionized by a metagalactic UV background. Therefore, we have developed an alternative model in which the LSSs are ionized by the radiation arising from the hot galactic halo (Viegas & Friaça 1995). The absorbers are identified with colder, denser clumps embedded in the hot halo. A chemo-dynamical model for the evolution of elliptical galaxies, predicts a large soft X-ray and EUV luminosity during the early phases (Friaça and Terlevich 1994). We choose as the reference of this paper, the gas distribution and the emergent radiation of the fiducial model of Friaça & Terlevich (1994) representing an elliptical with luminous mass of 2×10^{11} M$_\odot$, at an epoch $t = 2$ Gyr and a radius of 30 kpc. The ion column densities of the clumps, modeled as isobaric plane-parallel slabs, in pressure equilibrium with the hot halo, are obtained from the photoionization code AANGABA (Gruenwald & Viegas 1992). As illustrated in Fig. 1, our models can explain the large range of ionization stages in the clumps. Our spectrum has the optimal shape to explain the simultaneous presence of He I and O VI ions, in the sense that it is steep below the ionization edge of He I and flatter at higher energies.

In addition to explain the LLSs of the QSO HS1700+6416, the hot halos provide an additional ionization source required to explain other absorbers. The hot galactic halos would constitute the third ingredient in hybrid models

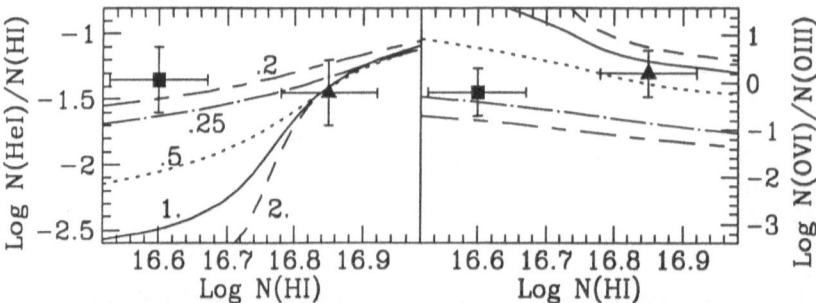

Fig. 1. Comparison between predicted and observed ion ratios of HeI/HI and OVI/OIII. The observational points are the LLSs at $z = 2.1678$ (triangles) and $z = 2.433$ (squares). The curves are labeled according to the $q(HI)/q(HI)_0$ ratio. The variation of $q(HI)$, the flux of photons photoionizing H I reflects the differences in density, temperature, and radiation intensity met by the actual clump and those of the reference – defining $q(HI)_0$ ($= 1.17 \times 10^9$ cm^{-2} s^{-1}).

for the ionizing radiation, the classical ingredients being AGN's and hot stars (Miralda-Escudé & Ostriker 1990; Madau 1991). Other possible origin for the ionizing radiation is the intracluster medium, however, its harder spectrum will make it more difficult to reproduce the ionization observed in LLLs. We do not find in our models that the observed column densities of O III to O VI could arise from the hot galactic halo, as proposed by Giroux, Sutherland & Shull (1994). In our models, the halo is considerably more ionized than in their model, and even for the O VI ion, the column densities integrated through the halo are at least two orders of magnitude lower than the observed values.

Acknowledgements. S.M.V. acknowledges support from the Brazilian funding agencies CNPq and Fapesp. A.C.S.F. acknowledges support from CNPq.

References

Friaça A.C.S, Terlevich R.J., 1994, in *Violent Star Formation from 30 Doradus to QSOs*, ed. Tenorio-Tagle G., (Cambridge: CUP), p.424.

Giroux M.L., Sutherland R.S., Shull J.M., 1994, ApJ, 435, L97.

Gruenwald R.B., Viegas S.M., 1992, ApJS, 78, 153.

Madau P., 1991, ApJ, 376, L33.

Miralda-Escudé, J., Ostriker, J.P., 1990, ApJ, 350, 1.

Reimers D., 1993, Vogel S., A&A, 276, L13.

Viegas S.M., Friaça A.C.S, 1995, MNRAS, in press.

Vogel S., Reimers D., 1993, A&A, 274, L5.

Lyα Clouds from Dwarf Outflows

Boqi Wang

The Johns Hopkins University, Baltimore, MD 21218, USA

Abstract. We propose that cooling outflows from star-forming dwarf galaxies may account for the QSO Lyα forests, providing a explanation for both the heavy elements and large sizes recently observed for them. Supernovae shock the interstellar gas to high temperature, and the resulting gas outflows from the dwarfs owing to their shallow potential wells. The gas cools radiatively as the radiative cooling time shortens because of adiabatic expansion. Subsequent thermal instability results in condensation of clouds. The clouds inherit the kinetic energy of the flow and coast to large distances, providing regions of the absorbing gas with sizes of up to hundreds of kpc. We calculate the Lyα absorption line profiles and the neutral column distribution, and show that they are consistent with observations. We suggest that the Lyα forests are caused by the faint blue galaxies found in the deep extragalactic surveys.

We consider radial, steady, outflows starting from an initial radius (r_i) with an initial temperature (T_i). Given r_i, T_i, and the mass loss rate λ (or equivalently the initial gas density), the outflow solutions are completely specified. The general solutions of the outflow problem have been obtained in Wang (1995a, b, to appear in ApJ), and we refer those papers for details. We adopt the cooling function for a gas with 1/100 solar abundance. We consider dwarf galaxies whose dominant dark matter is distributed as isothermal spheres but truncated at radius r_t; we take $r_t = 10$ kpc, the circular (rotation) velocity of the dark matter halo to be $v_{cir} = 50$ km/s, and $r_i = 2$ kpc, $T_i = 4 \times 10^5$ K and $\lambda = 0.5$ M$_\odot$/yr for the following calculation. In Figure 1, we show the Lyα line profiles for impact parameter $\rho = 2$, 3, and 4 kpc for $J_{21} = 1$. The contribution to the absorption in these cases comes mostly from the neutral hydrogen within the cooling radius.

At further large distances, if the line of sight (los) intercepts no clouds, the resulting line profile is a broad trough with a b-value comparable to the flow velocity (~ 170 km/s), shown in Figure 1 for $J_{21} = 1$ at $\rho = 20$ kpc. If the los intercepts a single cloud that dominates the absorption, one sees a narrow line with a width characteristic of the internal temperature 2×10^4 K. We show such an example in Figure 1, assuming that a cloud with 4 times the ambient density and a radius of 1 kpc is intercepted at $y = 0$ (y is the distance along the los measured from the pericenter). In practice, clouds at different y should also cause absorption, resulting in multiple lines with different strengths. However, the lines are likely to be dominated by clouds at $y = 0$ (moving perpendicular to the los) because the gas density is the highest here. This implies that the velocity difference in two los (e.g., toward a pair of QSOs) through the same galaxy should be small, consistent with recent observations.

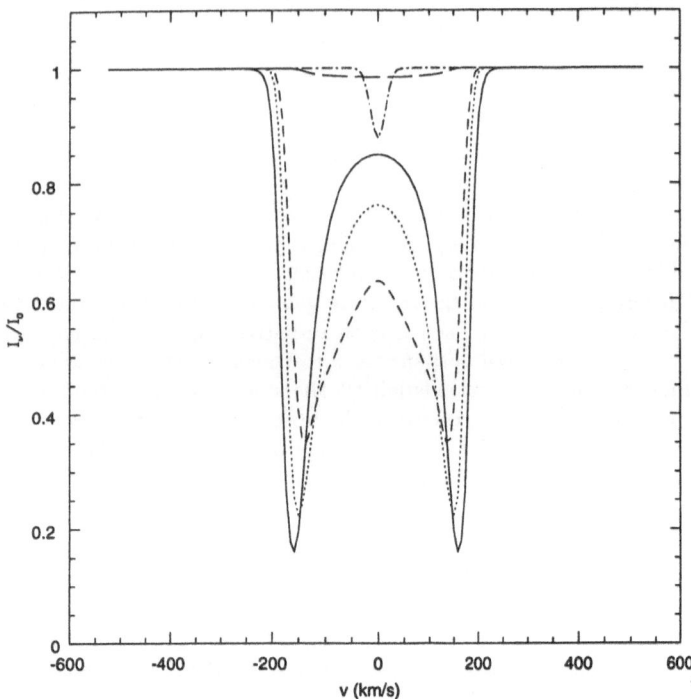

Fig. 1. The Lyα absorption line profiles, $I_\nu/I_0 = e^{-\tau_\nu}$, as a function of the velocity shift $v \equiv c\delta\nu$ for $J_{21} = 1$. The impact parameter to the line of sight is $\rho = 2$ (solid), 3 (dotted), 4 (dashed) kpc, respectively. For $\rho = 20$ kpc, we have assumed that the line of sight intercepts no clouds but only diffuse gas (long-dashed) or intercepts a single cloud with a radius of 1 kpc and an overdensity of 4 (dash-dotted).

We expect a strong evolution for the number of the absorbing dwarfs if they are responsible for most of the Lyα absorptions. For example, the dwarf comoving density is increased by about a factor of 2 from $z = 0$ to 0.4. Interestingly, deep extragalactic surveys show that the number of galaxies at apparent magnitude $b_J \sim 24$ in the blue band is larger than that expected from the local luminosity function by a similar factor, while spectroscopic observations of these faint galaxies show that they have a median z of 0.4, and most of them are sub-L_* galaxies experiencing active star formation. Thus we suggest that the faint galaxies in the deep surveys may be responsible for the Lyα clouds; this would also account for the heavy elements recently detected in the Lyα clouds (Tytler 1995; this conference).

Consequences of H II Region Model for Absorption Systems in QSOs

R. Srianand and Pushpa Khare

Department of Physics, Utkal University, Bhubaneswar - 751004, India

1 Introduction

It has been recently shown that the observed equivalent widths of various lines in the QSO absorption line systems (QSOALS) are consistent with their formation in the H II regions of O4 stars. These H II regions could in principle be present in low surface brightness or dwarf galaxies. However this is unlikely in view of the identification of absorbers with bright galaxies as well as presence of ionization sub-clustering among the components of the absorption lines. We thus assume that the H II regions are present in bright galaxies. Below we present some consequences of this hypothesis.

2 Star Formation Rate

The star formation rate (SFR) necessary to account for the observed frequency of occurrence of the QSOALS is given by

$$\text{SFR} = \int_0^\infty (m/M_\odot)\Phi(m)dm \quad M_\odot \ yr^{-1} \ pc^{-2},$$

over the entire area πr^2_{eff} of the galaxy. Here r_{eff} is the effective radius of galaxies for producing the QSOALS and $\phi(m)d(m)$ is the initial mass function (IMF), assumed to be

$$\Phi(m)dm = k \ m^{-2.35} \quad pc^{-2} \ yr^{-1},$$

where k is given by the assumption of unit covering factor of H II regions,

$$\int_0^\infty \Phi(m)t_{MS}(m)\pi r^2_{H\ II}(m)dm = 1.$$

Here t_{ms} yrs is the main sequence life time and $r_{H\ II}(m)$ pc is the radius of H II region of a star of mass m. The SFR for different values of lower and upper cutoff masses are higher by almost two orders of magnitude than the rates in the Milky way and in the star forming regions around the line of sight to the QSOs at the absorption redshift. The SFR also imply very short recycling time scales ($\sim Gyr$) for matter in the galaxies and thus imply too large a rate of chemical evolution.

3   Lyman-Alpha Surface Brightness

The Ly α luminosity emitted by number of H II regions estimated above is

$$L_{Ly\alpha} = 0.34h\nu_{Ly\alpha} \int_0^\infty \Phi(m)t_{MS}(m)n_{UV}(m)dm,$$

$n_{UV}(m)$ is the number of Ly α photons emitted per unit time by a star of mass m. Values of Ly α surface brightness are higher by an order of magnitude than the observed upper limits from the Ly α absorbers as well as from blank sky searches.

4   UV Flux from Galaxies

Number of UV ionizing photons emitted by the galaxy is given by

$$L_{UVG} = x\pi r_{eff}^2 \int_o^\infty dm\Phi(m)n_{UV}(m)t_{MS}(m) \ s^{-1},$$

where x is the average fraction of UV photons that escape the galaxy. For x between 0.1 and 0.5 the UV flux of galaxies will be greater than the inter-galactic background (10^{-21} ergs s^{-1} cm^{-2} Hz^{-1} Sr^{-1}) over distances ~ 0.3 - 1.6 Mpc from the galaxies. Thus proximity effect due to galaxies (metal line QSOALS) over these distances is expected. No such effect is however seen.

5   Fraction of Low Ionization Systems

Probability of occurrences of QSOALS without C IV and QSOALS with W(Si IV) > W(C IV) are given by

$$R = \frac{\int_{m_L}^{m_{min}} dm\Phi(m)R_{H\,II}^2(m)t_{MS}(m)}{\int_{m_L}^{m_U} dm\Phi(m)R_{H\,II}^2(m)t_{MS}(m)}$$

for m_{min} = 30 M$_\odot$ and 50 M$_\odot$. These are 0.15 and 0.4 respectively. The values obtained from the observations are 0.01 and 0.08 respectively.

6   Conclusions

The observed upper limits on SFR, Ly α surface brightness and the fraction of low ionization systems indicate that H II regions can at most produce 10% of the observed QSOALS

QSO Absorption Line Systems as Pressure-Confined Clouds in Galactic Haloes

H. J. Mo

Max-Planck-Institut für Astrophysik, 85748 Garching, Germany

Abstract. Recent observations show that some QSO absorption line systems, especially the MgII systems, are associated with very large galactic haloes. We propose that these absorbers are clouds confined by the pressure of ambient hot gas in galactic haloes. Such a model is based on the generally accepted picture that galaxies form from the cooling and condensation of gas in dark haloes. We determine the properties of this hot gas and of the absorption systems on the basis of observational and theoretical constraints. We show that a consistent model can be obtained in such a picture. Due to the drag force of the hot medium, clouds may move at a velocity that is lower than the circular velocity of the halo.

1 Model and Results

We model the halo of a normal galaxies by an isothermal sphere with a circular velocity $V_{\rm cir} = 200\,{\rm km\,s^{-1}}$. The hot gas is assumed to be at the virial temperature $T = 1.5 \times 10^6\,{\rm K}$. The pressure of the hot gas is assumed to be about $10^3 {\rm cm^{-3}K}$ around a radius $R \sim 70\,{\rm kpc}$ (with h=0.5). This pressure is obtained when the cooling time scale is comparable to the Hubble time at redshift $z \sim 2$. We have chosen a reference radius $R = 70\,{\rm kpc}$ which is about the halo virial radius at $z = 2$ and is approximately the radius at which the bulk of MgII line systems are observed (Steidel 1994). We assume that, due to thermal instability, a cold phase of gas exists as clouds heated and ionized by photoionization and in overall dynamic equilibrium with the hot medium. We model a cloud by a uniform sphere with radius R_c, proton number density n_c, and constant temperature T_c. The clouds are assumed to be optically thin and in thermal and ionization equilibrium with a constant UV flux $J(\nu) \sim 10^{-21}(\nu_{\rm T}/\nu)\,{\rm erg\,cm^{-2}sr^{-1}Hz^{-1}s^{-1}}$, where $\nu_{\rm T}$ is the threshold frequency of the Lyman continuum. In the density range concerned here, we estimated that the equilibrium temperature of the clouds is about $15000\,{\rm K}$. The density n_c of a cloud is obtained from pressure equilibrium. For a given cloud mass M_c, the radius R_c and the HI column density $N_{\rm HI}$ are determined.

When the mass of a cloud is too large and the external pressure too high, the cloud will become unstable as it sinks toward the center of a halo. This happens when the cloud mass exceeds the Jeans mass. Once a cloud is accelerated (by the gravitational field of dark halo) to a terminal velocity V at which the drag force of the hot medium is comparable to the self-gravitating force on its surface, the cloud will become hydrodynamically unstable. S-mall clouds will be evaporated by heat conduction. McKee and Cowie (1977) defined an evaporation parameter σ_0 which is essentially the ratio of the electron mean-free path to the cloud radius. They found that, when the magnetic field is negligible, clouds with $\sigma_0 \sim 0.03$ is radiatively stabilized. Clouds with larger σ_0 will evaporate, while those with smaller σ_0 will condensate.

Based on above considerations, we found that a consistent model can be obtained if the pressure of the hot gas is about $10^3 \, \mathrm{cm}^{-3} \, \mathrm{K}$ around a radius $R \sim 70 \, \mathrm{kpc}$. As shown in Mo (1994), such hot gas emits soft x-ray with a bolometric luminosity of the order of $10^{39} \, \mathrm{erg \, s}^{-1}$. In order for a absorbing cloud with $N_{\mathrm{HI}} \sim 10^{17.5} \, \mathrm{cm}^{-2}$ (typically that of MgII systems) to be accelerated to a velocity of about $100 \, \mathrm{km \, s}^{-1}$, to match the observed velocity structure of metal line systems, the clouds should have a mass $M_c \sim 10^6 \, M_\odot$. Such clouds are stable against both heat conduction and gravitational collapse. The radius and density of such clouds are $R_c \sim 1 \, \mathrm{kpc}$ and $n_c \sim 0.1 \, \mathrm{cm}^{-3}$, respectively. These results are in good agreement with those obtained by the study of the Lyman limited systems (a population similar to MgII systems, Steidel 1990). To have a unity covering factor within an impact parameter of about $70 \, \mathrm{kpc}$, the total mass in the clouds is about $5 \times 10^9 \, M_\odot$.

The other consequences of this model are: (1) Clouds with a lower HI column density $N_{\mathrm{HI}} \sim 10^{16} \mathrm{cm}^{-2}$ should be found at a larger radius ($R \sim 150 \, \mathrm{kpc}$, depending on the density profile of the hot gas) with lower density and lower velocity. Such clouds may produce Lyα forest systems associated with normal galaxies (Lanzetta et al. 1994), as suggested by Mo (1994). (2) Pressure-confined absorption systems with $N_{\mathrm{HI}} \gtrsim 10^{17.5} \mathrm{cm}^{-2}$ should not be found in haloes with circular velocities as low as $\sim 100 \, \mathrm{km \, s}^{-1}$, because the pressure in these haloes is too small. (3) Absorption systems with $N_{\mathrm{HI}} \sim 10^{14.5} \mathrm{cm}^{-2}$ can exist as pressure-confined clouds in small galactic haloes (with $V_{\mathrm{cir}} \lesssim 100 \, \mathrm{km \, s}^{-1}$). (4) Due to the drag force of the hot medium, clouds are moving with a velocity that can be much lower than the circular velocity of the halo. This may be relevant to the observations of Lyα forest in the spectra of quasar pairs (Bechtold et al. 1994; Dinshaw et al. 1994).

References

Bechtold J., et al., 1994, ApJ, 437, L83
Dinshaw N., et al., 1994, ApJ, 437, L87
Lanzetta K.M., et al., 1994, ApJ, in press
Mo H.J., 1994, MNRAS, 269, L49
Steidel C.C., 1990, ApJS, 74, 37
Steidel C.C., 1994, preprint

Jeans-Type Instability of the Reheated IGM

H.G. Bi[1][3], G. Börner[2], L.Z. Fang[1], and Q.B. Li[3]

[1] Dept. of Physics, Univ. of Arizona, Tucson, AZ85721 USA
[2] MPA 85740 Garching, Germany
[3] Beijing Observatory, Beijing, China

Abstract. The evolution of the baryonic intergalactic medium (IGM) between the epoch of QSO formation ($z \sim 6$) to the epoch of intense galaxy formation ($Z \sim 1$) is studied.

We consider first-order fluctuations of the intergalactic medium (IGM) plus CDM following the procedure of Field[1] The fluctuation quantities, $\delta = \rho/\rho_0 - 1$, $\tau = T/T_0 - 1$ and $\mathbf{v}$ are conveniently studied in k-space by using the Fourier transformation. As an important example, we assume that the initial heating happened at $z = 6$, when $T_0 = 100K$, $\tau = 0$ and the IGM had the same spatial distribution δ as the dark matter δ_{DM}. The follow-up evolution of the temperature is found surprisingly independent of the initial condition. We illustrate $\delta(\mathbf{k})$, $\tau(\mathbf{k})$ at $z = 1$ and $z = 0$ in Figs. 1a and 1b respectively. In the figures, the x-axis is k; the full curves are the ratio $\delta(\mathbf{k})/\delta_{DM}(\mathbf{k})$, the dashed curves for $\tau(\mathbf{k})/\delta_{DM}(\mathbf{k})$ and the dot-dashed curves for the function $(1 + x_b^2 k^2)^{-1}$. This formula gives the behaviour of δ/δ_{DM} at $k \to 0$ and $k \to \infty$ even to non-linear evolutionary states, and is also the exact solution of the $\gamma = 4/3$ polytropic gas of the equation of state $p \propto \rho^\gamma$ or a reasonable approximation of other γ[2,3]. Here the Jeans length x_b is defined as $x_b = \sqrt{3/2} v_s t/R(t)$ with the sound velocity $v_s = (5kT_0/3\mu m_p)^{1/2}$. We have the following two general conclusions from these numerical solutions:

1) The damping of δ relative to δ_{DM} on the scales $k^{-1} \leq x_b$ is due to the thermal pressure, since besides the gravity that affects equally all matter, the IGM is subject to another *expelling* thermal force. Therefore, the IGM distribution is filtered from, and smoother than, that of the dark matter. One can see this from the x-space representation of the formula $(1 + x_b^2 k^2)^{-1}$:
$\delta(\mathbf{x}_1) = \int \frac{1}{4\pi x_b^2 x} e^{-x/x_b} \delta_{DM}(x_{12}) d\mathbf{x}_2$.

2) The ratio $r(k)/\delta(k)$ is almost a constant, about 0.60, indicating that the density fluctuation has a proportional temperature fluctuation. If one approximates the IGM as $p \propto \rho^\gamma$, the index will be $\gamma = 1.60$ which is close to, but still differs from, and isotropic hydrogen gas of $\gamma = 5/3$.

In a related work about the $Ly\alpha$ forest, we studied the same IGM model except for assuming that the IGM follows a polytropic equation of state. The present IGM evolution is very close to the $\gamma = 5/3$ model discussed in somewhat detail in Ref. 2. Models of different γ are found to result in similar rectified absorpiton profiles, in which $\gamma > 1$ models display a positive correlation between line column densities and line temperatures, while $\gamma < 1$

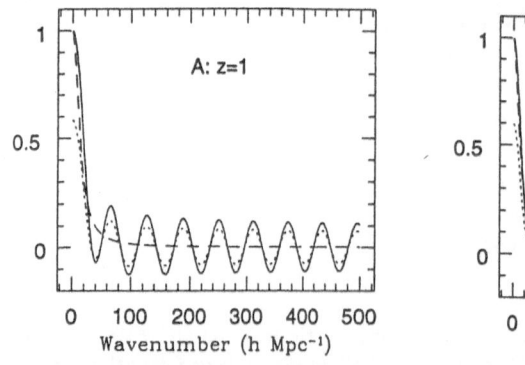 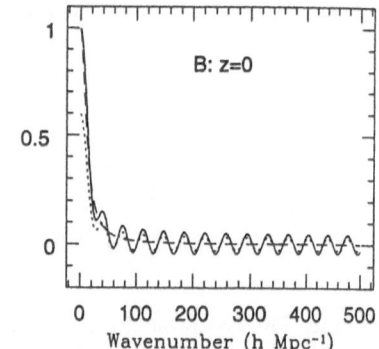

Fig. 1. a) The fluctuation quantities $\delta(k)/\delta_{DM}(k)$ (full line), $\tau(k)/\delta_{DM}(k)$ (dashed) and the approximate formula (dot–dashed) at the redshift $z = 1$. The x-axis is wavenumber k. b) the same as a) at $z = 0$.

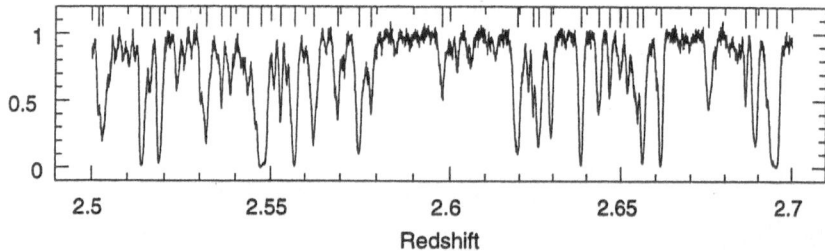

Fig. 2. A synthetic Lyα forest absorption spectrum of the IGM of the equation of state $p \propto \rho^{\gamma}(\gamma = 1)$ in the CDM cosmological model. Noise, instrumental point–spread–function and peculiar velocity are also simulated. Lyα lines having equivalent widths larger than 0.16Å are marked by bars.

models display an anti-correlation. Fig.2 illustrates a synthetic spectrum[5] of the isothermal $\gamma = 1$ model at the redshift range 2.5 to 2.7 in which instrumental effects (noise, point-spread-function etc.) and the peculiar velocity of the IGM have also been taken into account. The physical parameters used are as follows: $\Omega_H = 0.015h^{-2}$, $T_0 = 2 \times 10^4 K$, the square of variance of the IGM density = 1.72, and $J_c = 3.2 \times 10^{-21} ergs/s/cm^2/st$. The dark matter is assumed to be cold with a Gaussian power spectrum.

References

Field, G.B. 1965, ApJ, 142, 531

Bi, H.G. 1991, Ph.D. thesis (Universität München) or MPA preprint 620

Fang, L.Z., Bi, H.G., Xiang, S.P., Börner G. 1993, ApJ 413, 477

Bi, H.G, Börner, G., Chu, Y. 1992, A&A, 266, 1

Bi, H.G. 1993, ApJ, 405, 479

Ω_{baryon} and the Geometry of Intermediate Redshift Lyman α Absorption Systems

Michael Rauch[1] and Martin G. Haehnelt[2]

[1] Observatories of the Carnegie Institution of Washington, 813 Santa Barbara Street, Pasadena, CA 91101, USA
[2] Max-Planck Institut für Astrophysik, Karl-Schwarzschild-Straße 1, 85748 Garching, Germany

Abstract. Estimates of Ω_{bar} from primordial nucleosynthesis together with standard assumptions about the ionization state of low column density Lyman α forest clouds provide significant constraints on the axis ratio of the absorbers and on the overall fraction of the baryonic matter contained in these objects if the recently measured large values for their transversal size are representative.

Recent work (Petitjean et al. 1993, Meiksin & Madau 1993, Shapiro et al. 1994) has emphasized that low column density Lyman α absorption systems (10^{13}cm$^{-2} \leq N \leq 10^{15}$cm^{-2}) may contain a substantial fraction of the baryonic Ω, if the clouds are highly ionized.

The baryonic fraction of the closure density in these clouds at intermediate and high redshift redshifts can be computed from the neutral hydrogen (HI) column density distribution function (CDDF), f(N), as follows (Wolfe 1993):

$$\Omega_{Ly\alpha} = \frac{\mu m_H H_0}{c\rho_{0crit}} \int_{N_1}^{N_2} x^{-1}(N) N f(N) dN, \tag{1}$$

where N and x are the neutral hydrogen column density and the ratio of neutral to total hydrogen, respectively. H_0 and ρ_{0crit} are the Hubble constant and the critical density at the present epoch. μm_H is the mean molecular weight. We adopt the usual power-law fit to the lower column density range (between $N_1=10^{13}$ and $N_2=10^{15}$cm^{-2}) as e.g. given by Petitjean et al. (1993), f(N)=B$N^{-\beta}$, with B=7.76×10^{12} and β=1.83, and integrate between column densities N_1 and N_2. This parametrization was obtained from data with a mean redshift $\sim$ 2.5. Assuming that the neutral fraction x of hydrogen is determined by photoionization equilibrium, we obtain

$$x = 3.9 \times 10^{-6} \left(\frac{T}{3 \times 10^4}\right)^{-0.35} \left(\frac{I}{10^{-21}}\right)^{-0.5} \left(\frac{N}{10^{14}}\right)^{0.5} \left(\frac{D}{100kpc}\right)^{-0.5}, \tag{2}$$

where T is the gas temperature, I is the intensity of the ionizing UV background, N the HI column density, and D the thickness of the cloud (or the path length of our line-of-sight through it).

Taking D to be independent of the column density we obtain for the baryonic Ω:

$$\Omega_{bar} \approx 0.040 h_{75}^{-1} f_{Ly\alpha}^{-1} \left(\frac{T}{3 \times 10^4}\right)^{0.35} \left(\frac{I}{10^{-21}}\right)^{0.5} \left(\frac{D}{100 kpc}\right)^{0.5}. \quad (3)$$

Here $f_{Ly\alpha}$ is the fraction of baryons contained in Lyman α absorption systems. This value has to be compared to the range permitted by primordial nucleosynthesis arguments, $0.018 \leq \Omega_{nuc} h_{75}^2 \leq 0.027$ (Walker et al. 1990). The nucleosynthesis upper limit on Ω implies an upper limit on the size of the absorber along the line-of-sight, D.

Together with the transversal sizes derived from absorption towards multiple QSO images we can derive from this an upper limit on the axis ratio D/L (the ratio of thickness D to transversal length L) of the clouds: $\Omega_{nuc} \leq 0.027 h_{75}^{-2}$ implies

$$D/L \leq 0.045 h_{75}^{-2} f_{Ly\alpha}^2 \left(\frac{T}{3 \times 10^4}\right)^{-0.7} \left(\frac{I}{10^{-21}}\right)^{-1} \left(\frac{L}{1 h_{75}^{-1} Mpc}\right)^{-1}. \quad (4)$$

Two recent measurements indicate transversal "cloud diameters" (or coherence lengths) of order $1\ h_{75}^{-1}$ Mpc for absorbers in the redshift range 0.5 $\leq z \leq 0.9$ (Dinshaw et al. 1994), and $240\ h_{75}^{-1}$ kpc or larger at redshift ~ 1.8 (Bechtold et al. 1994).

These values suggest that low column density Lyman α absorption systems belong to *significantly flattened* structures even if the fraction of all baryons contained in these systems is close to one. To escape these conclusions we would have have to assume at least one of the following:

- the neutral fraction in the clouds is significantly larger due to clumpiness of the gas
- the standard assumption for the flux of ionizing background radiation overestimates the true value
- the nucleosynthesis constraint on Ω_{bar} is in error
- the recent size measurements of low column density Lyman α absorption systems at intermediate redshift are not representative.

References

Bechtold, J., Crotts, A.P.S., Duncan, R.S., Fang, Y., 1994, ApJ, 437, L83

Dinshaw, N.,Foltz, C.B., Impey,C.D., Weymann, R.J., Morris, S.L., (this workshop)

Meiksin, A., Madau, P., 1993, ApJ, 412, 34

Petitjean, P., Webb, J.K., Rauch, M., Carswell, R.F., Lanzetta, K.M., 1993, MN-RAS, 262, 499

Shapiro, P.R., Giroux, M.L., Babul, A., 1994, ApJ, 427, 25

Walker, T.P., Steigman, G., Schramm, D.N., Olive, K.A., Kang, H.S., 1990, ApJ, 376, 51

Wolfe, A.M., 1993, Annals of the New York Academy of Sciences, 688, 281

PART XI

WORKSHOP

SUMMARY AND CONCLUSION

Summary: An Absorbing Workshop

Virginia Trimble

Astronomy Department, University of Maryland, College Park, MD 20742
and
Physics Department, University of California, Irvine, CA 92717, USA

Abstract. After an initial, confusing epoch, in which astronomers disagreed vigorously about the locations of the gas clouds responsible for absorption lines, has come consensus that most, and the most interesting, are distributed through space between us and the source quasars/QSOs, at the distances indicated by their redshifts. Absorption lines thus provide a powerful probe of structure and conditions at redshifts $z = 1$–4 and beyond that are otherwise largely inaccessible. At least two important ideas crystallized at the workshop. First, categories of lines characterized by different column depths and (apparent) compositions are closely related, at least partly by evolutionary processes. And, second, the structures and conditions that the lines reveal are closely related, again at least partly by evolutionary processes, to the structures and conditions shown by QSOs themselves and by galaxies at smaller redshifts. The sheer richness of recent spectra from HST, Keck, and elsewhere was a third, striking feature.

1 Introduction

I do not know, but would very much like to find out, who gave the very first "concluding remarks" talk at a topical astronomical conference. The custom does, at least, predate quasars, which first appeared at the 1963 Texas symposium, with no less than three summaries, by P.G. Bergmann, R. Minkowski, and P. Morrison. The trail goes back to the first two IAU Symposia, in 1951, for which only A. Blaauw's summary was published, and 1953, with final thoughts contributed by J.M. Burgers (of the vector) and T. Gold. To look back further requires an older library than that of the University of Maryland. This is about my 16th effort and is probably beginning to approach the record set by L. Woltjer; we have, in recent years, occasionally been asked to share some such task, illustrating that the progress of astronomy is now so rapid that we can sit beside the giants on whose shoulders we stand.

Much better documented is the history of absorption lines in quasars and QSOs — the distinction between the former with strong radio emission and the latter without it is a useful one and should be maintained (though I will often fail to do so). The quasar "discovery package" consisted, of course, of four short letters to Nature in April, 1963: Hazard et al. reporting a lunar occultation position for 3C 273, Schmidt identifying its redshift from optical lines, Oke confirming this with Hα redshifted into the near infrared, and Greenstein and Matthews announcing the still-larger redshift of 3C 48. Soon after, Williams (1963) reported Ca II H and K and 21 cm absorption, caused by gas in our own galaxy, in spectra of 3C 273.

1965 was a banner year, beginning with Sandage's (1965) report of the discovery of "interlopers", radio quiet objects otherwise resembling quasars, and something like 100 times as common. Iriarte had already noted that some Tonanzintla blue "stars" resembled 3C 273. Sandage coined the abbreviation BSO, for bright stellar object, and BSO1 is described as having structured emission lines, which we would now record as narrow absorption cutting into the broad emission. This is arguably the real discovery of QSO absorption lines, though not generally so credited.

The same year, Schmidt, observing 3C 9, pushed redshifts above 2, thus moving Lyα into the wavelength range accessible from ground. His letter (Schmidt 1965) appears in the April 1st ApJ, but was not submitted until the 8th. We suspect that this issue was a little late reaching library shelves! Within minutes of his showing the spectrogram of 3C 9 at a Caltech seminar that spring, graduate student Jim Gunn recognized that the persistence of continuum emission blueward of Lyα in the quasar rest frame was, in his words, "important for cosmology". What we now call the Gunn-Peterson effect (Gunn and Peterson 1965) is, indeed, important for cosmology. Their paper, submitted in May, was not published until September, permitting Scheuer (1965) to be first into print. He had not, however, seen the actual spectrogram, and so could not provide any numbers to go with his deduction that high-redshift quasars would permit strict limits to be set on the density of intergalactic neutral hydrogen. For whatever reason, the Gunn and Peterson paper is still cited a dozen or more times a year, and the Scheuer paper almost never.

Bahcall and Salpeter (1965) brought the year to a fitting close by predicting that discrete absorption features, due to gas in intervening galaxies, ought to be detectable in QSO spectra.

The "official" and most-often cited discovery of quasar absorption lines came in 1966, with nearly-simultaneous reports on QSO 1116+12 (Lynds and Stockton 1966a, Lyα only), PHL 938 (Kinman 1966, CIV and Lyα, remarkably, on a spectrogram exposed and loaned by Sandage), and 3C 191 (Burbidge et al. 1966, Lynds and Stockton 1966b). All three had absorption redshifts smaller than, but very close to, the emission redshift. 3C 191, like most prototypes, is markedly untypical, with non-resonance lines, arising from excited levels. Burbidge et al. (1966) suggested immediately that the absorbing clouds might be intrinsic to the quasars.

2 Subtypes and Locations

Although some degree of unification was an important message of this workshop, the recognition of several rather different classes of absorption features was essential to early progress in the subject. PHL 5200 (Lynds 1967; Burbidge 1968) was the first of what we now call BAL objects, that is, ones showing several lines at z_a neither much smaller or much larger than z_e, whose widths exceed any reasonable thermal broadening.

Multiple absorption redshifts in a single source began with PKS 0237–23 (Greenstein and Schmidt 1967), followed closely by PHL 938 (Burbidge et al. 1968). These were also the first quasars with absorption at much smaller redshifts than the broad emission lines. Showing an enormous number of lines and (eventually) at least seven separate redshift systems, PKS 0237–23 required something beyond simple visual examination to choose likely values of z_a and establish their reality. Thus computers entered the picture (Bahcall 1968, Bahcall et al. 1968), and have never left it since.

Oke (1970) found the first Lyman limit system, meaning absorption beyond 912 Å at a redshift seen independently in metal lines. The target object was 4C 05.34, also the quasar that persuaded Lynds (1970, 1971) that most of the lines with wavelengths less than 1216 Å in the quasar's rest frame were attributable to Lyα absorption in large numbers of small clouds. The Lyman alpha forest rapidly took over much of the landscape.

Brown and Roberts (1973) found 21 cm absorption in 3C 286, and Haschick and Burke (1975) were the first to see the 21 cm feature at a velocity identical to that of a known galaxy along the line of sight. Their pair, 3C 232 and NGC 3067, still an interesting combination, also appeared at the workshop. Finally, damped Lyα systems (meaning ones with non-velocity broadening and only such ions as you would expect in an HI region) were shown to be a separable class by Strittmatter et al. (1973) and Carswell et al. (1975), both looking at PHL 957.

"Where are They?" was the question of the year in 1968, specifically addressed by Burbidge et al. (1968), Bahcall et al. (1968), and Grewing and Schmidt-Kaler (1968). An excess of both absorption and emission redshifts near $z = 1.95$ (Burbidge 1967) could have been due either to non-zero cosmological constant or to non-cosmological redshifts and so did not help much. But the redshift distribution was clearly the critical datum needed to distinguish between intervening and associated clouds. McCrea (1968) believed and Weymann and Wilcox (1968) denied that an excess of large z's was already detectable in the inventory then. Bahcall and Peebles (1969) put the redshift distribution test into quantitative form, for future use, when numbers of recognized z_a's should be large enough.

Between 1970 and 1979, a series of "state of the nucleus" talks at international conferences traced changing opinions. In 1970, E.M. Burbidge at the Vatican Conference and C.R. Lynds at IAU Symposium 44 were firmly on the side of intrinsic absorbers. P.A. Strittmatter, at the 7th Texas Symposium in 1974, agreed, but with less firm conviction. By the time of IAU Symposium 74, in 1976, the number of Lyα forest lines was large enough for A. Boksenberg (who had collected many of the lines himself) to conclude that their redshift distribution established them unequivocally as due to intervening clouds. The same year, M.J. Rees, summarizing at the 8th Texas Symposium, said that the issue could be regarded as still open in the case of metallic line systems (though his own opinion was pretty definitely on the intervening side).

Essentially our current view was in place by 1979, when R.A. Weymann, speaking at IAU Symposium 92, described four separate categories of absorption features. Of these, only the BALs with $z_a \approx z_e$ were produced in clouds associated with the quasars (etc.). The other three classes, narrow Lyα (forest) lines, metallic (+ Lyman limit) systems, and damped Lyα (+ 21 cm) absorption, all came from clouds along the lines of sight. In the course of this workshop, Weymann indicated that this had been his view from the first time he saw QSO absorption lines (presumably well before 1970), while E.M. Burbidge expressed, more quietly, the opinion that at least some absorption, even at z_a much less than z_e, is probably intrinsic to the sources.

It seems likely that the beliefs of every workshop participant can be accommodated somewhere between those two ideas. In the remainder of this discussion, I will adopt the convention of giving initials and surnames of speakers at the workshop and surnames only for the presenters of poster papers, in the hope that this may help readers a bit in locating items that interest them.

3   Broad Absorption Line QSOs

Just as redshift distributions demonstrated that the other categories of absorption lines are produced by intervening gas, the definitive proof that BALs are due to gas clouds belonging to the QSOs comes from the fact that the probability of seeing broad lines with $z_a \approx z_e$ is not a function of z. It is about 10% at all redshifts. The main physical question is, therefore, whether the BALs are a separate phenomenon or a ubiquitous one with a covering factor of about 10%. Clearly one needs to look for some systematic difference between quasars with and without BALs that cannot be explained by absorption or scattering in the BAL clouds themselves or by other related orientation effects.

Differences there clearly are, including a total absence of the radio-loudest quasars from the BAL class and other differences in the distribution of radio luminosity (Francis et al. 1993) and weakness of [OII] and other differences in the emission line spectra (Boroson and Meyers 1992). The richness of the spectrum of Q 0059–2735, shown by E.J. Wampler, suggests additional possible signatures without having definitely resolved this basic issue.

The poster by Campusano noted that essentially all QSOs have some absorption at $z_a \approx z_e$, and R. Weymann suggested that a subclass of these "associated absorption" features are closely related to BALs. This would seem to imply that the relevant clouds exist in more objects than the ones where we see them.

It turned out that most discussion of BALs occurred in the poster session, with models presented by Shchekinov, Vilkoviskij, and Mathur (this last focussing on ultraviolet and X-ray absorption in clouds that are highly ionized like the BAL ones) and a Keck high-resolution spectrum of one very broad and utterly structureless BAL shown by Burbidge. The concentration reflected the dominant interest of most speakers and participants in "applied lineology", studying the large scale structure and evolution of the universe with absorption lines as probes.

4   Intergalactic HI, HeII, and D/H

To this day, truly diffuse absorption by intergalactic HI (as opposed to a smear of forest lines that you have not done a good job of resolving, Cristiani et al. 1993) has never been seen. Repeated errors have driven τ down to 0.02 ± 0.03, at $z = 4$, according to E. Giallongo. This is, of course, consistent with zero. The corresponding limit on the diffuse baryon contribution to Ω depends very much upon ionization, but the assumption that the QSOs themselves (or something with the same UV spectral index) dominate leads to $\Omega \lesssim 0.01$ in diffuse baryons.

In contrast, absorption at the corresponding HeII edge at 304 Å has been seen, and at least part of it is genuinely diffuse, according to P. Jakobsen. The conversion of the observed optical depth, 1.7 (Jakobsen et al. 1994) to baryon density is exceedingly sensitive to the ionization level of the helium, and anything from 10^{-5} to 10^{-3} of closure density is possible.

The HeII absorption is interestingly weaker in a second, slightly lower redshift QSO, 1935–69, reported by D. Tytler. It has $\tau \lesssim 0.5$, though with quite large error bars. We will all wait most impatiently for additional examples! A careful interpretation of the HeII absorption requires allowance for the equivalent of the "proximity effect" in lines. That is, the quasar will itself ionize nearby gas to a higher level than the intergalactic average. This was discussed thoroughly by A. Davidsen and alluded to in the posters by Fardal, Meiksin, and Miralda-Escude.

The combination of detected 304 Å absorption and 1216 Å upper limits requires a very soft ionizing spectrum, such as might be produced by D. Sciama's decaying neutrinos, to the point where OVI then has to be collisionally ionized in the clouds where it is seen (Miralda-Escude and others). Finally, there was some discussion of the possibility that, at $z \gtrsim 3$, a given bit of intergalactic gas might be ionized by only a handful of the nearest QSOs, resulting in large place-to-place fluctuations and perhaps accounting for the different values of τ (304 Å) reported by P. Jakobsen and D. Tytler.

The ratio of D/H in distant gas is really associated with discrete lines not continuous absorption, but nevertheless somehow seems to belong here.

Carswell et al. (1994) and Songaila et al. (1994) recently created much consternation with a report that $D/H = 2.5 \times 10^{-4}$ in a single Lyman limit system. From the beginning, it was possible that they had really seen a stray Lyα line from a cloud with low column density, fortuitously placed to mimic deuterium at the redshift of the denser cloud. This is probably the correct explanation, given that D. Tytler finds $D/H = 2 \times 10^{-5}$ (to within 50%) from both Lyα and Lyβ in QSO 1937–1009.

5   Metallic Line (+ Lyman Limit) and Damped Lyα (+21 cm) Systems

Column depths in excess of about 2×10^{20} H/cm^2 are needed either to produce readily detected features of MgII, CIV, and OVI or to reach the damping part of the curve of growth in Lyα. The standard model, going into the workshop, was that these two categories of absorption line are produced in the same sorts of intervening galaxies, but the metallic lines arise in clouds of turbulent halo gas, while the damped systems (including metal lines characteristic of HI regions) arise in more quiescent gas.

I arrived at the workshop with a list of questions, of which the first was, naturally, whether this bipartite division is correct. The short answer seems to be "Yes, but...". Other questions were: What is the relationship of this absorbing gas to that in the Milky Way and other low-redshift sites? What are the mean values, ranges, and correlations of chemical composition, cloud size, degree of clustering, and ionization state of the clouds? Models of what we "ought" to expect for these correlations on evolutionary grounds are not very robust (Vogel and Reimers 1993; Khare and Rana 1993). Finally, a question from out of the past. An important early piece of evidence in favor of intrinsic or associated absorption clouds, even at large $z_e - z_a$, was an apparent excess of absorption redshifts seemingly locked to emission ones, in the sense of

$$\frac{1 + z_e}{1 + z_a} = 1.11 \text{ or } 1.32$$

corresponding to Lyβ or Lyα falling on the continuum edge. Is there any residual excess of such apparently-locked lines in the data base? Inevitably, none of these questions was entirely answered. The following subsections replace them with slightly different questions, to which an assortment of possible answers came from speakers and posters.

5.1   Where and What is the Absorbing Gas?

Many alternatives to the conventional map were propounded. The only one firmly ruled out by its proposer was HII regions around O4 stars, which would put Ly limit systems in galactic disks (Khare). Those not immediately ruled out included giant HI clouds (V. Khersonsky); small, low surface brightness and underluminous galaxies (Fried, M. Pettini, Rauch); pressure-confined clouds in halos (Mo), an idea which triggered much discussion of the lifetimes of such clouds; galaxies with high star formation rate and evidence for interactions (Caulet), which one can, however, imagine arising as a selection effect; and gas on the way to becoming a thick disk (A.M. Wolfe). The evidence for this last is a handful of damped systems where the lines show low density wings to red or blue, but not both (like a sight line cutting through a disk with differential rotation and a radial density gradient). R. Green, however, showed one example with low density wings on both sides.

In more general terms, E.B. Jenkins, B.D. Savage, and Saucedo drew analogies with Milky Way halo gas, in particular the high velocity clouds of high ionization (Savage), and with the expected appearance of pre-galactic, clumpy filaments for the damped systems (S. Warren). A particularly interesting datum, available only when you can see the galaxy responsible for the absorption, is that there are systematic differences in cloud composition and column depth as a function of impact parameter, in the expected sense (D. Bowen, S. di Serego Alighieri).

Another way to tackle the "what are they?" issue is to look for something else at the same place and redshift, starting with ordinary galaxies. The yield is about 50% for the low-redshift absorbers in the HST key project sample, according to R. Weymann, and essentially 100% for the $z \sim 1$ absorbers investigated by C. Steidel. One participant repeatedly emphasized that such detections of galaxies do not guarantee you have found the site of the absorber, which could be a smaller galaxy or isolated cloud in the same cluster (etc.) as the one you have seen. If this situation is common, then we are in the curious position of habitually having two galaxies at the same redshift, one of which cannot be seen, and one of which produces no absorption.

Lyman α emission in the cores of damped lines (for which there has been a countably infinite number of false alarms) apparently does not exist, and, according to M. Pettini, no more telescope time should be alloted to this project — a remarkable conclusion for someone who has been engaged in the search! Extended Lyα emission occurs, but is rare, according to Bergvall and S. Warren. I am inclined to regard this as part of the generic result that primeval galaxies cannot be found as Lyα sources. Hα emission, less vulnerable to dust, may yield more identifications (Bunker), and [OII] emission sometimes also turns up (Thimm).

Radio wavelengths provide another set of possibilities. Frayer's poster reported a firm detection of large quantities of CO associated with one $z = 0.4$ system toward 3C 196, though some discussion remarks made it clear that even 10^{11} $M_\odot$ of gas with a large line width is currently very close to detection limits and correspondingly uncertain. CO and other millimetric molecular features are clearly present in some cases (T. Wiklind).

R. Weymann reminded us of a statistical argument, dating back to the late 1960's, leading to the conclusion that, if the absorbers are all in galaxies, the galaxies must have radii of 100 kpc or so just to get as many lines as you see. This maximum impact parameter is one measure of cloud size. Another comes from dividing observed column density by space density, N_H/n. J. Bergeron pointed out that the latter is much smaller than the former, as a rule, and varies among ions, in the sense that individual MgII clouds are the smallest, CIV ones next, and OVI clouds the largest, perhaps as large as the impact parameter size.

5.2   Evolutionary Issues

In a slightly odd convention, shared with counters of radio galaxies, quasars, gamma ray bursts, and (for all I know) beans, the QSO absorption lines community tends to use "evolution" to refer to the circumstance where the numbers of something scale more steeply with redshift than $(1 + z)$, meaning that there must have been more of whatever you are counting (per comoving volume) in the past, independent of cosmological model. In this sense, the Lyman limit systems clearly evolve monotonically from $z = 4$ to 0 (L. Storrie-Lombardi and A. Boksenberg). A good deal of breath was expended over whether this evolution might best be described as a broken power law (with less-steep "evolution" at lower redshift), a continuous variable, or a single power law with large statistical errors. This will sort itself out with additional low-redshift data from HST.

At a somewhat lower confidence level, the numbers of CIV lines evolve positively at small redshift and negatively at high redshift, peaking some-where around $z = 2$–3. One does, however, have to be a little careful about what is being counted. The number of absorption lines will depend on a product of (a) numbers of clouds (possibly indeed largest at $z = 2$–3), (b) mean cloud size (decreasing monotonically down to the present, according to J. Bergeron), (c) the C/H ratio, reasonably assumed to peak at $z = 0$ (Stengler-Larrea), (d) the fraction of the element in the right ionization level (apparently highest at intermediate redshift, P. Madau), and (e) a depletion factor, arguably largest now, when there are the most other heavy atoms to form grains with, but not negligible at any z (M. Fall, B. Savage, E. Jenkins).

Quite a different sort of evolution might be expected in the properties of the galaxies hosting the absorption line clouds. A very exciting data set shown by C. Steidel includes a complete sample of $z \sim 1$ absorbing galaxies and reveals that their luminosity distribution, (B–K) colors, and sizes (based on maximum impact parameter) are almost indistinguishable from the corresponding properties of galaxies now. Only very bright galaxies are, obviously, present in the $z \sim 1$ sample, and these are the ones that (in many formation models) acquire their identities first. Is Steady State right after all?

Finally, one can ask about the total amount of gas present in the absorber clouds and how this has changed with time. There was real disagreement a-mong presenters on these points, beginning with which sorts of clouds contain most of the baryons, the Lyα forest (Rauch) or the highest N_H damped systems (A.M. Wolfe). Assuming that we decide to vote for the dense clouds, a discrepancy remains. L. Storrie-Lombardi concluded that the fraction of closure density in damped systems has always been small (0.001 or so) and has not changed systematically with redshift, while A. Wolfe showed results in which an Ω (gas) of at least 0.055 at $z = 3$ had declined to less than 0.001 at z less than 0.5. The baryon densities thus found are lower limits, because of the effects of dust in the absorbers, which will remove the background QSO from your sample completely. Adopting Wolfe's numbers, we conclude that the disappearing gas has obviously been turned into stars, and the amount is just about right to make the luminous parts of galaxies as we see them now. One is tempted to believe the result just because it is so exactly what we want and expect. In an case, the two speakers will be working together next year, so their differences should be resolved!

5.3 Ionization Issues

All absorption line systems, except damped Lyα ones, arise from quite highly ionized gas. The questions are pretty obvious: how many ionizing photons are out there; what is their spectrum; where do they come from; is there also a collisional ionization component; and to what extent do the answers vary with redshift or from place to place at a given redshift? The same questions, of course, arise for the ionization of the Lyα clouds, and the issues will be discussed together here.

From the discovery of the Gunn-Peterson limit in 1965, it was clear that the QSOs themselves would make a significant contribution to ionizing radiation, and, as Gunn then noted, if every place were as close as we are to a quasar as bright as 3C 273, the absence of 1216 Å absorption would be no surprise. Things turn out not to be quite that simple.

The best available measurement of the average ionizing flux far away comes from the incremental effect of each QSO on its own surroundings, the so-called proximity effect (which also occurs when there is a foreground QSO close to our line of sight, according to Fernandez-Soto, as, of course, it should). J. Bechtold and P. Madau reviewed recent, improved determinations of the intensity and spectrum of the ionizing flux and concluded that the sum of the contributions by identified AGNs falls short of the total by nearly an order of magnitude. The prediction of numbers of systems as a function of column density, if the intergalactic flux is due just to AGNs, may (Fardal) or may not (Levshakov) disagree with observations. P. Madau also noted that the assumption of a power law ionization flux is no longer a good enough approximation for calculating metal line ratios; thus further information about the nature of the sources could eventually come from spectral shape.

What are the alternatives? Friaça suggested that the gas in Ly limit systems might be partly ionized by UV and X-ray radiation from the host galaxies. Collisional ionization (in merging and starburst galaxy hosts, for instance) must be the main contributor to producing OVI if the intergalactic spectrum is as soft as is implied by the ratio of HeII to HI Gunn-Peterson absorption (Sect. 4). Finally, M. Fall believes that the AGN total may, in fact, be all that is required, after allowing for dust absorption on Lyman limit and other absorbers. The point is that the real luminosity function of AGNs is so steep, and surveys are so strictly magnitude limited, that even modest amounts of continuum absorption will remove many QSOs (and the densest absorbing clouds) from our inventories.

5.4   Composition Issues

The simplest question one can think to ask is, what is $Z[Z(z)]$? That is, how does the metallicity at a given position in a typical galaxy change with time? Even in this simplified guise, the aspiring answerer (whose task was described by E.J. Wampler and other speakers) must cope with correct placement of the continuum in spectra, the effects of saturated lines and blends, and the choice of velocity parameters and other inputs to curves of growth. The correct spectrum and intensity of the ionizing radiation, with which we ended the previous section, is another essential input. Depletion of metals on to dust grains adds to the sources of errors and confusion, especially for cool, Ly limit gas. Depletion factors can be as large as 100–1000, and we are saved from total ignorance only because they tend to occur in a fairly predictable pattern, correlated with condensation temperature, as described by E.B. Jenkins. Unfortunately, most of the dust you are ever going to get forms while gas abundances are falling by the first order of magnitude, and further decreases, to $[M/H] = -2$ or -3, make an enormous difference to the final metallicity you arrive at, but may not show up as additional reddening or absorption.

The next set of problems arises because my initial question is not really a well-posed one. First, what element(s) best probe metallicity? R.F. Green advocated zinc, because it depletes in lock-step with iron, and its lines are generally on the linear part of the curve of growth. Its production during chemical evolution of galaxies is, however, rather complex. Oxygen has been popular with people who look at gas in nearby metal-poor galaxies, and it is available in some QSO absorption line systems. It has the advantage of being produced mostly by Type II (core collapse) supernovae, so that it traces a definite aspect of chemical evolution. Unfortunately, stellar oxygen abundances are very difficult to come by, making comparison of the absorption line gas with galaxies difficult.

Second, the criteria for choosing "an absorption system" to analyze are not entirely self-evident when the velocity structure of lines is a strong function of ionization potential, as in many of the systems shown by A.M. Wolfe and Lu and in the SII line profile analyzed by R.F. Green.

Third, additional thought is needed when you go to compare gas and stars, because there are good reasons to suspect that star formation may not select an unbiased sample of the average interstellar medium. Because metals help to cool gas, the idea of metal-enhanced star formation (MESF) is attractive, and, indeed, there seem to be no HII regions in our neighborhood as metal rich as the sun. On the other hand, some stars (e.g. the λ Boo subset of Ap stars) give the impression that dust was preferentially excluded when they formed, at least from their surface layers. This will distort element ratios as well as absolute metallicity.

Perhaps surprisingly, in light of this multiplicity of difficulties, there was reasonable agreement among speakers and poster presenters (including A.M. Wolfe, R.F. Green, Lu, and Saucedo) on the following issues:

a. "Most" metal line systems have [Fe/H] in the range -1 to -2.

b. Some of the traditional anomalies seen in galactic population II stars appear, including [O/Fe] > 0, [N/O] < 0 (Lipman), and the odd-even effect, e.g. [Al/Fe] > 0. These anomalies can be modeled without doing grave violence to other things we think we know about chemical evolution (U. Fritze – v. Alvensleben).

c. The scatter at each redshift is comparable with the full range of [Fe/H] seen; this is reminiscent of the situation in the disk of the Milky Way, where metallicity is nearly uncorrelated with ages of stars locally, but there is a radial gradient.

d. If the damped Lyα systems, most of which have [Fe/H] < -1, turn into stars (only 2% of which in our neighborhood have [Fe/H] < -1), then there is a sort of G dwarf problem. There are also differences from Milky Way star properties in some of the element ratios and in mean metallicity vs. look-back time.

e. The lowest metallicity found is [Fe/H] between -2.5 and -3.0 (Vladilo), while galactic halo stars extend down to -4.0 and beyond. This may be a mere selection effect; stellar spectroscopists had to look at nearly 10^5 stars to find one that metal poor. Alternatively, it may say that absorption line clouds are more like halo globular clusters, for which the minimum metallicity is about -2.5.

f. Some BAL and other associated line systems have heavy element abundances as large as or larger than that of our sun (P. Petitjean, D. Turnshek).

6   The Lyα Forest

As with the metallic line (etc.) systems, I arrived in Garching with a set of
questions about the Lyα forest clouds, not always closely related to the questions that the speakers (and the data) were ready to answer. These included:
What is the relationship between absorbers at high and low redshift? What
is the connection between the forest clouds and the ones of higher column
density? Are the clouds held together primarily by external pressure or by
gravity (and pressure and/or gravity of what)? And, what are the mean values, redshift dependence, and correlations among Fe/H (if non-zero), cloud
size, ionization level, line velocity widths and structures, and clustering properties?

6.1   Intrinsic Cloud Properties

No very simple description of means and variances is possible, because, as
emphasized by R.F. Carswell, $N(z, N_H, b)$, where N is (comoving) number
density of clouds, z is redshift, N_H is column density, and b is the velocity
width parameter, is not a separable function. The implication seems to be
either that there are several populations with discrete distributions of properties, or that some high redshift clouds turn into other things that are more
"like" galaxies at low redshift.

The question of whether column depth and velocity parameter are correlated is an ancient and honorable one. Yes votes came from Rodriguez-Pascual
and de la Fuente; no votes from Carballo, D. Tytler, and R.F. Carswell. If
you look at a given source a second time with improved wavelength resolution, you inevitably find more lines with smaller N_H and b than you did the
first time. As a result, all nominal average values are really upper limits, and
artificial correlations are likely to appear in data that have not been carefully
corrected for selection effects. Two arguably real effects are (a) a good many
lines have $b = 20$–25 km/sec, rather less than the expected thermal width
at 10^5 K, and (b) the power-law distribution for $N(N_H)$ at large redshift
apparently turns over below 10^{13} H/cm^2.

Some information on cloud sizes (or anyhow sizes of complexes into which
clouds are grouped) comes from finding, or not finding, the same absorption
redshift in spectra of lensed quasar components and close quasar pairs. Distance in the sky to the nearest detectable galaxy at the same redshift as an
absorption feature also provides some size information. The numbers that
come out (K. Lanzetta, N. Dinshaw, A. Smette) range from 10's of kpc up
to a megaparsec or more, and are disconcertingly large in some cases. M.
Disney, in several discussion remarks, emphasized that this could mean that
the absorbing clouds are not actually all part of a single galaxy (either in the
Lyα case or for metallic line systems looked at in the same way). Instead,
we are merely learning that the absorber clouds live in the same large scale
structures that galaxies do. The analyses have generally been done assuming

rather simple, sharp-edged structures for the clouds, and M.J. Rees pointed out that this can lead to artificially large "best fit" radii.

6.2 Evolutionary Perspective

In a neat reversal of the usual English-language cliché, A. Boksenberg noted that part of the problem in studying Lyα only lines is that you can't see the trees for the forest. In any case, it does seem to make sense to try to connect properties of the clouds responsible for these lines with properties of other entities — in composition, degree of clustering, and so forth. A little fuzzing at the edges turns out to be necessary to fit everything into a single picture!

Lyα forest lines, more or less by definition, have no associated metal lines. It is quite difficult to push limits for individual clouds below the [Fe/H] < -1.5 to -2.0 reported for two objects by D. Womble and S. Savaglio. This is, in some sense, a high metallicity for the context (to paraphrase Maurice Goldhaber's remark on a double-beta-decay experiment that had reported no counts). D. Tytler reported a positive detection, but at a lower level [C/H] $= -2.5$, for the median value of a large number of Lyα clouds, based on the presence of a CIV feature in the summation of their spectra, shifted in wavelength to line up properly.

Whether Lyα forest lines are clustered in redshift space has been long debated. Results reported at the workshop at first seemed to be adding to the confusion. K. Lanzetta set an upper limit of less clustering at $z \gtrsim 2$ than is shown by dwarf galaxies (at $z = 0$, of course). But he also reported a correlation of Lyα equivalent width with impact parameter for cases where the responsible galaxies have apparently been seen. This indicates clustering around galaxies, at least. "Associated systems" show no such correlation. R.F. Carswell, in contrast, found some clustering at $z \lesssim 2$. And N. Dinshaw described the situation as showing no correlation along the line of sight and lots in the plane of the sky.

I think one can summarize and reconcile the previous few paragraphs by saying that, as you go up in column density, toward the values found for Ly limit systems, which really are clustered (according to Francis), and down in redshift, toward times when galaxies are known to exist and be clustered, and when you focus on velocity separations in the range 50–150 km/sec, then you find at least as much structure (reported by V.V. Chernomordik, J. Stocke, and S. Cristiani, as well as others already mentioned) as you would expect (according to calculations by S. Bajtlik). In addition, the lines are at least sporadically associated with detectable (often faint) galaxies and are not pure big bang gas. Thus they are not totally disjoint from the rest of the universe (D. Tytler, K. Lanzetta, Lebrun).

In a couple of cases reported by K. Lanzetta, the associated galaxy is only about 1% as bright as the Milky Way, but, based on the velocity separation of components on different sight lines, it has a mass of about 10^{12} M$_\odot$, implying a remarkably high M/L ratio.

Finally, it is possible to describe an otherwise plausible universe in which Lyα forest clouds form and evolve to yield the statistics seen, without the need to invoke any additional "tooth fairies" beyond the ones already needed to make galaxies (J. Charleton, M.J. Rees, and J. Miralda-Escude).

7   Applied Lineology

Not infrequently, someone claims that, with a sufficiently large supply of QSO absorption lines (or gamma ray bursts, or gravitational lenses, or whatever the speaker works on) you can measure H_0, q_0, Λ, or Ω. There was, mercifully, very little of this at the workshop.

Some of the interesting applications of line statistics or individual cloud properties that were presented include (a) setting a limit on temporal evolution of the fine structure constant (Levshakov; it is not as tight as previously advertized), (b) distinguishing real gravitational lenses from QSO pairs (A. Smette), (c) limiting, and eventually measuring, our peculiar velocity from the excess of features in whichever direction we happen to be travelling (Rauch), and (d) perhaps most obviously, testing various models for the formation of very large scale structure, streaming, and the nature of dark matter (Doroshkevich, Francis, Khersonsky, Mücket, Petitjean).

One can also say a few specific things about some aspects of large scale structure based on absorption lines. First, the voids are genuinely regions of low baryon density, but are not entirely empty (J. Stocke). Second, large scale structure, based on QSOs themselves and absorption redshifts in the range $z \gtrsim 2$–3, exists on scales up to at least 10 Mpc (P. Møller). Over the range $z = 1.8$ to 0.7, the autocorrelation function of the absorbers has an amplitude $\xi(r) = 0.6 \pm 0.2$ between 15 and 100 Mpc, with a possible periodicity at 30 Mpc, according to C.D. Impey. And, over that redshift range, the large scale structure thus revealed seems to be nearly in free expansion and to show very little deviation from smooth Hubble flow.

8   L'Envoi

It is the traditional privilege of the last speaker to express the collective gratitude of conference participants to those who made it all possible, beginning, in this case, with the host organization, ESO, and its director, Riccardo Giacconi. Enthusiastic thanks go also to the Scientific Organising Committee for organizing the provocative program, and to C. Stoffer and the others who handled the local arrangements so expeditiously. I am personally grateful to Tom Kinman and Margaret Burbidge for sharing some first-hand memories of the paleo-line period.

This conference was described by many speakers as the successor to one held in 1987 at STScI. Simple subtraction indicates that we will next meet in 2001, somewhere near latitude 55° N and longitude 101° E. This puts us somewhere between Irkutsk and Krasnoyarsk, and close enough to Kyzyl (former capital of Tana Tuva, or Tuvanian SSR) that it cannot be a coincidence. And surely no one would venture to predict firmly that Giacconi would not be the director there! Until then, auf wiedersehen, au revoir, hasta la vista, aloha, and, as the ancient Egyptians said: Ankh, wdh, snb — may you live, prosper, and be healthy.

References

Bahcall J.N., 1968, ApJ, 153, 679
Bahcall J.N., Peebles P.J.E., 1969, ApJ, 156, L7
Bahcall J.N., Salpeter E.E., 1965, ApJ, 142, 1677
Bahcall J.N., Greenstein J.L., Sargent W.L.W., 1968, ApJ, 153, 686
Boroson T.A., Meyer K.A., 1992, ApJ, 397, 442
Brown R.L., Roberts M.S., 1973, ApJ, 184, L7
Burbidge E.M., 1968, ApJ, 152, L111
Burbidge G.R., 1967, ApJ, 147, 851
Burbidge E.M., Lynds C.R., Burbidge G.R., 1966, ApJ, 144, 447
Burbidge E.M., Lynds C.R., Stockton A.N., 1968, ApJ, 152, 1077
Carswell R.F., et al., 1975, ApJ, 196, 351
Carswell R.F., et al., 1994, MNRAS, 268, L1
Cristiani S., et al., 1993, A&A, 268, 86
Francis P.J., et al., 1993, AJ, 106, 417
Greenstein J.L., Matthews T.A., 1963, Nature, 197, 1041
Greenstein J.L., Schmidt M., 1967, ApJ, 148, L13
Grewing M., Schmidt-Kaler Th., 1968, A. f. Ap., 69, 247
Gunn J.E., Peterson B.A., 1965, ApJ, 142, 1633
Haschick A.D., Burke B.F., 1975, ApJ, 200, L137
Hazard C., Mackey M.B., Shimmins A.J., 1963, Nature, 197, 1037
Jakobsen P., et al., 1994, Nature, 370, 35
Khare P., Rana N.C., 1993, J. Astrophys. Astron., 14, 83
Kinman T.R., 1966, ApJ, 144, 1232
Lynds C.R., 1967, ApJ, 147, 396
Lynds C.R., 1970, IAU Symp. 44
Lynds C.R., 1971, ApJ, 164, L73
Lynds C.R., Stockton A.N., 1966a, ApJ, 144, 445
Lynds C.R., Stockton A.N., 1966a, ApJ, 144, 451
McCrea W.H., 1968, Nature, 218, 257
Oke J.B., 1963, Nature, 197, 1040
Oke J.B., 1970, ApJ, 161, L17
Sandage A., 1965, ApJ, 141, 1560
Scheuer P.A.G., 1965, Nature, 207, 963
Schmidt M., 1963, Nature, 197, 1040
Schmidt M., 1965, ApJ, 141, 1295
Songaila A., et al., 1994, Nature, 368, 599
Strittmatter P.A., et al., 1973, ApJ, 183, 767

Vogel S., Reimers D., 1993, A&A, 274, L5

Weymann R.A., Wilcox R.C., 1968, Nature, 219, 103.

Williams D.R.W., 1963, Talk at Texas Symp., publishing 1965 in I. Robinson et
    al. (eds.), Quasi Stellar Sources and Gravitational Collapse. U. Chicago Press,
    p. 213

Author Index